Petrological Evolution of the European Lithospheric Mantle

Geological Society books refereeing procedures

The Society makes every effort to ensure that the scientific and production quality of its books matches that of its journals. Since 1997, all book proposals have been refereed by specialist reviewers as well as by the Society's Books Editorial Committee. If the referees identify weaknesses in the proposal, these must be addressed before the proposal is accepted.

Once the book is accepted, the Society Book Editors ensure that the volume editors follow strict guidelines on refereeing and quality control. We insist that individual papers can only be accepted after satisfactory review by two independent referees. The questions on the review forms are similar to those for *Journal of the Geological Society*. The referees' forms and comments must be available to the Society's Book Editors on request.

Although many of the books result from meetings, the editors are expected to commission papers that were not presented at the meeting to ensure that the book provides a balanced coverage of the subject. Being accepted for presentation at the meeting does not guarantee inclusion in the book.

More information about submitting a proposal and producing a book for the Society can be found on its web site: www.geolsoc.org.uk.

It is recommended that reference to all or part of this book should be made in one of the following ways:

Coltorti, M., Downes, H., Grégoire, M. & O'Reilly, S. Y. (eds) 2010. *Petrological Evolution of the European Lithospheric Mantle*. Geological Society, London, Special Publications, **337**.

Tabor, F. A., Tabor, B. E. & Downes, H. 2010. Quantitative characterization of textures in mantle spinel peridotite xenoliths. *In*: Coltorti, M., Downes, H., Grégoire, M. & O'Reilly, S. Y. (eds) *Petrological Evolution of the European Lithospheric Mantle*. Geological Society, London, Special Publications, **337**, 195–211.

GEOLOGICAL SOCIETY SPECIAL PUBLICATION NO. 337

Petrological Evolution of the European Lithospheric Mantle

EDITED BY

M. COLTORTI
University of Ferrara, Italy

H. DOWNES
Birkbeck College, University of London, UK

M. GRÉGOIRE
University of Toulouse III, France

and

S. Y. O'REILLY
Maquarie University, Australia

2010
Published by
The Geological Society
London

THE GEOLOGICAL SOCIETY

The Geological Society of London (GSL) was founded in 1807. It is the oldest national geological society in the world and the largest in Europe. It was incorporated under Royal Charter in 1825 and is Registered Charity 210161.

The Society is the UK national learned and professional society for geology with a worldwide Fellowship (FGS) of over 9000. The Society has the power to confer Chartered status on suitably qualified Fellows, and about 2000 of the Fellowship carry the title (CGeol). Chartered Geologists may also obtain the equivalent European title, European Geologist (EurGeol). One fifth of the Society's fellowship resides outside the UK. To find out more about the Society, log on to www.geolsoc.org.uk.

The Geological Society Publishing House (Bath, UK) produces the Society's international journals and books, and acts as European distributor for selected publications of the American Association of Petroleum Geologists (AAPG), the Indonesian Petroleum Association (IPA), the Geological Society of America (GSA), the Society for Sedimentary Geology (SEPM) and the Geologists' Association (GA). Joint marketing agreements ensure that GSL Fellows may purchase these societies' publications at a discount. The Society's online bookshop (accessible from www.geolsoc.org.uk) offers secure book purchasing with your credit or debit card.

To find out about joining the Society and benefiting from substantial discounts on publications of GSL and other societies worldwide, consult www.geolsoc.org.uk, or contact the Fellowship Department at: The Geological Society, Burlington House, Piccadilly, London W1J 0BG: Tel. +44 (0)20 7434 9944; Fax +44 (0)20 7439 8975; E-mail: enquiries@geolsoc.org.uk.

For information about the Society's meetings, consult *Events* on www.geolsoc.org.uk. To find out more about the Society's Corporate Affiliates Scheme, write to enquiries@geolsoc.org.uk.

Published by The Geological Society from:
The Geological Society Publishing House, Unit 7, Brassmill Enterprise Centre, Brassmill Lane, Bath BA1 3JN, UK

(*Orders*: Tel. +44 (0)1225 445046, Fax +44 (0)1225 442836)
Online bookshop: www.geolsoc.org.uk/bookshop

British Library Cataloguing in Publication Data

A catalogue record for this book is available from the British Library.
ISBN 978-1-86239-304-2

Typeset by Techset Composition Ltd, Salisbury, UK
Printed by CPI Antony Rowe, Chippenham, UK

Distributors

North America
For trade and institutional orders:
The Geological Society, c/o AIDC, 82 Winter Sport Lane, Williston, VT 05495, USA
Orders: Tel. +1 800-972-9892
Fax +1 802-864-7626
E-mail: gsl.orders@aidcvt.com

For individual and corporate orders:
AAPG Bookstore, PO Box 979, Tulsa, OK 74101-0979, USA
Orders: Tel. +1 918-584-2555
Fax +1 918-560-2652
E-mail: bookstore@aapg.org
Website: http://bookstore.aapg.org

India
Affiliated East-West Press Private Ltd, Marketing Division, G-1/16 Ansari Road, Darya Ganj, New Delhi 110 002, India
Orders: Tel. +91 11 2327-9113/2326-4180
Fax +91 11 2326-0538
E-mail: affiliat@vsnl.com

Contents

Petrological evolution of the European lithospheric mantle: introduction

MASSIMO COLTORTI[1]*, HILARY DOWNES[2], MICHEL GRÉGOIRE[3] & SUZANNE Y. O'REILLY[4]

[1]*University of Ferrara, Department of Earth Sciences, Polo Scientifico-Tecnologico, Via Saragat 1 Ferrara, 44100 Italy*

[2]*School of Earth Sciences, Birkbeck College, Malet Street, London WC1E 7HX, UK*

[3]*UMR5562-Terrestrial and Planetary Dynamics, University Toulouse III, Observatoire Midi-Pyrenees, 14 Avenue E. Belin, Toulouse 31320, France*

[4]*Department of Earth and Planetary Sciences, Macquarie University, Sydney, NSW 2109, Australia*

**Corresponding author (e-mail: clt@unife.it)*

This volume, together with its companion volume in *Journal of Petrology* (Volume 50, No. 7), is the result of the EMAW (European MAntle Workshop: Petrological evolution of the European Lithospheric Mantle: from Archean to Present Day) held in Ferrara from 29 to 31 August 2007. The meeting was organized by M. Coltorti (Earth Sciences Department, University of Ferrara), H. Downes (Birkbeck College, London University), M. Grégoire (Observatoire Midi Pyrénées, CNRS, Toulouse) and S. Y. O'Reilly (ARC National Key Centre, GEMOC, Macquarie University), and was sponsored by the University of Ferrara, the Istituto Universitario di Studi Superiori (IUSS) of the same university, the Gruppo Nazionale di Petrografia (GNP) and the Federazione Italiana di Scienze della Terra (FIST). The organizers would like to express their deep satisfaction with the success of the meeting and the enthusiasm it provoked, as well as a sincere thanks to all participants for their contributions. Almost 100 researchers participated in the meeting, coming from most European countries, China, Japan and Australia.

The meeting was an attempt to homogenize the different databases and models that have been developed from many years of study on European mantle xenoliths, peridotite massifs, ophiolites and mafic magmas spanning in age from Archaen to Recent times. Xenoliths from Europe are mostly entrained in Cenozoic mafic magmas, and the imprints of older events may be difficult to recognize in these materials. On the other hand, ophiolites and peridotite massifs record events confined to the Mesozoic history of the upper mantle, while the mafic magmas carrying the mantle xenoliths were generated beneath terrains that have been involved in different orogenic processes. The last major geological process was the Caledonian orogeny on Spitsbergen, the Hercynian orogeny at Calatrava (Spain) and the French Massif Central, and the Alpine *sensu lato* geological cycle for the Hungarian and Serbian localities.

The EMAW brought together different approaches in an attempt to integrate the findings of geophysical and geochemical investigations of the European lithospheric mantle. Most of the broad studies have already been published in *Journal of Petrology*, while those presented here provide particular emphasis on regional petrological studies of the European lithospheric mantle, from Spain to the Pannonian Basin, from Corsica and Serbia as far north as Svalbard. Six contributions are based on studies of mantle xenoliths, while the remaining four deal with ophiolitic and peridotitic complexes. A further article provides an update on the textural classification of mantle rocks initially proposed by Mercier & Nicolas (1975), using a computer-aided approach.

From subcontinental to suboceanic lithospheric mantle

The two papers by **Piccardo & Guarnieri** and **Piccardo** deal with two sectors of SCLM (subcontinental lithospheric mantle) that outcrop in ultramafic complexes in northern Corsica (Mt Maggiore Massif) and in NW Italy (Lanzo massif). These are believed to represent SCLM that was exhumed and exposed on the sea floor during the Late Jurassic opening of the Ligurian–Piedmont Basin, the westernmost branch of the Jurassic Tethys, between the Europe and Adria (Africa) margins. The peridotite massifs record composite histories of subsolidus

From: Coltorti, M., Downes, H., Grégoire, M. & O'Reilly, S. Y. (eds) *Petrological Evolution of the European Lithospheric Mantle*. Geological Society, London, Special Publications, **337**, 1–5.
DOI: 10.1144/SP337.1 0305-8719/10/$15.00

exhumation and melt–peridotite interaction, depending on their location with respect to the axial zone of the extensional system, where MORB (mid-ocean ridge basalt) melts were formed by decompression melting in the upwelling asthenosphere and percolated the overlying lithospheric mantle by porous flow. Peridotites exposed at more marginal settings (i.e. the North Lanzo massif) escaped significant melt–rock interaction, contain widespread Sp(Gnt)-pyroxenites and preserve significant relicts of the lithospheric evolution (lithospheric Sp lherzolites). Peridotites exposed at more internal settings (i.e. South Lanzo and Mt Maggiore massifs) show extreme compositional heterogeneity induced by MORB-type melt–peridotite interaction (reactive Sp peridotites, impregnated Plg peridotites, replacive Sp harzburgites and dunites). Piccardo & Guarnieri show that the presence of decametric–hectometric remnants of pyroxenite-bearing fertile lherzolites in the melt-modified peridotites indicates that the melt–peridotite reaction processes affected SCLM protoliths and not refractory residua of the asthenospheric mantle after oceanic partial melting.

The Lanzo massif exposes, in particular, a transect from more marginal, pyroxenite-bearing lithospheric Sp lherzolites (the North Lanzo massif) to more distal strongly melt-modified, extremely heterogeneous peridotites (the South Lanzo massif). Piccardo proposes that the marginal North Lanzo peridotites were exhumed from shallower lithospheric levels and escaped the significant melt–rock interaction that affected the more internal South Lanzo peridotites which were exhumed from deeper lithospheric levels. Field relationships in the more distal, strongly heterogeneous peridotites reveal the evolution of progressive exhumation and melt–rock interaction events in these peridotites, which were percolated during continental extension and stretching. Inception of asthenospheric melting during adiabatic upwelling caused porous-flow diffuse percolation of pyroxene-undersaturated MORB-type melts that reacted (dissolving mantle pyroxenes and precipitating olivine) with the percolated SCLM under Sp peridotite-facies conditions and transformed it to depleted reactive Sp peridotites. At shallower depth (Plg-facies conditions), under increasing conductive heat loss, the percolating MORB-type melts were silica-saturated, reacted with the percolated peridotite (dissolving olivine and clinopyroxene/forming orthopyroxene (+Plg)) and underwent interstitial crystallization, enriching the peridotites in plagioclase and gabbroic microgranular aggregates. Melt impregnation and interstitial crystallization clogged the melt pathways and further melt migration was focused into structural and compositional discontinuities, giving rise to replacive olivine-rich, harzburgite and dunite channels. The replacive channels were locally (e.g. at South Lanzo) exploited for upwards migration of aggregated MORB melts and sporadic alkaline melts. Alkaline melts in Lanzo post-date the onset of MORB magmatism and mark an abrupt compositional change in the melting source: a mechanism of delamination of sectors of the lower-mantle lithosphere is proposed by Piccardo to explain sinking of Gnt-pyroxenite-bearing lithospheric mantle into the upwelling mantle asthenosphere. Partial melting of these pyroxenite-bearing sections under the appropriate pressure conditions could have formed the alkaline melts, which migrated upwards and percolated through the extending lithospheric mantle after the MORB percolation and impregnation events.

Although the sizes of mantle xenoliths limit structural studies of the reaction and refertilization processes like those described above, **Grégoire *et al.*** propose a similar model for composite xenoliths from Spitsbergen Island (Svalbard archipelago). These xenoliths have websterite veins cutting lherzolite protoliths. Unusually, major element analyses of minerals and trace element analyses of clinopyroxene are similar in both lithotypes. Grégoire *et al.* propose a tholeiitic (MORB-like) metasomatism that, taking also into account the higher melt/rock ratio involved, resembles the refertilization process invoked by Piccardo in this volume for Lanzo. The LREE-depleted clinopyroxenes of the composite websterite–lherzolite are comparable to those that are often interpreted as residues of mantle melting (Ionov *et al.* 2002). However, in this case the presence of websterite veins, which could not survive the homogenization caused by a melting process, requires a different mechanism. This LREE-depleted pattern could be caused by MORB-like tholeiitic melts circulating within the upper-mantle peridotites. As extensive tholeiitic (MORB-like) magma percolation would be expected during extension and rifting, it is highly probable that the LREE-depleted patterns usually found in Cpx could simply be the result of this metasomatic–refertilization process. The influx of relatively large volumes of magma would also change the thermal and rheological properties of the SCLM, strongly constraining the geometry and composition of the lithosphere during the opening of the basin. The finding of portions of SCLM left stranded within clearly oceanic settings (Delpech 2004; Bonadiman *et al.* 2006; Coltorti *et al.* 2008, 2010; O'Reilly *et al.* 2009) are consistent with this scenario.

Eclogites and pyroxenites in the SCLM

Garnet- and spinel-bearing pyroxenites are relatively abundant in the lithospheric mantle of both the European and Adriatic plates (Montanini *et al.*

2006; Piccardo & Vissers 2007; Piccardo *et al.* 2007). A rare case of graphite-bearing garnet clinopyroxenite occurs in an ultramafic complex in the External Ligurides. **Montanini *et al.*** suggest that this body equilibrates within the SCLM at approximately 2.8 GPa and 1100 °C. Carbon phases (graphite or diamond) typically occur in cratonic SCLM (Pearson *et al.* 1994; Haggerty 1995), and this rare occurrence extends the debate on the nature of C within the mantle, that is, deep provenance v. crustal recycling. Graphite-bearing mantle rocks have been reported in the orogenic peridotite massifs of Ronda (Davies *et al.* 1993) and Beni Bousera (Pearson *et al.* 1993). Carbon in the pyroxenites of the External Ligurides has a typical mantle isotopic signature (-4.5 ± 0.5 ‰: Thomassot *et al.* 2007) and the associated sulphide assemblage (Ni-free pyrrhotite, pentlandite, Cu–Fe sulphides) indicates exsolution from a high-temperature monosulphide solution (MSS). The crystallization of elemental carbon from a silicate melt is not considered a suitable mechanism owing to the low solubility of C in silicate melts (Bulanova 1995). Recent experimental results on graphite/diamond precipitation from alkaline carbonate melts (e.g. Arima *et al.* 2002; Palyanov *et al.* 2007) do not seem appropriate for the system under investigation. Thus, on the basis of the graphite–sulphide association, the authors suggest that C was reduced by the interaction of CO_2-rich fluids and S-rich immiscible melt (see also Haggerty 1986; Bulanova 1995; Bulanova *et al.* 1998; Luque *et al.* 1998).

The Cenozoic Calatrava volcanic field comprises more than 200 volcanic centres in an area of about 5500 km^2 (Ancochea & Nixon 1987). Strombolian cones, tuff rings and maars transport abundant mantle xenoliths to the surface, providing a splendid opportunity for studying the lithospheric mantle beneath central Spain. Lherzolites are the most abundant lithotypes; wehrlites, websterites and dunites are very subordinate (**Bianchini *et al.*** and **Villaseca *et al.***). Some of these lherzolites are anomalously rich in Fe (and Ti), which Bianchini *et al.* attribute to Fe–(Ti)-rich metasomatizing agents, derived from eclogite material present beneath the lithospheric mantle of central Spain. The HIMU signature of the Cpx appears to be the result of long-term recycling of oceanic basalts/gabbros (or their eclogitic equivalent) via ancient subduction. According to the authors, the recent subduction along the Betic collisional belt would have remobilized these old relicts from the lower–upper mantle transition zone (410–660 km). Fe–Ti rich melts characterized by a HIMU isotopic signature would be generated, and infiltrate and metasomatize the shallower lithospheric mantle. Villaseca *et al.* focus their study on Mg-('normal') lherzolites and wehrlites, some of them bearing amphibole and phlogopite. Based on the chemistry of minerals and glasses, they envisage at least two different metasomatic agents: the first is a subduction-related one, which in this case refers to the Tertiary subduction; the second one, just before xenolith entrainment, is related to undersaturated silicate alkaline melt. The latter may also be responsible for the origin of the Fe-rich wehrlite, suggesting a bridge toward the Fe-lherzolite enrichment model of Bianchini *et al.*

Fingerprints of subduction and intraplate metasomatism

The paper of **Yoshikawa *et al.*** attempts to precisely constrain the temperature and depth of derivation of xenoliths from the Massif Central (France), using Raman-based spectroscopy geobarometry. On the basis of mineral composition (both major and trace element) they also identify two different metasomatizing agents: a carbonatitic subduction-related one that affected the shallower, northern part of the lithosphere; and an asthenospheric melt that modified its deeper southern portion. The silicate-rich carbonatite melt may be derived from carbonate sediments brought into the lithospheric mantle of the French Massif Central via subduction. The age of this subduction is questionable, but geological reconstruction (Shaw *et al.* 1993; Pin & Paquette 1997), Hf model ages (Wittig *et al.* 2006) and a DM (depleted mantle) model age determined by Sr–Nd on Cpx favour a Hercynian event.

Dobosi *et al.* analysed clinopyroxenes and orthopyroxenes in anhydrous mantle peridotite xenoliths from the western part of the Pannonian Basin. HREE contents of the clinopyroxenes suggest that most of the xenoliths experienced less than 15% partial melting. The lowest degrees of melting occur in the LREE-depleted xenoliths, and the highest degree in the LREE-enriched xenoliths, although the refertilization caused by the interaction of large volumes of magma with the lithosphere (Piccardo, Piccardo & Guarnieri and Grégoire *et al.* in this volume) has to be kept in mind. The authors reconstruct the composition of the asthenosphere-derived metasomatizing agents and suggest that they are similar to alkali lamprophyres that were emplaced in the area during Cretaceous time. They also exclude any evidence for subduction-related melt, except for a family of Cpx with U-shaped REE patterns. These findings appear to contrast with the hypothesis put forward by Coltorti *et al.* (2007) to explain amphibole-bearing mantle xenoliths from Kapfenstein (Styria Basin), where a subduction-related, adakite-like metasomatic agent was inferred. It also suggests that the western Pannonian lithospheric mantle experienced repeated intraplate alkaline metasomatism from Cretaceous to Quaternary times despite the Neogene calc-alkaline (*sensu lato*) magmatism that has been described in the area (Seghedi *et al.* 2004, 2005). Particular attention in this paper is also given to textural features, which, according to Downes *et al.* (1992), are also correlated with enrichment style and extent.

Mantle textures revisited

Tabor *et al.* focus on the microstructures of mantle rocks, and propose a method for quantitative characterization of grain size by means of optical scanning and computer measurement of individual grain areas. They apply this technique to spinel peridotite xenoliths from the Massif Central and Eifel regions, and note that: (i) the three main groups of textures, that is, protogranular, porphyroclastic and equigranular, proposed by Mercier & Nicolas (1975) form a continuous series rather than discrete groups; and (ii) protogranular samples have standard deviations higher than porphyroclastic samples. Thus, large crystals embedded in a finer grain matrix do not uniquely characterize the porphyroclastic microstructure. The study of Tabor *et al.* is based only on peridotite xenoliths without obvious modal metasomatism, and further measurements on deeper (garnet-bearing) samples and xenoliths from different tectonic settings are needed to fully understand the geological and petrological meaning of these results.

Relationships between metasomatim and magmatism

Cvetković *et al.* discuss the origin of enrichment in East Serbian mantle xenoliths, reversing the concept that metasomatism could be related to magmatism as proposed by many authors, at least for Na-alkaline products (Wulff-Pedersen *et al.* 1999; Coltorti *et al.* 2004). According to these authors, metasomatism represents the infiltration of a small volume of magma rising up to the surface, thus metasomatism would be successive to magmatism. Most of the Serbian mantle xenoliths are harzburgites with abundant both modal and cryptic metasomatic features, while lherzolites, which are devoid of secondary modal enrichment, appear to have Al_2O_3, Cr_2O_3 and FeO content higher than undepleted peridotites. The petrological study of these mantle xenoliths, together with the inversion modelling applied on the most uniform and least contaminated basanite from the area, suggests that the enrichment in the mantle source of the basanite is similar to that observed in the metasomatised harzburgite. The model proposed by Cvetković *et al.* involves the percolation of CO_2- and H_2O-rich fluids that strongly metasomatized the lithosphere prior to it being uplifted and heated by asthenospheric upwelling. In this way metasomatism represents a precursor to alkaline magmatism.

References

ANCOCHEA, E. & NIXON, P. H. 1987. Xenoliths in the Iberian peninsula. *In*: NIXON, P. H. (ed.) *Mantle Xenoliths*. Wiley, Chichester, 119–124.

ARIMA, M., KOZAI, Y. & AKAISHI, M. 2002. Diamond nucleation and growth by reduction of carbonate melts under high-pressure and high-temperature conditions. *Geology*, **30**, 691–694.

BONADIMAN, C., COLTORTI, M., SIENA, F., O'REILLY, S. Y., GRIFFIN, W. L. & PEARSON, N. J. 2006. Archean to Proterozoic depletion in Cape Verde lithospheric mantle. *In*: *Abstracts of 16th Goldschmidt Conference*, 27 August–1 September 2006, Melbourne, Australia.

BULANOVA, G. P. 1995. The formation of diamond. *Journal of Geochemical Exploration*, **53**, 1–23.

BULANOVA, G. P., GRIFFIN, W. L. & RYAN, C. G. 1998. Nucleation environment of diamonds from Yakutian kimberlites. *Mineralogical Magazine*, **62**, 409–419.

COLTORTI, M., BECCALUVA, L., BONADIMAN, C., FACCINI, B., NTAFLOS, T. & SIENA, F. 2004. Amphibole genesis via metasomatic reaction with clinopyroxene in mantle xenoliths from Victoria Land, Antarctica. *Lithos*, **75**, 115–139.

COLTORTI, M., BONADIMAN, C., FACCINI, B., NTAFLOS, T. & SIENA, F. 2007. Slab melt and intraplate metasomatim in Kapfenstein mantle xenoliths (Styrian Basin, Austria). *In*: COLTORTI, M., DOWNES, H. & PICCARDO, G. B. (eds) *Melting, Metasomatism and Metamorphic Evolution in the Lithospheric Mantle. Lithos*, **94**, Special Issue 66–89.

COLTORTI, M., BONADIMAN, C., O'REILLY, S. Y., GRIFFIN, W. L. & PEARSON, N. 2008. Heterogeneity in the oceanic lithosphere as evidenced by mantle xenoliths from Sal island (Cape Verde Archipelago). *In*: *Abstracts of 33rd International Geological Congress*, 6–14 August 2008, Oslo.

COLTORTI, M., BONADIMAN, C., O'REILLY, S. Y., GRIFFIN, W. L. & PEARSON, N. J. 2010. Buoyant ancient continental mantle embedded in oceanic lithospheric (Sal Island, Cape Verde Archipelago). *Lithos*, doi: 10.1016/j.lithos.2009.11.005.

DAVIES, G. R., NIXON, P. H., PEARSON, D. G. & OBATA, M. 1993. Tectonic implications of graphitized diamonds from the Ronda peridotite massif, southern Spain. *Geology*, **21**, 471–474.

DELPECH, G. 2004. *Trace Element and Isotopic Fingerprints in Ultramafic Xenoliths From the Kerguelen Arcipelago (South Indian Ocean)*. PhD thesis, Macquarie–St. Etienne Universities.

DOWNES, H., EMBEY-ISZTIN, A. & THIRLWALL, M. F. 1992. Petrology and geochemistry of spinel peridotite xenoliths from the western Pannonian Basin (Hungary): evidence for an association between enrichment and texture in the upper mantle. *Contributions to Mineralogy and Petrology*, **109**, 340–354.

HAGGERTY, S. E. 1986. Diamond genesis in a multiply constrained model. *Nature*, **320**, 34–38.

HAGGERTY, S. E. 1995. A diamond trilogy: superplumes, supercontinents, and supernovae. *Science*, **285**, 851–860.

IONOV, D. A., BODINIER, J. L., MUKASA, S. B. & ZANETTI, A. 2002. Mechanisms and sources of mantle metasomatism: major and trace element compositions of peridotite xenoliths from Spitsbergen in the context of numerical modelling. *Journal of Petrology*, **43**, 2219–2259.

LUQUE, F. J., PASTERIS, J. D., WOPENKA, B., RODAS, M. & BARRENECHEA, J. F. 1998. Natural fluid-deposited graphite: mineralogical characteristics and mechanisms of formation. *American Journal of Science*, **298**, 471–498.

MERCIER, J.-C. C. & NICOLAS, A. 1975. Textures and fabrics of upper-mantle peridotites as illustrated by xenoliths from basalts. *Journal of Petrology*, **16**, 454–487.

MONTANINI, A., TRIBUZIO, R. & ANCZKIEWICZ, R. 2006. Exhumation history of a garnet pyroxenite bearing mantle section from a continent–ocean transition (Northern Apennine ophiolites, Italy). *Journal of Petrology*, **47**, 1943–1971.

O'REILLY, Y. S., MING, Z., GRIFFIN, W. L., BEGG, G. & HRONSKY, J. 2009. Ultradeep continental roots and their oceanic remnants: a solution to the geochemical 'mantle reservoir' problem? *Lithos*, **112**, 1043–1054.

PALYANOV, YU. N., BORZDOV, YU. M., BATALEVA, YU. V., SOKOL, A. G., PALYANOVA, G. A. & KUPRIYANOV, I. N. 2007. Reducing role of sulfides and diamond formation in the Earth's mantle. *Earth and Planetary Science Letters*, **260**, 242–256.

PEARSON, D. G., BOYD, F. R., HAGGERTY, S. E., PASTERIS, J. D., FIELD, S. W., NIXON, P. H. & POKHILENKO, N. P. 1994. The characterisation and origin of graphite in cratonic lithospheric mantle: a petrological carbon isotope and Raman spectroscopic study. *Contributions to Mineralogy and Petrology*, **115**, 449–466.

PEARSON, D. G., DAVIES, G. R. & NIXON, P. H. 1993. Geochemical constraints on the petrogenesis of diamond facies pyroxenites from the Beni Bousera peridotite massif, North Morocco. *Journal of Petrology*, **34**, 125–172.

PICCARDO, G. B. & VISSERS, R. L. M. 2007. The pre-oceanic evolution of the Erro-Tobbio peridotite (Voltri Massif–Ligurian Alps, Italy). *Journal of Geodynamics*, **43**, 417–449.

PICCARDO, G. B., ZANETTI, A., PRUZZO, A. & PADOVANO, M. 2007. The North Lanzo peridotite body (NW Italy): lithospheric mantle percolated by MORB and alkaline melts. *Periodico di Mineralogia*, **76**, 175–196.

PIN, C. & PAQUETTE, J.-L. 1997. A mantle-derived bimodal suite in the Hercynian Belt: Nd isotope and trace element evidence for a subduction-related rift origin of the Late Devonian metavolcanics, Massif Central (France). *Contributions to Mineralogy and Petrology*, **129**, 222–238.

SEGHEDI, I., DOWNES, H., VASELLI, O., HARANGI, S., MASON, P. R. D. & PÉCSKAY, Z. 2005. Geochemical response of magmas to Neogene-Quaternary continental collision in the Carpathian–Pannonian region. A review. *Tectonophysics*, **410**, 485–499.

SEGHEDI, I., DOWNES, H., VASELLI, O., SZAKÁCS, A., BALOGH, K. & PÉCSKAY, Z. 2004. Post-collisional Tertiary–Quaternary mafic alkalic magmatism in the Carpathian–Pannonian region: a review. *Tectonophysics*, **393**, 43–62.

SHAW, A., DOWNES, H. & THIRLWALL, M. F. 1993. The quartz-diorites of Limousin: Elemental and isotopic evidence for Devono-Carboniferous subduction in the Hercynian belt of the French Massif Central. *Chemical Geology*, **107**, 1–18.

THOMASSOT, E., CARTIGNY, P., HARRIS, J. W. & VILJOEN, K. S. 2007. Methane-related diamond crystallization in the Earth's mantle: stable isotope evidences from a single diamond-bearing xenolith. *Earth and Planetary Science Letters*, **257**, 362–371.

WITTIG, N., BAKER, J. A. & DOWNES, H. 2006. Dating the mantle roots of young continental crust. *Geology*, **34**, 237–240.

WULFF-PEDERSEN, E., NEUMAN, E. R., VANNUCCI, R., BOTTAZZI, P. & OTTOLINI, L. 1999. Silicic melts produced by reaction between peridotite and infiltrating basaltic melts: ion probe data on glasses and minerals in veined xenoliths from La Palma Canary Islands. *Contributions to Mineralogy and Petrology*, **137**, 59–82.

The Monte Maggiore peridotite (Corsica, France): a case study of mantle evolution in the Ligurian Tethys

GIOVANNI B. PICCARDO* & LUISA GUARNIERI

Dipartimento per lo Studio del Territorio e delle sue Risorse, Universita' di Genova, Corso Europa 26, I-16132 Genova, Italy

**Corresponding author (e-mail: piccardo@dipteris.unige.it)*

Abstract: The Monte Maggiore peridotite represents subcontinental mantle that underwent tectonic and magmatic evolution during the rifting stage of the Jurassic Ligurian Tethys oceanic basin. Pristine garnet peridotites were first equilibrated under spinel-facies conditions. During continental extension they were diffusely infiltrated by asthenospheric melts that consisted of single fractional melt increments (6% melting degree) showing depleted MORB (mid-ocean ridge basalt) signature. Diffuse melt migration of undersaturated melts at spinel-facies conditions formed reactive spinel peridotites, and melt impregnation at plagioclase-facies conditions formed impregnated plagioclase peridotites. Further focused melt migration occurred within high-porosity dunite channels.

Subsequently, the single melt fractions underwent coalescence to form aggregate MORB melts that were intruded into shallow magma chambers. They underwent fractional crystallization and formation of variably evolved Mg-rich and Fe–Ti-rich magmas. Mg- and Fe–Ti-gabbroic dykes were formed by intrusion along fractures of these magmas. Melt-percolated peridotites and gabbroic rocks are isotopically homogeneous, suggesting that melts which percolated and intruded the mantle lithosphere derived from isotopically homogeneous asthenospheric mantle sources.

The magmatic cycle, that is, asthenosphere partial melting, lithosphere diffuse melt percolation and dyke intrusion, occurred during Late Jurassic times (163–150 Ma) and represents the youngest events of lithosphere–asthenosphere interaction so far documented in ophiolitic peridotites from the Ligurian Tethys. The Ligurian Tethys basin never reached a mature oceanic stage, that is, the genetic link between exposed oceanic crustal rocks and refractory mantle peridotites.

Ophiolite sequences from the Western Alps (WA) and the Northern Apennines (NA) derive from sectors of the oceanic lithosphere of the Ligurian or Western Tethys basin (also known as the Ligurian–Piemontese basin) that separated the Europe and Adria plates during Late Jurassic–Cretaceous times. Opening of the Ligurian Tethys basin was kinematically related to pre-Jurassic rifting and Late Jurassic spreading in the Central Atlantic, and was a consequence of the passive extension of the Europe–Adria continental lithosphere. Mesozoic rifting and opening of the Ligurian Tethys ocean is generally thought to have occurred in response to the break-up of Gondwana, involving eastwards motion of the African plate relative to America and Eurasia, and the related northwards opening of the Atlantic since the early Mesozoic (Dewey *et al.* 1973).

The peculiar stratigraphy of the Alpine–Apennine (A–A) ophiolites (i.e. the association of mantle peridotites, basaltic pillow lavas and oceanic sediments) was first described by Steinmann (1927; translation 2003) in the Central Eastern Alps and by Decandia & Elter (1969, 1972) in the Northern Apennines. Further studies evidenced that the Ligurian ophiolite sequences are significantly different from the ophiolite stratigraphy as defined by the Penrose Conference on Ophiolites in 1972 (Anon. 1972). In fact, serpentinized mantle peridotites underlie both MORB basaltic lava flows and oceanic sediments, whereas sheeted dyke complexes are lacking and km-scale gabbroic bodies are intruded into the mantle peridotites. Accordingly, the Ligurian Tethys basin was floored by a peridotite basement (Bezzi & Piccardo 1971; Piccardo 1976; Lagabrielle *et al.* 1984; Lemoine *et al.* 1987; Abbate *et al.* 1994; Marroni *et al.* 1998, 2002).

Petrological studies of the Ligurian ophiolitic mantle rocks revealed the fertile lherzolite composition of peridotites from the External Ligurides and the depleted composition of peridotites from the Internal Ligurides that have been considered similar to present-day abyssal peridotites (e.g. Piccardo 1976, 1977; Beccaluva & Piccardo 1978; Beccaluva *et al.* 1984; Piccardo *et al.* 1990; Rampone *et al.* 1996; Rampone & Piccardo 2000). It was inferred that the External Liguride peridotites

From: Coltorti, M., Downes, H., Grégoire, M. & O'Reilly, S. Y. (eds) *Petrological Evolution of the European Lithospheric Mantle*. Geological Society, London, Special Publications, **337**, 7–45.
DOI: 10.1144/SP337.2 0305-8719/10/$15.00

derive from the subcontinental mantle, and that the Internal Liguride peridotites could represent depleted oceanic mantle residua after Jurassic oceanic partial melting. On the other hand, the isotopic contrast between some refractory peridotites (Mt Fucisa, Internal Ligurides) and the basaltic–gabbroic rocks from the Internal Ligurides ophiolites allowed the theory to be sustained that mantle and crustal rocks in the Internal Liguride ophiolites are not linked by a residue–melt relationship and, accordingly, do not represent mature oceanic lithosphere, cogenetic and coeval (Rampone *et al.* 1998).

Recent studies on the A–A ophiolitic peridotites promoted significant deepening in the structural and compositional knowledge, and unravelled the mantle processes responsible for their extreme variability. Mantle peridotites from the ocean–continent transition (OCT) zones of the Late Jurassic Ligurian Tethys basin revealed their long-term residence in the subcontinental lithosphere and the subsolidus tectonic–metamorphic exhumation during lithospheric extension. Mantle peridotites from the more internal oceanic (MIO) settings of the Ligurian basin revealed their extreme compositional heterogeneity that was induced on pristine subcontinental peridotites by diffuse percolation and melt–peridotite interaction of MORB-type asthenospheric melts during the rifting stages of the basin (e.g. Müntener & Piccardo 2003; Piccardo 2003, 2008; Piccardo *et al.* 2004, 2007*a*; Piccardo & Vissers 2007; and references therein).

The Monte Maggiore peridotite body, which crops out within the Alpine Corsica metamorphic belt, has been considered to derive from a MIO setting of the Ligurian Tethys basin. Previous studies (e.g. Rampone *et al.* 1997) described the presence of depleted, clinopyroxene (Cpx)-poor spinel lherzolites that were interpreted as mantle residua after MORB-type partial melting processes, and plagioclase-bearing peridotites that were interpreted as formed by melt entrapment and melt impregnation. Moreover, on the basis of isotopic data on peridotites and gabbroic dykes, it has been determined (Rampone 2004) that isotopic compositional similarities exist between depleted mantle peridotites (interpreted as the melting refractory residua) and associated magmatic crustal rocks (interpreted as crystallization products of the produced melts), and that the Monte Maggiore spinel peridotites are consistent with a Jurassic age of partial melting. According to Rampone (2004, p. 226), 'the Monte Maggiore gabbro–peridotite association constitutes the first record of the attainment of a mature oceanic stage (i.e. genetic crust–mantle link) of the Ligurian Tethys ocean'.

Recent papers on the Monte Maggiore peridotite body (e.g. Müntener & Piccardo 2003; Rampone *et al.* 2008) aimed at presenting the multistage melt–rock interaction in the Monte Maggiore peridotite. They were focused on the variably depleted, pyroxenite-poor spinel peridotites and the plagioclase peridotites. The depleted spinel peridotites have been recognized as reactive peridotites, formed by melt–rock interaction, on the basis of the 'diffuse melt–rock interaction microtextures' and the 'contrasting bulk v. mineral chemistry features which cannot be simply reconciled with partial melting' (Rampone *et al.* 2008, p. 466). The plagioclase peridotites have been recognized, on the basis of microtextural and chemical evidences, as impregnated peridotites formed by melt–rock interaction and interstitial crystallization of trapped melt (Müntener & Piccardo 2003; Rampone *et al.* 2008).

A recent paper (Piccardo 2007) provided evidenced at Monte Maggiore of: (1) the presence and local abundance of pyroxenite-bearing, Cpx-rich spinel-facies lherzolites; and (2) the replacement relationships of the depleted spinel peridotites at the expense of these pyroxenite-bearing lherzolite rock types. Field and structural relationships reveal that the pyroxenite-veined spinel lherzolites are the pristine mantle protoliths that were transformed by diffuse melt–rock interaction to pyroxenite-free, Cpx-poor depleted spinel peridotites (see below).

As we will discuss in the following, the recognition that the mantle protoliths pre-existing melt–peridotite interaction were veined fertile lherzolites of the ancient subcontinental lithospheric mantle is crucial in reconstructing the geodynamic evolution of this section of oceanic lithosphere and, accordingly, on the evolution stages of the Ligurian Tethys basin that are recorded in the Monte Maggiore peridotites.

In this paper we present new field, structural and compositional data on mantle peridotites and associated mafic rocks from the Monte Maggiore peridotite body that evidence, in particular, the existence of widespread remnants of pristine pyroxenite-bearing spinel lherzolites. Then we discuss the whole available field, structural, petrological–geochemical and geochronological data on this peridotite mass in order to present the composite structural–petrological scenario for the evolution of a sector of lithospheric mantle, from its exhumation from subcontinental mantle depths to exposure on the sea floor of the Late Jurassic Ligurian Tethys basin, that was characterized by percolation, impregnation and intrusion of Late Jurassic MORB-type melts.

The Monte Maggiore peridotite

The Monte Maggiore peridotite massif is located in the NE sector of the Alpine Corsica, at the northernmost end of Cap Corse (Fig. 1).

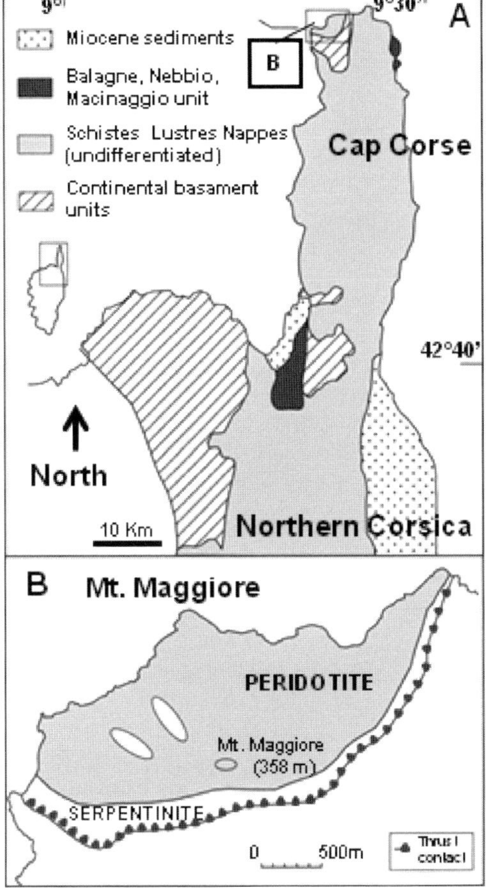

Fig. 1. (**a**) Tectonic sketch map of northern Corsica (redrawn after Rampone *et al.* 2008). (**b**) The Monte Maggiore peridotite body (redrawn after Jackson & Ohnenstetter 1981; Rampone *et al.* 2008).

These mantle peridotites constitute the uppermost tectonic unit of a pile of tectonic nappes, which consist of (from the bottom):

- schists and calc-schists with intercalations of metabasites (prasinites), followed by alternance of prasinitic and serpentinitic levels;
- Centuri gneisses;
- prasinites (greenschist metabasites), probably deriving from gabbros;
- peridotites, mantle ultramafics partly serpentinized, intruded by mafic dykes.

In the peridotite body of Monte Maggiore different rock types are present, consisting of:

- mantle peridotites, showing in places pyroxenite banding;
- mafic–ultramafic cumulate pods;
- gabbroic veins and cm-wide dykelets;
- metre-wide gabbroic dykes;
- a few mafic porphyritic dykes.

Mantle peridotites and related rocks

Previous studies on the Monte Maggiore peridotites (e.g. Rampone *et al.* 1997, 2008; Müntener & Piccardo 2003) have investigated spinel and plagioclase peridotites, which show widespread structural and compositional features indicating melt–peridotite interaction and melt impregnation. A recent contribution (Piccardo 2007) reveals the presence and local abundance of spinel pyroxenite-bearing lherzolites, which are recognized as the pristine mantle rocks preceding melt–rock interaction. In fact, they are replaced in the field by the other peridotite rock types and are transformed into pyroxene-depleted spinel peridotites and plagioclase-enriched peridotites (Piccardo 2007).

The present paper, on the basis of an accurate field and petrographical–structural work, identifies the main rock types and their mutual relationships allowing a better understanding of the mantle processes that are recorded in the Monte Maggiore peridotite body. Mantle peridotites show structural–compositional features (described in the following) that indicate their strong heterogeneity, that is, they record different mantle processes and their evolution with time. The main rock types cropping out in the Monte Maggiore body are: (i) fertile spinel lherzolites, veined by widespread pyroxenite banding (referred to as *lithospheric spinel lherzolites*); (ii) pyroxene-depleted spinel harzburgites and Cpx-poor lherzolites, lacking pyroxenite banding (referred to as *reactive spinel peridotites*); (iii) plagioclase-enriched peridotites (referred to as *impregnated plagioclase peridotites*); and (iv) pyroxene-free spinel dunites, with interstitial plagioclase and sporadic gabbroic veins (referred to as *replacive spinel dunites*).

Field, petrographical and structural features

Lithospheric spinel lherzolites. In the field they are characterized by abundant pyroxenes and by widespread parallel, coarse-grained pyroxenite bands (Fig. 2a), formed by up to 1 cm-size pyroxene and spinel grains, locally showing tight similar folding. Centimetre-size rounded Opx + Sp (+Cpx) clusters are frequent in both lherzolites and pyroxenite bands. These lherzolites generally show a porphyroclastic and tectonitic structure preserving less deformed areas with protogranular structure. They have a spinel-facies [olivine (Ol) + orthopyroxene (Opx) + clinopyroxene (Cpx) + spinel (Sp)] assemblage with relatively high Cpx (average 10% vol.) content.

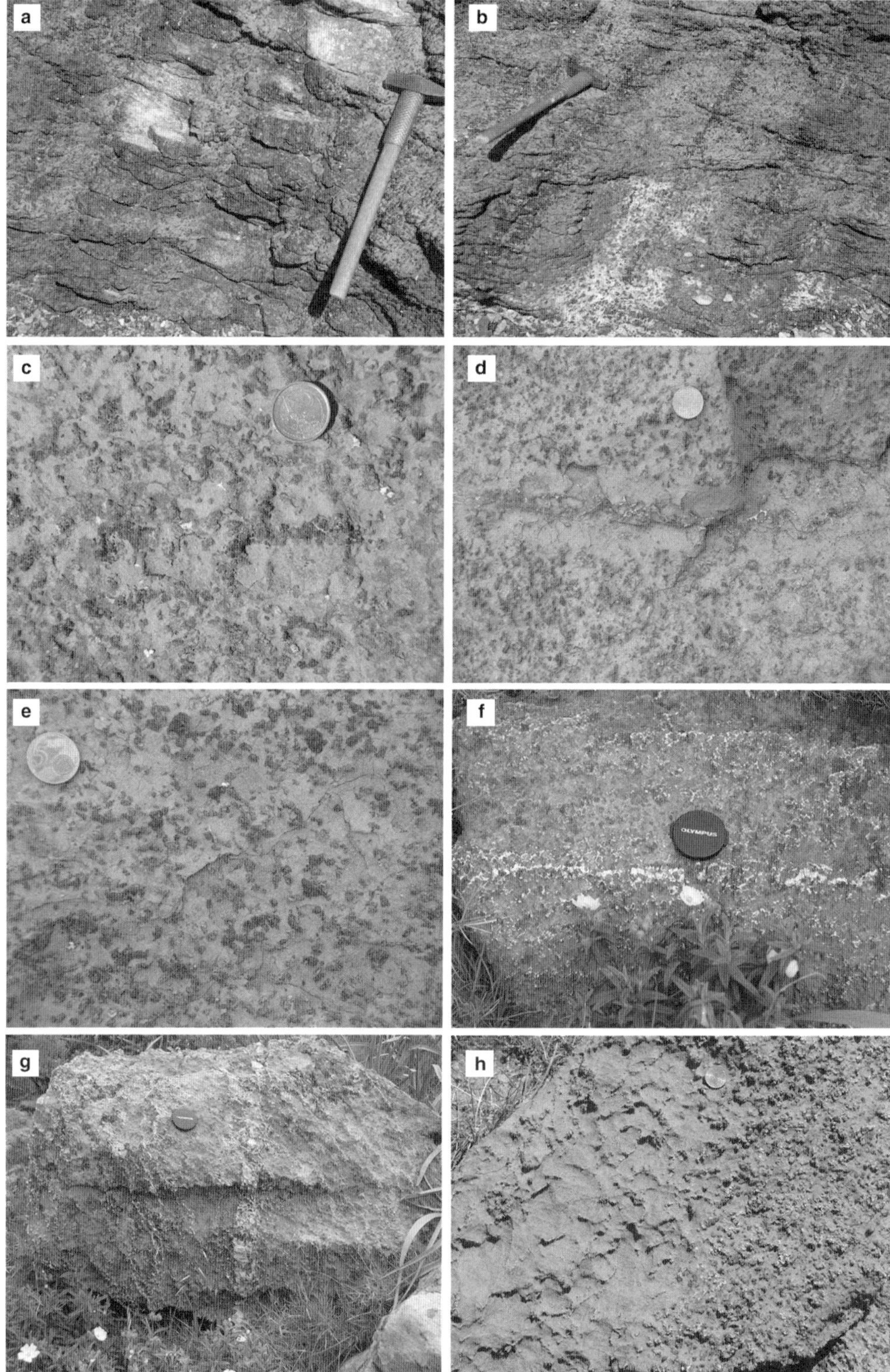

Fig. 2.

This rock type crops out as hm–km-wide masses (where 1 hm = 1 × 10^2 m) embedded within the reactive and impregnated peridotites, and showing rather sharp transitions (a few decimetres (dm)) to the rock types formed by melt–rock interaction. In places, across the transition between pristine lithospheric lherzolites and reactive spinel peridotites, spinel pyroxenite bands in the former continue as spinel dunite bands in the latter. This indicates that the spinel dunite bands, preserving widespread pyroxene and spinel relicts, were formed by pyroxene dissolution and olivine precipitation processes at the expense of pre-existing spinel pyroxenite bands (see later). Frequently, this rock type forms decametric-wide relicts within the reactive spinel peridotites, clearly indicating that they are uncompletely modified remnants of the pristine lithospheric lherzolite protolith. Field evidence and mutual relationships, thus, clearly indicate that reactive and impregnated peridotites replace pristine spinel lherzolites, and, moreover, melt-reactive percolation and impregnation occurred on pyroxenite-veined fertile lherzolites of the extending lithosphere (see below).

In thin section they are characterized by peculiar microstructures: (i) the already mentioned rounded Opx + Sp clusters; and (ii) thin vermicular exsolutions of spinel at the borders of the Opx porphyroclasts, that indicate their subsolidus evolution. In fact, the Opx + Sp clusters represent the complete recrystallization to spinel-facies conditions of precursor garnets; thus indicating that the pristine mantle rock underwent exhumation from garnet-facies conditions and recrystallization at shallower spinel-facies conditions. The vermicular spinels represent the exsolution of Mg-tschermakitic ($MgAlAlSiO_6$) components, in the form of spinel, from pre-existing Al-richer Opx, stable at higher-temperature conditions under spinel-facies conditions. This microstructure indicates that the mantle protolith, once recrystallized at spinel-facies conditions, underwent significant cooling under spinel-facies conditions, that is, under almost isobaric conditions. Geothermometric estimates of the spinel-facies equilibration gave temperatures of about 1000–1100 °C (Müntener & Piccardo 2003; Piccardo 2003, 2007).

Plastic deformation and microstructural features furnish evidence for the subsolidus evolution of these mantle lherzolites that upwelled from garnet-facies conditions and were recrystallized to spinel-facies assemblages at continental lithosphere conditions (i.e. *the lithospheric evolution*). The presence of pristine garnet pyroxenite banding, that is, high-pressure mafic intercalations, suggest that the host peridotite did not undergo partial melting events subsequent to the pyroxenite formation and preceding the subsolidus evolution (i.e. garnet- to spinel-facies decompression recrystallization).

Accordingly, it can be speculated that the pyroxenite-bearing mantle protoliths of the Monte Maggiore peridotites were isolated from the convective mantle under garnet-facies conditions, and were accreted to the thermal lithosphere where they were finally equilibrated and cooled under spinel-facies conditions at an intermediate continental geotherm.

In samples not affected by subsequent plagioclase-forming melt impregnation, a small rim of olivine + plagioclase is locally developed between pyroxenes and spinel, according to the reaction pyroxenes + spinel = olivine + plagioclase (as already evidenced by Müntener & Piccardo 2003). This microstructure provides rare evidence for a metamorphic origin of some plagioclase in

Fig. 2. (*Continued*) Field aspects of the Monte Maggiore peridotites and mafic rocks. (**a**) Lithospheric spinel lherzolites showing dm-wide spinel pyroxenite bands. Pyroxenites are characterized by coarse granular (up to a few cm in size) pyroxenes and cm-size rounded Opx + Sp clusters, suggesting pristine garnet-bearing assemblages. (**b**) Lithospheric spinel lherzolites showing parallel relicts of pristine pyroxenite bands. The host peridotite is strongly depleted in pyroxenes, and the pyroxenite bands are strongly reduced by progressive replacement of new olivine on previous coarse granular pyroxenes. This structural features is particularly evident in thin section. (**c**) Centimetre-wide spinel pyroxenite band significantly replaced along the borders by new olivine, forming two symmetrical cm-wide bands of dunite. The centre of the band is formed by corroded skeletal pyroxene crystals, while the outer borders are mostly composed by new olivine, forming two dunitic bands. Sp + Opx clusters are concentrated in the centre of the band, indicating a compositional banding in the previous garnet-bearing pyroxenite. (**d**) Centimetre-wide spinel pyroxenite band significantly replaced along the borders by new olivine, forming two symmetrical cm-wide bands of dunite bounding the central zone formed by corroded skeletal Cpx and spinel, frequently in Sp + Opx aggregates. Note that Opx + Sp clusters are preserved in both the pyroxenite relict and the host peridotite, indicating precursor garnet-bearing peridotites and pyroxenites. (**e**) Field aspect of a reactive spinel harzburgite. Note the coarse granular, isotropic texture, the abundance of olivine and the presence of some relict Opx + Sp clusters. (**f**) Inpregnated plagioclase-rich peridotite, showing oriented veins rich in plagioclase (from Müntener & Piccardo 2003). (**g**) Impregnated plagioclase-rich peridotites showing oriented parallel plagioclase-rich veins. A later vertical gabbroic dykelet related to the dunite channels discordantly cuts the early plagioclase vein orientation (from Müntener & Piccardo 2003). (**h**) Contact between a strongly impregnated plagioclase peridotite (on the right) and a pyroxene-free replacive dunite channel (on the left). Note the cm-size dimensions of the olivine crystals in the dunite and the presence of thin plagioclase films (in relief) in-between the olivine megacrystals.

the Monte Maggiore peridotites and indicates that pristine subcontinental mantle peridotites underwent decompression from the spinel peridotite to the plagioclase peridotite field during exhumation.

Accordingly, the lithospheric spinel lherzolites preserve records of ancient enrichment events (i.e. veining of pyroxenite bands) at high pressure–high temperature conditions and of different steps of subsolidus evolution under decreasing pressure (P) and temperature (T) conditions (i.e. plastic deformation and garnet- to spinel-facies recrystallization), leading to the thermal equilibration at a subcontinental geothermal gradient. Sporadic growth of plagioclase + olivine assemblages by reaction between spinel-facies pyroxenes and spinel suggests decompression exhumation towards shallower, and most probably colder, lithospheric levels ($P < 1.0$ GPa).

Reactive spinel harzburgites. In the field the investigated rocks are characterized by: (i) highly variable but rather low Cpx (<8% vol.) and high olivine (70–90% vol.) content; (ii) coarse granular textures; (iii) lack of pyroxenite bandings (Fig. 2b); and (iv) presence of cm–dm-wide spinel dunite banding and aligned trains of spinel grains. In places aligned relicts of pristine pyroxenites bands are still preserved, marked by parallel thin bands of corroded clinopyroxene crystals and spinel grains (Fig. 2b).

Spinel dunite bands mostly consist of olivine and spinel, showing relicts of corroded and dissolved pyroxene crystals. Relict coarse-grained pyroxenes (up to 1 cm in size), spinel grains and Sp + Opx clusters are preserved in the centre of the band (Fig. 2c, d). In places, cm–dm-wide patches of partially replaced, coarse pyroxene crystals are preserved and surrounded by cm–dm-wide olivine-rich dunitic zones. These patches are sometimes folded, recalling the tight similar folding of pyroxenite bands in lithospheric spinel lherzolites. Sporadically, across the transition zone between pristine lithospheric lherzolites and reactive spinel harzburgites, spinel pyroxenite bands in the former propagate as spinel dunite bands in the latter. This indicates that the spinel dunite bands, preserving widespread pyroxene and spinel relicts, were formed by pyroxene dissolution and olivine precipitation processes at the expense of pre-existing spinel pyroxenite bands.

In thin section the spinel harzburgites are characterized by the presence of reaction microstructures on the pyroxene crystals, that is, the presence of coronas of new unstrained olivine surrounding the deformed and exsolved pyroxene porphyroclasts. These microstructures are interpreted as having been formed by the interaction of the mantle protolith with a melt that percolated by diffuse porous flow mechanisms. The reaction microstructures on pyroxenes (i.e. pyroxene dissolution and olivine precipitation) suggest that the percolating melts were silica(-pyroxenes)-undersaturated.

Frequently, the reactive harzburgites preserve structural relicts (i.e. Opx + Sp clusters and Sp vermicular exsolution in relict Opx porphyroclasts) of the pristine lithospheric lherzolites, supporting field evidence that reactive harzburgites formed at the expense of pre-existing lithospheric lherzolites. Accordingly, field and microstructural evidence indicates that the mantle protoliths were diffusely percolated by silica-undersaturated melts, which caused their depletion of pyroxenes by dissolution and enrichment of olivine by precipitation. The almost complete disappearance of the pyroxenite bands and the formation of spinel dunite bands pertains to the reactive interaction with the percolating melts that dissolved the pyroxenes and precipitated abundant olivine, leaving solely the alligned trains of spinel grains of the pristine pyroxenites. The reactive spinel peridotites are recognizable in the outcrop by their coarse-granular structure, the rather low pyroxene (Cpx) modal content and, accordingly, the rather high olivine content (Fig. 2e), the lack of pyroxenite bands, and the presence of spinel dunite bands and linear trains of spinel grains.

In summary, the reactive spinel peridotites show structural and petrographical features which indicate that they were formed by the reactive porous flow percolation and reactive interaction of silica-undersaturated melts with the pristine lithospheric lherzolites. Absence of plagioclase in the magmatic assemblages and in the products of the melt–peridotite reactions suggests that the melt-reactive percolation occurred under spinel-facies conditions.

Impregnated plagioclase peridotites. These peridotites are characterized by the presence of plagioclase (Plg), which reaches relatively high modal contents (up to 20% vol.), and by orthopyroxene, which replaces mantle olivine. Plagioclase is present as: (i) microgranular and symplectitic aggregated of Opx + Plg surrounding and replacing spinel-facies Cpx porphyroclasts; (ii) isolated crystals cutting kinked mantle olivine; (iii) granular aggregates of Opx + Plg (microgabbro–noritic aggregates) interstitial to the mantle olivine porphyroclasts; and (iv) mm- to cm-size Plg-rich veins cutting through the peridotite and showing fuzzy contacts with the host rock (Fig. 2f, g). The new Plg and Opx grains in these microstructures are significantly undeformed. These microstructures are interpreted as having been formed by interaction and interstitial crystallization of a percolating melt. Widespread Opx replacement on mantle olivine indicates the silica(-orthopyroxene)-saturated nature of the percolating melts.

In summary, the impregnated plagioclase peridotites are characterized by structural features which indicate that they were formed by porous flow percolation and interstitial crystallization of silica-saturated melts through pre-existing spinel lherzolites and reactive spinel peridotites. The presence of plagioclase in the magmatic assemblages and in the products of the melt–peridotite reactions suggests that the melt-reactive percolation and interstitial crystallization occurred under plagioclase-facies conditions.

Replacive spinel dunites. These rocks consist of olivine crystals, frequently more than 1 cm in size, that enclose single grains and aggregates of spinel (Figs 2h & 3a, b). Sporadic relicts of orthopyroxene almost completely replaced by olivine suggest a melt–pyroxene reaction and indicate that the pristine rock underwent almost complete dissolution of their pyroxenes. Olivine enlargement is, most probably, related to the action of percolating strongly undersaturated melts, which caused olivine overgrowth by continuous olivine precipitation around previous mantle olivine, and the enclosure of previous spinel grains within the growing olivine crystals. Frequently, thin mm-size films of plagioclase are present surrounding the olivine megacrystals (Figs 2h & 3a). In some places new magmatic pyroxenes crystallize together with plagioclase at olivine triple junctions and along olivine grain boundaries; locally, these films merge to form mm- to cm-wide gabbroic veins showing fuzzy contacts with the surrounding dunite (Fig. 3b). Gabbro–noritic veins and dykelets are formed that are rooted in the dunites and intrude the host peridotite (Fig. 3c).

These dunites are, accordingly, interpreted as the products of reactive and focused percolation through spinel and plagioclase peridotites of silica-undersaturated melts that completely dissolved pyroxenes and plagioclase, and caused olivine crystallization and overgrowth to cm-size dimensions. The plagioclase films and gabbro–noritic veins and dykelets represent the incipient interstitial crystallization of the melts migrating within the dunite channels. These replacive channels were, later on, exploited by the upward-focused migration of saturated melts, which formed gabbro–noritic dykelets and cumulate pods.

Mafic–ultramafic cumulates. In places, decametre-scale pods of coarse-granular–pegmatoid magmatic rocks intrude the impregnated plagioclase peridotites: they consist of pyroxene-rich mafic–ultramafic cumulates showing olivine gabbro–noritic composition (Fig. 3d, e). Pyroxenes are dominant on olivine and Opx is very abundant. Olivine and pyroxenes are generally euhedral, frequently reaching cm-size dimensions, whereas plagioclase is interstitial: accordingly, the crystallization order is olivine > pyroxenes > plagioclase (Fig. 3e). In places, a broad dm-wide compositional layering is evident, going from (bottom) ultramafic olivine + pyroxene-rich cumulates to plagioclase-rich gabbro–noritic cumulates to (top) almost pure plagioclase cumulates.

Gabbroic bodies and dykes. The melt percolation and impregnation stages were followed by the intrusion of variably fractionated magmas, which formed gabbroic dykes (Fig. 3f) and decametre-scale cumulate intrusive masses (Fig. 3g). These intrusive rocks mostly consist of Mg gabbros (olivine and pyroxenes gabbros) and Fe–Ti gabbros (Fe–Ti oxide gabbros). The gabbroic dykes show porphyritic tectures made up of cm-size Ol, Cpx and Plg phenocrysts in a medium-grained matrix formed by the same minerals. Decametric intrusive bodies are composed of olivine cumulates, olivine gabbros and Fe–Ti gabbros. They show the following crystallization order: olivine > plagioclase > clinopyroxene > orthopyroxene > Fe–Ti oxides and titanian–pargasite amphibole, which is typical of MORB gabbros from modern oceans and from the ophiolitic sequences of the Alpine–Apennine system.

The different gabbroic rock types crystallized from magmas variably evolved, as indicated by the olivine composition that ranges from Fo81 (in the less evolved Mg gabbros) to Fo68 (in the more evolved Fe gabbros).

Basaltic dykes. All previous structures are cut by dm-wide medium- to fine-grained basaltic dykes, which sporadically show cm-size plagioclase and, rarely, olivine phenocrysts in a fine-grained ophitic (clinopyroxene + plagioclase) groundmass (Fig. 3h).

Field relationships

In the field, the mutual relationships between the different ultramafic rock types are clearly visible. The pyroxenite-bearing lithospheric spinel lherzolites are the oldest rock types as they are replaced by hm-scale masses of coarse-granular, pyroxenite-free reactive spinel harzburgites, frequently showing the transition in contacts a few cm wide, demonstrating the abrupt modification of both composition and structure. The pyroxenite bands of the pristine lherzolites are progressively dissolved in the reactive harzburgites, and spinel dunite bands and aligned trains of spinel grains in the spinel harzburgites testify to the pre-existing pyroxenite bands in

Fig. 3.

the spinel lherzolite protolith. Both lithospheric lherzolites and reactive peridotites are locally enriched in plagioclase, frequently reaching 15% modal content, and transformed into impregnated plagioclase peridotites. Plagioclase is distributed in the rock along the contacts of the mineral grains of the previous spinel-facies assemblage and, in places, it is concentrated in veins, showing fuzzy contacts with the host peridotite. Metre- to decametre-wide bands of replacive spinel dunites cut across the pre-existing spinel harzburgites and plagioclase peridotites. Millimetre-size gabbroic dykelets and decametre-size mafic–ultramafic cumulates, showing fuzzy contacts with the host peridotites, cut all of the previous rock types. The youngest magmatic event is represented by the intrusion of decameter-size gabbroic bodies, metre-wide gabbroic dykes and dm-wide fine-grained porphyritic mafic dykes. These intrusive bodies show sharp contacts and sporadically chilled margins against the host peridotite.

Composition

New bulk-rock and mineral major and trace element compositions data are reported for spinel peridotites, mafic–ultramafic cumulates, gabbroic dykelets and gabbro dykes, whereas additional compositional data on spinel and plagioclase peridotites are taken from the literature (Müntener & Piccardo 2003; Piccardo *et al.* 2004; Rampone *et al.* 2008).

Analytical methods

Bulk rock major element compositions were determined using X-ray fluorescence (XRF) (Philips PW1480) at the Dipartimento di Scienze della Terra dell'Università degli Studi di Modena e Reggio Emilia, Italy, on pressed pellets, using the methods of Leoni & Saitta (1976). Bulk-rock trace element compositions were analysed using inductively coupled plasma mass spectroscopy (ICP-MS) at the Activation Laboratories (ACTLABS, Toronto, Canada). Major element compositions of minerals were determined by electron microprobe JEOL JXA-8600, at IGG-CNR, Florence section, according to the procedure described by Vaggelli *et al.* (1999). Trace element compositions of minerals were analysed using *in situ* laser ablation-ICP-MS (LA-ICP-MS) techniques at IGG-CNR, Pavia (Italy). Precision and accuracy (both better than 10% for concentrations at ppm levels) were assessed from repeated analyses of SRN NIST 612 and BCR-2 standards. Full details of the analytical parameters and quantification procedures can be found in Tiepolo *et al.* (2003).

Peridotites

Lithospheric spinel lherzolites. Major element bulk-rock compositions recalculated on a loss on ignition (LOI)-free basis (Table 1) are characterized by relatively low Mg (MgO 39.8–40.9 wt%), high Si (SiO_2 45.4–45.5 wt%), Al (Al_2O_3 2.73–3.12 wt%) and Ca (CaO 2.59–3.09 wt%) approaching fertile primordial mantle (PM) compositions (Hofmann 1988). The calculated modal compositions (Table 1), accordingly, have relatively high pyroxene (Opx 30.6–31.5% vol., Cpx 9.1–11.4% vol.) and low olivine (Ol 54.4–58.2% vol.) content.

Available mineral chemistry data show that spinel in the cluster structure is rather rich in Al (Al_2O_3 44.7 wt%) and significantly poor in Ti (TiO_2 0.12 wt%) (Table 2, Fig. 4). Spinel of the vermicular exsolution in orthopyroxene porphyroclasts is relatively rich in Al (Al_2O_3 51.7 wt%) (Fig. 4), confirming their origin as exsolution of the Mg-tschermakitic components. Spinel-facies porphyroclastic Cpx preserves rather high Na and Al ($Na_2O > 1.0$ wt%, $Al_2O_3 > 7.0$ wt%) and low Si ($SiO_2 < 50.51$ wt%) content (Table 3).

Reactive spinel harzburgites. Major element bulk-rock compositions recalculated on a LOI-free basis (Table 1) (data from Romairone 1996) are characterized relatively high Mg (MgO 44.4–48.1 wt%), and low Si (SiO_2 41.0–43.4 wt%), Al (Al_2O_3 1.0–1.6 wt%) and Ca (CaO 0.7–1.9 wt%),

Fig. 3. (*Continued*) Field aspects of the Monte Maggiore peridotites and mafic rocks. (**a**) Particular of replacive dunite channel. Note the presence of interstitial pyroxenes, accompanying plagioclase, between the huge olivine megacrystals. (**b**) Replacive dunite channel showing anastomosing mm-size plagioclase films to form a cm-size gabbroic dykelet. Note that pyroxene crystals tend to be euhedral in the centre of the dykelet. (**c**) A cm-size gabbroic dykelet, showing rounded euhedral pyroxene crystals, that cuts across the contact between impregnated plagioclase peridotite (right up) and a replacive dunite channel (left down) (from Müntener & Piccardo 2003). (**d**) Field aspect of the mafic–ultramafic cumulates. In this outcrop the ultramafic layers are dominant, whereas plagioclase is concentrated along cm-size layers. (**e**) Sample of a mafic cumulate (the fly is given as scale). Note the pegmatoid cm-wide grain size and the euhedral habits of olivine and pyroxenes (from Müntener & Piccardo 2003). (**f**) Plagioclase-rich, metre-wide pegmatoid gabbroic dyke (from Müntener & Piccardo 2003). (**g**) Decametre-size pod of olivine cumulate. Note the huge size of the olivine crystals and their perfect euhedral shape. Plagioclase and, rarely, pyroxenes are interstitial. (**h**) Decimeter-size medium-grained basaltic dyke showing porphyritic texture and widespread plagioclase phenocrysts.

Table 1. *Major element bulk-rock composition of Mt Maggiore peridotites*

Sample	MM01	MM02	MM1/5†	MM1/9†	MM2/10†	MM2/11†	MM2/12†	MM1/10*	MM1/11*	MM1/4*	MM2/5A*
Rock type	Lith. spinel lherzolite	Lith. spinel lherzolite	Reactive Sp-perid.	Reactive Sp-perid.	Reactive Sp-perid.	Reactive Sp-perid.	Reactive Sp-perid.	Impr. Plg-perid.	Impr. Plg-perid.	Impr. Plg-perid.	Impr. Plg-perid.
SiO_2	45.52	45.42	43.17	43.80	42.80	41.42	42.83	44.60	44.29	44.30	44.76
TiO_2	0.08	0.07	0.04	0.04	0.04	0.02	0.03	0.04	0.04	0.04	0.05
Al_2O_3	3.12	2.73	1.14	1.57	1.44	0.98	1.30	3.22	3.05	2.73	4.94
FeO^T	8.29	8.13	7.62	7.60	8.12	8.18	7.91	8.06	8.18	8.25	7.90
MnO	0.12	0.12	0.14	0.14	0.15	0.14	0.14	0.14	0.14	0.14	0.14
MgO	39.78	40.94	46.37	44.88	45.84	48.54	46.46	41.95	42.51	42.62	39.42
CaO	3.09	2.59	1.53	1.97	1.61	0.71	1.33	1.99	1.79	1.92	2.63
Na_2O	n.d.	n.d.	n.d.	n.d.	n.d.	n.d.	n.d.	n.d.	n.d.	n.d.	n.d.
K_2O	n.d.	n.d.	n.d.	n.d.	n.d.	n.d.	n.d.	n.d.	n.d.	n.d.	n.d.
Total	100.00	100.00	100.00	100.00	100.00	100.00	100.00	100.00	100.00	100.00	99.85
Modes (wt%) calculated with Petmix4											
Ol	54.4	58.2	82.5	76.5	81.6	93.4	83.5	68.3	70.8	71.1	65.1
Cpx	11.4	9.1	5.4	7.5	6.0	2.3	4.8	2.7	2.1	3.3	1.3
Opx	31.5	30.6	11.7	15.4	11.4	3.4	11.0	23.3	21.5	21.5	22.3
Sp	2.7	2.1	0.4	0.6	1.1	0.9	0.7				
Plg								5.7	5.4	4.1	11.2

*Data from Rampone *et al.* (2008).
†Data from Romairone (1996).
Abbreviations: Lith., lithospheric; Sp-perid., spinel peridotite; Plg-perid., plagioclase peridotite; Impr., impregnated; n.d., not determined.

Table 2. *Major element composition of spinels in Mt Maggiore peridotites*

Type	Spinel (1) Cluster in lith.	Spinel (4) Exsolution in lith.	Spinel (1) Reactive	Spinel (4) Impregnated
SiO_2	0.05	0.09	0.06	0.10
TiO_2	0.12	0.13	0.24	0.52
Al_2O_3	44.66	51.78	42.16	31.24
Fe_2O	1.46	2.33	2.41	4.16
FeO	11.63	10.29	10.78	14.66
MnO	0.18	0.31	0.37	0.31
MgO	17.92	19.47	18.21	14.78
CaO	0.05	0.09	0.02	0.07
Cr_2O_3	24.03	15.72	25.96	34.60
Total	99.95	99.96	99.96	100.21
Cr#	26.52	16.92	29.23	42.62

Abbreviation: Lith., lithospheric lherzolites.

with respect to fertile PM compositions (Hofmann 1988). Al and Ca are slightly enriched (Al_2O_3 2.0–2.4 wt% and CaO 1.4–1.6 wt%) in samples that show a small percentage of plagioclase (Plg $< 5\%$ vol.). The calculated modal compositions (Table 1), accordingly, are very rich in olivine (Ol 76.5–93.4% vol.) and poor in pyroxenes (Opx 3.4–15.4% vol., Cpx 2.3–7.5% vol.).

Available mineral major and trace element (Tables 3–5) data (data from Romairone 1996; Müntener & Piccardo 2003; Piccardo *et al.* 2004; Rampone *et al.* 2008) indicate that clinopyroxenes show high Mg# (91–94), relatively low incompatibile element content (Ti, Zr and LREE), and Sr (0.8–2.1 ppm) and normalized REE (rare earth element) patterns rather flat in the MREE–HREE region (at $<10 \times$ C1), showing a strong LREE negative fractionation (Ce_N/Sm_N 0.02–0.06) (Fig. 5). Orthopyroxenes show high Mg# (91–93) and relatively low incompatible trace element content: their normalized REE patterns are progressively fractionated from HREE to LREE (Ce_N/Yb_N 0.001–0.004), starting from Lu_N at 2–3 $\times$ C1 (Fig. 5). Granular spinels maintain a relatively high Al content (Al_2O_3 42.2 wt%), while their Ti content is significantly increased (TiO_2 0.24 wt%) with respect to spinels in clusters from the spinel lherzolite protoliths (Table 2). The spinel representative composition falls, in the TiO_2 v. Cr# diagram of Figure 4, into the field of spinels in equilibrium with MORBs (Dick & Bullen 1984). The Ti increase in spinel is, most probably, owing to Ti equilibrium partitioning with the percolating basaltic melt.

Impregnated plagioclase peridotites. Available major element bulk-rock compositions (Table 1) (data from Rampone *et al.* 2008), recalculated on a LOI-free basis, demonstrate their significantly different compositional characteristics with respect to the reactive spinel peridotites. In fact, they show lower Mg (MgO 39.11–42.23 wt%) and higher Si (SiO_2 43.89–44.41 wt%), Al (Al_2O_3 2.71–4.90 wt%) and Ca (CaO 1.68–2.73 wt%) content. The calculated modal compositions (Table 1), accordingly, are relatively lower in olivine (Ol 65.1–71.1% vol.) and clinopyroxene (Cpx 1.3–3.3% vol.) and higher in orthopyroxene (Opx 21.5–23.3% vol.). Plagioclase content varies in the range 4–11% vol. Bulk-rock major element and modal compositions are consistent with field and petrographical–microstructural evidence, which shows that impregnated plagioclase peridotites mostly derive from pristine reactive spinel peridotites that underwent significant replacement of olivine by orthopyroxene and of clinopyroxene by orthopyroxene + plagioclase aggregates, and the interstitial formation of plagioclase, during percolation of silica-saturated melts.

Available mineral major and trace element (Tables 4–6) compositions (data from Romairone 1996; Müntener & Piccardo 2003; Piccardo *et al.* 2004; Rampone *et al.* 2008) indicate that clinopyroxenes show high Mg# (91.2–91.5) and are significantly enriched in many trace elements (i.e. REE, Sc, V, Ti, Zr, Y), with respect to the Cpx of the reactive harzburgites. Their normalized REE patterns are convex-upwards with MREE–HREE greater than 10 $\times$ C1, and show a strong LREE negative fractionation (Ce_N/Sm_N 0.03–0.06) and an evident Eu negative anomaly, suggesting equilibration with plagioclase (Fig. 5). Orthopyroxenes, both mantle porphyroclasts and new magmatic grains, show high Mg# (92–93) and greater trace element content with respect to the orthopyroxene of reactive peridotites (Fig. 5).

Plagioclases are relatively rich in Ca (An89). They show relatively low Sr content

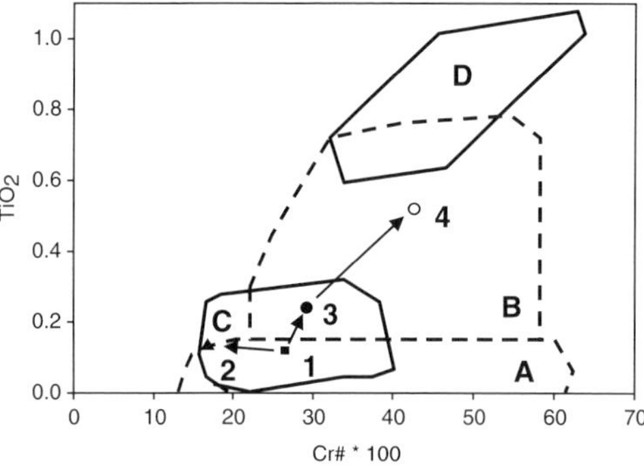

Fig. 4. TiO_2 v. Cr# × 100 diagram for spinels in spinel and plagioclase-bearing spinel peridotites of Monte Maggiore. (1) Porphyroclastic spinel (average value of very similar compositions) (Al_2O_3 44.7 wt%) in Opx + Sp clusters from spinel peridotites. They represent the breakdown product of pristine mantle garnets that are remnants of the pre-percolation mantle protoliths. Note that the TiO_2 content (TiO_2 0.12 wt%) is significantly small. (2) Vermicular spinel exsolution (average value of very similar compositions) (Al_2O_3 51.7 wt%) in mantle orthopyroxene porphyroclasts from spinel peridotites. They represent subsolidus exsolutions of Mg-tschermakite components, in the form of Mg–Al-rich spinel, from pristine Al-rich spinel-facies orthopyroxene during cooling under spinel-facies conditions. (3) Porphyroclastic spinel (average value of very similar compositions) (Al_2O_3 42.2 wt%) in reactive spinel harzburgites. Note that the Ti content is significantly increased (TiO_2 0.24 wt%) with respect to pre-percolation spinel (1), most probably due to equilibration with the percolating basaltic melt. The representative point falls, accordingly, in the field (B) of spinel compositions in equilibrium with MORBs. (4) Porphyroclastic spinel (average value of very similar compositions) (Al_2O_3 32.2 wt%) in plagioclase-bearing spinel peridotites. Note that TiO_2 and Cr# are greatly increased and the representative point falls inside the field (B) of spinel compositions in equilibrium with MORBs. Also reported for comparison: (**a**) the field of spinel compositions in refractory abyssal peridotites (Dick & Bullen 1984); (**b**) the field of spinel compositions in equilibrium with MORB melts (Dick & Bullen 1984); (**c**) the field of spinel compositions in reactive spinel peridotites of the South Lanzo Massif (Piccardo *et al.* 2007*a*); and (**d**) the field of spinel compositions in impregnated plagioclase peridotites of the South Lanzo Massif (Piccardo *et al.* 2007*a*).

(6.8–28.6 ppm), with respect to plagioclase in equilibrium with a MORB melt, and have normalized REE patterns characterized by LREE negative fractionation (Fig. 5; Ce_N/Sm_N <0.3–0.4).

Porphyroclastic spinels have relatively low Al content (Al_2O_3 average composition 32.2 wt%), consistent with equilibration with plagioclase, whereas Ti and Cr# are strongly increased (TiO_2 average composition 0.52, Cr# average composition 42.6; Table 2) with respect to spinel in reactive spinel peridotites. The average composition of spinel falls in the TiO_2 v. Cr# diagram of Figure 4 inside the field (B) of spinel compositions in equilibrium with MORBs (Dick & Bullen 1984). Ti increase in spinel is, most probably, owing to Ti equilibrium partitioning with the percolating basaltic melt.

Mafic–ultramafic cumulate and gabbroic dykelets. Available mineral major and trace element (Tables 7–9) compositions indicate that these rocks are characterized by olivine relatively rich in forsterite (Fo90), plagioclase very rich in Ca (An86.6–91.5) and pyroxenes high in Mg# (in the range 90.0–93.7).

Clinopyroxenes are rich in Cr (Cr_2O_3 1.20–1.40 wt%) and relatively poor in Ti (TiO_2 0.28–0.44 wt%). Their normalized REE patterns are rather flat in the HREE–MREE region (at about 10 × C1), do not show any Eu anomalies and are strongly LREE fractionated (Fig. 5; La_N/Sm_N 0.003–0.015). They have, moreover, very low La_N (0.02–0.16), Sr (0.71–2.06 ppm) and Zr (2.29–6.63 ppm) content. Orthopyroxenes are relatively rich in Cr (Cr_2O_3 0.57–0.97 wt%). Plagioclases are characterized by very small content of Ce_N (0.013–0.03) and Sr (15.6–30.4 ppm): their normalized REE patterns show a strong negative LREE fractionation (Fig. 5; La_N/Sm_N 0.06–0.48).

Gabbroic dykes. Available mineral major and trace element (Tables 10–13) compositions indicate that these rocks are characterized by olivine (Fo81.4–67.9), plagioclase (An55.8–41.4) and clinopyroxene (Mg# 85.3–74.1).

Olivine Fo, clinopyroxene Mg# and plagioclase An compositions are positively correlated (Fig. 6). Ti-magnetite crystals are present in samples showing lower Fo, Mg# Cpx and An contents, indicating the transition to oxide gabbroic compositions.

The more primitive dykes (showing the higher Fo content in olivine) have bulk-rock normalized REE patterns characterized by flat patterns in the MREE–HREE region, an evident LREE negative fractionation and a significant Eu_N positive anomaly, suggesting a cumulus phase of plagioclase, according to the crystallization order (Fig. 7a).

Clinopyroxenes have compositions that vary from diopside (in the more primitive rocks) to augite (in the more evolved rocks), and the overall normalized REE patterns show a significant LREE negative fractionation. The normalized concentration range from MREE to HREE is about 10 × C1 in Cpx from the more primitive samples and about 50 × C1 in Cpx from the more evolved samples. The Eu_N anomaly becomes progressively negative with increasing fractionation (Fig. 7b).

Table 3. *Major element compositions of mineral in Mt Maggiore reactive spinel peridotites*

Sample	MM1/5	MM1/5	MM1/5	MM1/5	MM2/10	MM2/10	MM2/10	MM2/10	MM2/10
Rock type	Reactive Sp-perid.	Reactive Sp-perid.	Reactive Sp-perid.	Reactive Sp-perid.	Reactive Sp-perid.	Reactive Sp-perid.	Reactive Sp-perid.	Reactive Sp-perid.	Reactive Sp-perid.
Mineral	Cpx pc	Cpx pr	Opx pc	Opx pr	Cpx pc	Cpx pr	Opx pc	Opx pr	Ol
SiO_2	51.29	52.20	54.81	54.34	50.88	51.80	54.47	55.25	40.63
TiO_2	0.34	0.29	0.06	0.09	0.29	0.23	0.14	0.06	0.09
Cr_2O_3	1.26	1.00	0.53	0.47	0.94	0.70	0.47	0.29	0.04
Al_2O_3	5.78	4.62	4.05	3.62	6.53	4.54	4.60	3.62	0.00
FeO^T	2.80	2.58	5.74	5.65	3.05	2.62	5.79	6.08	8.97
MnO	0.60	0.07	0.17	0.15	0.06	0.11	0.21	0.09	0.21
MgO	16.74	17.34	33.49	33.69	16.59	17.50	32.81	33.75	49.41
CaO	21.37	21.81	1.07	0.96	21.88	22.38	1.53	0.86	0.09
Na_2O	0.30	0.10	0.03	0.00	0.11	0.09	0.05	0.00	0.00
K_2O	0.00	0.00	0.00	0.00	0.00	0.00	0.00	0.00	0.00
NiO	n.d.	n.d.	n.d.	n.d.	n.d.	n.d.	n.d.	n.d.	n.d.
Mg#	93.15	92.10	93.10	92.40	90.70	94.00	92.60	92.50	
An%									
Fo%									90.60

Abbreviations: Sp-perid., spinel peridotite; pc, phenocryst core; pr, phenocryst rim; n.d., not determined.
Data from Romairone (1996).

Table 4. *Major element compositions of mineral in Mt Maggiore impregnated plagioclase peridotites*

Sample	MM1/10	MM1/10	MM1/10	MM1/10	MM1/10	MM2/5A	MM2/5A	MM2/5A	MM2/5A	MM2/5A	MM2/5A	MM2/5A
Rock type	Impr. Plg-perid.	Impr. Plg-perid.	Impr. Plg-perid.	Impr. Plg-perid.	Impr. Plg-perid.	Impr. Plg-perid.	Impr. Plg-perid.	Impr. Plg-perid.	Impr. Plg-perid.	Impr. Plg-perid.	Impr. Plg-perid.	Impr. Plg-perid.
Mineral	Cpx pc	Cpx pr	Opx pc	Opx nmr	Ol	Cpx pc	Cpx pr	Opx pc	Opx pr	Opx nmr	Plg	Ol
SiO_2	50.81	51.94	54.57	55.45	40.96	52.72	52.25	55.50	55.74	56.48	46.08	40.88
TiO_2	0.31	0.36	0.14	0.12	0.07	0.32	0.47	0.17	0.27	0.21	0.09	0.12
Cr_2O_3	1.01	1.06	0.62	0.61	0.03	1.12	1.23	0.74	0.81	0.63		0.08
Al_2O_3	6.18	4.08	4.06	2.77	0.00	4.78	4.37	4.50	3.53	2.61	34.73	0.00
FeO^T	3.27	2.67	5.90	6.16	8.70	3.96	2.99	6.65	6.47	6.54	0.29	9.97
MnO	0.08	0.08	0.18	0.23	0.23	0.12	0.17	0.13	0.20	0.22		0.22
MgO	17.51	17.75	32.78	33.03	49.77	20.86	16.89	33.52	33.39	33.44	0.15	48.76
CaO	20.63	21.97	1.31	1.29	0.09	17.48	23.06	1.16	2.37	1.22	18.17	0.18
Na_2O	0.11	0.13	0.03	0.01	0.00	0.12	0.15	0.09	0.11	0.06	1.10	0.00
K_2O	0.00	0.00	0.00	0.00	0.00	0.00	0.00	0.00	0.00	0.00	0.16	0.00
NiO	n.d.	n.d.	n.d.	n.d.	n.d.	n.d.	n.d.	n.d.	n.d.	n.d.	n.d.	n.d.
Mg#	91.20	92.65	91.90	92.55		91.50	91.52	92.30	93.70	90.43		
An%											89.30	
Fo%					90.90							89.50

Sample	MM2/6A	MM2/6A	MM2/6A	MM2/6A	MM2/6B	MM2/6B	MM2/6B	MM2/6B
Rock type	Impr. Plg-perid.	Impr. Plg-perid.	Impr. Plg-perid.	Impr. Plg-perid.	Impr. Plg-perid.	Impr. Plg-perid.	Impr. Plg-perid.	Impr. Plg-perid.
Mineral	Cpx pc	Cpx pr	Opx pc	Ol	Cpx pc	Cpx pr	Opx pc	Opx nmr
SiO_2	51.13	51.78	55.02	40.63	51.35	52.93	54.58	55.51
TiO_2	0.42	0.46	0.18	0.07	0.39	0.38	0.19	0.19
Cr_2O_3	1.20	1.13	0.64	0.04	1.24	0.87	0.64	0.64
Al_2O_3	5.64	4.43	3.02	0.00	5.43	2.58	3.73	2.34
FeO^T	2.93	2.67	6.18	9.05	2.92	2.19	6.22	6.16
MnO	0.11	0.11	0.18	0.15	0.07	0.09	0.17	0.25
MgO	16.89	17.13	33.90	24.77	17.46	17.88	33.41	34.06
CaO	21.85	22.50	0.83	24.96	21.37	22.90	1.03	0.77
Na_2O	0.13	0.14	0.04	0.00	0.10	0.09	0.02	0.03
K_2O	0.01	0.00	0.00	0.00	0.00	0.00	0.00	0.00
NiO	n.d.	n.d.	n.d.	n.d.	n.d.	n.d.	n.d.	n.d.
Mg#	90.95	92.25	93.30		91.53	93.70	92.95	92.70
An%								
Fo%				90.60				

Sample	MM2/5B	MM2/5B	MM2/5B	MM2/5B	MM2/5B	MM2/6	MM2/6	MM2/6	MM2/6	MM2/6
Rock type	Gabbroic vein	Gabbroic vein	Gabbroic vein	Gabbroic vein	Gabbroic vein	Gabbroic vein	Gabbroic vein	Gabbroic vein	Gabbroic vein	Gabbroic vein
Mineral	Cpx pc	Cpx pr	Opx pc	Opx pr	Plg	Cpx pc	Opx pc	Opx nmr	Ol	Plg
SiO_2	51.17	52.00	55.54	55.88	46.29	51.40	55.09	55.46	40.77	45.98
TiO_2	0.38	0.45	0.26	0.25	0.08	0.36	0.18	0.21	0.02	0.04
Cr_2O_3	1.25	1.31	0.65	0.67		1.18	0.54	0.59	0.00	
Al_2O_3	5.02	3.90	3.12	2.64	34.61	5.33	3.25	2.47	0.00	33.98
FeO^T	3.09	3.07	6.62	6.79	0.28	3.02	6.13	6.28	9.29	0.19
MnO	0.16	0.13	0.23	0.20		0.09	0.18	0.20	0.21	
MgO	16.53	17.14	32.72	33.35	4.82	17.11	33.65	34.00	49.18	0.44
CaO	22.59	22.76	1.15	1.06	13.99	21.40	0.92	0.87	0.08	16.94
Na_2O	0.16	0.19	0.09	0.05	0.87	0.12	0.01	0.03	0.00	1.68
K_2O	0.00	0.00	0.00	0.00	0.14	0.00	0.00	0.00	0.00	0.12
NiO	n.d.	n.d.	n.d.	n.d.	n.d.	n.d.	n.d.	n.d.	n.d.	n.d.
Mg#	91.75	93	89.9	90.7		90.86	92.5	92.15		
An%					89.45					84.2
Fo%									90.2	

Abbreviations: Impr., impregnated; Plg-perid., plagioclase peridotite; pc, pshenocryst core; pr, phenocryst rim; nmr, new magmatic rim; n.d., not determined.
Data from Romairone (1996).

Table 5. *Trace element composition of minerals in Mt Maggiore spinel peridotites*

Sample	MM1/5*	MM1/5*	MM1/9*	MM1/9*	MM2/10*	MM2/10*	MM2/12*	MM2/12*	C221/2*	CC1[†]	CC1[†]
Rock type	Reactive Sp-perid.	Reactive Sp-perid.	Reactive Sp-perid.	Reactive Sp-perid.	Reactive Sp-perid.	Reactive Sp-perid.	Reactive Sp-perid.	Reactive Sp-perid.	Reactive Sp-perid.	Reactive Sp-perid.	Reactive Sp-perid.
Mineral	Cpx	Opx	Cpx (2)	Opx	Cpx (2)	Opx	Cpx (2)	Opx	Cpx (2)	Cpx	Opx
Sc	56	27	53	32	58	37.0	55	34	44	na	na
Ti	1502	557	1444	498	1421	646	1596	687	1269	1678	779
V	294	135	300	155	286	171	306	173	234	na	na
Sr	2.1	0.02	0.89	0.02	0.92	0.05	0.80	0.20	0.83	na	na
Y	11.7	1.60	12.3	1.66	13.3	2.4	12.2	2.6	9.8	12.5	1.64
Zr	2.8	0.48	1.70	0.24	1.48	0.38	1.38	0.46	1.64	na	na
La	0.008	bd	0.005	bd	0.005	bd	0.002	bd	bd	bd	bd
Ce	0.12	0.002	0.073	0.001	0.058	0.004	0.067	0.007	0.063	0.071	bd
Pr	0.069	bd	0.048	bd	0.043	bd	0.046	bd	bd	0.047	bd
Nd	0.75	0.015	0.76	0.007	0.66	0.027	0.65	0.038	0.51	0.74	0.012
Sm	0.79	0.019	0.70	0.012	0.68	0.032	0.65	0.043	0.48	0.68	0.012
Eu	0.32	0.013	0.31	0.008	0.25	0.022	0.29	0.023	0.23	0.31	0.008
Gd	1.32	0.095	1.37	0.041	1.29	0.114	1.32	0.137	1.09	1.24	0.040
Tb	0.27	0.033	0.28	0.015	0.30	0.027	0.28	0.032	bd	0.26	0.015
Dy	1.85	0.26	2.0	0.20	2.2	0.29	2.1	0.38	1.54	1.92	0.20
Ho	0.41	bd	0.44	0.060	0.48	0.084	0.48	0.098	bd	0.42	0.059
Er	1.31	0.21	1.26	0.23	1.40	0.32	1.38	0.36	0.86	1.23	0.23
Tm	0.175	0.041	0.192	0.046	0.210	0.057	0.209	0.059	bd	0.187	0.043
Yb	1.26	0.36	1.25	0.33	1.41	0.41	1.20	0.44	0.94	1.22	0.25
Lu	0.16	0.048	0.17	0.059	0.20	0.073	0.18	0.058	bd	0.17	0.0591
Hf	0.27	0.027	0.175	0.033	0.22	0.039	0.23	0.074	bd	bd	bd

*Data from Rampone *et al.* (2008).

[†]Data from Müntener & Piccardo (2003).

Abbreviations: Sp-perid., spinel peridotite; bd, below detection; na, not analysed.

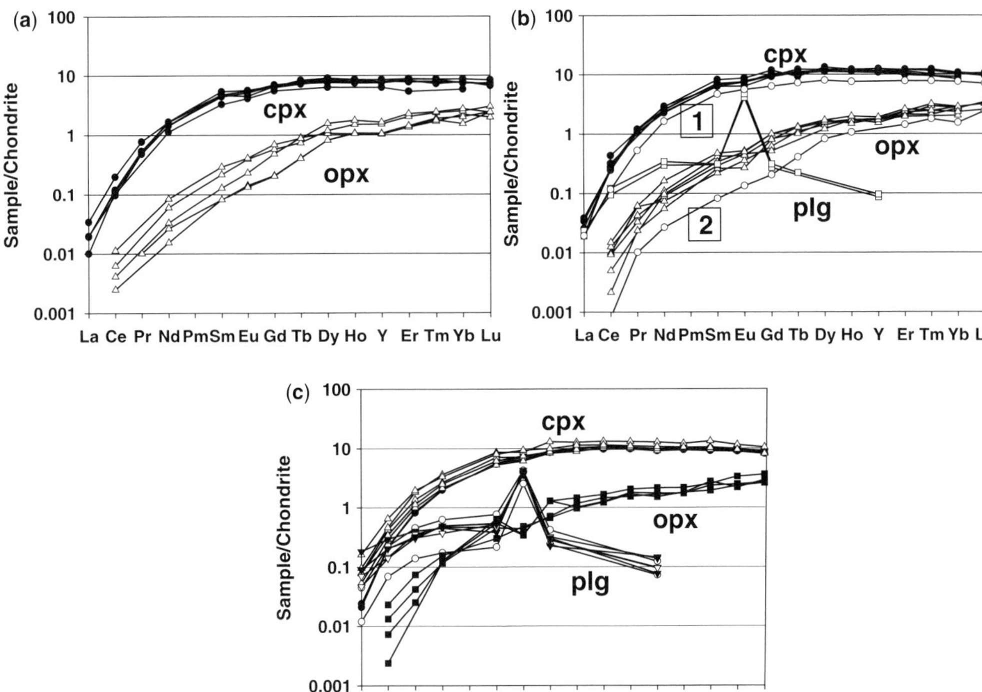

Fig. 5. C1-normalized REE patterns of mineral in Monte Maggiore peridotites, cumulates and gabbroic dykelets. (**a**) Clinopyroxenes (Cpx) and orthopyroxenes (Opx) from some reactive spinel peridotites (data from Müntener & Piccardo 2003; Rampone *et al.* 2008). Note the strong LREE fractionation of the clinopyroxene REE patterns. (**b**) Clinopyroxenes (Cpx), orthopyroxenes (Opx) and plagioclases (Plg) from some plagioclase-bearing peridotites (data from Müntener & Piccardo 2003; Rampone *et al.* 2008). Note the strong LREE fractionation of clinopyroxenes and the negative LREE fractionation of plagioclases. Also reported for comparison: [1] the REE pattern of a representative clinopyroxene of a reactive spinel peridotite; and [2] the REE pattern of a representative orthopyroxene of a reactive spinel peridotite. (**c**) Clinopyroxenes (Cpx), orthopyroxenes (Opx) and plagioclases (Plg) from the studied mafic–ultramafic cumulates and dykelets. Note the strong LREE fractionation of clinopyroxenes and the negative LREE fractionation of plagioclases. The cumulus clinopyroxenes show a slightly increasing absolute content and parallel Eu negative anomaly, most probably related to slight melt evolution and concomitant plagioclase crystallization during intrusion.

Plagioclases are relatively rich in Sr (314–349 ppm) and have normalized REE patterns showing positive LREE fractionation, similar to plagioclase in equilibrium with MORB: similarly to clinopyroxenes, the absolute REE concentrations increase with the increasing degree of evolution of the host rock (Fig. 7b).

Discussion

The mantle protolith

The presence and local abundance of masses of pyroxenite-bearing spinel lherzolites showing evident structural and petrological records of subsolidus evolution in the lithosphere and the clear evidence that the reactive spinel peridotites formed at the expense of these lithospheric protoliths shed new light on the early evolution stages of the Monte Maggiore peridotites.

The above evidence indicates that the pyroxenite-bearing spinel lherzolites were the mantle protoliths existing prior to the formation of reactive spinel peridotites by reactive melt percolation. Geothermometric estimates (Piccardo 2007) yield temperatures in the range of 1000–1100 °C for the spinel-facies recrystallization of the lherzolite protoliths, whereas the spinel-facies cooling structures (i.e. vermicular spinel exsolution in Opx porphyroclasts) yielded temperatures in the range 970–1100 °C (Rampone *et al.* 2008). These protoliths record a composite lithospheric evolution, that is, pyroxenite veining, subsolidus decompression from garnet- to spinel-facies conditions and spinel-facies recrystallization under cooling. Accordingly, these pyroxenite-bearing

Table 6. *Trace element composition of minerals in Mt Maggiore plagioclase peridotites*

Sample	MM1/4*	MM1/4*	MM1/10*	MM1/10*	MM1/11*	MM1/11*	MM2/5*	MM2/5*	C36*	C36*	C36*	CC2†	CC2†	CC2†
Rock type	Impr. Plg-perid.	Impr. Plg-perid.	Impr. Plg-perid.	Impr. Plg-perid.	Impr. Plg-perid.	Impr. Plg-perid.	Impr. Plg-perid.	Impr. Plg-perid.	Impr. Plg-perid.	Impr. Plg-perid.	Impr. Plg-perid.	Impr. Plg-perid.	Impr. Plg-perid.	Impr. Plg-perid.
Mineral	Cpx (3)	Opx (2)	Cpx (2)	Opx (2)	Cpx (2)	Opx (3)	Cpx (2)	Plg	Cpx (3)	Opx	Plg (2)	Cpx	Opx	Plg
Sc	58	34.0	63	29.0	59	30.3	65	na	53	28	na	na	na	na
Ti	1992	759	1845	642	1910	740	1761	100	1620	761	58	1618	1678	na
V	362	188	349	186	331	176	285	3	277	166	2.0	na	na	na
Sr	0.328	0.0125	1.00	bd	0.445	0.04	1.16	28.6	2.8	0.19	6.8	na	na	na
Y	17.7	2.8	17.1	2.4	16.8	2.7	16.9	0.18	16.7	2.7	0.135	19.4	3.0	0.15
Zr	3.1	0.48	2.9	0.90	2.8	0.60	5.2	bd	4.1	1.01	na	na	na	na
La	0.008	bd	0.009	bd	0.006	bd	bd	0.011	bd	bd	0.017	0.006	bd	bd
Ce	0.149	0.006	0.153	0.007	0.169	0.005	0.193	0.060	0.26	0.003	0.058	0.15	bd	0.074
Pr	0.107	0.005	0.093	0.004	0.104	bd	bd	bd	bd	bd	bd	0.095	bd	bd
Nd	1.28	0.034	1.02	0.041	1.15	0.050	1.17	0.160	1.32	0.044	0.135	1.06	0.025	0.16
Sm	1.20	0.033	0.94	0.043	1.04	0.061	1.02	0.1	0.93	0.050	0.045	0.93	0.039	0.046
Eu	0.48	0.02	0.41	0.026	0.42	0.025	0.42	0.2	0.43	0.028	0.23	0.36	0.015	0.26
Gd	2.3	0.122	1.87	0.102	1.93	0.162	2.01	0.050	1.92	0.197	0.053	1.79	0.158	0.061
Tb	0.33	0.040	0.44	bd	0.40	0.037	bd	bd	na	na	na	0.39	0.047	0.008
Dy	3.2	0.34	2.9	0.29	2.8	0.38	2.78	0.05	2.7	0.38	bd	2.9	0.41	bd
Ho	0.68	0.096	0.67	0.096	0.62	0.096	bd	bd	na	na	na	0.68	0.110	bd
Er	1.91	0.37	1.93	0.33	1.81	0.34	1.56	bd	1.68	0.31	bd	1.90	0.42	bd
Tm	0.28	0.059	0.30	0.071	0.25	0.051	bd	bd	na	na	na	0.29	0.060	bd
Yb	1.7	0.46	1.7	0.45	1.6	0.43	1.48	bd	1.35	0.33	bd	1.7	0.38	bd
Lu	0.23	0.078	0.25	0.080	0.24	0.084	bd	bd	na	na	na	0.26	0.062	bd
Hf	0.41	0.058	0.34	0.064	0.31	0.077	bd	bd	na	na	na	bd	bd	bd

*Data from Rampone *et al.* (2008).
†Data from Müntener & Piccardo (2003).
Abbreviations: Impr. Plg-perid., impregnated plagioclase peridotite; bd, below detection; na, not analysed.

Table 7. *Major element compositions of minerals from gabbroic dykelets and cumulates in Mt Maggiore peridotites*

Sample	CR0	CR0	CR0	CR33	CR33	CR33	CR9-1	CR9-1	CR9-1	CR5-2	CR5-2	CR5-2	CR10	CR10	CR10	CR10
Rock type	Dykelet	Dykelet	Dykelet	Dykelet	Dykelet	Dykelet	Cumulate	Cumulate	Cumulate	Cumulate	Cumulate	Cumulate	Cumulate	Cumulate	Cumulate	Cumulate
Mineral	Cpx pc	Opx pc	Ol	Cpx pc	Cpx pr	Plg	Cpx pc	Cpx pr	Plg	Cpx pc	Cpx pr	Plg	Cpx pc	Cpx pr	Opx	Plg
SiO_2	52.19	55.28	41.00	50.66	51.35	46.18	51.22	52.02	46.98	51.07	51.89	46.40	50.60	52.16	53.89	46.15
TiO_2	0.44	0.16	0.02	0.28	0.37	0.00	0.31	0.35	0.00	0.29	0.28	0.00	0.31	0.43	0.21	0.00
Cr_2O_3	1.20	0.88	0.05	1.18	1.41	0.00	1.34	1.41	0.00	1.41	1.21	0.00	1.30	1.18	0.76	0.00
Al_2O_3	3.97	3.35	0.05	6.26	4.89	34.64	5.61	4.31	34.16	5.44	3.56	33.99	5.53	3.26	3.15	34.79
FeO^T	2.93	6.57	9.63	3.09	2.76	0.29	2.92	2.81	0.08	2.78	2.36	0.22	2.80	2.43	4.86	0.17
MnO	0.07	0.15	0.18	0.12	0.09	0.00	0.13	0.08	0.05	0.09	0.10	0.03	0.09	0.08	0.09	0.02
MgO	17.06	32.09	49.08	15.68	15.54	0.11	15.68	15.65	0.05	15.82	15.95	0.05	15.90	16.77	30.52	0.06
CaO	21.51	2.20	0.06	21.61	22.30	18.07	21.63	22.45	17.36	22.12	22.70	17.53	21.80	22.50	3.74	17.59
Na_2O	0.48	0.07	0.00	0.14	0.16	0.93	0.46	0.47	1.49	0.27	0.26	1.39	0.30	0.17	0.06	1.30
K_2O	0.00	0.00	0.00	0.01	0.00	0.00	0.01	0.02	0.00	0.01	0.00	0.04	0.00	0.02	0.00	0.00
NiO	n.d.	n.d.	n.d.	n.d.	n.d.	n.d.	n.d.	n.d.	n.d.	n.d.	n.d.	n.d.	n.d.	n.d.	n.d.	n.d.
Mg#	91.22	89.69		90.06	90.94		90.53	90.85		93.74	92.34		91.00	92.48	91.80	
An%						91.50			86.60			87.20				88.20
Fo%			90.08													

Abbreviations: pc, phenocryst core; pr, phenocryst rim; n.d., not determined.

Table 8. *Trace element composition of minerals in Mt Maggiore dykelets*

Sample	CR33	CR33	CR34	CR34
Rock type	Dykelets	Dykelets	Dykelets	Dykelets
Mineral	Cpx (3)	Plg (2)	Cpx (3)	Plg (3)
Sc	54	0.27	54	0.28
TiO_2	2041	86	1978	180
V	357	3.7	341	3.4
Sr	0.71	14.7	0.79	15.6
Y	14.1	0.113	15.4	0.192
Zr	2.5	bd	2.3	bd
La	0.006	0.003	0.005	0.011
Ce	0.132	0.042	0.130	0.136
Pr	0.077	0.012	0.073	0.041
Nd	0.91	0.079	0.89	0.28
Sm	0.80	0.032	0.84	0.113
Eu	0.37	0.140	0.40	0.24
Gd	1.67	0.047	1.70	0.082
Tb	0.34	0.005	0.37	0.012
Dy	2.3	0.025	2.6	0.048
Ho	0.54	0.004	0.57	0.005
Er	1.5	bd	1.6	0.022
Tm	0.22	bd	0.24	0.004
Yb	1.44	bd	1.6	bd
Lu	0.20	bd	0.22	bd
Hf	0.20	bd	0.22	bd
Ta	bd	0.101	bd	0.009
Pb	0.16	4.1	bd	bd
Th	bd	0.013	bd	bd
U	bd	bd	0.25	0.003

Abbreviation: bd, below detection.

spinel lherzolite protoliths were resident in the subcontinental mantle under lithospheric conditions where they were equilibrated at an average continental geothermal gradient.

The Monte Maggiore pyroxenite-veined mantle protoliths represent the subcontinental mantle lithosphere of the Ligurian lithospheric domain prior to exhumation during lithosphere extension (i.e. the transition from spinel- to plagioclase-facies conditions) and MORB-type melt percolation. Structural–compositional relicts derived from these lithospheric protoliths (i.e. Opx + Sp clusters, vermicular spinel exsolutions in Opx porphyroclasts and high Al core compositions in porphyroclastic pyroxenes) are preserved within the reactive spinel peridotites. This evidence reinforces the inference that the pristine subcontinental pyroxenite-bearing, lithospheric lherzolites were transformed by reactive interaction (pyroxene dissolution and olivine precipitation) of migrating silica-undersaturated melts into pyroxene-poor reactive spinel peridotites.

Pyroxenite-veined fertile spinel lherzolites characterize the mantle rocks from the subcontinental lithosphere that were exhumed and exposed on the sea floor in the OCT zones. They are preserved as remnants within the melt-modified reactive and impregnated peridotites that were exposed at the MIO setting of the Ligurian Tethys basin (Piccardo 2008 and references therein). These peridotites show remarkably similar structural and compositional characteristics: that is (i) records of pristine garnet-facies assemblages; (ii) widespread veining of spinel(-garnet)-pyroxenites; and (iii) records of subsolidus tectonic–metamorphic evolution during exhumation. Available isotope data from some External Ligurides and North Lanzo peridotites (data from Bodinier *et al.* 1991; Rampone *et al.* 1995) indicate Proterozoic depleted mantle (DM) and CHUR (chondritic uniform reservoir) Nd model ages that have been interpreted as the ages of isolation from the convective mantle asthenosphere and accretion to the subcontinental thermal lithosphere (Piccardo 2008).

Origin of the melt-modified peridotites and gabbroic intrusions

The reactive spinel peridotites. The analysed reactive spinel peridotites have highly variable and significantly low pyroxene (Cpx 2.3–7.5% vol., Opx 3.4–15.4 vol.) content, coupled with significantly high olivine (Ol 76.5–93.4% vol.) content, in the different samples (Table 1). Increasing Ol and decreasing Cpx and Opx with respect to the pristine spinel lherzolites are in good agreement with petrographical evidence indicating significant pyroxene dissolution and olivine precipitation during melt–rock reactive interaction. Notwithstanding the highly variable clinopyroxene content in the different samples, Cpx of the reactive spinel peridotites show remarkably similar trace element (REE) content (Fig. 5). Clinopyroxene REE patterns from the whole investigated samples are, accordingly, closely similar and indicate that they could correspond with Cpx compositions in refractory residua after 6% fractional melting of a spinel-facies DM asthenospheric mantle source. A strong decoupling is evident between the highly variable modal composition and the homogeneous Cpx trace element content in the different samples.

According to Rampone *et al.* (2008), the depleted chemical signature of the Monte Maggiore spinel peridotites, although recognized as affected by melt–rock interaction by reactive porous flow, is similar to many abyssal peridotites and indicates that 'they likely experienced MORB-type partial melting' and 'were variably modified by reactive porous flow'. As evidenced above, the spinel peridotites show contrasting compositional features, both depleted (i.e. Al–Ca–Si depletion, Mg enrichment) and enriched (i.e. Ti enrichment in spinel) chemical

Table 9. *Trace element composition of minerals in Mt Maggiore cumulates*

Sample	CR5/2	CR5/2	CR5/2	CR7/2	CR7/2	CR7/2	CR9/1	CR9/1	CR9/1	CR9/2	CR9/2	CR9/2	CR9/3	CR9/3	CR10/1	CR10/1	CR10/1
Rock type	Cumulate	Cumulate	Cumulate	Cumulate	Cumulate	Cumulate	Cumulate	Cumulate	Cumulate	Cumulate	Cumulate	Cumulate	Cumulate	Cumulate	Cumulate	Cumulate	Cumulate
Mineral	Cpx (5)	Opx (2)	Plg (3)	Cpx (4)	Opx (1)	Plg (3)	Cpx (2)	Opx (1)	Plg (3)	Cpx (4)	Opx (1)	Plg (3)	Cpx (4)	Plg (3)	Cpx (3)	Opx (2)	Plg (1)
Sc	60	30	0.60	61	28	0.37	61	26	1.02	67	32.2	0.60	55	0.39	61	27.6	0.51
Ti	1928	769	97	2047	802	100	2054	764	113	2471	932	102	1917	122	2303	978	121
V	365	169	2.93	365	167	2.61	344	173	2.25	398	196	3.08	335	3.80	382	180	2.49
Sr	1.9	0.18	30	2.1	0.06	28	1.26	0.10	24.1	1.45	bd	26.6	1.21	26.7	1.75	0.02	22.87
Y	14.3	2.6	0.128	15.4	2.6	0.152	14.2	2.7	0.22	19.9	3.32	0.227	15.4	0.148	16.6	2.35	0.120
Zr	3.1	0.97	0.52	3.3	0.73	0.101	3.0	0.66	bd	6.6	1.15	bd	3.44	bd	4.9	1.01	bd
La	0.020	bd	0.023	0.019	bd	0.043	0.013	bd	0.017	0.0226	bd	0.0212	0.0184	0.0110	0.038	bd	0.0214
Ce	0.21	0.018	0.098	0.26	0.008	0.178	0.177	bd	0.089	0.279	bd	0.123	0.208	0.085	0.40	0.0140	0.124
Pr	0.091	0.0071	0.0176	0.121	0.0038	0.036	0.090	0.002	0.032	0.163	bd	0.0272	0.102	0.0277	0.173	0.0066	0.0290
Nd	0.95	0.067	0.120	1.20	0.053	0.20	0.96	0.057	0.21	1.65	0.056	0.223	1.12	0.168	1.51	0.071	0.224
Sm	0.80	0.091	0.030	1.04	0.067	0.056	0.78	0.094	0.068	1.27	0.080	0.078	0.91	0.077	1.18	0.044	0.079
Eu	0.37	0.028	0.187	0.40	0.025	0.24	0.35	0.0195	0.22	0.49	0.0192	0.211	0.42	0.208	0.52	0.027	0.200
Gd	1.48	0.166	bd	1.77	0.140	0.056	1.64	0.25	0.063	2.5	0.25	0.044	1.72	0.056	1.95	0.132	0.060
Tb	0.32	0.051	bd	0.37	0.043	bd	0.32	0.037	bd	0.46	0.053	bd	0.35	bd	0.41	0.035	bd
Dy	2.3	0.39	bd	2.5	0.35	bd	2.5	0.31	0.054	3.2	0.40	bd	2.6	0.039	2.8	0.30	bd
Ho	0.53	0.086	bd	0.56	0.085	bd	0.55	0.099	bd	0.72	0.113	bd	0.57	bd	0.61	0.093	bd
Er	1.45	0.29	bd	1.57	0.28	bd	1.65	0.27	0.034	1.92	0.34	bd	1.63	bd	1.66	0.29	bd
Tm	0.22	0.051	bd	0.23	0.066	bd	0.25	0.057	bd	0.32	0.061	bd	0.25	bd	0.25	0.047	bd
Yb	1.43	0.45	0.030	1.49	0.35	bd	1.6	0.40	0.044	1.9	0.54	bd	1.5	bd	1.5	0.38	bd
Lu	0.20	0.068	bd	0.20	0.071	bd	0.21	0.064	bd	0.25	0.089	bd	0.20	bd	0.20	0.063	bd
Hf	0.25	0.055	bd	0.28	0.071	0.034	0.28	0.049	0.032	0.495	0.118	0.0239	0.257	0.0261	0.295	0.0583	bd
Ta	bd	bd	0.070	bd	bd	0.53	bd	bd	bd	bd	bd	bd	bd	bd	bd	bd	bd
Pb	bd	bd	0.044	0.093	bd	0.099	bd	bd	bd	0.040	bd	0.041	bd	0.033	0.32	bd	bd
Th	bd	bd	0.02	0.008	bd	0.09	bd	bd	bd	bd	0.0028	0.005	0.0021	0.004	0.0025	bd	bd
U	0.001	bd	0.027	0.38	bd	0.39	bd	bd	0.003	0.0026	0.0011	bd	bd	0.025	0.0017	bd	bd

Abbreviation: bd, below detection.

Table 10. *Trace element bulk-rock composition of Mt Maggiore gabbros*

Sample	C17	C20	C26	M21	M23	M25	M26	M28	M33	M35	M40	M42	M46
Rock type	Gabbro	Gabbro	Gabbro	Gabbro	Gabbro	Gabbro	Gabbro	Gabbro	Gabbro	Gabbro	Gabbro	Gabbro	Gabbro
Rb	0.19	0.099	0.43	0.070	0.110	0.124	0.102	0.102	0.53	0.21	0.25	0.28	0.17
Sr	176	177	77	180	186	218	211	160	264	191	188	147	210
Y	8.0	8.3	3.2	7.4	10.7	10.4	6.0	6.4	12.2	5.4	10.7	14.1	5.2
Zr	10.5	17	11.6	16	23	26	12.0	9.2	18	5.7	11.9	15	9.8
Nb	0.089	0.059	0.22	0.116	0.41	0.38	0.41	0.050	0.44	0.044	0.078	0.088	0.16
Cs	0.176	0.011	0.016	0.005	0.01	0.01	0.02	0.02	0.22	0.01	0.05	0.17	0.09
Ba	2.1	2.6	10.3	1.18	3.0	2.7	1.7	2.7	10.2	2.4	3.1	2.1	3.0
La	0.56	0.58	0.51	0.56	0.83	0.96	0.46	0.45	0.70	0.43	0.59	0.58	0.68
Ce	1.9	2.0	1.5	2.0	2.9	3.2	1.5	1.5	2.3	1.3	2.0	2.2	2.0
Pr	0.35	0.38	0.23	0.36	0.52	0.56	0.27	0.28	0.44	0.22	0.38	0.46	0.33
Nd	2.1	2.3	1.21	2.1	3.1	3.2	1.6	1.7	2.8	1.3	2.4	3.1	1.8
Sm	0.80	0.85	0.32	0.78	1.11	1.09	0.60	0.64	1.11	0.47	0.95	1.3	0.60
Eu	0.59	0.51	0.28	0.47	0.63	0.63	0.44	0.44	0.86	0.50	0.69	0.75	0.62
Gd	1.08	1.14	0.38	1.02	1.4	1.4	0.76	0.83	1.5	0.63	1.3	1.8	0.74
Tb	0.22	0.23	0.069	0.21	0.29	0.28	0.16	0.17	0.31	0.13	0.27	0.38	0.15
Dy	1.4	1.4	0.45	1.3	1.8	1.7	0.95	1.04	1.9	0.85	1.6	2.3	0.90
Ho	0.33	0.34	0.118	0.30	0.42	0.41	0.23	0.26	0.47	0.21	0.41	0.57	0.22
Er	0.93	0.96	0.40	0.85	1.18	1.15	0.65	0.72	1.3	0.62	1.14	1.6	0.60
Tm	0.13	0.13	0.064	0.12	0.17	0.16	0.092	0.101	0.19	0.092	0.16	0.23	0.087
Yb	0.82	0.84	0.52	0.74	1.05	1.00	0.56	0.63	1.25	0.60	1.03	1.4	0.55
Lu	0.13	0.13	0.099	0.118	0.17	0.16	0.088	0.099	0.20	0.100	0.17	0.22	0.089
Hf	0.37	0.55	0.30	0.47	0.72	0.70	0.32	0.33	0.63	0.21	0.42	0.59	0.33
Ta	0.006	0.005	0.01	0.009	0.03	0.04	0.03	0.005	0.04	0.003	0.007	0.007	0.02
Pb	0.26	1.14	0.27	0.17	0.20	2.6	0.30	0.49	0.44	0.17	0.23	0.27	0.23
Th	0.005	bd	0.007	0.008	0.026	0.03	0.02	0.003	0.002	0.003	0.005	0.005	0.004
U	0.002	bd	0.007	0.004	0.011	0.011	0.008	0.004	0.001	0.001	0.003	0.002	0.002

Abbreviation: bd, below detection.

Table 11. *Major element mineral compositions of the Mt Maggiore gabbros*

Sample	M21	M21	M21	M21	M21	M21	M21	M26	M26	M26	M26	M26	M28	M28	M28	M28	M28
Rock type	Gabbro	Gabbro	Gabbro	Gabbro	Gabbro	Gabbro	Gabbro	Gabbro	Gabbro	Gabbro	Gabbro	Gabbro	Gabbro	Gabbro	Gabbro	Gabbro	Gabbro
Mineral	Cpx pc	Cpx pr	Cpx gr	Plg pc	Plg gr	Ol pc	Ol gr	Cpx pc	Cpx gr	Plg pc	Plg gr	Ol pc	Cpx gr	Cpx pc	Plg pc	Plg gr	Ol pc
SiO_2	50.59	50.56	50.67	54.48	53.88	38.49	38.09	50.79	50.72	53.72	53.43	38.78	50.92	51.43	53.48	53.66	38.30
TiO_2	1.08	1.07	1.07	0.25	0.20	0.15	0.15	0.75	0.69	0.25	0.21	0.20	1.00	0.66	0.17	0.26	0.17
Cr_2O_3	0.52	0.48	0.48	0.00	0.00	0.14	0.10	0.47	0.50	0.00	0.00	0.15	0.54	0.49	0.00	0.00	0.15
Al_2O_3	2.55	2.21	2.41	28.53	28.53	0.00	0.00	2.95	2.17	28.33	28.89	0.00	2.50	2.56	28.95	29.08	0.00
FeO^T	0.23	5.92	5.60	0.34	0.31	17.86	18.33	5.23	5.17	0.31	0.24	17.42	5.40	5.61	0.40	0.61	17.76
MnO	0.28	0.31	0.29	0.00	0.00	0.36	0.38	0.25	0.22	0.00	0.00	0.40	0.27	0.27	0.00	0.00	0.42
MgO	16.43	17.28	17.01	0.00	0.00	43.50	42.83	16.81	16.75	0.00	0.00	42.50	16.54	16.73	0.00	0.00	43.67
CaO	21.78	20.07	21.02	10.91	10.85	0.15	0.15	21.85	22.57	11.07	11.51	0.13	21.68	21.57	11.40	11.27	0.15
Na_2O	0.40	0.38	0.38	4.99	5.38	0.00	0.00	0.44	0.47	5.12	4.91	0.00	0.40	0.26	5.16	5.55	0.00
K_2O	0.15	0.18	0.18	0.20	0.20	0.00	0.00	0.17	0.16	0.19	0.20	0.00	0.18	0.16	0.20	0.19	0.00
NiO	n.d.	n.d.	n.d.	n.d.	n.d.	n.d.	n.d.	n.d.	n.d.	n.d.	n.d.	n.d.	n.d.	n.d.	n.d.	n.d.	n.d.
Mg#	84.70	83.90	84.40					85.13	85.20				84.50	84.20			
An%				54.08	52.20					53.80	55.80				54.40	52.30	
Fo%						81.30	80.70					81.20					81.40

Sample	C17	C17	C17	C17	C17	C20	C20	C20	C20	C20	M23	M23	M23	M23	M23	M23	M25
Rock type	Gabbro	Gabbro	Gabbro	Gabbro	Gabbro	Gabbro	Gabbro	Gabbro	Gabbro	Gabbro	Gabbro	Gabbro	Gabbro	Gabbro	Gabbro	Gabbro	Gabbro
Mineral	Cpx pc	Cpx gr	Plg pc	Plg gr	Ol gr	Cpx pc	Cpx gr	Plg pc	Plg pr	Ol gr	Cpx pc	Cpx gr	Plg pc	Plg pr	Plg gr	Ol gr	Cpx pc
SiO_2	52.62	51.80	54.86	54.95	38.91	53.04	53.09	55.48	55.76	39.29	50.45	50.05	54.86	54.79	54.48	37.55	51.31
TiO_2	0.43	0.94	0.14	0.13	0.07	0.52	0.98	0.14	0.10	0.05	1.29	1.23	0.22	0.29	0.21	0.22	0.80
Cr_2O_3	0.55	0.61	0.09	0.07	0.03	0.40	0.40	14.11	0.12	0.00	0.44	0.51	0.00	0.00	0.00	0.18	0.44
Al_2O_3	0.22	2.98	27.62	27.53	0.10	3.12	2.53	14.15	27.92	0.16	2.44	2.54	28.15	28.21	28.45	0.00	2.61
FeO^T	5.40	6.80	0.27	0.30	22.02	5.76	6.23	0.32	0.22	21.65	6.06	6.57	0.31	0.48	0.48	20.55	6.04
MnO	0.12	0.21	0.00	0.00	0.39	0.17	0.21	0.02	0.02	0.33	0.28	0.26	0.00	0.00	0.00	0.45	0.21
MgO	17.62	16.12	0.00	0.00	40.06	16.33	16.76	0.00	0.00	40.79	16.23	16.18	0.00	0.00	0.00	41.16	17.31
CaO	19.07	19.39	10.25	10.19	0.05	21.20	20.42	7.97	10.23	0.07	21.66	21.11	10.52	10.48	10.79	0.17	20.80
Na_2O	0.36	0.23	5.65	5.68	0.00	0.24	0.15	2.61	6.05	0.00	0.18	0.48	5.76	5.76	5.54	0.00	0.24
K_2O	0.08	0.08	0.15	0.13	0.00	0.11	0.08	0.16	1.14	0.00	0.15	0.16	0.17	0.20	0.21	0.00	0.19
NiO	n.d.	n.d.	n.d.	n.d.	n.d.	n.d.	n.d.	n.d.	n.d.	n.d.	n.d.	n.d.	n.d.	n.d.	n.d.	n.d.	n.d.
Mg#	85.30	8.9				83.50	82.50				82.72	81.48					83.65
An%			49.70	49.50				53.15	47.90				49.80	49.60	51.27		
Fo%					76.50					77.00						78.12	

(*Continued*)

Table 11. *Continued*

Sample	M25	M25	M25	M25	M25	M25	M33	M33	M35	M35	M35	M35	M35	M35	M40	M40	M40	M40
Rock type	Gabbro	Gabbro	Gabbro	Gabbro	Gabbro	Gabbro	Gabbro	Gabbro	Gabbro	Gabbro	Gabbro	Gabbro	Gabbro	Gabbro	Gabbro	Gabbro	Gabbro	Gabbro
Mineral	Cpx gr	Plg pc	Plg pr	Plg gr	Ol gr	Ol gr	Cpx pr	Cpx gr	Cpx pc	Cpx pr	Cpx gr	Plg pc	Plg pr	Plg gr	Cpx pc	Cpx gr	Plg gr	Ol pc
SiO_2	50.73	54.43	53.69	54.99	38.09	38.39	50.05	50.92	50.56	50.89	50.36	55.98	56.74	56.04	51.14	51.13	56.94	37.00
TiO_2	1.02	0.21	0.26	0.24	0.25	0.18	1.07	0.99	1.02	0.88	1.11	0.27	0.17	0.22	1.07	1.05	0.18	0.20
Cr_2O_3	0.36	0.00	0.00	0.00	0.20	0.16	0.39	0.27	0.27	0.36	0.33	0.00	0.00	0.00	0.30	0.33	0.00	0.06
Al_2O_3	2.39	28.34	28.43	28.46	0.00	0.00	2.30	2.03	2.27	2.43	2.48	26.55	27.13	27.32	2.36	2.30	26.64	0.00
FeO^T	5.91	0.26	0.53	0.36	18.47	18.83	8.30	9.44	7.32	7.10	7.01	0.45	0.35	0.33	7.64	8.21	0.35	24.86
MnO	0.22	0.00	0.00	0.00	0.41	21.26	0.44	0.33	0.29	0.35	0.28	0.00	0.00	0.00	0.35	0.36	0.00	0.51
MgO	17.19	0.00	0.00	0.00	43.20	21.43	15.00	16.11	15.70	15.27	15.51	0.00	0.00	0.00	15.74	15.13	0.00	37.09
CaO	15.63	10.79	10.86	10.72	0.18	0.07	20.96	19.15	21.26	21.97	21.63	9.11	8.95	9.32	21.51	21.16	8.84	0.18
Na_2O	0.29	5.25	5.09	5.61	0.00	0.00	0.24	0.46	0.68	0.28	0.17	6.60	6.89	6.37	0.19	0.22	6.52	0.00
K_2O	0.10	0.17	0.24	0.19	0.00	0.00	0.20	0.16	0.18	0.20	0.06	0.18	0.18	0.17	0.00	0.17	0.21	0.00
NiO	n.d.	n.d.	n.d.	n.d.	n.d.	n.d.	n.d.	n.d.	n.d.	n.d.	n.d.	n.d.	n.d.	n.d.	n.d.	n.d.	n.d.	n.d.
Mg#	83.85						76.30	75.45	79.73	79.30	79.80				78.60	76.80		
An%		52.70	53.30	50.82								42.85	41.40	44.30			42.34	
Fo%					80.65	79.40												72.70

Sample	M40	M40	M42	M42	M42	M42	M42	M42	M46	M46	M46	M46	M46	M46	M46	C26	C26	C26
Rock type	Gabbro	Gabbro	Gabbro	Gabbro	Gabbro	Gabbro	Gabbro	Gabbro	Gabbro	Gabbro	Gabbro	Gabbro	Gabbro	Gabbro	Gabbro	Cumulate	Cumulate	Cumulate
Mineral	Ol pr	Ol gr	Cpx pc	Cpx gr	Plg gr	Ol pr	Ol pc	Ol gr	Cpx pc	Cpx gr	Plg pc	Plg gr	Ol pc	Ol pr	Ol gr	Cpx pc	Cpx gr	Ol gr
SiO_2	37.24	37.08	50.63	50.21	56.25	36.50	36.52	36.72	50.78	50.00	55.11	56.32	36.79	36.56	36.92	52.11	51.83	36.88
TiO_2	0.14	0.21	0.98	1.04	0.22	0.14	0.21	0.22	1.19	1.17	0.20	0.28	0.21	0.12	0.21	1.07	0.98	0.05
Cr_2O_3	0.16	0.18	0.32	0.29	0.00	0.14	0.22	0.24	0.33	0.28	0.00	0.00	0.17	0.16	0.18	0.41	0.42	0.05
Al_2O_3	0.00	0.00	2.21	2.38	27.00	0.00	0.00	0.00	2.56	2.55	27.77	27.27	0.00	0.00	0.00	2.81	2.56	0.10
FeO^T	24.53	25.68	7.19	8.33	0.40	26.29	26.70	26.13	8.95	7.72	0.43	0.48	25.58	25.12	24.60	9.36	9.58	28.61
MnO	0.43	0.50	0.29	0.33	0.00	0.48	0.58	0.56	0.32	0.33	0.00	0.00	0.55	0.58	0.55	0.23	0.26	0.57
MgO	37.38	36.06	16.53	15.68	0.00	36.41	36.92	37.60	17.10	15.43	0.00	0.00	37.84	37.81	38.53	15.35	15.40	33.99
CaO	0.15	0.24	20.06	19.61	9.06	0.16	0.11	0.18	18.86	21.15	10.26	9.38	0.17	0.15	0.17	19.02	18.69	0.13
Na_2O	0.00	0.00	0.37	0.32	6.49	0.00	0.00	0.00	0.37	0.29	5.79	6.54	0.00	0.00	0.00	0.06	0.49	0.00
K_2O	0.00	0.00	0.14	0.18	0.20	0.00	0.00	0.00	0.24	0.19	0.20	0.20	0.00	0.00	0.00	0.06	0.11	0.00
NiO	n.d.	n.d.	n.d.	n.d.	n.d.	n.d.	n.d.	n.d.	n.d.	n.d.	n.d.	n.d.	n.d.	n.d.	n.d.	n.d.	n.d.	n.d.
Mg#			80.40	77.03					77.30	78.09						74.50	74.10	
An%					43.02						48.92	43.58						
Fo%	73.10	71.45				71.15	71.10	71.53					72.50	72.70	73.63			67.90

Abbreviations: pc, phenocryst core; pr, phenocryst rim; gr, groundmass grain; n.d., not determined.

Table 12. *Trace element composition of plagioclase in Mt Maggiore gabbros*

Sample	M21	M23	M26	M28	M35	M42	M46	C20
Rock type	Gabbro	Gabbro	Gabbro	Gabbro	Gabbro	Gabbro	Gabbro	Gabbro
Mineral	Plg (1)	Plg (3)	Plg (2)	Plg (2)	Plg (2)	Plg (2)	Plg (2)	Plg (2)
Sc	bd	bd	bd	0.63	bd	bd	bd	bd
Ti	539	539	413	420	779	509	599	479
V	1.70	1.59	2.6	3.0	2.9	1.79	3.3	2.5
Sr	349	341	313	325	320	329	341	314
Y	0.25	0.27	0.194	0.22	0.23	0.26	0.38	0.192
La	0.75	0.98	0.32	0.34	0.43	0.57	0.58	0.62
Ce	1.79	2.03	1.00	0.88	1.25	1.36	1.63	1.52
Pr	0.20	0.25	0.101	0.48	0.161	0.182	0.22	0.182
Nd	0.91	1.05	0.46	0.43	0.64	0.73	0.94	0.66
Sm	0.135	0.096	0.077	0.065	0.127	0.155	0.166	0.093
Eu	0.49	0.43	0.32	0.33	0.53	0.60	0.60	0.47
Gd	0.128	bd	0.119	0.112	0.107	0.127	0.163	bd
Tb	0.011	0.006	bd	0.012	0.010	0.011	0.020	0.009
Dy	0.070	0.056	0.060	0.030	0.070	0.046	0.098	0.057
Ho	0.008	0.010	bd	0.009	0.008	0.012	0.024	bd
Er	bd	bd	0.024	bd	0.027	0.043	bd	bd
Yb	0.017	bd	0.034	bd	0.015	bd	0.022	bd
Lu	bd	0.008	bd	bd	bd	bd	bd	bd
Th	bd	bd	0.033	bd	bd	bd	bd	bd

Abbreviation: bd, below detection.

signatures, and contrasting bulk and mineral chemistry characteristics. The highly variable Cpx modal compositions and the close similarity of the Cpx trace element composition highlight the facts that: (1) the percolating melts dissolved the mantle pyroxenes and precipitated new magmatic olivine in variable proportions in the different samples; and (2) the peridotite mantle minerals (i.e. clinopyroxenes) underwent equilibrium trace element redistribution with the percolating melts that were characterized by very similar compositions.

To test the partial melting residua hypothesis, different estimates of the hypothetical partial melting degrees have been performed that furnish different indications. According to the MgO bulk-rock contents (Niu 1997), the partial melting estimates vary in the range 21–32%, whereas according to the Cr# value of spinel (Hellebrand *et al.* 2001) the estimates vary in the range 6–11%. When the bulk-rock compositions are plotted on a MgO v. SiO_2 diagram (Fig. 8), the representative points fall outside the compositional trends as calculated by Niu (1997) for refractory residua after any kind of partial melting. In fact, they fall at significantly lower values of SiO_2 content, at the corresponding MgO content values, than the partial melting refractory residua.

Accordingly, the field, petrological and geochemical data available on the Monte Maggiore spinel peridotites do not confirm that 'the Cpx-poor spinel lherzolites are consistent with mantle residues after low degrees fractional melting' and that 'these peridotites experienced an "oceanic-type" evolution, namely asthenospheric upwelling and MORB-type melting', as hypothesized by Rampone *et al.* (2008). The above arguments support the theory that the depleted compositions of the investigated spinel peridotites and the whole of their compositional and structural features are related to melt–rock interaction processes. Moreover, neither compositional nor structural characteristics allow for speculation that the multistage melt migration and interaction recorded by these reactive peridotites was preceded by an 'oceanic-type' MORB-type partial melting, as put forward by Rampone *et al.* (2008).

The reactive spinel harzburgites are, thus, products of melt–peridotite interaction at the expense of pristine subcontinental lithospheric spinel lherzolites, as recognized in the reactive spinel harzburgites of other A–A ophiolitic peridotites (e.g. South Lanzo and Erro-Tobbio: Piccardo 2003; Müntener & Piccardo 2003; Piccardo *et al.* 2004, 2007*a*; Rampone *et al.* 2004; Piccardo & Vissers 2007).

Information on the compositional characteristics of the percolating melts has been obtained from the Cpx trace element compositions, assuming that the mantle Cpx and percolating liquid attained the trace element equilibration during melt–peridotite interaction. REE compositions and patterns have been calculated for the liquids in equilibrium with

Table 13. *Trace element composition of clinopyroxenes in Mt Maggiore gabbros*

Sample	M21	M23	M26	M28	M33	M35	M42	M46	C20	C26
Rock type	Gabbro	Gabbro	Gabbro	Gabbro	Gabbro	Gabbro	Gabbro	Gabbro	Gabbro	Gabbro
Mineral	Cpx (2)	Cpx (3)	Cpx (4)	Cpx (2)	Cpx (2)	Cpx (3)	Cpx (2)	Cpx (3)	Cpx (3)	Cpx (2)
Sc	102	116	110	94	126	121	122	100	100	121
Ti	4555	7312	3972	3177	5574	5933	4875	4435	5274	4375
V	365	490	382	337	520	489	449	442	395	458
Cr	1478	1151	1151	1717	631	839	643	194	1651	171
Sr	12.8	12.7	14.3	15.0	17.5	13.6	12.3	11.7	13.7	13.9
Y	26	38	23	18.6	50	36	31	32	28	69
Zr	41	75	32	14.1	43	34	22	36	51	56
Nb	0.23	0.63	0.190	0.060	0.027	0.062	0.054	0.110	0.104	0.023
La	0.68	0.82	0.41	0.31	0.82	0.62	0.43	0.58	0.62	1.17
Ce	3.8	5.0	2.3	1.79	5.5	3.9	2.7	3.9	3.8	8.0
Pr	0.93	1.29	0.63	0.47	1.50	0.96	0.73	0.99	0.95	2.0
Nd	6.0	7.9	4.3	3.2	9.5	6.2	5.0	6.2	6.2	14.3
Sm	2.5	3.6	2.0	1.58	4.5	3.1	2.7	3.0	2.9	6.6
Eu	0.84	1.16	0.67	0.68	1.58	1.17	0.93	0.96	0.89	2.25
Gd	4.0	5.8	3.5	2.7	6.6	5.2	4.3	4.6	4.2	11.1
Tb	0.69	1.06	0.63	0.51	1.28	0.93	0.86	0.91	0.70	2.06
Dy	4.6	6.6	3.8	3.2	8.0	5.9	5.1	5.7	5.0	13
Ho	1.05	1.45	0.91	0.72	1.76	1.41	1.19	1.26	1.01	2.8
Er	3.0	4.2	2.5	2.0	5.3	4.0	3.2	3.6	3.0	7.4
Tm	0.43	0.58	0.37	0.31	0.70	0.58	0.47	0.49	0.43	0.98
Yb	2.9	3.9	2.4	1.8	4.9	3.8	3.3	3.5	2.9	6.2
Lu	0.42	0.57	0.34	0.28	0.70	0.53	0.47	0.47	0.38	0.96
Hf	1.22	2.0	1.03	0.62	1.62	1.29	0.91	1.18	1.41	2.1
Ta	0.015	0.136	0.028	bd	bd	bd	bd	bd	0.022	bd
Th	0.025	0.18	0.011	0.008	0.010	0.018	0.010	0.031	0.011	bd
U	0.016	0.044	bd	bd	bd	0.012	bd	bd	0.036	bd

Abbreviation: bd, below detection.

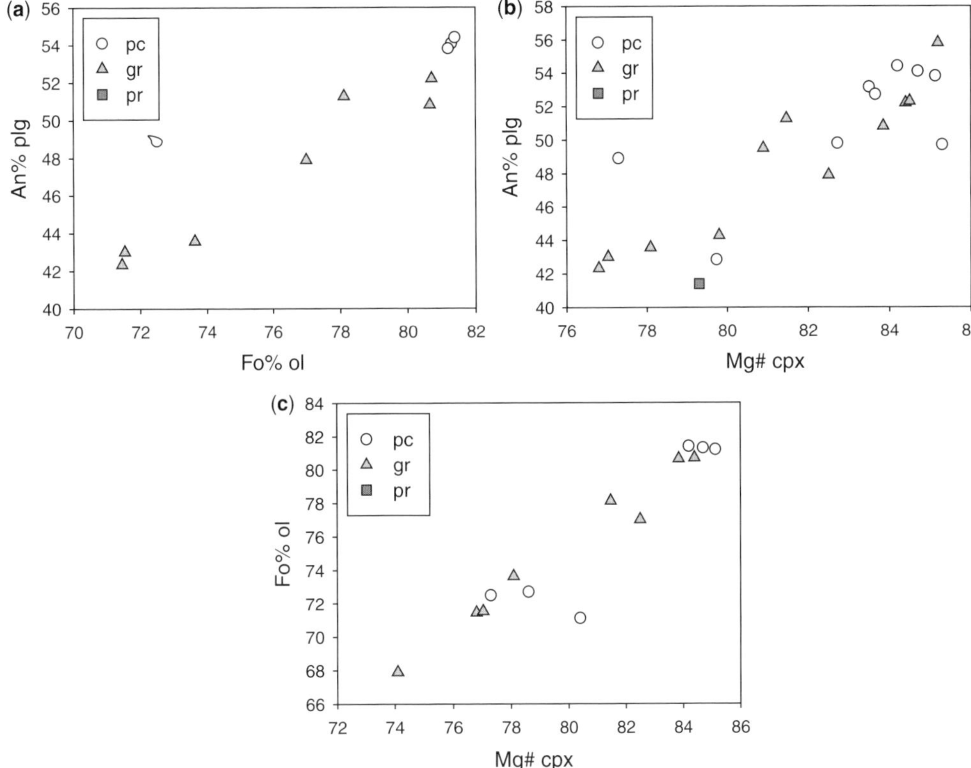

Fig. 6. Mg# (clinopyroxene) v. Fo (olivine) v. An (plagioclase) in gabbroic dykes and cumulate pods. Note the co-variations of these compositional parameters from olivine gabbros (olivine Fo81.4, plagioclase An55.8 and clinopyroxene Mg# 85.3) to Fe–Ti oxide gabbros (olivine Fo67.9, plagioclase An41.4 and clinopyroxene Mg# 74.1).

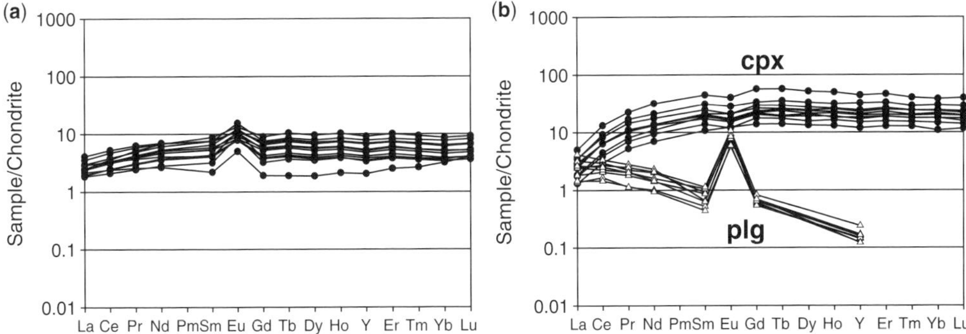

Fig. 7. C1-normalized REE patterns of gabbroic dykes from Monte Maggiore. (**a**) Bulk rocks. Note the slight LREE fractionation and the evident positive Eu anomaly, suggesting that they are plagioclase-rich cumulates. (**b**) Clinopyroxenes (Cpx) and plagioclases (Plg) of gabbroic dykes. Note the slight LREE negative fractionation of clinopyroxenes and the positive LREE fractionation of plagioclase, very similar to the REE patterns of minerals in MORBs, and the evident positive Eu anomaly, suggesting that they are plagioclase-rich cumulates.

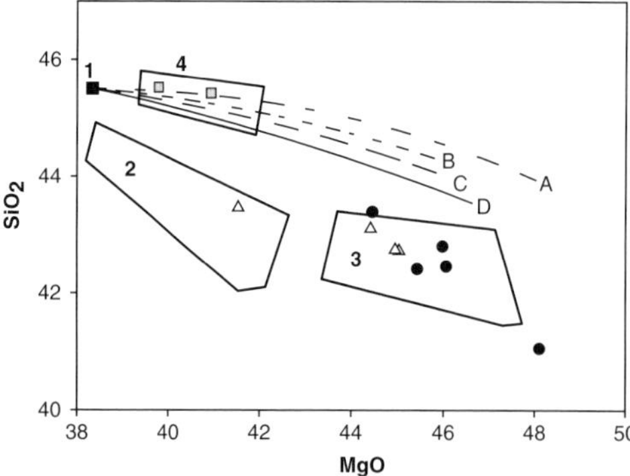

Fig. 8. SiO_2 v. MgO diagram of the bulk compositions of some plagioclase-free and plagioclase-bearing spinel peridotites from Monte Maggiore. (1) The composition of the fertile starting composition: (**a**) the compositional trend of refractory residua after polybaric melting at 25–8 kbar; (**b**) the compositional trend of refractory residua after polybaric melting at 15–8 kbar; (**c**) the compositional trend of refractory residua after batch melting at 20 kbar; (**d**) the compositional trend of refractory residua after batch melting at 10 kbar. Compositional trends from Niu (1997). (2) The field (3) of reactive spinel peridotites of the South Lanzo Massif (from Piccardo *et al.* 2007*a*). (3) The field (2) of impregnated plagioclase peridotites of the South Lanzo Massif (from Piccardo *et al.* 2007*a*). (4) The field (4) of subcontinental lithospheric lherzolites of North Lanzo and Erro-Tobbio (Piccardo *et al.* 2007*b*; Piccardo & Vissers 2007).

Note that: (1) the lithospheric peridotites lie along the melting trend curves, consistent with the subcontinental lithospheric lherzolites of other OCT ophiolitic peridotite bodies (i.e. North Lanzo, Erro-Tobbio) of the Ligurian Tethys (field 4); (2) the spinel peridotites have significantly lower SiO_2 content than the compositions of refractory residua after partial melting, at the corresponding MgO content, consistent with reactive pyroxene depletion and olivine addition; and (3) the plagioclase peridotites show decreasing MgO and slightly increasing SiO_2 content, with respect to the reactive peridotites, consistent with addition of gabbroic material.

clinopyroxene core compositions in the different rock types (using Cpx/liquid K_Ds of Hart & Dunn 1993). Calculated REE composition of liquids are quite similar to those of liquids produced by 6% fractional melting of spinel-facies DM asthenospheric mantle sources (Fig. 9a). Accordingly, it can be speculated that these spinel harzburgites were formed by the reactive percolation of fractional melt increments showing MORB affinity.

The impregnated plagioclase peridotites. Available field, petrographical–microstructural and compositional data show that impregnated plagioclase peridotites mostly derive from pristine reactive spinel peridotites that underwent significant replacement of olivine by orthopyroxene and of clinopyroxene by Opx + Plg aggregates, and interstitial crystallization of plagioclase, during percolation and interstitial crystallization of basaltic melts.

The analysed impregnated plagioclase peridotites have variable plagioclase (Plg 4.1–11.2% vol.) content and significantly low levels of clinopyroxene (Cpx 2.1–3.3% vol.). Orthopyroxene content is significantly higher (Opx 21.5–23.3 vol.), coupled with significantly lower olivine (Ol 65.1–71.14% vol.) content with respect to the reactive spinel peridotites. Increasing Opx, and decreasing Cpx and Ol, is in good agreement with petrographical evidence indicating significant olivine and clinopyroxene replacement by orthopyroxene during melt–rock interaction.

The porphyroclastic mantle pyroxenes in the impregnated Monte Maggiore peridotites have trace element compositions unlike residues of fractional melting and are enriched in many trace elements (i.e. REE, Ti, Sc, V, Zr, Y) with respect to porphyroclastic pyroxenes from pristine reactive spinel peridotites (see Fig. 5). Clinopyroxene core compositions shows convex-upward REE spectra, with a significant MREE enrichment, and both pyroxenes generally show a negative Eu_N anomaly. This indicates that: (1) porphyroclastic pyroxenes in the impregnated peridotites attained trace element equilibrium with the migrating liquid; and (2) the impregnating liquids were significantly enriched in many trace elements (see also Müntener & Piccardo 2003).

Rampone *et al.* (2008) demonstrated the existence of within-mineral core-rim chemical zoning in clinopyroxene showing trace element enrichment at the rims of reacted clinopyroxene porphyroclasts, which is interpreted as *in situ* melt–rock interaction and the crystallization of trapped melts.

Orthopyroxene saturation of the infiltrating melts can have been attained by the interaction of the primary melts with the ambient peridotite during upwards migration in the mantle column, where they dissolved mantle pyroxenes and crystallized olivine (Müntener & Piccardo 2003). In fact, liquids rising by diffuse porous flow through the mantle and reacting with the percolated mantle rock would dissolve pyroxenes and crystallize olivine. As discussed by Kelemen *et al.* (1995), the composition of liquids that reacted with mantle peridotite depends on the relative effects of reaction with the surrounding host peridotite and cooling. For relatively rapid reactions and slow cooling, liquids might quickly become saturated in orthopyroxene, whereas more rapid cooling, or slower pyroxene dissolution, would produce less silica-rich liquid compositions. Continued melt–rock reaction and/or cooling finally led to saturation in two pyroxenes.

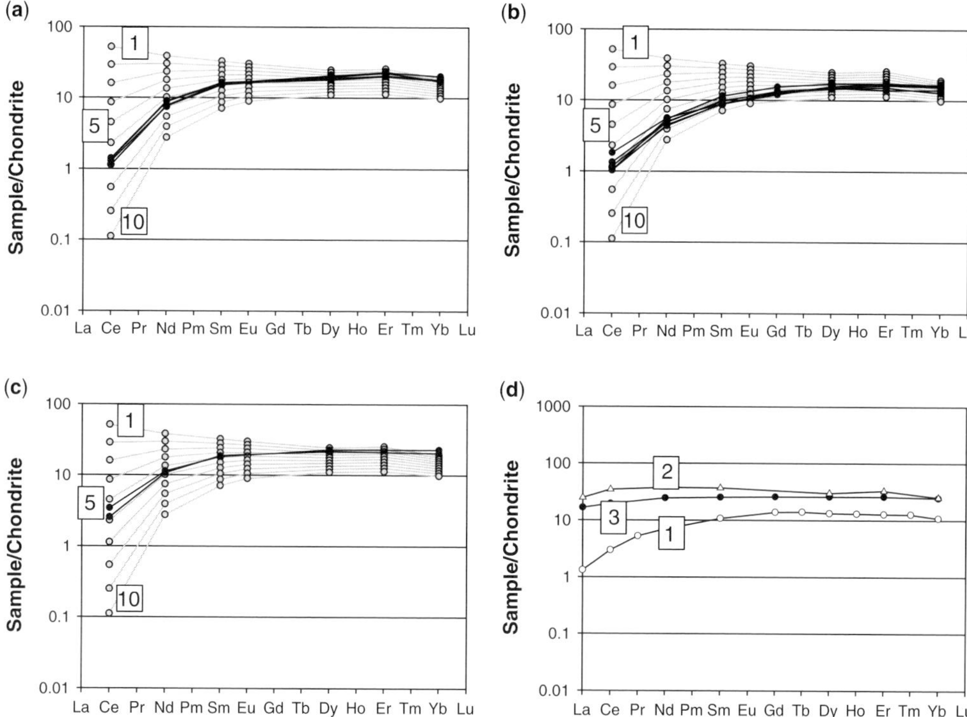

Fig. 9. C1-normalized REE patterns of liquids calculated in equilibrium with Cpx in Monte Maggiore peridotites and gabbros, using the K_D Cpx/liquid values of Hart & Dunn (1993) (Fig. 9a) and of Vannucci *et al.* (1998) (Fig. 9b, c). Liquid REE patterns calculated at variable degrees of fractional melting, according to the fractional melting model of Johnson *et al.* (1990), starting from the source Cpx of Ionov *et al.* (2002), under spinel-facies conditions, are reported for comparison. (**a**) The REE patterns of liquids calculated in equilibrium with clinopyroxenes of reactive spinel peridotites. It can be deduced that the reactively percolating melts were very similar to single-melts fractions after 6% of fractional melting of a DM mantle source at spinel-facies condition. [1], [5] and [10] represent the REE patterns of liquids obtained at the different degrees of melting. (**b**) The REE patterns of liquids calculated in equilibrium with clinopyroxenes of plagioclase-bearing peridotites. It can be deduced that the reactively percolating melts were very similar to single-melts fractions after 6% of fractional melting of a DM mantle source at spinel-facies condition. [1], [5] and [10] represent the REE patterns of liquids obtained at the different degrees of melting. (**c**) The REE patterns of liquids calculated in equilibrium with clinopyroxenes of representative mafic–ultramafic cumulates and dykelets. It can be deduced that the reactively percolating melts were very similar to single-melts fractions after 5–6% of fractional melting of a DM mantle source at spinel-facies condition. [1], [5] and [10] represent the REE patterns of liquids obtained at the different degrees of melting. (**d**) C1-normalized average REE pattern of the aggregate MORB melt [2] in equilibrium with clinopyroxene [1] of the less evolved gabbroic dyke, having an olivine composition of Fo81. The C1-normalized REE pattern of N-MORB [3] (data from Hoffman 1988) is also reported for comparison.

Magmatic plagioclases and trace element melt-equilibrated clinopyroxenes are strongly LREE depleted and have very low Sr content, significantly different from corresponding minerals in the primary aggregate MORB. Accordingly, the impregnating melts were strongly depleted in the most incompatible elements, most probably corresponding to depleted single-melt increments that were produced during the fractional melting of an asthenospheric mantle source (Rampone *et al.* 1997; Piccardo *et al.* 2007*a*).

Information on the compositional characteristics of the percolating melts has been obtained from the Cpx trace element compositions, assuming that the Cpx and liquid attained the trace element equilibration during percolation and/or crystallization, using the Cpx/liquid K_Ds of Vannucci *et al.* (1998) that are useful for saturated basaltic systems. The REE patterns of the calculated equilibrium liquids are quite similar to those of liquids produced by 6% fractional melting of spinel-facies DM asthenospheric mantle sources (Fig. 9b). Accordingly, it can be speculated that the impregnated plagioclase peridotites were formed by percolation and interstitial crystallization of asthenospheric fractional melt increments showing MORB affinity.

The replacive spinel dunites. The formation of discordant dunite channels marks the transition from diffuse porous flow to focused porous flow. Textural evidence for a replacive origin of the discordant dunite bodies was first given by Boudier & Nicolas (1972) in the Lanzo peridotite. Dunite channels locally replace early pyroxenite banding, as indicated by trains of Cr spinel in dunite that are continuous with the pyroxenite bands in the host peridotite. Melt percolating in the dunite channels sporadically crystallized plagioclase films surrounding olivine megacrystals and interstitial clinopyroxenes and, later, trails of clinopyroxene + plagioclase megacrysts (up to 2 cm in size), which are precursors to the early gabbroic veins and dykelets.

The mafic–ultramafic cumulate pods and gabbroic dykelets. The mafic–ultramafic cumulates and gabbroic dykelets have abundant euhedral Opx, and are characterized by minerals showing incompatible trace element compositions significantly depleted with respect to the corresponding minerals in equilibrium with aggregate MORB melts, suggesting crystallization from depleted and silica-saturated melts fractions. The REE patterns obtained for liquids in equilibrium with clinopyroxenes from these rock types, using the Cpx/liquid K_Ds of Vannucci *et al.* (1998), useful for saturated basaltic systems, are quite similar to those of liquids produced by 5–6% fractional melting of spinel-facies DM asthenospheric mantle sources (Fig. 9c). Accordingly, it can be speculated that the liquids which migrated upwards within the dunite channels and intruded as dykelets and cumulate pods consisted of fractional melt increments showing MORB affinity.

The gabbro–noritic cumulates are very similar, as far as their modal and mineral compositions, to the gabbro–noritic mafic–ultramafic cumulates, rich in orthopyroxene and strongly depleted in incompatible trace elements, found at Site 334 of DSDP along the Mid-Atlantic Ridge (MAR) (Ross & Elthon 1993).

The gabbroic dykes. Gabbroic dykes represent cumulates from variably evolved MORB-type melts. On the basis of mineral compositions and crystallization order it can be deduced that primary melts underwent evolution by fractional crystallization, becoming progressively enriched in fusible and incompatible elements, reaching Fe-basaltic compositions.

Geochemical modelling of the REE composition of melts in equilibrium with clinopyroxenes of the more primitive Mg-gabbroic sample (having olivine with Fo81 composition) indicate that the equilibrium melts have a rather flat REE pattern (at about 30 × C1) showing slight negative LREE fractionation (Fig. 9d). The REE patterns is similar to those of N-MORBs and of liquids in equilibrium with the more primitive Mg-gabbros of the Apennine ophiolites and the MAR, although they have slightly higher REE concentrations because they are slightly evolved (Rampone *et al.* 2004; Tribuzio *et al.* 2004 and references therein).

It may be inferred that the primary melts of the gabbroic dykes at Monte Maggiore were aggregate N-MORB melts, similar to those of the other ophiolitic gabbroic rocks of the Ligurian ophiolites.

Rampone (2004) presented and discussed isotope data for a limited set of samples from the Monte Maggiore spinel and plagioclase peridotites and gabbroic rocks. Clinopyroxene separates from spinel, and plagioclase peridotites display relatively high $^{147}Sm/^{144}Nd$ ratios (0.49–0.59) and $^{143}Nd/^{144}Nd$ values in the range 0.513367–0.513551. The associated gabbroic rocks have Nd isotopic compositions typical of MORB ($^{143}Nd/^{144}Nd$ = 0.513122–0.513138). Moreover, plagioclase whole-rock clinopyroxene Sm/Nd data for a gabbro–noritic dykelet, related to the melt impregnation, and an olivine gabbro dyke define internal isochrons that yield Late Jurassic ages (155 ± 6 and 162 ± 10 Ma, respectively) and initial ε_{Nd} values of 9.7 and 8.9, respectively (Rampone 2004). In a $^{147}Sm/^{144}Nd$ v. $^{143}Nd/^{144}Nd$ diagram, the peridotite data conform to the linear array defined by the gabbroic rocks, their initial (160 Ma) ε_{Nd} values varying in the range 7.6–8.9. The Sm/Nd isotopic compositions of the Monte Maggiore peridotites have been interpreted by Rampone (2004) as being consistent with a Jurassic age of partial melting. Available data point, moreover, to isotopic compositional similarities between depleted peridotites and associated magmatic rocks. This homogeneity has been regarded by Rampone (2004) as evidence of a melt–source isotopic equilibration in this sector of the ancient Ligurian Tethys basin. Accordingly, the Monte Maggiore gabbro–peridotite association has been considered by Rampone (2004) to represent the first record of the attainment of a mature oceanic stage in the Ligurian Tethys ocean.

Noticeably, our field, structural and compositional results indicate that the whole Monte Maggiore depleted spinel and plagioclase peridotites formed by interaction–impregnation processes involving MORB-type melts on the pristine fertile (veined) lithospheric lherzolites. On this basis, we infer that the overall isotopic homogeneity shown by the Monte Maggiore peridotites (as reported by Rampone 2004) is related to the melt–peridotite interaction processes and were induced by isotopic equilibration with the percolating melts.

Accordingly, the overall isotopic homogeneity documented by the Monte Maggiore peridotites and gabbros indicate that the lithospheric mantle protoliths were isotopically homogenized by extensive melt–rock interaction that erased any preexisting isotopic heterogeneity. The percolating MORB melts were isotopically similar to the parental melts of the gabbroic intrusions, and the asthenospheric mantle sources of the percolating and intruding melts were isotopically homogeneous.

As a whole, the isotope characteristics of the spinel and plagioclase peridotites and gabbros, along with geochronological constraints, support the following scenario. During Late Jurassic times an isotopically homogeneous asthenosphere underwent adiabatic upwelling and decompression melting. The asthenospheric melts migrated by porous flow percolation through the mantle lithosphere, thus erasing any pre-existing isotopic heterogeneity and imparting an overall isotopic homogenization to the percolated–impregnated mantle lithosphere. Finally, after the upwelling of mantle peridotites to more cold and rigid shallow lithospheric levels, similar melts rose up from the same asthenospheric sources, underwent variable evolution, and intruded as gabbroic pods and dykes.

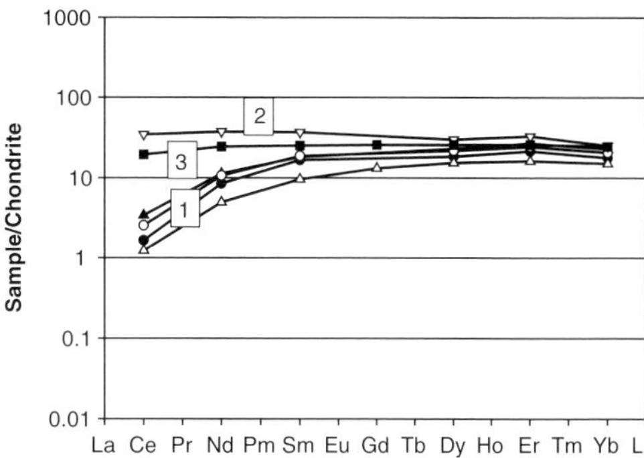

Fig. 10. C1-normalized average REE patterns of the fractional MORB-type melts in equilibrium with the clinopyroxenes of spinel peridotites, plagioclase-bearing peridotites, mafic–ultramafic cumulates and dykelets [1], and the aggregate MORB melt in equilibrium with clinopyroxene of the less evolved gabbroic dyke [2], having an olivine composition of Fo81. The C1-normalized REE pattern of N-MORB [3] (data from Hoffman 1988) is also reported for comparison.

Melt dynamics and thermal evolution of the mantle lithosphere

Significant compositional differences exist between magmatic clinopyroxenes and plagioclases from early stages of percolation–intrusion (i.e. gabbro-noritic veins, dykelets and mafic–ultramafic cumulate pods) and later gabbroic intrusion in dykes. Considering clinopyroxenes showing similar Mg#, clinopyroxenes of the early dykelets and cumulates have REE patterns showing strong LREE fractionation and small content of incompatible elements (e.g. Sr 0.7–2.0 ppm), whereas clinopyroxenes of the late gabbroic dykes have REE patterns showing moderate LREE fractionation and a significant content of highly incompatible elements (e.g. Sr 11.7–17.5 ppm). In accordance, plagioclases show negative LREE fractionation and relatively small content of incompatible elements (e.g. Sr 14.7–30.4 ppm) in the former, and positive LREE fractionation and large content of incompatible elements (e.g. Sr 310–350 ppm) in the latter. These features suggest that the early percolating–intruding melts had significantly depleted MORB-type characteristics, whereas later melts intruding as gabbroic dykes were quite similar to aggregate N-MORB.

As previously stated, early percolating melts that formed reactive spinel peridotites, impregnated plagioclase peridotites and mafic–ultramafic cumulates, most probably consisted of depleted single-melt increments with MORB affinity deriving from 5–6% fractional melting of a spinel-facies DM asthenospheric mantle source. By contrast, the late intrusive melts were represented by aggregate N-MORB melts (Fig. 10). Available isotope data indicate an overall Sm/Nd isotopic homogeneity between a gabbro–noritic veinlet related to early impregnation and the late gabbroic dyke (Rampone 2004), suggesting provenance of their parental melts from similar isotopically homogeneous asthenospheric mantle sources.

This indicates that the melt dynamics in the melting asthenosphere changed during the evolution of the melting process. At the inception of the melting process the single-melt fractions survived unmixed and migrated isolated through the overlying spinel–plagioclase-facies mantle lithosphere. Subsequently, the single fractional melt increments were progressively aggregated and the aggregate MORB liquids were delivered to shallow crustal levels. There, they underwent magmatic evolution by low-pressure fractional crystallization, most probably within ephemeral magma chambers, and the variably evolved magmas were intruded into fractures.

The pyroxenite-bearing spinel lherzolite protoliths were completely equilibrated in the subcontinental mantle lithosphere at spinel-facies conditions and temperatures (T) in the range 970–1100 °C. During subsequent exhumation from spinel- to plagioclase-facies conditions these mantle sections were infiltrated by melt fractions via diffuse porous flow percolation. Melt percolation should have occurred at temperature conditions higher than the liquidus conditions of the percolating melt (i.e. at $T > 1250\,^{\circ}$C). In fact, geothermometric estimates based on trace element (Sc, V) distribution between coexisting ortho- and clinopyroxene (method of Seitz *et al.* 1999) indicate $T > 1250$–$1300\,^{\circ}$C in the melt-percolated spinel and plagioclase peridotites (Müntener & Piccardo 2003; Piccardo *et al.* 2004). Rampone *et al.* (2008) obtained equilibration temperatures ranging from 1200 to 1300 °C preserved in pyroxene porphyroclasts of a few samples of (reactive) spinel peridotites. The above estimates are consistent with conditions for diffuse porous flow migration of basaltic melts and indicate the attainment of asthenospheric thermal conditions in the mantle lithosphere during the melt diffuse percolation. Accordingly, it may be speculated that asthenosphere upwelling and melt percolation during lithosphere extension caused the thermomechanical and chemical erosion of the extending mantle lithosphere.

Subsequently, progressive upwelling under increasing conductive heat loss caused, first, the interstitial crystallization of the percolating melts that formed the impregnated plagioclase peridotites. Melt impregnation should have occurred at temperature conditions close to the liquidus conditions of the percolation melts (i.e. 1150–1250 °C, as obtained in plagioclase peridotites by Rampone *et al.* 2008), which infiltrated the peridotite by diffuse porous flow and underwent later interstitial crystallization of early cumulus minerals in the form of microgabbroic aggregates.

Interstitial crystallization in plagioclase peridotites clogged the infiltration pathways of the migrating melts. Further melt migration was only allowed by focused percolation within compositional and/or structural discontinuities. Dunite channels were formed by intense melt–rock interaction (pyroxene dissolution and olivine precipitation) during open-system melt migration within these pathways. The interconnected melt network in dunite channels became progressively clogged with crystallization products. Because of cooling and crystallization, melt migration changed from focused percolation in dunite channels to intrusion into narrow conduits, most probably formed by substantial hydrostatic overpressure that initiated the formation of cracks and expelled liquids into veins, dykelets and dykes (Müntener & Piccardo 2003 and references therein).

Geochronological data

Available isotope data on the Monte Maggiore peridotites and gabbroic dykes (Rampone 2004) provide reliable geochronological information (i.e. Sm–Nd clinopyroxene–plagioclase–whole-rock isochron ages and Sm–Nd model ages). Plagioclase–whole rock–clinopyroxene (Plg–wr–Cpx) data obtained from a gabbroic rock defines an internal isochron yielding Late Jurassic age (162 ± 10 Ma) (Fig. 11). The initial ε_{Nd} value of 8.9 is consistent with a MORB affinity. The Plg–wr–Cpx isochron for a gabbro–noritic veinlet formed by plagioclase-bearing melt impregnation yields an age of 155 ± 6 Ma; the initial ε_{Nd} value is 9.7, also indicative of a MORB signature (Rampone 2004).

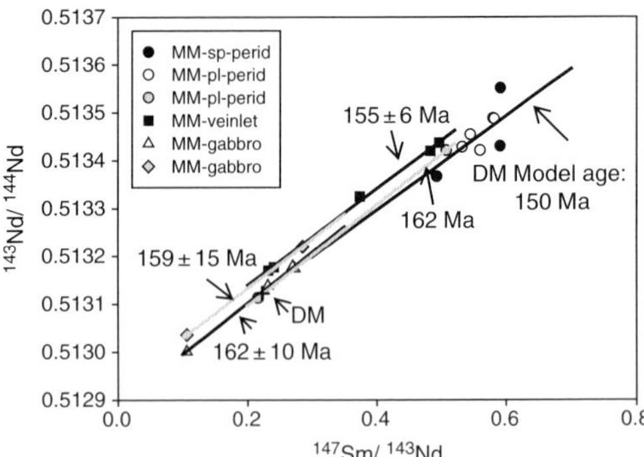

Fig. 11. ^{143}Nd/^{144}Nd v. ^{147}Sm/^{144}Nd diagram of clinopyroxenes, plagioclase and whole rocks of spinel peridotites (MM Sp-perid.), plagioclase peridotites (MM Plg-perid.), and gabbroic veinlet and dykes (MM-veinlet and MM-gabbros) from Monte Maggiore (from Rampone 2004). Note that the Cpx–Plg–wr isochrons of the gabbroic rocks (veinlet and dykes) point to ages in the range of 155–162 Ma, while the spinel and plagioclase peridotite DM model ages (150 Ma) conform to these isochrons.

Peridotite data conform, in a ^{147}Sm/^{144}Nd v. ^{143}Nd/^{144}Nd diagram, to the linear array defined by the gabbroic rocks, extending this correlation to high Sm–Nd isotope ratios (Fig. 11). Their initial (160 Ma) ε_{Nd} values vary in the range 7.6–8.9.

As our field, structural and compositional data indicate that both depleted spinel peridotites and impregnated plagioclase peridotites were formed by melt–rock interaction processes induced by MORB melt percolation on pristine lithospheric lherzolites, we infer that the isotopic signatures of peridotites were induced by melt–rock isotopic equilibration. Accordingly, the Sm/Nd isotopic signatures of the Monte Maggiore peridotites reflect the isotopic features of the percolating melts and indicate a Jurassic age for the isotopic equilibration, that is, the MORB melt percolation. Accordingly, MORB melts deriving from isotopically homogeneous asthenospheric mantle sources infiltrated the extending lithospheric mantle during Late Jurassic times.

Available geochronological data from the whole ophiolitic peridotites from the Alpine–Apennine system support the time–space evolution of the subcontinental mantle during lithosphere extension and continental failure leading to the opening of the Ligurian Tethys basin (see the discussion in Piccardo *et al.* 2009).

Sm–Nd isotopic data on clinopyroxenes from some OCT External Ligurides (Suvero) and North Lanzo lithospheric peridotites (data from Bodinier *et al.* 1991; Rampone *et al.* 1995) yielded Proterozoic CHUR and DM model ages, suggesting that they were isolated from the convective mantle asthenosphere and accreted to the subcontinental lithosphere during Proterozoic times (Piccardo 2008 and references therein) (Fig. 12a).

Plagioclase-facies peridotite mylonites in shear zones of some OCT External Ligurides lherzolites yielded Lu–Hf isochrons indicating minimum ages of 220 ± 13 Ma (Montanini *et al.* 2006), and hydrated shear zones (tremolite–chlorite–peridotite mylonites) in the Malenco peridotites

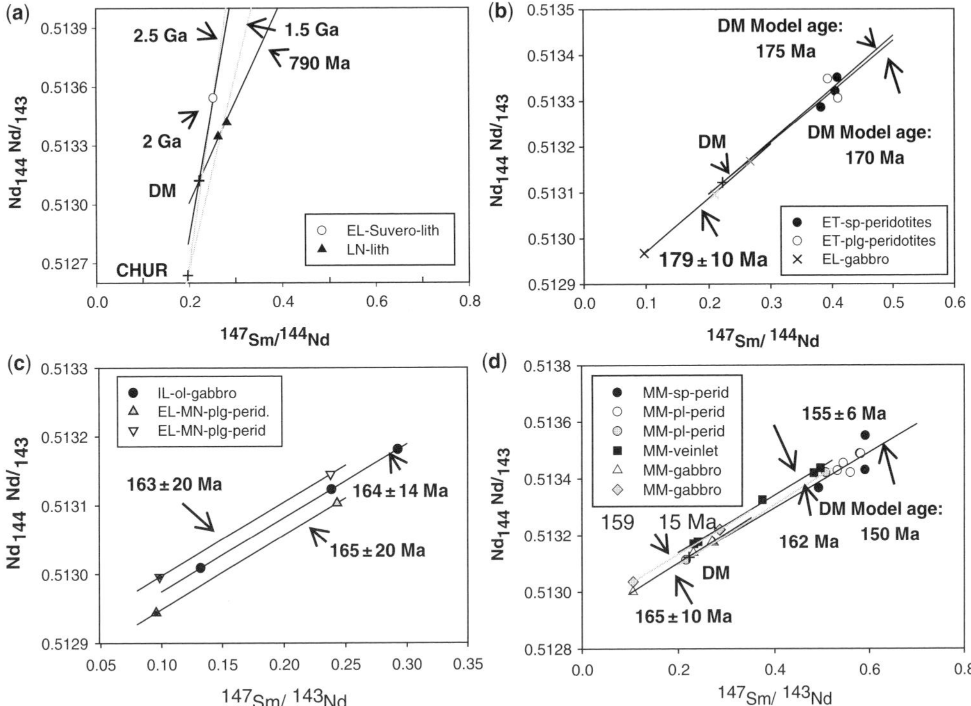

Fig. 12. Geochronological data from ophiolitic peridotites and gabbros of the Ligurian Tethys. (**a**) $^{143}Nd/^{144}Nd$ v. $^{147}Sm/^{144}Nd$ diagram of clinopyroxenes from OCT lithospheric lherzolites from the External Ligurides (EL Suvero lith.) and North Lanzo (LN lith.) (data from Bodinier *et al.* 1991; Rampone *et al.* 1995). CHUR and DM model ages indicate Proterozoic ages that have been interpreted as the ages of accretion from the convecting mantle to the lithospheric mantle. (**b**) $^{143}Nd/^{144}Nd$ v. $^{147}Sm/^{144}Nd$ diagram of clinopyroxenes from OCT reactive (ET Sp-peridotites) and impregnated (ET Plg-peridotites) peridotites of Erro-Tobbio (ET) (data from Rampone *et al.* 2005) and from a synrift gabbro of the OCT External Ligurides (EL-gabbro) (data from Tribuzio *et al.* 2004). Note that the Cpx–Plg–wr isochrons of the EL-gabbro point to an age of 179 Ma, while the DM model ages of ET spinel and plagioclase peridotites conform to these isochrons giving ages of 170–175 Ma. These Early Jurassic ages are interpreted as the age of inception of asthenosphere decompression melting and melt infiltration–intrusion in the OCT subcontinental lithospheric mantle when the OCT peridotite protoliths were still located along the axial zone of the Ligurian extensional system. (**c**) $^{143}Nd/^{144}Nd$ v. $^{147}Sm/^{144}Nd$ diagram of clinopyroxenes, plagioclases and whole rock from EL impregnated peridotites (EL MN Plg-peridotites) and from a IL gabbroic body from the Internal Liguride Units (IL-Ol-gabbro) (data from Rampone *et al.* 1998). Note that the Cpx–Plg–wr isochrone of the IL-gabbro points to an age of 164 Ma, while the Cpx–Plg isochrons of the EL impregnated plagioclase peridotites conform to this isochron giving 163–165 Ma ages. These Middle–Late Jurassic ages are interpreted as the age of asthenosphere decompression melting and infiltration of the melt fractions in the subcontinental lithospheric mantle when the OCT peridotite protoliths were deplaced to more pericontinental positions and MIO peridotite protoliths upwelled along the axial zone of the Ligurian extensional system. (**d**) The same as Figure 11.

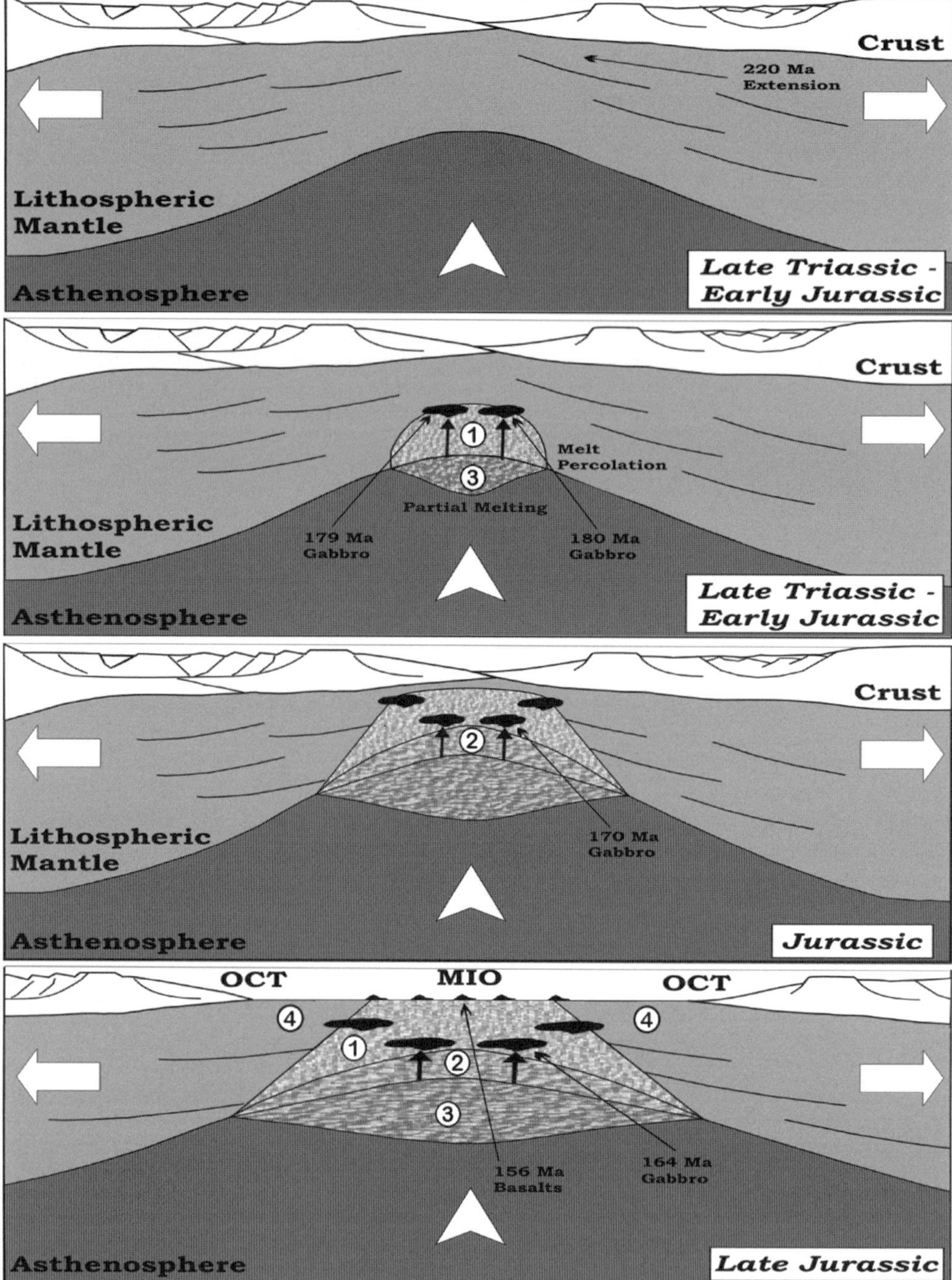

Fig. 13. The geodynamic evolution of the Ligurian Tethys (from top to bottom). Late Triassic–Early Jurassic times – extension of the continental lithosphere, tectonic exhumation of lithospheric mantle and adiabatic upwelling of the asthenosphere. Lithosphere extension was already active during Triassic (220 Ma in the EL mantle: Montanini *et al.* 2006), and was accomodated by km-scale extensional shear zones (e.g. Drury *et al.* 1990; Vissers *et al.* 1991; Hoogerduijn Strating *et al.* 1993). Early Jurassic times – the adiabatically upwelling asthenosphere underwent inception of decompressional partial melting (3). The asthenospheric melts infiltrated by porous flow mechanisms through the

(Central Alps) yielded $^{40}Ar/^{39}Ar$ amphibole age of about 225 Ma (Müntener & Hermann 2001). These data have been related to the onset of continental rifting, indicating that the subcontinental mantle lithosphere underwent subsolidus exhumation by extensional shear zones during Triassic times.

An Early Jurassic age for the onset of asthenosphere decompression partial melting can be deduced from the oldest melt percolation and intrusion events in the OCT peridotites. Available isotope data from the OCT Erro-Tobbio and External Liguride peridotites yielded Nd DM model ages and Sm–Nd Cpx–Plg isochron ages ranging from 180 to 170 Ma (Fig. 12b). Relatively younger Middle Jurassic ages are yielded by melt impregnation in more internal peridotite bodies (Mt Nero) of the External Ligurides (Sm–Nd Cpx–Plg isochron ages of 163–165 $\pm$ 20 Ma) (Fig. 12c) and by some gabbroic bodies from the Internal Liguride Units (Rampone *et al.* 1998; Tribuzio *et al.* 2004 and references therein). The geochronological data from the Monte Maggiore peridotites and gabbros (Fig. 12d) indicate that MORB melt percolation and intrusion in this more internal MIO peridotite massif represent the youngest, Late Jurassic, magmatic events recorded by the Ligurian lithospheric mantle.

Accordingly, available geochronological data support the hypothesis that asthenosphere partial melting, lithosphere melt percolation and intrusion in the Europe–Adria lithospheric mantle, prior to the opening of the Ligurian Tethys basin, occurred starting from Early Jurassic times, when the continental lithosphere underwent significant stretching and thinning during lithosphere extension. MORB melt percolation–intrusion was slightly older (Early Jurassic (180–170 Ma)) in the peridotites that were, later on, confined to more 'external' OCT settings of both the Europe and Adria margins, and slightly younger (Late Jurassic (162–150 Ma)) in the peridotites that upwelled later along the axial zone of the extensional system and were exposed on the sea floor at more internal MIO settings of the basin.

The 'oceanic stage' in the Ligurian Tethys

Field, structural and stratigraphic data on A–A ophiolites describe two different types of ophiolite sequences, depending on presence or absence of crustal material from the continental margins. The more 'external' OCT-type sequences are characterized by the presence of continental crustal remnants and continental crust-derived sediments associated with mantle peridotites, basaltic lava flow and oceanic sediments, suggesting a close relationship with the continental margin. The more 'internal' MIO-type sequences are, however, characterized by a lack of crustal remnants, and a discontinuous cover of basaltic volcanites and oceanic sediments (Abbate *et al.* 1994; Marroni *et al.* 1998).

Apparently, the 'oceanic stage' seems to correspond to the complete removal of the continental crust, the lack of crust-derived sediments and the deposition of oceanic sediments, besides the discontinuous basaltic extrusions (see Piccardo 2008 for a more detailed discussion).

From the mantle perspective, the OCT settings are characterized by sea-floor exposure of 'cold' mantle exhumed from the shallow subcontinental lithosphere, whereas the MIO settings are characterized by sea-floor exposure of 'hot' asthenospherized mantle exhumed from deeper lithospheric levels where it was significantly modified by MORB melt interaction (see Piccardo *et al.* 2007*b* for a more detailed discussion). In any case, A–A ophiolites consist of the stratigraphic association of MORB volcanites and oceanic sediments with subcontinental lithospheric mantle that either escaped melt infiltration (the OCT peridotites) or was deeply modified by melt percolation and interaction (the MIO peridotites) (Piccardo 2008) (Fig. 13).

This opens the question about the meaning of 'oceanic stage' in the Ligurian Tethys. Oceanic

Fig. 13. (*Continued*) extending lithospheric mantle (1) and formed sporadic gabbroic intrusions (Piccardo *et al.* 2007*a*; Piccardo & Vissers 2007). Melt percolation was facilitated by the network of shear zones driving lithosphere extension. Available ages of MORB gabbroic intrusion (179–180 Ma of the oldest MORB gabbros in OCT peridotites (Tribuzio *et al.* 2004; Rampone *et al.* 2005)) indicate that asthenosphere partial melting started, most probably, no later than Early Jurassic times. Jurassic times – ongoing extension caused further lithosphere exhumation and significant asthenosphere upwelling. Refractory residua after Jurassic partial melting, coeval and cogenetic with the Jurassic MORB melts, were accreted to the base of the lithosphere (2). Late Jurassic times – the continental crust underwent complete failure and the passive margins were formed. The entrapment of asthenospheric melts by percolation–impregnation in the extending lithospheric mantle prevented significant melt extrusion, and the passive margins were typically non-volcanic, although magmatic. The subcontinental lithospheric mantle was exhumed and exposed on the sea floor. The lithospheric mantle peridotites, deriving from shallower mantle levels, were exhumed at more external OCT zones of the basin close to the continental margins (4), whereas lithospheric peridotites deriving from deeper mantle levels that underwent significant melt percolation–impregnation processes were exhumed and exposed on the sea floor at MIO settings of the basin.

stage should mean that the oceanic lithosphere is produced from the products of asthenosphere partial melting, that is, basalts and gabbros, formed by the melt fraction, and abyssal peridotites, formed by the refractory residua. In the case of the Ligurian Tethys, the attainment of a true oceanic stage should consist of: (1) the break-up and complete removal of the whole continental lithosphere (both crust and mantle) between newly formed continental margins and ocean–continent transition zones; and (2) the formation of new oceanic lithosphere, that is, the close association of oceanic gabbros + basalts and refractory peridotites, that should be broadly coeval and cogenetic.

In this case, no one of the known A–A ophiolite sequences represents true oceanic lithosphere. In fact, the depleted spinel peridotites of Monte Maggiore and the other MIO massifs were formed by reactive percolation of infiltrating asthenospheric melts through the mantle lithosphere. They do not represent refractory residua after Jurassic oceanic-type partial melting of the mantle asthenosphere (e.g. Piccardo *et al.* 2004, 2007*a*). From a mantle perspective, the transition from the continent-ward OCT peridotites to the ocean-ward MIO peridotites in the Jurassic Ligurian Tethys did not represent the transition from subcontinental mantle to oceanic mantle (i.e. Jurassic oceanic refractory residua), as MIO peridotites were pristine subcontinental mantle modified by MORB melt interaction (Fig. 13). Available isotopic data on the Monte Maggiore peridotites indicate that the subcontinental mantle underwent Jurassic '*oceanization*', in the sense that it was significantly depleted and attained MORB isotopic signatures by interaction with Jurassic MORB melts rising from the melting asthenosphere.

The lack of evidence of true refractory residua on the sea floor, that is, within the known A–A ophiolite sequences, indicates that, most probably, the refractory residua after Jurassic partial melting did not reach the sea floor and the whole peridotite basement of the oceanic lithosphere was constituted by exhumed subcontinental lithospheric mantle, significantly modified by melt percolation along the more internal settings of the embryonic basin.

Accordingly, the 'oceanic stage', frequently related in the past to the internal 'more oceanic' ophiolites of the Ligurian Tethys basin, consisted of the complete failure of the continental crust but not of the formation of true new oceanic lithosphere, showing genetic mantle–crust relationships.

On the basis of present knowledge on the studied A–A ophiolites, it can be sustained that the complete break-up and failure of the subcontinental mantle lithosphere did not occur in the Ligurian Tethys basin, as testified by the known A–A ophiolite sequences. Lithosphere extension by far-field tectonic forces and melt entrapment in the extending lithospheric mantle caused the formation of passive, non-volcanic margins, large ocean–continent transition zones characterized by the sea-floor exposure of the subcontinental mantle, variably modified by MORB melt interaction. Accordingly, the Ligurian Tethys basin, as evidenced by the known A–A ophiolites, was an 'embryonic' oceanic basin most probably formed by ultra-slow continental extension and oceanic spreading (Piccardo 2008; Piccardo *et al.* 2009) that never reached a complete 'mature' oceanic stage. It was, most proberly, similar to the 120 Ma reconstruction of the Newfoundland–Iberia system in the North Atlantic (Whitmarsh *et al.* 2001).

Conclusion

The Monte Maggiore peridotite body is derived from lherzolite protoliths that were accreted to the subcontinental mantle lithosphere and were refertilized by pyroxenite veining under garnet-facies conditions. These pyroxenite-bearing lherzolites were subsequently equilibrated under spinel-facies conditions at a continental geothermal gradient. These subcontinental lithospheric lherzolites underwent exhumation towards plagioclase-facies conditions during lithosphere extension and thinning, which accommodated rifting and drifting in the Ligurian Tethys domain (Fig. 13). Lithosphere thinning and stretching induced adiabatic upwelling and decompression melting of the asthenosphere, and upwards migration of the asthenospheric melts.

The lithospheric mantle was infiltrated via diffuse porous flow by single-melt increments showing depleted MORB characteristics, which were formed by about 6% degree of fractional melting of a DM spinel-facies asthenospheric mantle and survived unmixed and migrated isolated through the overlying mantle lithosphere. Diffuse melt percolation through the extending mantle lithosphere initiated at spinel-facies conditions, forming reactive spinel peridotites, and ceased at plagioclase-facies conditions forming impregnated plagioclase peridotites. Further focused upwards migration of MORB-type fractional melts was only allowed within high-porosity dunite channels. The strongly transformed, melt-reacted lithospheric mantle peridotites were further exhumed to shallow levels where conductive heat loss became dominant on convective heat transport, and the rheology of the upwelling lithosphere progressively changed towards more rigide and cold conditions. Partial melting and melt dynamics in the asthenosphere gradually evolved and the single-melt fractions underwent coalescence to form aggregate N-MORB melts.

Diffuse and focused percolation in the more plastic levels of the lower lithosphere was dominated by single fractional MORB-type melts, whereas shallow intrusion in more rigid and fragile shallower lithosphere was characterized by aggregate MORBs that underwent magmatic evolution by fractional crystallization, and variably evolved Mg-rich and Fe–Ti-rich magmas were formed.

Available isotope data indicate isotopic homogeneity between melt-percolated peridotites and gabbroic rocks, suggesting that melts percolating and intruding the mantle lithosphere derived from isotopically homogeneous asthenospheric mantle sources. Geochronological information suggests that the magmatic cycle, consisting of decompression melting of the asthenosphere, early melt percolation and late melt intrusion in the mantle lithosphere, occurred during Late Jurassic times and thus represents the youngest percolation–intrusion events so far documented in ophiolitic peridotites from the Ligurian Tethys.

The tectonic and magmatic evolution recorded by the Monte Maggiore peridotites is well representative of the composite rifting history of the subcontinental lithospheric mantle of the Europe–Adria system from passive lithosphere extension to asthenosphere decompression melting and 'oceanization' of the subcontinental lithospheric mantle by interaction and isotopic equilibration with Late Jurassic MORB-type melts.

The Italian MIUR (Ministero dell'Istruzione, dell'Universita' e della Ricerca) (PRIN-COFIN2005: 'Lithosphere evolution induced by migration of mantle-derived melts at different geodynamic settings') and the University of Genova are acknowledged for their financial support. An early version of this paper was revised by Y. Dilek and an anonymous referee; their suggestions greatly improved the final version. Discussions and suggestions by R. Vannucci are greatly acknowledged. The participants of the International Corsica Field Trip, Working Group on Mediterrranean Ophiolites (G.L.O.M.), 16–21 September 2007, are acknowledged for their fruitful discussions at Monte Maggiore.

Thanks are due also to M. Piccardo for his technical assistance.

References

ABBATE, E., BORTOLOTTI, V., PASSERINI, P., PRINCIPI, G. & TREVES, B. 1994. Oceanisation processes and sedimentary evolution of the Northern Apennine ophiolite suite: a discussion. *Memorie della Società Geologica Italiana*, **48**, 117–136.

ANON. 1972. Penrose field conference on ophiolites. *Geotimes*, **17**, 24–25.

BECCALUVA, L. & PICCARDO, G. B. 1978. Petrology of the Northern Apennine ophiolites and their significance in the Western Mediterranean area. *In*: CLOSS, H., ROEDER, D. & SCHMIDT, K. (eds) *Alps, Apennines, Hellenides*. Inter-Union Commission on Geodynamics Scientific Report, **38**, E. Schweizerbart'sche, Stuttgart, 243–254.

BECCALUVA, L., MACCIOTTA, G., PICCARDO, G. B. & ZEDA, O. 1984. Petrology of lherzolitic rocks from the Northern Apennine ophiolites. *Lithos*, **17**, 299–316.

BEZZI, A. & PICCARDO, G. B. 1971. Structural features of the Ligurian ophiolites: petrologic evidence for the 'oceanic' floor of northern Apennines geosyncline. *Memorie della Società Geologica Italiana*, **10**, 53–63.

BODINIER, J. L., MENZIES, M. A. & THIRWALL, M. F. 1991. Continental to oceanic mantle transition – REE and Sr–Nd isotopic geochemistry of the Lanzo Lherzolite Massif. *In*: MENZIES, M. A., DUPUY, C. & NICOLAS, A. (eds) *Orogenic Lherzolites and Mantle Processes. Journal of Petrology*, Special Lherzolite Issue, 191–210.

BOUDIER, F. & NICOLAS, A. 1972. Fusion partielle gabbroique dans la lherzolite de Lanzo (Alpes Peimontaises). *Schweizerische Mineralogische und Petrographische Mittelungen*, **52**, 39–56.

DECANDIA, F. A. & ELTER, P. 1969. Riflessioni sul problema delle ofioliti nell'Appennino Settentrionale (nota preliminare). *Atti della Società Toscana di Scienze Naturali*, **75**, 1–9.

DECANDIA, F. A. & ELTER, P. 1972. La 'zona' ofiolitifera del Bracco nel settore compreso fra Levanto e la Val Graveglia (Appennino ligure). *Memorie della Società Geologica Italiana*, **11**, 503–530.

DEWEY, J. F., PITTMAN, W. C., RYAN, W. B. F. & BONNIN, J. 1973. Plate tectonics and the evolution of the alpine system. *Geological Society of America Bulletin*, **84**, 3137–3180.

DICK, H. J. B. & BULLEN, T. 1984. Chromian spinel as a petrogenetic indicator in abyssal and alpine-type peridotites and spatially associated lavas. *Contributions to Mineralogy and Petrology*, **86**, 54–76.

DRURY, M. R., HOOGERDUIJN STRATING, E. H. & VISSERS, R. L. M. 1990. Shear zone structures and microstructures in mantle peridotites from the Voltri Massif, Ligurian Alps, NW Italy. *Geologie en Mijnbouw*, **69**, 3–17.

HART, S. R. & DUNN, T. 1993. Experimental cpx/melt partitioning of 24 trace elements. *Contributions to Mineralogy and Petrology*, **113**, 1–8.

HELLEBRAND, E., SNOW, J. E., DICK, H. J. B. & HOFMANN, A. W. 2001. Coupled major and trace elements as indicators of the extent of melting in mid-ocean-ridge peridotites. *Nature*, **410**, 677–681.

HOFMANN, A. W. 1988. Chemical differentiation of the earth: the relationship beween mantle, continental crust and oceanic crust. *Earth and Planetary Science Letters*, **90**, 297–314.

HOOGERDUIJN STRATING, E. H., RAMPONE, E., PICCARDO, G. B., DRURY, M. R. & VISSERS, R. L. M. 1993. Subsolidus emplacement of mantle peridotites during incipient oceanic rifting and opening of the Mesozoic Tethys (Voltri Massif, NW Italy). *Journal of Petrology*, **34**, 901–927.

IONOV, D. A., BODINIER, J. L., MUKASA, S. B. & ZANETTI, A. 2002. Mechanisms and sources of mantle metasomatism: major and trace element conditions of peridotite xenoliths from Spitzbergen in the context of numerical modelling. *Journal of Petrology*, **43**(12), 2219–2259.

JACKSON, M. D. & OHNENSTETTER, M. 1981. Peridotite and gabbroic structures in the Monte Maggiore massif, Alpine Corsica. *Journal of Geology*, **89**, 703–719.

JOHNSON, K. T. M., DICK, H. J. B. & SHIMIZU, N. 1990. Melting in the oceanic upper mantle: an ion microprobe study of diopsides in abyssal peridotites. *Journal of Geophysical Research*, **95**, 2661–2678.

KELEMEN, P. B., WHITEHEAD, J. A., AHARONOV, E. & JORDAHL, K. A. 1995. Experiments on flow focusing in soluble porous media, with applications to melt extraction from the mantle. *Journal of Geophysical Research*, **100**, 475–496.

LAGABRIELLE, Y., POLINO, R. & AUZENDE, J. M. 1984. Les témoins d'une tectonique intra-océanique dans le domaine téthysien: analyse des rapports entre les ophiolites et leur couverture métasédimentaire dans la zone piémontaise des Alpes franco-italiennes. *Ofioliti*, **9**, 67–88.

LEMOINE, M., TRICART, P. & BOILLOT, G. 1987. Ultramafic and gabbroic ocean floor of the Ligurian Tethys (Alps, Corsica, Apennines): in search of a genetic model. *Geology*, **15**, 622–625.

LEONI, L. & SAITTA, M. 1976. X-ray fluorescence analysis of 29 trace elements in rock and mineral standards. *Rendiconti della Società Italiana di Mineralogia e Petrologia*, **32**, 497–510.

MARRONI, M., MOLLI, G., MONTANINI, A. & TRIBUZIO, R. 1998. The association of continental crust rock with ophiolites in the Northern Apennine (Italy): implication for the continent–ocean transition in the Western Tethys. *Tectonophysics*, **292**, 43–66.

MARRONI, M., MOLLI, G., MONTANINI, A., OTTRIA, G., PANDOLFI, L. & TRIBUZIO, R. 2002. The External Ligurian units (Northern Apennine, Italy): from rifting to convergence of a fossil ocean–continent transition zone. *Ofioliti*, **27**, 119–131.

MONTANINI, A., TRIBUZIO, R. & ANCZKIEWICZ, R. 2006. Exhumation history of a garnet pyroxenite-bearing mantle section from a continent–ocean transition (Northern Apennine ophiolites, Italy). *Journal of Petrology*, **47**(10), 1943–1971.

MÜNTENER, O. & HERMANN, J. 2001. The role of lower crust and continental upper mantle during formation of nonvolcanic passive margins: evidence from the Alps. *In*: WILSON, R. C. L., WITHMARSH, R. B., TAYLOR, B. & FROITZHEIM, N. (eds) *Non-volcanic Rifting of Continental Margins: A Comparison of Evidence From Land and Sea*. Geological Society, London, Special Publications, **187**, 267–288.

MÜNTENER, O. & PICCARDO, G. B. 2003. Melt migration in ophiolites: the message from Alpine–Apennine peridotites and implications for embryonic ocean basins. *In*: DILEK, Y. & ROBINSON, P. T. (eds) *Ophiolites in Earth History*. Geological Society, London, Special Publications, **218**, 69–89.

NIU, Y. 1997. Mantle melting and melt extraction processes beneath ocean ridges: evidence from abyssal peridotites. *Journal of Petrology*, **38**, 1047–1074.

PICCARDO, G. B. 1976. Petrologia del massiccio lherzolitico di Suvero (La Spezia). *Ofioliti*, **1**, 279–317.

PICCARDO, G. B. 1977. Le ofioliti dell'areale ligure: petrologia ed ambiente geodinamico di formazione. *Rendiconti della Società Italiana di Mineralogia e Petrologia*, **33**, 221–252.

PICCARDO, G. B. 2003. Mantle processes during ocean formation: petrologic records in peridotites from the Alpine–Apennine ophiolites. *Episodes*, **26**(3), 193–200.

PICCARDO, G. B. 2007. Evolution of the ultra-slow spreading Jurassic Ligurian Tethys: view from the mantle. *Periodico di mineralogia*, **76**, 67–80.

PICCARDO, G. B. 2008. The Jurassic Ligurian Tethys, a fossil ultraslow spreading ocean: the mantle perspective. *In*: COLTORTI, M. & GROGOIRE, M. (eds) *Metasomatism in Oceanic and Continental Lithospheric Mantle*. Geological Society, London, Special Publications, **293**, 11–34.

PICCARDO, G. B. & VISSERS, R. L. M. 2007. The preoceanic evolution of the Erro-Tobbio peridotite (Voltri Massif–Ligurian Alps, Italy). *Journal of Geodynamics*, **43**, 417–449.

PICCARDO, G. B., MÜNTENER, O., ZANETTI, A. & PETTKE, T. 2004. Ophiolitic peridotites of the Alpine–Appenine system: mantle processes and geodynamic relevance. *International Geology Review*, **46**, 1119–1159.

PICCARDO, G. B., RAMPONE, E. & VANNUCCI, R. 1990. Upper mantle evolution during continental rifting and ocean formation: evidence from peridotite bodies of the Western Alpine–Northern Apennine system. *Mémoirs de la Société Géologique de la France*, **156**, 323–333.

PICCARDO, G. B., VANNUCCI, R. & GUARNIERI, L. 2009. Evolution of the lithospheric mantle in an extensional settino: Insights from ophiolitic peridotites. *Lithosphere*, **1–2**, 1–7.

PICCARDO, G. B., ZANETTI, A. & MÜNTENER, O. 2007*a*. Melt/peridotite interaction in the Lanzo South peridotite: field, textural and geochemical evidence. *Lithos*, **94**(1–4), 181–209.

PICCARDO, G. B., ZANETTI, A., PRUZZO, A. & PADOVANO, M. 2007*b*. The North Lanzo peridotite body (NW-Italy): lithospheric mantle percolated by MORB and alkaline melts. *Periodico di Mineralogia*, **76**(2–3), 199–221.

RAMPONE, E. 2004. Mantle dynamics during Permo-Mesozoic extension of the Europe-Adria lithosphere: insights from the Ligurian Ophiolites. *Periodico di Mineralogia*, **73**, 215–230.

RAMPONE, E. & PICCARDO, G. B. 2000. The ophiolite-oceanic lithosphere analogue: new insights from the Northern Apennines (Italy). *In*: DILEK, Y., MOORES, E. M., ELTHON, D. & NICOLAS, A. (eds) *Ophiolites and Oceanic Crust: New Insights From Field Studies and the Ocean Drilling Program*. Geological Society of America, Special Paper, **349**, 21–34.

RAMPONE, E., HOFMANN, A. W., PICCARDO, G. B., VANNUCCI, R., BOTTAZZI, P. & OTTOLINI, L. 1995. Petrology, mineral and isotope geochemistry of the External Liguride peridotites (Northern Apennine, Italy). *Journal of Petrology*, **36**, 81–105.

RAMPONE, E., HOFMANN, A. W., PICCARDO, G. B., VANNUCCI, R., BOTTAZZI, P. & OTTOLINI, L. 1996. Trace element and isotope geochemistry of depleted peridotites from an N-Morb type ophiolite (Internal

Liguride, N-Italy). *Contributions to Mineralogy and Petrology*, **123**, 61–76.

Rampone, E., Hofmann, A. W. & Raczek, I. 1998. Isotopic contrasts within the Internal Liguride ophiolite (N. Italy): the lack of a genetic mantle–crust link. *Earth and Planetary Science Letters*, **163**, 175–189.

Rampone, E., Piccardo, G. B. & Hofmann, A. W. 2008. Multi-stage melt–rock interaction in the Mt. Maggiore (Corsica, France) ophiolitic peridotite: microstructural and geochemical evidence. *Contributions to Mineralogy and Petrology*, **156**, 453–475, doi: 10.1007/s00410-008-0296-y.

Rampone, E., Piccardo, G. B., Vannucci, R. & Bottazzi, P. 1997. Chemistry and origin of trapped melts in ophiolitic peridotites. *Geochimica et Cosmochimica Acta*, **61**, 4557–4569.

Rampone, E., Romairone, A., Abouchami, W., Piccardo, G. B. & Hofmann, A. W. 2005. Chronology, petrology and isotope geochemistry of Erro-Tobbio peridotites (Ligurian Alps, Italy): record of late Paleozoic lithospheric mantle. *Journal of Petrology*, **46**, 799–827.

Rampone, E., Romairone, A. & Hofmann, A. W. 2004. Contrasting bulk and mineral chemistry in depleted mantle peridotites: evidence for reactive porous flow. *Earth and Planetary Science Letters*, **218**, 491–506.

Romairone, A. 1996. *Petrologia e geochimica delle peridotiti di Monte Maggiore (Corsica) [Petrology and Geochemistry of the Monte Maggiore Peridotites]*. Graduate thesis, University of Genova.

Ross, K. & Elthon, D. 1993. Cumulates from strongly depleted mid-ocean-ridge basalt. *Nature*, **365**, 826–829.

Seitz, H. M., Altherr, R. & Ludwig, T. 1999. Partitioning of transition elements between orthopyroxene and clinopyroxene in peridotitic and websteritic xenoliths: New empirical geothermometers. *Geochimica et Cosmochimica Acta*, **63**, 3967–3982.

Steinmann, G. 1927. Die ophiolitischen Zonen in den mediterranen Kettengebirgen. *In*: *14th International Geological Congress*, Volume 2, 637–668.

Steinmann, G. 2003. Die ophiolitischen Zonen in den mediterranen Kettengebirgen (The ophiolitic zones in the mediterranean mountain chains), Bernoulli, G. and Friedman, G. M. translators. *In*: Dilek, Y. & Newcomb, S. (eds) *Ophiolite Concept and the Evolution of Geologic Thought*. Geological Society of America, Special Paper, **373**, 77–91.

Tiepolo, M., Bottazzi, P., Palenzona, M. & Vannucci, R. 2003. A laser probe coupled with ICP Double-Focusing Sector-Field Mass Spectrometer for *in situ* analysis of geological samples and U–Pb dating of zircon. *The Canadian Mineralogist*, **41**, 259–273.

Tribuzio, R., Thirlwall, M. F. & Vannucci, R. 2004. Origin of the gabbro–peridotite association from the Northern Apennine ophiolites (Italy). *Journal of Petrolgy*, **45**(6), 1109–1124.

Vaggelli, G., Olmi, F. & Conticelli, S. 1999. Quantitative electron microprobe analysis of reference silicate mineral and glass samples. *Acta Vulcanologica*, **11**(2), 297–303.

Vannucci, R., Bottazzi, P., Wulff-Pedersen, E. & Neumann, E.-R. 1998. Partitioning of REE, Y, Sr, Zr and Ti between clinopyroxene and silicate melts in the mantle under La Palma (Canary Islands): implications for the nature of the metasomatic agents. *Earth and Planetary Science Letters*, **158**, 39–51.

Vissers, R. L. M., Drury, M. R., Hoogerduijn Strating, E. H. & Van Der Wal, D. 1991. Shear zones in the upper mantle: a case study in an Alpine type lherzolite massif. *Geology*, **19**, 990–993.

Whitmarsh, R. B., Manatschal, G. & Minshull, T. A. 2001. Evolution of magma-poor continental margins from rifting to seafloor spreading. *Nature*, **413**, 150–154.

The Lanzo peridotite massif, Italian Western Alps: Jurassic rifting of the Ligurian Tethys

GIOVANNI B. PICCARDO

Dipartimento per lo Studio del Territorio e delle sue Risorse, Universita' di Genova, Corso Europa 26, I-16132 Genova, Italy (e-mail: piccardo@dipteris.unige.it)

Abstract: The Lanzo Massif in the Western Alps consists of three bodies (North, Central and South) of mantle peridotites that were exhumed from the subcontinental mantle lithosphere to the sea floor during lithosphere extension related to the formation of the Jurassic Ligurian Tethys oceanic basin. The North Lanzo protoliths were located at shallower lithospheric levels than the South Lanzo protoliths. During exhumation, early MORB-type fractional melts from the asthenosphere infiltrated and modified the South Lanzo protoliths. Later on, aggregate MORB melts passed through the South Lanzo peridotites, migrating within replacive peridotite channels, and impregnated the North Lanzo peridotites. Ongoing lithosphere extension and stretching caused break-up of the continental crust and sea-floor exposure of the Lanzo peridotites. The North Lanzo peridotites, deriving from shallower lithospheric levels, were exhumed and exposed at more external ocean–continent transition (OCT) zones of the basin, whereas the South Lanzo peridotites, deriving from deeper lithospheric levels, were exhumed and exposed at more internal oceanic (MIO) settings of the basin. Field, petrographical–structural and petrological–geochemical studies on the Lanzo mantle peridotites provide mantle constraints regarding the geodynamic evolution of the Europe–Adria extensional system during the rifting and opening of the Ligurian Tethys basin.

The Lanzo peridotite body belongs to the internal part of the high-pressure–low-temperature metamorphic belt of the Western Alps. The Lanzo peridotite body (*c.* 150 km^2) is located approximately 30 km NW of Torino (NW Italy) and is bounded by Po plain sediments to the east, high-pressure ophiolites and schistes lustrés to the west, and continental units of the Sesia–Lanzo Zone to the north (Nicolas 1974). The Lanzo Massif has been subdivided into the South (*c.* 55 km^2), Central (*c.* 90 km^2) and North (*c.* 5 km^2) bodies, separated by mylonitic shear zones (Boudier 1976, 1978; Kaczmarek & Müntener 2008) (Fig. 1). The massif is characterized by an unusually large proportion of fresh mantle peridotites that consist dominantly of plagioclase peridotites, surrounded and partially replaced by serpentinites (Boudier 1978). Early studies stressed the chemical and structural variation between the South–Central and the North bodies, evidencing that the North peridotites display, as a whole, lower LREE (light rare earth elements) depletion and Sr–Nd isotope ratios than the South peridotites, which show an isotope composition similar to those of Atlantic MORB (mid-ocean ridge basalts) (Bodinier 1988). According to Bodinier *et al.* (1991), peridotites from the Central body show geochemical features transitional between those of the peridotites from the South and the North bodies. It was suggested (Bodinier 1988; Bodinier *et al.* 1991) that the Lanzo peridotites represent a residual piece of upper mantle after MORB extraction: the North body was residual after melt extraction lower than 6% and was equilibrated with T-MORB (transitional MORB), whereas the Central and South bodies underwent variable partial melting (from 6% in the Central body to 12% in the South body) showing affinity with both T- and N-MORB (normal MORB).

Nicolas (1986) and Bodinier *et al.* (1991) considered the Lanzo Massif as an asthenospheric diapir that was emplaced at shallow levels in the early Mesozoic, during the opening of the Ligurian–Piemontese basin. On the basis of Sr–Nd isotope data, the North body has been considered by Bodinier *et al.* (1991) as a fragment of subcontinental mantle lithosphere consisting of older depleted MORB mantle (DMM), which became isolated by the convective mantle 400–700 Ma ago, and was accreted to the subcontinental lithosphere. The oceanic evolution of the Lanzo peridotite was suggested early on (Bodinier 1988) on the basis of presence of MORB-type gabbroic–basaltic dykes, which have been dated as Middle Jurassic (U–Pb zircon ages of 158–163 Ma: Kaczmarek *et al.* 2008). Lithostratigraphical studies indicate that the Lanzo peridotites are overlain by metabasites, Mn-rich metaquartzites and calc-schists (e.g. Lagabrielle *et al.* 1989), closely corresponding to the oceanic volcanic and sedimentary cover of other ophiolite sequences in the Western Alps and Northern Apennines. Accordingly, the Lanzo

From: Coltorti, M., Downes, H., Grégoire, M. & O'Reilly, S. Y. (eds) *Petrological Evolution of the European Lithospheric Mantle*. Geological Society, London, Special Publications, **337**, 47–69.
DOI: 10.1144/SP337.3 0305-8719/10/$15.00

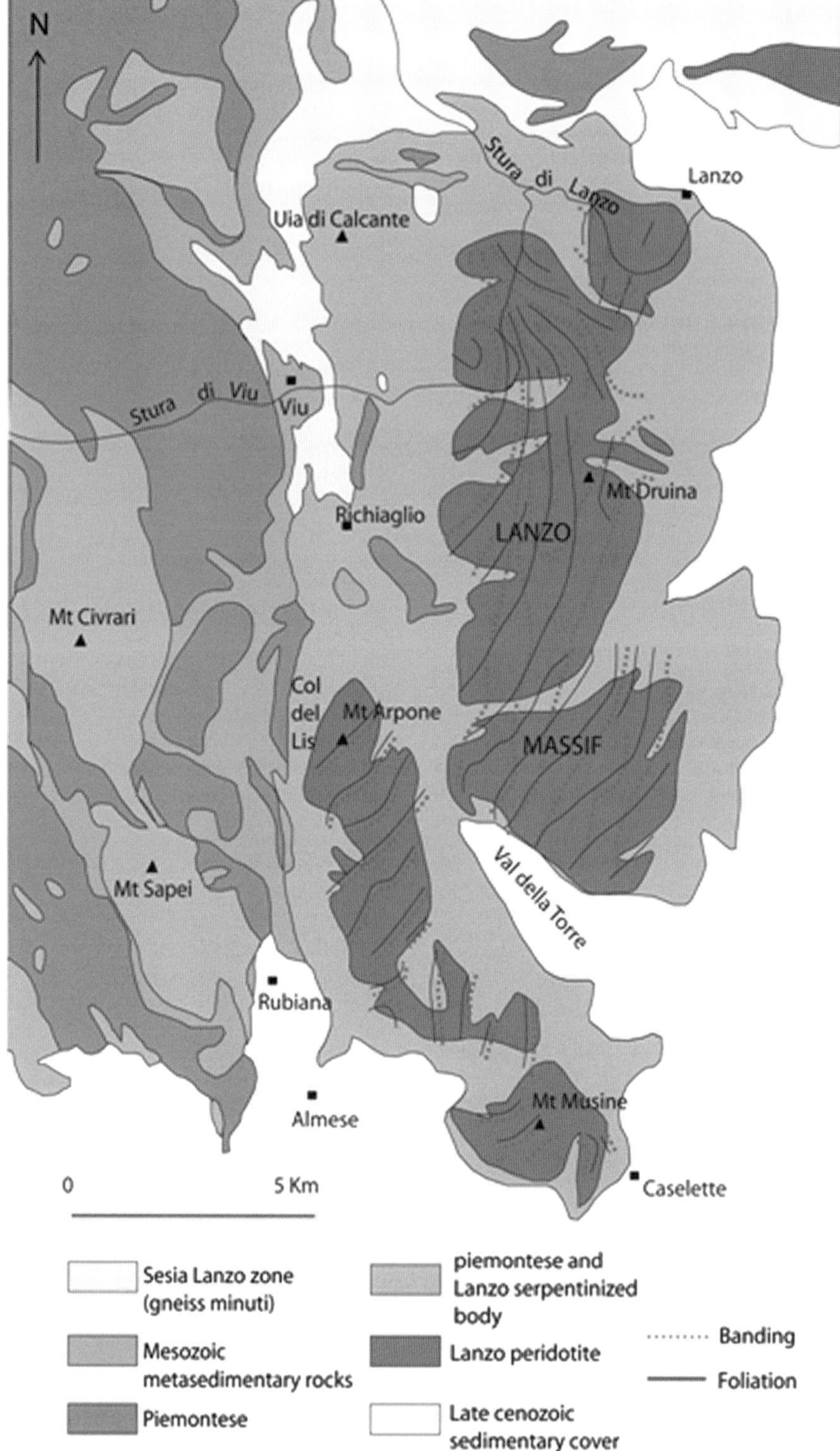

Fig. 1. Sketch map of the Lanzo peridotite Massif (Western Alps) (from Kaczmarek & Müntener 2008, modified).

peridotites were exhumed and exposed on the sea floor of the Jurassic Ligurian–Piemontese (or Ligurian Tethys) oceanic basin.

Recent investigations on the Lanzo peridotites (Piccardo *et al.* 2007*a*, *b* and references therein) document the structural and compositional complexity of the rock types, and the related mantle processes, recorded in each single body of the Lanzo Massif, which do not allow each peridotite mass to be identified with a single petrogenetic process. These studies describe composite tectonic–metamorphic and magmatic histories for the South and North peridotite bodies that are, as a whole, not consistent with the asthenospheric diapir model proposed by previous authors. The wealth of field, petrological and geochemical data indicate that the Lanzo peridotites: (1) were derived from the Europe–Adria subcontinental lithospheric mantle; (2) were progressively exhumed towards shallow levels during passive lithospheric extension driving the pre-oceanic rifting in the Jurassic Ligurian Tethys system, as recorded mostly by the North Lanzo peridotites; and (3) were strongly modified by melt–rock interaction operated by MORB-type melts during exhumation, as recorded mostly by the Central–South Lanzo peridotites.

This paper discusses recent field, structural and compositional data on the North and South Lanzo peridotite bodies and presents a new scenario for evolution of the Lanzo mantle during rifting of the Late Jurassic Ligurian Tethys basin.

The Lanzo Peridotite

North Lanzo body

Rock types. The North body of the Lanzo peridotite massif in the Western Alps is a small mass (5 km^2), separated by mylonitic shear zones from the main massif. It was first considered as part of an asthenospheric mantle diapir emplaced during the early stages of the opening of the Jurassic Ligurian Tethys (Nicolas 1974, 1984). Bodinier *et al.* (1991) interpreted it as DMM material separated from the convective mantle and accreted into the subcontinental lithosphere near the Proterozoic–Phanerozoic boundary.

Recent field and structural–petrological investigations (Piccardo *et al.* 2007*b*) have documented the presence of: (1) fertile lherzolites, showing abundant spinel pyroxenite banding, referred to as *lithospheric lherzolites*; (2) 10–100 m-wide bodies of strongly plagioclase-enriched peridotite, characterized by plagioclase-rich pyroxenite layers, referred to as *impregnated plagioclase peridotites*; (3) 1–10 m-wide bodies and channels of coarse granular pyroxene-depleted spinel harzburgites, referred to as *replacive spinel peridotites* (Figs 2 & 3).

On the basis of mutual field relationships, it has been shown that the lithospheric lherzolites represent the oldest rock types that were locally transformed to plagioclase-impregnated peridotites, whereas both lithospheric and impregnated peridotites have been subsequently replaced along channels by pyroxene-depleted harzburgites and dunites (Piccardo *et al.* 2007*b*).

Lithospheric lherzolites. They consist of clinopyroxene (Cpx) rich (12.0–14.5% vol.) porphyroclastic lherzolite, with spinel-facies mineral assemblage and abundant cm- to dm-wide spinel pyroxenite banding. Both lherzolites and pyroxenites show widespread rounded Opx + Sp (orthopyroxene + spinel) clusters, suggesting complete spinel-facies breakdown and recrystallization of precursor garnet. Taking into account present knowledge of the stability fields of peridotites and pyroxenites (e.g. Simon & Podladchikov 2008 and references therein), these microstructures suggests that the lherzolite–pyroxenite association underwent decompressional exhumation from P (pressure) > 2.5 GPa (i.e. garnet-facies assemblages in peridotites) to $P < 1.5$ GPa (garnet breakdown to spinel-facies assemblages in pyroxenites). Spinel-facies minerals and microtextures are overprinted by plagioclase-bearing microtextures. In fact, the coarse-granular spinel-facies clinopyroxene porphyroclasts show abundant exsolution lamellae of orthopyroxene and plagioclase, whereas spinel grains are rimmed by plagioclase + olivine coronas against spinel-facies pyroxenes. These latter microstructures suggest further decompressional exhumation to $P < 1.0$ GPa (i.e. plagioclase peridotite-facies conditions). Thus, following Bodinier *et al.* (1991), these rocks will be referred in the following to as spinel–plagioclase lherzolites, to indicate that their plagioclase was mainly formed by subsolidus recrystallization.

Impregnated plagioclase peridotites. They form hectometre (hm)-scale masses that are strongly enriched in plagioclase (up to 15% vol.): they commonly preserve pyroxenite bands that are strongly enriched in plagioclase and transformed to 'gabbroic' rocks. Both peridotites and pyroxenites are characterized by the presence of relict Opx + Sp clusters, and large clinopyroxene porphyroclasts showing abundant exsolution lamellae of plagioclase and orthopyroxene. Occurrence of widespread pyroxenite remnants demonstrate that plagioclase enrichment, and the related process, occurred at the expense of pristine lithospheric lherzolites with pyroxenite bands, which escaped important effects of melt–peridotite reactive percolation (i.e.

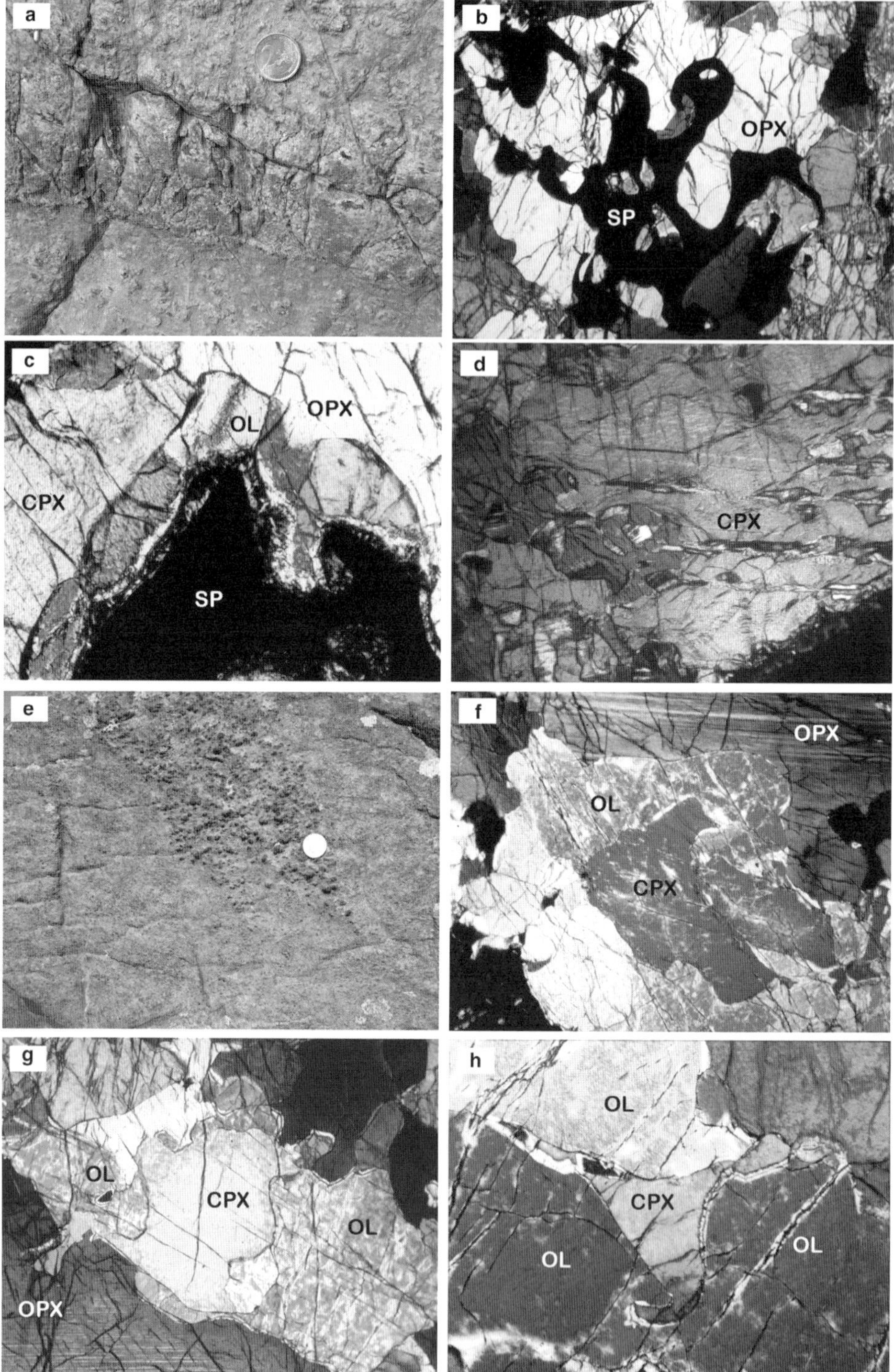

Fig. 2. Field and petrographical features of the Lanzo peridotites. (**a**)–(**d**) *Lithospheric lherzolites from North Lanzo.* (a) Spinel pyroxenite band in spinel lherzolite: note the presence of Sp + Opx clusters in the pyroxenite, indicating complete spinel-facies recrystallization of precursor garnet. (b) Rounded Sp + Opx cluster in lherzolite, indicating

pyroxene dissolution) by undersaturated melts. The exsolved clinopyroxene porphyroclasts are typically corroded and partially replaced by plagioclase + olivine (Plg + Ol) symplectites and granular aggregates, and olivine-bearing, plagioclase-rich microgabbroic patches interstitial between the large mantle porphyroclasts are widespread. These microstructures suggest that pristine lithospheric lherzolites underwent diffuse percolation, interaction and incipient interstitial crystallization (i.e. impregnation and refertilization) by olivine-oversaturated melts at plagioclase-facies conditions (Piccardo *et al.* 2007*b*).

Replacive spinel peridotites. They form metre- to decametre-wide masses and elongated bands of spinel harzburgites that are strongly depleted in pyroxenes. They are characterized by a coarse granular isotropic texture mostly formed by rounded orthopyroxene aggregates surrounded by abundant coarse granular olivine that replaces the pristine textures of the spinel and plagioclase peridotites that they cut. Their modal composition (i.e. Cpx 6.7% vol., Opx 7.1% vol., Ol 82.5% vol., Spl 3.7% vol.) points to a strong olivine enrichment coupled with a significant depletion in both pyroxenes with respect to both lithospheric and impregnated peridotites. In thin section the replacement structures of new, unstrained olivine on exsolved orthopyroxene porphyroclasts are widespread, suggesting that these refractory channels were formed by focused and reactive percolation of pyroxene(-silica)-undersaturated melts that dissolved pyroxenes and precipitated olivine. Clinopyroxene is mostly present in the form of small unstrained interstitial grains that show no reaction replacement by olivine. These microstructures indicate the late interstitial crystallization of magmatic clinopyroxene from the percolating melt.

Some compositional features. Lithospheric lherzolites have rather fertile compositions, characterized by relatively high SiO_2 (45.3 wt%), Al_2O_3 (2.8 wt%) and CaO (2.5 wt%), and relatively low MgO (40.7 wt%). Impregnated plagioclase peridotites have comparable SiO_2 (45.1–45.4 wt%), basically higher Al_2O_3 (2.5–4.1 wt%) and CaO (2.5–3.6 wt%), and lower MgO (37.8–40.2 wt%), content that is consistent with significant plagioclase addition. Replacive spinel harzburgites have relatively lower SiO_2 (43.0–43.3 wt%), Al_2O_3 (1.0–1.3 wt%) and CaO (0.65–0.84 wt%), and higher MgO (45.4–46.0 wt%), content that is consistent with their significantly low pyroxene content. In the MgO v. SiO_2 diagram (Fig. 4) the representative lithospheric lherzolite composition falls along the melting trends calculated by Niu (1997) for refractory peridotite residua after any kind of partial melting, suggesting a rather primitive composition that accords with its relatively high SiO_2 and low MgO content. Impregnated peridotites follow an 'enrichment trend', characterized by decreasing MgO and slightly increasing SiO_2 content, consistent with the petrographical evidence of the addition of silica-saturated minerals (i.e. plagioclase and pyroxenes). Replacive spinel harzburgites fall outside the melting trends, at lower SiO_2 content at the corresponding MgO content, suggesting that their strongly refractory compositions cannot be derived by a melting process. Therefore, according with the petrographical evidence, a more correct process should have been a melt–peridotite reaction process that depleted pyroxenes and enriched olivine.

Lithospheric peridotites show bulk-rock trace element compositions (and C1-normalized REE patterns) rather fertile, sometimes exceeding primordial mantle (PM) compositions. Reactive spinel harzburgites show sinusoidal C1-normalized REE content, being characterized by significantly low HREE content and relatively enriched M- and LREE content, which cannot be formed by any kind of partial melting but instead suggest transient geochemical features caused by chromatographic effects (Fig. 5a).

Clinopyroxenes of the lithospheric lherzolites show C1-normalized REE patterns slightly

Fig. 2. (*Continued*) complete spinel-facies recrystallization of the precursor garnet. Petrographical evidence indicates that both peridotites and pyroxenites were exhumed from garnet-facies conditions ($P > 2.5$ GPa, for peridotites) and recrystallized at spinel-facies conditions. (c) Rim of Ol + Plg (altered) between spinel-facies pyroxenes and spinel in spinel lherzolite, indicating incipient recrystallization to plagioclase-facies conditions during exhumation. (d) Plg + Opx exsolution lamellae in spinel-facies clinopyroxene porphyroclast, indicating transition to shallower (plagioclase-facies) P–T (pressure–temperature) conditions during exhumation. (**e**)–(**h**) *Reactive spinel peridotites from South Lanzo.* (e) Field aspect of reactive peridotites: note the abrupt change in grain and modal composition, from coarse granular Cpx-bearing harzburgite (in the centre) to fine-grained Opx-bearing dunite. (f) Reaction microstructure: unstrained Ol border replaces exsolved and deformed Cpx and Opx porphyroclasts, indicating peridotite interaction with a silica-undersaturated melt. (g) Reaction microstructure: unstrained Ol crystals replace exsolved and deformed Cpx porphyroclasts, and small Ol crystals form between Cpx and Opx porphyroclasts, indicating peridotite interaction with a silica-undersaturated melt. (h) New unstrained and interstitial magmatic Cpx grain formed at the triple junction between mantle Ol porphyroclasts, indicating that the percolating melt reached silica-saturation and started crystallizing pyroxenes along the percolation pathways.

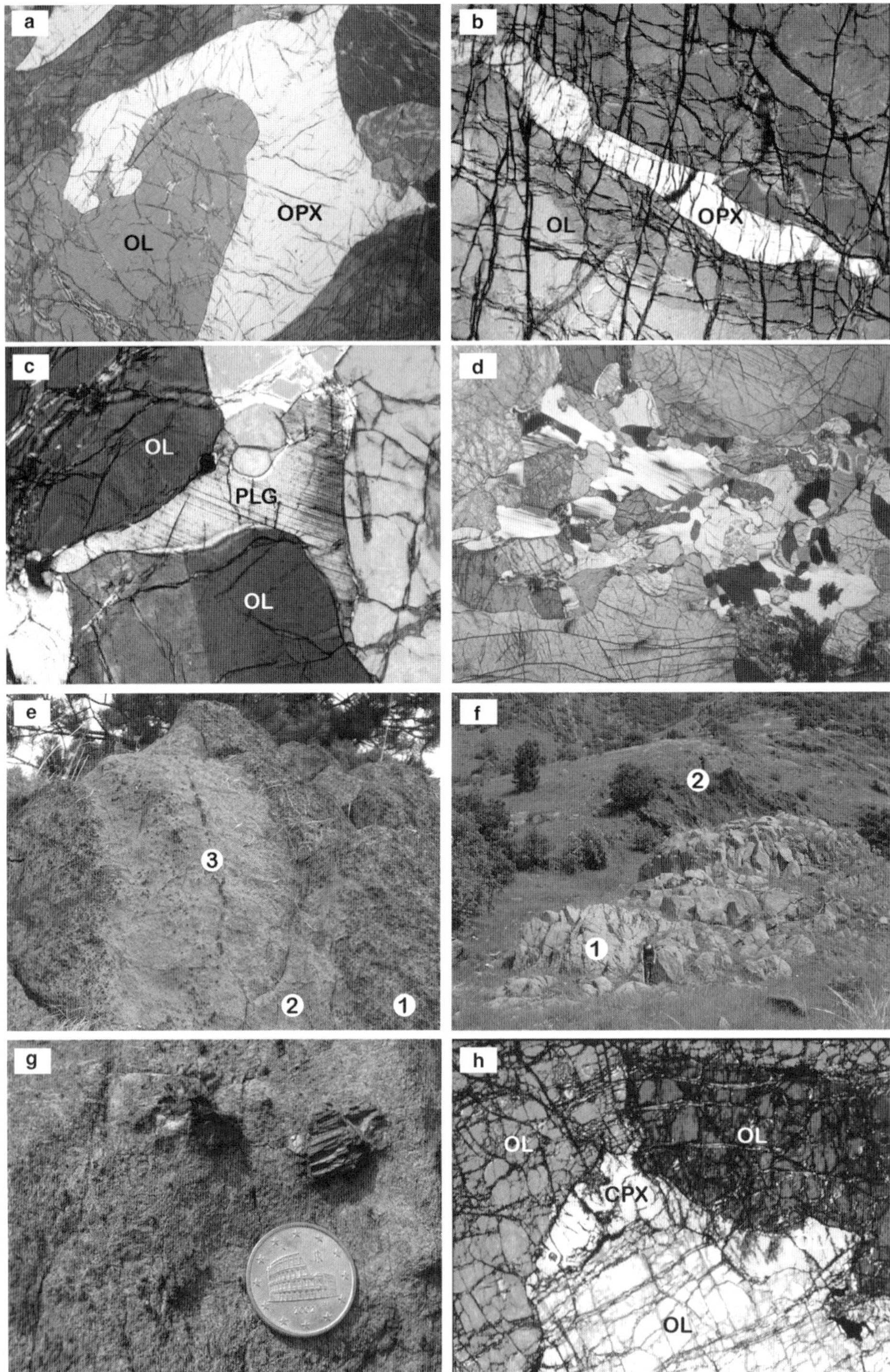

Fig. 3. Field and petrographical features of the Lanzo peridotites. (**a**)–(**d**) *Impregated plagioclase peridotites from South-Central Lanzo.* (a) Unstrained Opx replaces kinked Ol porphyroclast. This indicates the peridotite interaction with a silica-saturated melt. (b) Opx vein cuts across kinked Ol porphyroclasts. This indicates the peridotite interaction

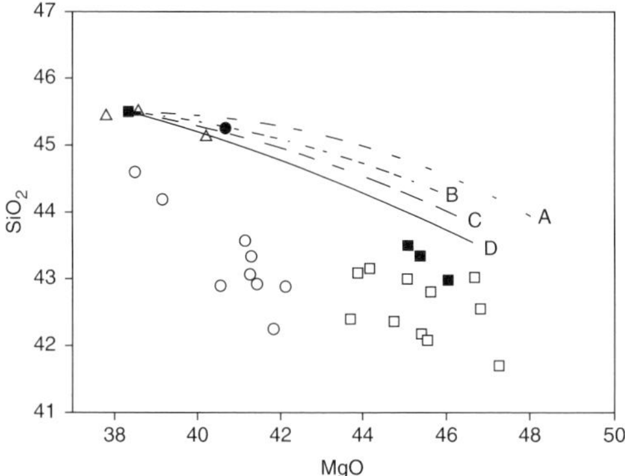

Fig. 4. SiO_2 v. MgO diagram of the bulk compositions of available peridotites from Lanzo. South and Central body: reactive spinel peridotites (open squares) and impregnated plagioclase peridotites (open circles) (note that melt impregnation causes significant MgO decrease and SiO_2 increase starting from previous reactive peridotites). North body: lithospheric lherzolite (full circle), impregnated plagioclase peridotites (open triangles), replacive spinel harzburgites (full squares) (note that melt impregnation cause slight SiO_2 enrichment and significant MgO decrease starting from the fertile lherzolite composition).

Also reported is the composition of the fertile starting composition (full square at left of compositional trends):

(A) the compositional trend of refractory residua after polybaric melting at 25–8 kbar;
(B) the compositional trend of refractory residua after polybaric melting at 15–8 kbar;
(C) the compositional trend of refractory residua after batch melting at 20 kbar;
(D) the compositional trend of refractory residua after batch melting at 10 kbar. Compositional trends from Niu (1997).

Note that the spinel peridotites (both the reactive peridotites of South–Central Lanzo and the replacive harzburgites of North Lanzo) do not fit with refractory residua compositions (having significantly lower SiO_2 content than the compositions of refractory residua after partial melting, at the corresponding MgO content). The lithospheric lherzolites fit with low-degree refractory residua, suggesting that they escaped melt–rock reaction processes.

LREE-depleted (Sm_N/Ce_N 1.12) and almost flat in the MREE–HREE region (at 7–9 × C1). Clinopyroxenes of the impregnated plagioclase peridotites show C1-normalized REE patterns that are moderately LREE-depleted (Sm_N/Ce_N 2.35–3.25), being flat to slightly convex-upwards in the MREE–HREE region (10–30 × C1) (Fig. 6a).

Clinopyroxenes of the replacive spinel harzburgites show Cl-normalized patterns characterized by enrichments in the LREE–MREE region (maximum at Pr 13.6 × Cl; Sm_N/Yb_N 2.40), and significantly fractionated HREE patterns (i.e. significantly low absolute HREE content: Lu 4.5 × C1) (Fig. 6c). The Nb and Ta content is significantly higher than in the other clinopyroxenes from Lanzo North ultramafics, being up to 2.8 and 7 × C1, respectively. Sr results in marked negative spikes. Clinopyroxenes with similar C1-normalized REE patterns have been found by Bodinier *et al.* (1991) in refractory peridotites (sample L212) enclosed in impregnated plagioclase peridotites of the Central Lanzo body.

Migrating melts. The similarity between the REE and HFSE (high field strength elements) fractionations of clinopyroxenes from the impregnated peridotites and those shown by clinopyroxenes in equilibrium with aggregate N-MORB (see Piccardo *et al.* 2007*a* for a more detailed discussion) allow for the possibility that they are segregated from aggregate N-MORB. Accordingly, it can be inferred that in early stage of melt diffuse percolation through the plagioclase-facies North Lanzo lherzolites the process involved aggregate N-MORB magmas (Fig. 6b).

Interstitial magmatic clinopyroxenes in replacive harzburgites shows strongly LREE-enriched and HREE-depleted REE patterns. The REE composition estimated for the liquid in equilibrium with these clinopyroxene (see Piccardo *et al.* 2007*b* for a more detailed discussion) shows a strong LREE–HREE fractionation ($La_N/Yb_N > 11$), very high $LREE_N$ ($La_N > 140$) and notably low $HREE_N$ values, pointing to an overall

Fig. 3. (*Continued*) with a silica-saturated melt. (c) An unstrained magmatic Plg crystal cuts across kinked Ol porphyroclasts. This indicates plagioclase crystallization (impregnation) from a percolating melt. (d) Millimetre-size patch of Plg-rich gabbroic material interstitial between mantle porphyroclasts. This indicated the interstitial crystallization (impregnation) of the percolating melt. (**e**)–(**h**) *Replacive spinel harzburgites and dunites from South Lanzo.* (e) Replacive dunite channel (2) cutting through replacive coarse granular spinel harzburgite (1). Note the vein of magmatic Cpx (3) running in the centre of the channel, which indicates the inception of crystallization of the focused migrating melt. The Cpx trace element composition indicates a clear MORB affinity. (f) Decametric–hectometric replacive spinel dunite body (1) enclosed in impregnated plagioclase peridotites (2). (g) Centimetre-size euhedral magmatic Cpx megacrystal in replacive spinel dunite, indicating the crystallization of the channelled migrating melt. The Cpx trace element composition indicates a clear MORB affinity. (h) Magmatic Cpx formed at the triple junctions between Ol porphyroclasts in replacive spinel dunite. The Cpx trace element composition indicates a clear MORB affinity.

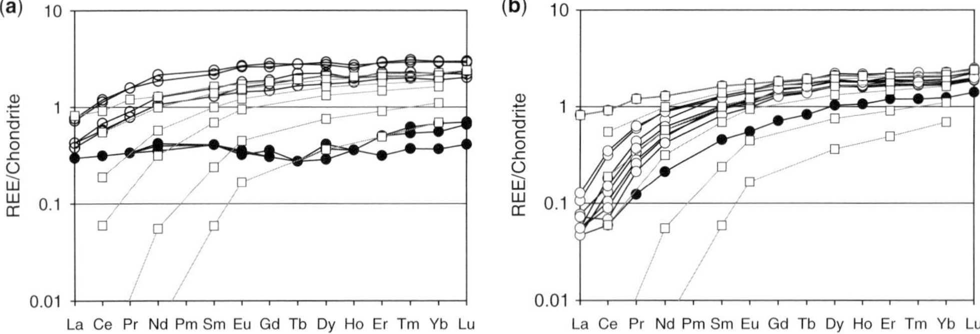

Fig. 5. (**a**) Bulk-rock C1-normalized REE patterns of North Lanzo peridotites: (1) replacive spinel harzburgites (full circles); and (2) impregnated plagioclase peridotites (open circles). (**b**) Bulk-rock C1-normalized REE patterns of South Lanzo peridotites: (1) reactive spinel peridotites (full circles); and (2) impregnated plagioclase peridotites (open circles). The chondrite-normalized REE abundance calculated for refractory peridotites after variable degrees ($F = 1-15\%$) of spinel-facies fractional melting of DMM are reported for comparison in (a) and (b). (Fractional melting equation is from Shaw (1970), modal composition from Hirschmann & Stolper (1996), geochemical composition of DMM from Workman & Hart 2005 and solid–liquid partition coefficients from Johnson *et al.* 1990.)

alkaline affinity (Fig. 6d). Sample L212 clinopyroxene, studied by Bodinier *et al.* (1991), has significantly high $^{87}Sr/^{86}Sr$ ratios (>0.70453) and low $^{143}Nd/^{144}Nd$ ratios (<0.512796) (E_{Sr} +3.4, E_{Nd} +0.9) that plot close to CHUR (chondritic uniform reservoir) values: it is characterized by much higher Sr and lower Nd isotopic ratios than that of the host plagioclase peridotites. According to Bodinier *et al.* (1991), it shows a 'plume-type' signature. Accordingly, the available trace element and isotopic data indicate that melts with alkaline affinity and OIB (oceanic island basalt) isotopic signature percolated within the replacive spinel harburgite channels.

The chemical data indicate that, at North Lanzo, a strong variation in the geochemical affinity of the migrating melts existed between the early pervasive percolation and impregnation, and the late channeled migration stages, pointing to significant changes in terms of composition and physicochemical characteristics of the mantle sources. In particular, if the partial melting of a spinel-facies DM source can reconcile the MORB geochemical features deduced for the parental melts of the plagioclase-bearing impregnation, a more fertile garnet-bearing mantle source must be considered for the alkaline-type melts migrating through the replacive conduits.

South–Central Lanzo bodies

Rock types. The South and Central bodies of the Lanzo peridotite massif consist of large masses (55 and 90 km^2, respectively) separated by mylonitic and partially serpentinized shear zones (Boudier & Nicolas 1972; Boudier 1978). They mostly consist of plagioclase peridotites and subordinated refractory harzburgites and dunites. They were first considered as part of the mantle asthenosphere that rose from the garnet stability field as a high-temperature diapir, accompanied by a large degree of melt extraction (Nicolas 1984, 1986; Bodinier *et al.* 1991). However, Pognante *et al.* (1985) suggested that the Lanzo peridotite represented subcontinental lithosphere that underwent decompression and partial melting during continental rifting, and was later intruded by asthenospheric melts with N-MORB affinity.

The large compositional variability – from plagioclase-enriched, refertilized peridotites to depleted refractory peridotites – and the significant REE variations in the Lanzo peridotites were related by Bodinier *et al.* (1991) to different processes, such as: (1) primary mantle heterogeneity; (2) polybaric melting; (3) melt extraction processes; and (4) secondary plagioclase crystallization. Recently, Bodinier & Godard (2003) proposed that the high-temperature plagioclase peridotites of Lanzo represent strongly 'asthenospherized' mantle containing only sparse relicts of lithospheric origin.

New field and structural–petrographical investigations (Piccardo *et al.* 2007*a* and references therein) show that the South Lanzo massif consists of: (1) 10–20 m-wide remnants of spinel pyroxenite-bearing fertile lherzolites, referred to as *lithospheric lherzolites*; (2) 100 m-wide bodies of pyroxene-depleted spinel harzburgites and dunites, referred to as *reactive spinel peridotites*; (3) large, km-scale masses of plagioclase-enriched peridotites, referred to as *impregnated plagioclase peridotites*; and (4) 10–20 m-wide elongated bodies and channels of pyroxene-free spinel dunites, and subordinated

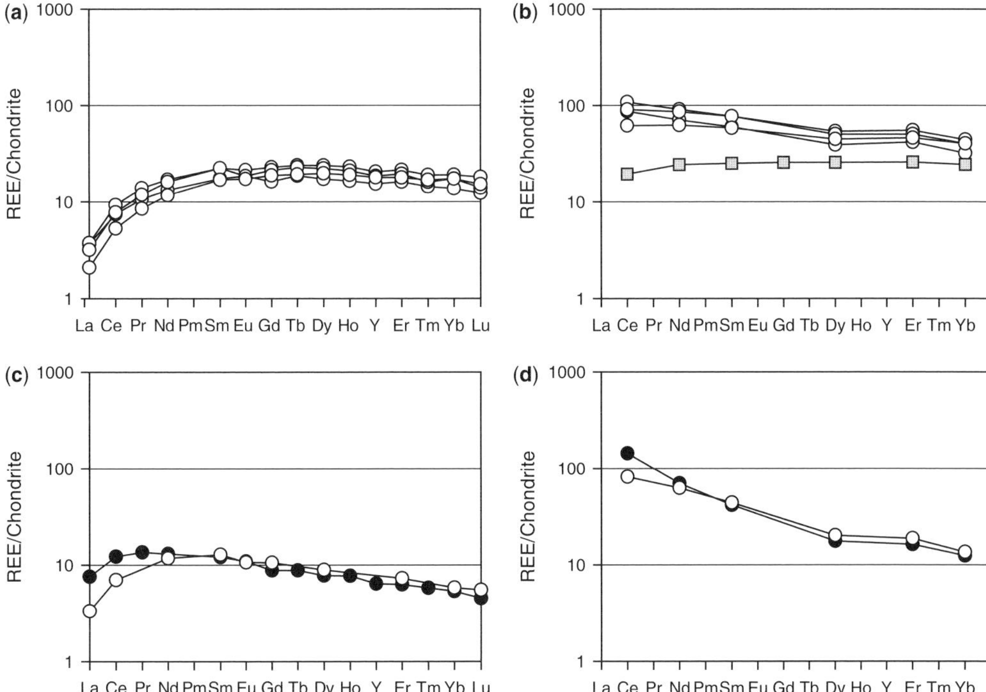

Fig. 6. (**a**) C1-normalized REE patterns of clinopyroxenes from representative impregnated plagioclase peridotites of North Lanzo (open circles). (**b**) C1-normalized REE patterns of liquids calculated in equilibrium with magmatic clinopyroxenes of representative North Lanzo impregnated plagioclase peridotites (open circles). The C1-normalized REE pattern of N-MORB (grey squares) (data from Hofmann 1988) is also reported for comparison. (**c**) C1-normalized REE patterns of representative clinopyroxenes from Lanzo peridotites: (1) magmatic interstitial Cpx in replacive harzburgite channels in North Lanzo (full circles); and (2) refractory peridotite channel in Central Lanzo (open circles) (sample L212 of Bodinier *et al.* 1991). L212 clinopyroxene has significantly high $^{87}Sr/^{86}Sr$ ratios (>0.70453) and low $^{143}Nd/^{144}Nd$ ratios (<0.512796) (E_{Sr} +3.4, E_{Nd} +0.9) and plots close to CHUR values: it is characterized by much higher Sr and lower Nd isotopic ratios than the host plagioclase peridotites. According to Bodinier *et al.* (1991), it shows a ‘plume-type’ signature. (**d**) C1-normalized REE patterns of liquids calculated in equilibrium with magmatic clinopyroxenes from: (1) representative North Lanzo replacive spinel harzburgite channels (full circles); and (2) representative Central Lanzo refractory spinel peridotites (open circles) (sample 212 of Bodinier *et al.* 1991). Note the strong enrichment in LREE and depletion in HREE of the C1-normalized REE patterns, indicating their alkaline affinity.

harzburgites, referred to as *replacive spinel peridotites* (Figs 2 & 3).

The Central Lanzo body in composed of the same rock types, although the pyroxenite-bearing lithospheric lherzolite remnants are more abundant, the impregnated plagioclase peridotites are more widespread and the replacive ‘refractory’ peridotites are less common (e.g. Bodinier *et al.* 1991). The rock-type distribution, in some way, supports the theory that the Central body represents the transition between the North and the South bodies (Bodinier *et al.* 1991).

In both the Central and the South Lanzo bodies, on the basis of mutual field relationships, it has been proved that the oldest rock type is represented by the lithospheric peridotites, preserved as decametre-scale remnants within the reactive spinel peridotites, that have been in their turn transformed into impregnated peridotites by significant plagioclase enrichment. Both reactive and impregnated peridotites are, subsequently, cross-cut by the spinel harzburgite–dunite channels (Piccardo *et al.* 2007*a*).

Lithospheric lherzolites. They consist of foliated, porphyroclastic spinel lherzolites showing a tectonite fabric and widespread pyroxenite banding. In thin section they show spinel-facies assemblage and are characterized by: (i) cm-scale exsolved and deformed pyroxene porphyroclasts; (ii) subsolidus microtextures consisting of broadly rounded intergrowths (clusters) of Opx + Sp (±Cpx); and (iii) micro-exsolutions of vermicular Sp at the outer border of huge Opx porphyroclasts. The above microstructures indicate: (ii) the breakdown products of complete spinel-facies recrystallization of a precursor garnet; and (iii) cooling (i.e. exsolution of Al-rich components from Opx) under spinel-facies conditions (Piccardo *et al.* 2007*a*).

Foliated lithospheric lherzolites are chiefly transformed to pyroxene-poor reactive spinel harzburgites, characterized by coarse granular structure and low clinopyroxene content compared to the pre-existing foliated peridotites. The contact between foliated and granular rock types is generally sharp. In the granular reactive peridotites, the pyroxenite bands of the pre-existing lithospheric spinel peridotites are almost completely dissolved and are only recorded by aligned concentrations of spinel grains.

Reactive spinel peridotites. They consist of coarse granular, Cpx-poor spinel hazburgites and subordinated dunites, which preserve, in places, Opx + Sp clusters and spinel trains, suggesting that they formed is at the expense of pristine pyroxenite-bearing lithospheric lherzolites. In thin section the coarse granular spinel harzburgites have microstructures indicating the growth of new olivine on the spinel-facies pyroxene porphyroclasts, that is: (i) coarse rims of unstrained olivine surrounding and partially replacing the exsolved pyroxene porphyroclasts; (ii) crystallization of granular aggregates of undeformed olivine; and (iii) coarsening of the pre-existing mantle olivine, which encloses broadly rounded spinel crystals and fragments of pre-existing Opx porphyroclasts still preserving vermicular spinel exsolutions. A subsequent stage is marked by the interstitial crystallization of small crystals of both pyroxenes at triple junctions or along contacts between huge olivine porphyroclasts.

The structural and compositional characteristics of these rock types suggest structural recovering and mineral modal modification (i.e. pyroxene depletion and olivine enrichment) that has been referred to the action of reactively percolating melts (Piccardo *et al.* 2007*a*). The mode of melt–rock interaction (i.e. pyroxene dissolution and olivine precipitation) indicates the silica-undersaturated nature of the percolating melts, which possibly attained pyroxene saturation at the end of the process and, thus, crystallizing interstitial pyroxene grains. The absence of plagioclase in the magmatic assemblages and in the products of the melt–peridotite reactions suggests that the melt-reactive percolation occurred under spinel-facies conditions.

Impregnated plagioclase peridotites. The plagioclase peridotites have high proportions of modal plagioclase (up to 20% vol.), typically concentrated in mm-scale aggregates and veins. Most plagioclase-enriched peridotites preserve a former spinel-facies foliation. Plagioclase is diffuse in the granular rock types or aligned along the foliation in the penetratively deformed units. In places the plagioclase-rich peridotites are in direct contact with the plagioclase-free reactive spinel peridotites; the transition is rather sharp and can be followed continuously for tens of metres. In thin section the plagioclase peridotites consist of an older, deformed and exsolved spinel-facies assemblage, and a new, mostly unstrained, mm-size granular aggregate of plagioclase-rich gabbroic material. Plagioclase peridotites are characterized by orthopyroxene-forming, olivine-dissolving microstructures. The most common microstructures are: (i) widespread undeformed plagioclase crystals that are interstitial or cross-cutting deformed and exsolved mantle minerals; (ii) the replacement of kinked mantle olivine by undeformed orthopyroxene patches; (iii) mm-size plagioclase-rich gabbroic veins and pods, and symplectitic pyroxene + plagioclase interstitial patches; and (iv) rims of plagioclase surrounding spinel. In places these microstructures are coupled with clinopyroxene-dissolving, orthopyroxene + plagioclase-forming reactions (i.e. symplectitic Opx + Plg coronas surrounding corroded Cpx) and interstitial crystallization of Cpx-free noritic microgranular aggregates.

The structural and compositional characteristics of these rocks suggest infiltration, melt–rock interaction and the interstitial crystallization of melts migrating through pre-existing lithospheric lherzolites or reactive spinel peridotites (Piccardo *et al.* 2007*a*). The type of melt–rock interaction (i.e. olivine replacement by orthopyroxene) indicates the silica-saturated nature of the percolating melts. The presence of plagioclase in the magmatic assemblages and in the products of the melt–peridotite reactions suggests that the melt-reactive percolation and interstitial crystallization occurred under plagioclase-facies conditions.

Replacive spinel peridotites (dunites and harzburgites). In places plagioclase peridotites are deformed along decametre–hectometre-wide shear zones, where they acquire a strong foliation and tectonite fabric that are marked by strongly parallel pyroxenes + plagioclase aggregates and thin mylonitic bands. In thin section the mm-size shear bands are composed of very fine-grained Plg + Ol + Px granoblastic aggregates that sporadically grade to extremely fine-grained bands and to ultra-fine-grained and glassy pseudotachylytes. Plagioclase peridotite tectonites and mylonites are locally transformed to granular spinel harzburgites and dunites (the replacive spinel peridotites). The replacive spinel peridotites occur as: (i) elongated, metre-wide bands within the tectonite–mylonite shear zones, concordant with the main foliation of the country rock; and (ii) huge, decametre-wide, discordant dunite masses cross-cutting both the peridotite foliation and the concordant spinel harzburgite and dunite bands. The concordant spinel harzburgites are characterized by recovery of the tectonite fabrics to coarse granular textures, and by the disappearance of Plg and Cpx. Metre- to decametre-wide bodies and channels of spinel dunites (discordant dunites) cut, at high angles, the foliation of the plagioclase peridotites, the strike of the plagioclase tectonite shear zones and the concordant harzburgite–dunite bands. Spinel dunites often preserve oriented trains of Cr-spinel that are continuous with the foliation of the host peridotite. These features indicate the replacive origin of these spinel dunites (Boudier & Nicolas 1972; Boudier 1978), that is they have been formed by focused and reactive migration of pyroxene-undersaturated melts (e.g. Kelemen *et al.* 1995*a*, *b*). Replacive spinel dunites and harzburgites in some cases contain both small interstitial and undeformed grains of clinopyroxene, mostly crystallized at triple junctions between olivine crystals, and cm-scale euhedral clinopyroxene megacrystals commonly concentrated in irregular 'pyroxenite' bands. Euhedral clinopyroxene and interstitial plagioclase form locally microgabbroic aggregates and veins a few centimetres thick.

Some compositional features. Reactive spinel peridotites have a relatively low content of fusible components (Al_2O_3 0.8–2.1 wt%, CaO 0.8–2.2 wt%, TiO_2 0.03–0.08 wt%), and the highest MgO (43.2–45.0 wt%) and lowest SiO_2 (41.8–43.0 wt%) content, which is consistent with their relatively low pyroxene content. Impregnated plagioclase peridotites are enriched in fusible components (Al_2O_3 3.1–4.0 wt%, CaO 2.3–3.3 wt%, TiO_2 0.08–0.14 wt%) and silica (SiO_2 42–44.5 wt%), and depleted in Mg (MgO 38.4–41.7 wt%), with respect to the reactive peridotites, which is consistent with significant addition of plagioclase and micro-gabbroic material. Some strongly impregnated plagioclase peridotites attain gabbroic compositions (Al_2O_3 9.3–12.8 wt%, CaO 8.6–8.9 wt%, TiO_2 0.18–0.34 wt%, SiO_2 45.2–45.9 wt%, MgO 22.5–28.4 wt%).

In the MgO v. SiO_2 diagram (Fig. 4) the replacive spinel harzburgites fall outside the melting trends calculated by Niu (1997) for refractory peridotite residua after any partial melting, showing a lower SiO_2 content at the corresponding MgO content. This suggests that their strongly refractory compositions cannot have been derived by a melting process. According to petrographical evidence, a melt–peridotite reaction process, dissolving pyroxenes and precipitating olivine, should have been the more correct process. Impregnated peridotites follow an 'enrichment trend', characterized by decreasing MgO and increasing SiO_2 content, starting from the compositions of the reactive spinel peridotites, consistent with the petrographical evidence of the addition of silica-saturated (i.e. pyroxenes) and Mg-free (i.e. plagioclase) minerals.

The bulk-rock C1-normalized REE pattern of the representative reactive spinel peridotite LAS3 is almost flat in the MREE–HREE region with significant LREE depletion, which could correspond to that of a refractory residuum after small degree of fractional melting. Impregnated plagioclase peridotites shows bulk-rock C1-normalized REE patterns to be almost flat in the MREE–HREE region with variable LREE depletion, which could correspond to those of a refractory residua after a small degree of fractional melting (Fig. 5b).

Clinopyroxenes of the reactive peridotites show variable C1-normalized REE patterns, varying from LREE–MREE-enriched sinusoidal patterns with low HREE contents to strongly LREE-depleted patterns with a rather flat HREE pattern, at about $10 \times$ chondrite (i.e. sample LAS3) (Fig. 7a). Clinopyroxenes of the impregnated plagioclase peridotites show C1-normalized REE patterns usually characterized by high MREE–HREE content, showing flat or slightly humped patterns, associated with a strong–moderate LREE depletion (Fig. 7c). Most clinopyroxenes from replacive spinel harzburgites and dunites show almost flat C1-normalized REE patterns in the MREE–HREE region (at 8–$9 \times$ chondrite) with variable, but low, LREE fractionation (Fig. 7e). The trace element composition and C1-normalized REE patterns are consistent with those of clinopyroxenes crystallized by N-MORBs to slightly LILE (large-ion lithophile element)-enriched MORBs.

Migrating melts. Clinopyroxene trace element content has been used to investigate the geochemical affinity of the percolating melts in the hypothesis that trace element solid–liquid equilibrium partitioning was attained during melt–peridotite interaction. Structural and compositional features suggest that the reactive peridotites were formed by porous flow percolation through the spinel-facies mantle lithosphere of silica-undersaturated melts. Clinopyroxenes from most of the reactive peridotite samples show C1-normalized patterns that are in equilibrium with depleted single-melt increments formed by relatively low (about 5%) degrees of fractional melting of a spinel-facies DM asthenospheric source showing MORB affinity (Fig. 7b). Some clinopyroxenes show sinusoidal REE patterns which have been interpreted (Piccardo *et al.* 2007*a*) as reflecting transient geochemical gradients that affected the asthenospheric melts during melt–rock interaction.

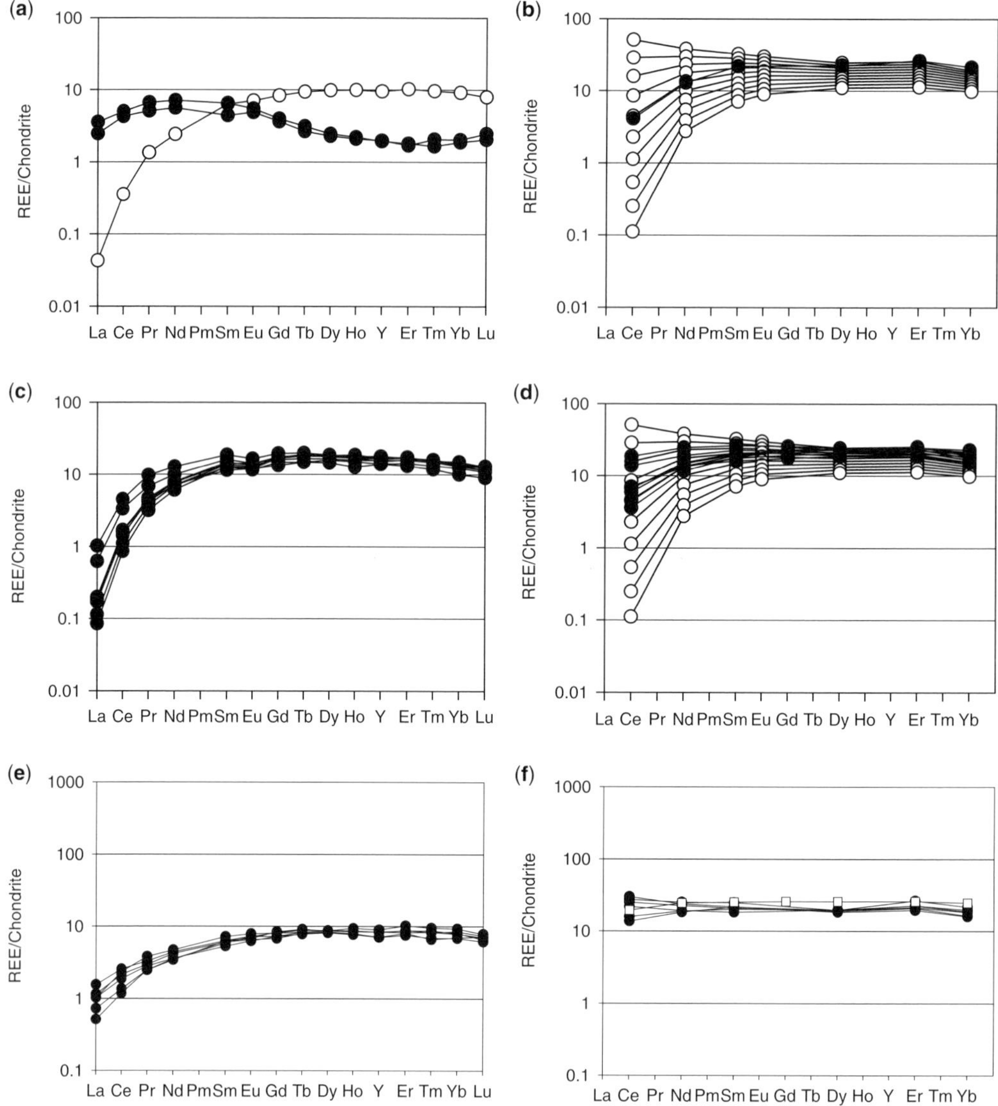

Fig. 7. (**a**) C1-normalized REE patterns of clinopyroxenes from representative reactive spinel harzburgites of South Lanzo: 1) Cpx showing trace element equilibration with depleted fractional melt (sample LAS3) (open circles); and (2) Cpx showing sinusoidal patterns recording transient geochemical gradients in the equilibrium melts during melt–rock interaction (full squares). (**b**) C1-normalized REE patterns of the liquid calculated in equilibrium with magmatic clinopyroxenes of the representative South Lanzo reactive spinel harzburgite LAS3 (full circles), which are representative of the reactive spinel peridotites that attained trace element equilibration with the reactively percolating melt. The calculated REE pattern of the equilibrium liquid (full circles) matches a fractional melt increment formed by 5% of fractional melting of a spinel-facies DM asthenospheric mantle source. Accordingly, it has been postulated that the pristine lithospheric lherzolites were transformed to reactive spinel peridotites via the reactive percolation of small degrees of fractional melt increments with MORB affinity.

The fractional melting model of Johnson *et al.* (1990), starting from the source Cpx of Ionov *et al.* (2002), under spinel-facies conditions using the $K_D^{cpx/liquid}$ values of Hart & Dunn (1993), which are useful for silica-undersaturated basaltic compositions. The REE patterns (grey, open circles) of melts formed by 1% (top) to 10% (bottom) of fractional melting are reported. (**c**) C1-normalized REE patterns of clinopyroxenes from representative impregnated plagioclase peridotites of South Lanzo (full circles). (**d**) C1-normalized REE patterns of liquids calculated in equilibrium with clinopyroxenes of the representative impregnated plagioclase peridotites of South Lanzo that are believed to have attained the trace element equilibration with the percolating melts or were crystallized from them. The

Clinopyroxenes from impregnated peridotites have C1-normalized REE patterns similar to those of Cpx in equilibrium with fractional melt increments generated by 1–5% fractional melting of a spinel-facies DM asthenospheric source, analogous to what has been argued for the reactive peridotites (Fig. 7d).

The trace element compositions of interstitial magmatic clinopyroxenes in replacive peridotites indicate that they attained the chemical equilibrium with aggregate MORBs, both N-MORB and slightly LILE-enriched MORB (Piccardo *et al.* 2007*a*) (Fig. 7f). The above data indicate that the melt dynamics changed during asthenosphere partial melting and melt percolation. The early percolation events, that is the reactive percolation and melt impregnation, involved single-melt fractions that survived unmixed and migrated isolated after their formation in the asthenosphere, whereas the later focused migration within replacive channels took place by aggregate MORB transportation; most probably, these liquids were aggregated in the asthenosphere prior to migration.

Discussion

Peridotite petrology and geodynamics in the Ligurian Tethys

It has recently been shown (Piccardo 2008) that the Western Alpine–Northern Apennine ophiolitic peridotites record and document different stages of lithosphere–asthenosphere evolution during the pre-oceanic rifting stages of the Late Jurassic Ligurian Tethys basin (Fig. 8).

- Lithosphere extension caused the tectonic–metamorphic evolution of the subcontinental mantle lithosphere, starting from spinel-facies conditions, accommodated by the formation of lithosphere-scale tectonite–mylonite shear zones (e.g. Montanini *et al.* 2006; Piccardo & Vissers 2007).
- Lithosphere extension and thinning caused adiabatic upwelling and decompressional melting of the underlying asthenosphere; the asthenospheric melts migrated via diffuse porous flow through the extending mantle lithosphere (Piccardo *et al.* 2007*a*), most probably exploiting the network of extensional shear zones.
- Deformation and melt-related processes in the mantle lithosphere were interdependent and mutually enhancing during lithosphere exhumation (Piccardo & Vissers 2007).
- Heating by asthenosphere upwelling and reactive percolation of asthenospheric melts through the mantle lithosphere caused diffuse structural, compositional and rheological modification of the mantle lithosphere, leading to its thermochemical and thermomechanical erosion (Piccardo 2003; Corti *et al.* 2007; Ranalli *et al.* 2007).

Moreover, it has been recognized (Piccardo 2008) that the petrological characteristics of mantle peridotites are significantly different in the various ophiolitic peridotite bodies and are strictly correlated to the inferred palaeogeographical settings of the ancient Ligurian Tethys. In fact, peridotites deriving from the OCT zones of the basin (i.e. the more continentward sequences) consist of spinel to plagioclase lherzolites recording tectonic–metamorphic exhumation from the subcontinental mantle, whereas mantle peridotites from the MIO settings of the basin (i.e. the more axial zones) consist of depleted and enriched peridotites, which derived from pristine subcontinental mantle strongly modified by melt–rock interaction and melt impregnation by MORB melts.

Mantle processes in the Lanzo peridotites

The North Lanzo body mostly consists of lherzolites and harzburgites, characterized by spinel-facies assemblage. These lithospheric peridotites show: (1) widespread effects of subsolidus recrystallization to plagioclase-facies assemblages; (2) limited effects of melt percolation by variably evolved aggregate MORB-type melts that formed

Fig. 7. (*Continued*) calculated REE pattern of the equilibrium liquids (full circles) match the fractional melt increments formed by 3–6% of fractional melting of a spinel-facies DM asthenospheric mantle source. Accordingly, it has been determined that the pre-existing reactive spinel peridotites were transformed to impregnated plagioclase peridotites via the percolation and interstitial crystallization of small degrees of fractional melt increments with MORB affinity.

The fractional melting model of Johnson *et al.* (1990), starting from the source Cpx of Ionov *et al.* (2002), under spinel-facies conditions using the $K_D^{cpx/liquid}$ values of Vannucci *et al.* (1998), which is useful for silica-saturated basaltic compositions. (**e**) C1-normalized REE patterns of magmatic interstitial and megacrystic clinopyroxenes of representative replacive harzburgite–dunite channels in South Lanzo (full circles). (**f**) C1-normalized REE patterns of liquids calculated in equilibrium with the magmatic clinopyroxenes of representative South Lanzo replacive harzburgite–dunite channels (full circles). The C1-normalized REE pattern of N-MORB (open squares) (data from Hofmann 1988) is also reported for comparison. REE concentrations and patterns of these liquids stress that magmatic clinopyroxenes within the replacive channels crystallized from MORBs.

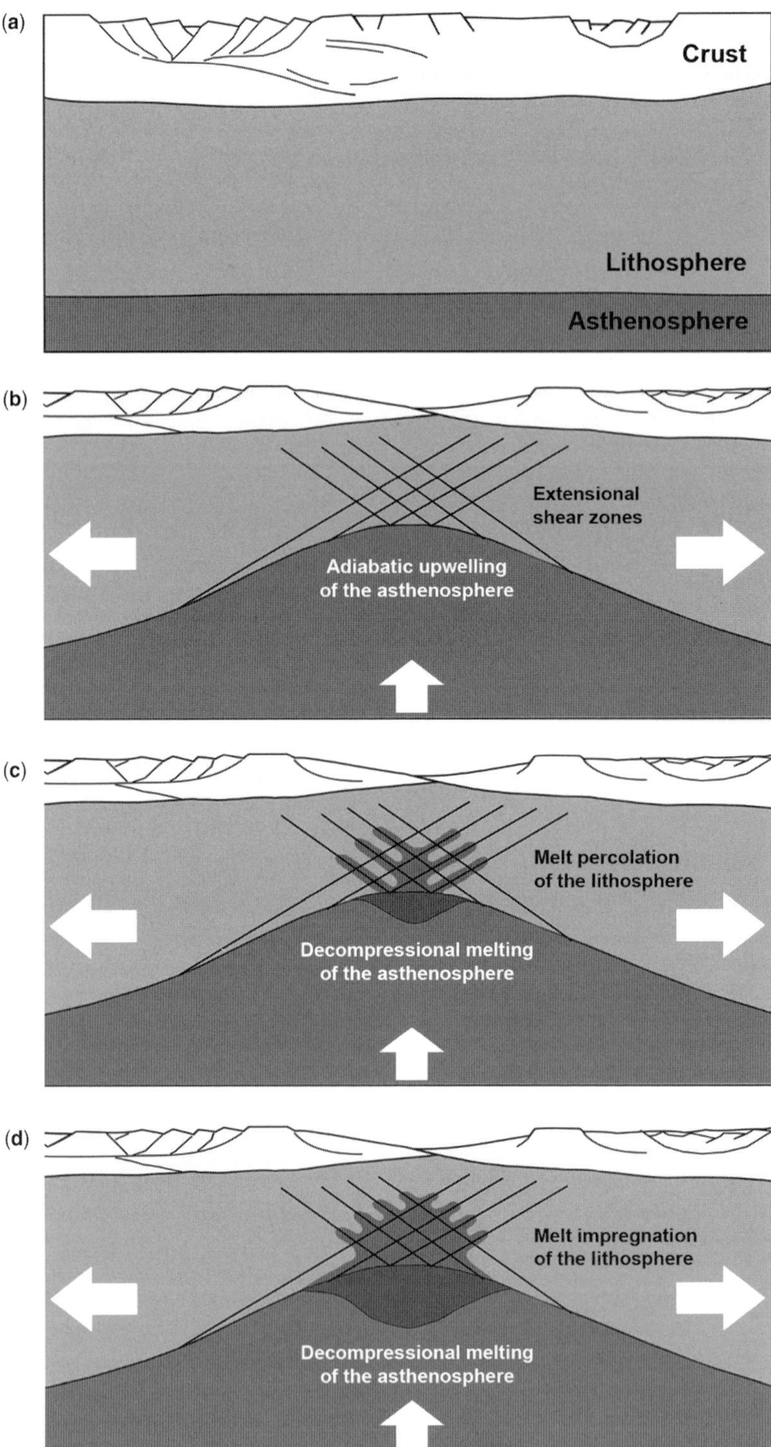

Fig. 8. Schematic evolution of the Europe–Adria continental system during pre-oceanic rifting in the Ligurian Tethys realm (from top to bottom). (**a**) The pre-Triassic situation. (**b**) Far-field tectonic forces caused lithosphere extension and thinning accommodated by a network of lithosphere-scale extesional shear zones. The underlying asthenosphere

hectometre-scale masses of impregnated plagioclase peridotites; (3) localized focused infiltration of alkaline melts that formed metre- to decametre-scale channels of strongly pyroxene-depleted, replacive spinel harzburgites (Piccardo *et al.* 2007*b*).

The diffuse partial recrystallization occurring at plagioclase-facies conditions suggests that these lithospheric peridotites were involved in the processes of lithospheric mantle exhumation related to the lithosphere extension. The subsequent localized impregnation by aggregate MORB melts indicates that asthenospheric melts percolated through the North Lanzo peridotites when they had been already exhumed to plagioclase-facies conditions, that is, at relatively shallow levels in the mantle lithosphere. The late formation of replacive spinel harzburgite channels, which have interstitial magmatic pyroxenes showing clear alkaline affinity, indicates that after pervasive percolation and impregnation by MORB-type melts these peridotites underwent focused migration along strongly pyroxene-depleted channels of alkaline melts, deriving from garnet-bearing mantle sources. Accordingly, it can be inferred that the North Lanzo spinel-facies lherzolite protoliths are the remnants of the subcontinental lithospheric mantle of the Europe–Adria system involved in the tectonic and magmatic evolution linked to the opening of the Ligurian Tethys. The lithospheric mantle underwent subsolidus exhumation, whereas the underlying asthenosphere underwent partial melting under decompression in response to adiabatic upwelling related to extension and thinning of the lithosphere.

Previous work suggested the provenance of the North Lanzo peridotites from the subcontinental mantle lithosphere of the Adria block (Bodinier *et al.* 1991; Müntener *et al.* 2005). Our structural and petrological–geochemical data provide important similarities with other ophiolitic peridotite massifs from the OCT zones of the Adria block in the Jurassic Ligurian Tethys and, particularly, with External Liguride peridotites of the Northern Apennines (e.g. Piccardo 1976, 2007; Rampone *et al.* 1995; Marroni *et al.* 1998; Montanini *et al.* 2006 and references therein). In fact, they are commonly characterized by: (i) Proterozoic ages of isolation from the convective mantle and accretion to the subcontinental thermal lithosphere; (ii) early spinel-facies assemblage and abundance of spinel-(and garnet-) pyroxenite bands; (iii) subsequent subsolidus evolution towards shallower, plagioclase-facies conditions; (iv) decametre–hectometre-scale areas showing effects of percolation and impregnation by aggregate MORB-type melts; and (v) late focused percolation of alkaline melts along replacive spinel harzburgite channels.

The South Lanzo body mostly consists of peridotites profoundly modified by interaction with MORB-type fractional melts (Müntener & Piccardo 2003; Piccardo 2003; Piccardo *et al.* 2004, 2007*a* and references therein). Sporadic decametre-scale remnants of pyroxenite-bearing peridotites indicate that subcontinental lithospheric lherzolites were the precursors of the melt-modified peridotites. As a whole, pristine lithospheric lherzolites underwent melt–peridotite interaction (both reactive depletion and refertilization) related to porous flow melt percolation operated by MORB-type fractional melt increments. Melt diffuse percolation started under spinel-facies conditions and ceased under plagioclase-facies conditions.

Field and structural data indicate that the South Lanzo peridotites underwent early reactive melt percolation at spinel-facies conditions, melt impregnation at plagioclase-facies conditions and, lastly, focused percolation within plagioclase-facies shear zones (Piccardo *et al.* 2007*b*). The late focused migration allowed the shallow-level delivery of aggregate MORB melts (Piccardo *et al.* 2007*a* and references therein). Geochemical data indicate that the migrating melts formed by partial melting of spinel-facies DM asthenospheric mantle (Piccardo *et al.* 2007*a*, *b*), which underwent adiabatic upwelling and decompression melting during lithosphere extension and thinning (Piccardo 2003).

The composite pattern of melt–peridotite interaction and the extreme compositional heterogeneity of the South Lanzo peridotites reflect important similarities with other ophiolitic peridotite massifs from the MIO settings of the Ligurian Tethys basin (Rampone *et al.* 1997, 1998, 2004; Piccardo *et al.* 2004; Piccardo 2007, 2008; Piccardo & Vissers 2007). As previously mentioned, lithostratigraphical data demonstrate that the Lanzo body was exhumed to the sea floor of the Jurassic Ligurian Tethys ocean, and was covered by volcanics and oceanic sediments, similar to the ophiolite sequences deriving from the Ligurian Tethys

Fig. 8. (*Continued*) underwent almost adiabatic upwelling, approaching solidus conditions. (**c**) Upwelling asthenosphere underwent decompressional partial melting and melts from the asthenosphere migrated via focused percolation within the pre-existing shear zones and infiltrated the host lithospheric mantle. Reactive melt migration caused melt–peridotite interaction (i.e. pyroxene dissolution and olivine precipitation) forming pyroxene-depleted reactive spinel peridotites. (**d**) Increasing asthenosphere partial melting and lithosphere porous flow percolation caused the compositional and rheological modification of significant sectors of the extending mantle lithosphere.

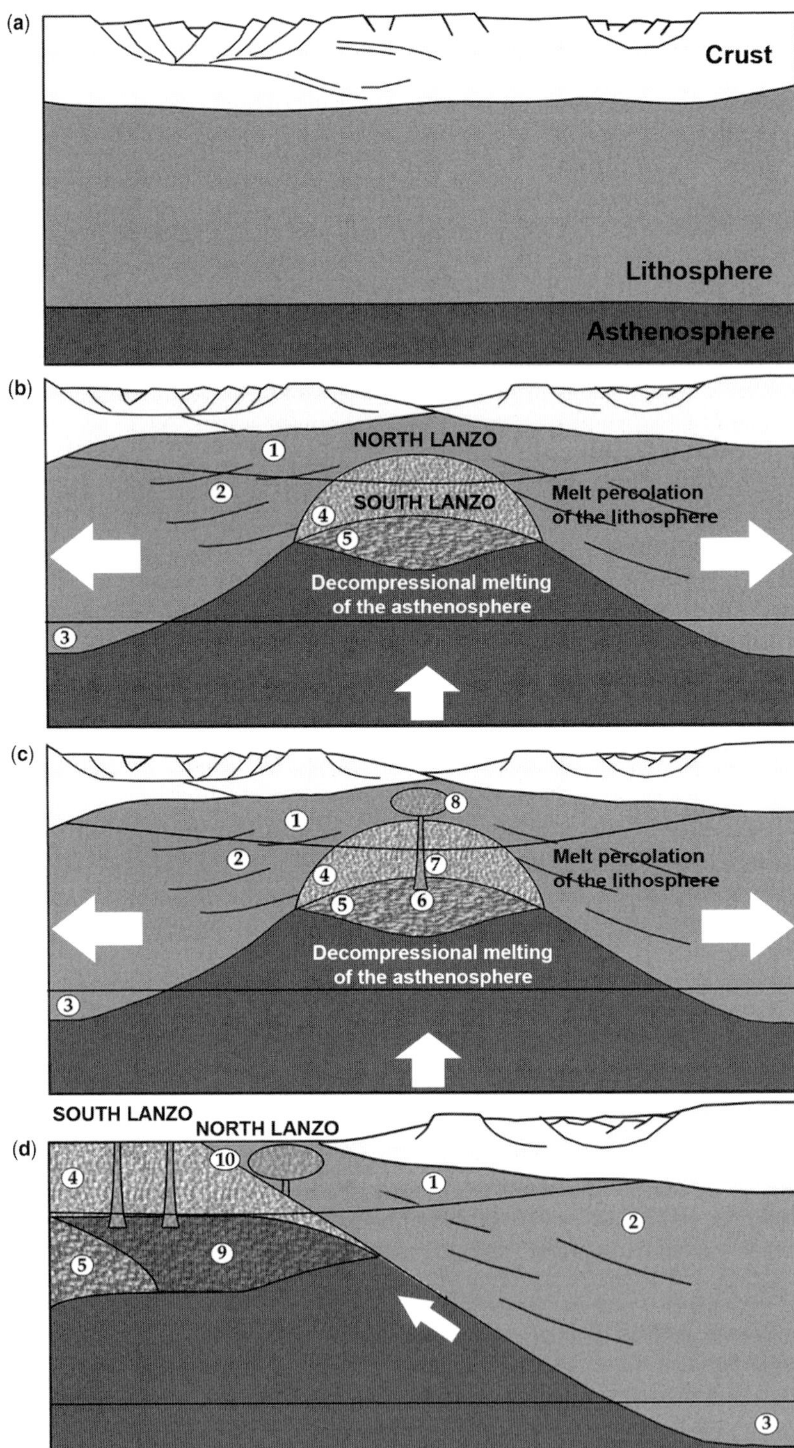

Fig. 9. The evolution of the Lanzo Massif (Italian Western Alps) in the frame of rifting of the Jurassic Ligurian Tethys. (**a**) The pre-Triassic setting (not to scale). (**b**) Extension driven by far-field tectonic forces led to a significant thinning of the continental lithosphere, causing: (i) exhumation of the shallower lithospheric mantle to plagioclase-facies conditions

basin. Present knowledge of the different bodies of the Lanzo massif supports the conclusion that both the OCT-type (i.e. North Lanzo body) and MIO-type peridotites (i.e. South–Central Lanzo body), derived from different palaeogeographical settings of the Western Alpine sector of the Jurassic Ligurian Tethys, are preserved in the Lanzo Massif.

The presence of alkaline melts in the Lanzo peridotites

The presence of alkaline melts in the OCT peridotites from the Jurassic Ligurian Tethys has recently been confirmed (Piccardo *et al.* 2007*b*; Piccardo 2008). As alkaline magmas are generated by the partial melting of garnet pyroxenites (Hirschmann *et al.* 2003 and references therein), the change in the geochemical affinity of the percolating melts from MORB to alkaline substantiates the entrainment of garnet pyroxenites in the melting source. Experimental data indicate that pyroxenites melt at significantly lower temperatures and/or at greater depths than peridotites (Petermann & Hirschmann 2003). This implies that if the garnet-pyroxenite bearing asthenospheric mantle is upwelling and melting under decompression, then alkaline liquids deriving from the melting of garnet pyroxenites should form earlier and rise prior to MORB melts deriving from the melting of host peridotites. Field evidence from Lanzo and other ophiolitic peridotites from the Ligurian Tethys indicates that the alkaline melts ascent *not preceding* but *following* the diffuse percolation of MORB-type melts (Piccardo 2008). This indicates that garnet pyroxenites did not rise from deeper asthenospheric levels and, accordingly, did not undergo melting prior to asthenospheric spinel-facies DM peridotites, but were entrained into the melting asthenosphere after inception of MORB-forming partial melting of peridotites.

Garnet pyroxenites and spinel (ex-garnet) pyroxenites are present in the lithospheric mantle peridotites from the Jurassic Ligurian Tethys (e.g. Montanini *et al.* 2006; Piccardo *et al.* 2007*b*; Piccardo & Vissers 2007; this work), thus indicating that the deep levels ($P > 1.5$ GPa) of the Europe–Adria subcontinental lithospheric mantle was veined by garnet pyroxenite bands. Field and petrographical evidence suggests that the earliest MORB-type melts infiltrated the extending lithosphere when peridotites and garnet pyroxenites of the lower lithosphere were completely recrystallized to spinel-facies assemblages, that is, had already been exhumed to $P < 1.5$ GPa (Piccardo *et al.* 2007*a*; Piccardo & Vissers 2007). Moreover, geochemical data indicate that the MORB-type asthenospheric melts that percolated through the spinel-facies mantle lithosphere were formed by fractional melting at relatively shallow levels ($P < 2.5$ GPa) of spinel-facies DM asthenospheric sources (Piccardo *et al.* 2007*a*; Piccardo & Vissers 2007).

Accordingly, it is highly improbable that the geochemical signature of these alkaline melts was derived by garnet pyroxenites of the lithospheric mantle, which have undergo thermomechanical erosion by the upwelling asthenospheric melts (as suggested by Bodinier & Godard (2003) and by Montanini & Tribuzio (2007*a*, *b*) for the garnet signature in the Ligurian MORB crustal rocks (see later)). As demonstrated above, the thermomechanical/thermochemical erosion was caused by MORB melts formed under spinel-facies conditions that infiltrated spinel-facies, spinel pyroxenite-bearing mantle lithosphere.

Garnet pyroxenites evidently were introduced into the melting asthenosphere at pressures greater than 1.5 GPa, to preserve their garnet-bearing assemblages. A suitable mechanism for their entrainment in the mantle asthenosphere is the early

Fig. 9. (*Continued*) (i.e. Lanzo North protoliths); (ii) adiabatic upwelling and inception of decompressional melting of the asthenosphere; and (iii) porous flow percolation through the lower mantle lithosphere (i.e. Lanzo South protoliths) of MORB-type fractional melts that escaped aggregation, survived unmixed and migrated isolated, forming reactive and impregnated peridotites. (1) plagioclase-facies lithosphere; (2) spinel-facies lithosphere; (3) garnet-facies lithosphere; (4) melt percolated lithosphere; and (5) melting asthenosphere. (**c**) Increase in partial melting during progressive decompression, where the single-melt fractions were more efficiently mixed and completely aggregated to form MORB magmas. Aggregate MORB magmas passed through the South Lanzo peridotites by focused migration within replacive harzburgite–dunite channels, undergoing variable evolution by fractional crystallization and infiltrated the North Lanzo peridotites; refractory residua after asthenosphere partial melting were accreted to the thermal lithosphere. (6) Aggregation of MORB melts in the melting asthenosphere; (7) aggregate MORB migration through the South Lanzo melt-modified peridotites within replacive channels; (8) MORB impregnation of the plagioclase-facies North Lanzo lherzolites; and (9) refractory residua after MORB melting were accreted to modified mantle lithosphere. (**d**) Ongoing lithosphere extension and stretching by means of km-scale shear zones caused the break-up of the continental crust and the sea-floor exposure of the subcontinental lithospheric mantle. The North Lanzo peridotites, deriving from shallower lithospheric mantle levels, were exhumed and exposed at more external OCT zones of the basin, close to the Adria continental margin. The strongly modified South Lanzo peridotites, deriving from deeper lithospheric mantle levels, were exhumed and exposed on the sea floor at more MIO settings of the basin.

delamination of the deep, garnet pyroxenite-bearing sectors of the mantle lithosphere during inception of lithosphere extension and preceding the decompressional subsolidus transition of garnet pyroxenites to spinel-facies assemblages (i.e. at $P > 1.5$ GPa). It can be speculated that segments of the lower lithospheric mantle were tectonically detached and sank into the upwelling hot mantle asthenosphere, where they underwent heating and partial melting under the appropriate pressure conditions to forming alkaline melts. These melts migrated upwards and percolated through the extending lithospheric mantle *subsequent* to the MORB percolation and impregnation.

It has recently been stressed by Montanini & Tribuzio (2007*a*, *b*) and Montanini *et al.* (2007) that the subsequent oceanic magmatism, which formed the crustal rocks (gabbroic intrusives and basaltic volcanites) of the Ligurian ophiolites, although showing MORB affinity, is characterized by peculiar trace element fingerprints. According to these authors, the Zr enrichment, the relatively high Sm/Yb ratios and the anomalous composition of the less radiogenic Nd isotope of the Ligurian ophiolitic crustal rocks is not compatible with the small extent of fractional melting of a depleted spinel peridotite but, however, can be explained by the small degree of melting of a mixed source of spinel peridotite with small amounts of garnet pyroxenite.

It has been speculated (Montanini & Tribuzio 2007*a*, *b*; Montanini *et al.* 2007) that the interaction between the uprising MORB melts and the pyroxenite bands of the lithospheric mantle played a role in generating the peculiar garnet signatures of the MORB-type parental melts of the Ligurian ophiolitic crustal rocks. On the basis of the above arguments (i.e. the spinel-facies assemblage of pyroxenites in the extending lithospheric mantle), this seems to be highly unlikely. It is more plausible that these signatures were produced by garnet pyroxenite relicts left in the DM melting sources after delamination and sinking of sections of the deep garnet pyroxenite-bearing mantle lithosphere.

According to the above scenario: (i) MORB melts were generated and migrated earlier through the extending lithosphere; (ii) alkaline melts subsequently formed by the melting of the sinking garnet pyroxenites; and (iii) garnet pyroxenite relicts were left in the melting asthenospheric source to generate the enriched signature of the subsequent magmatism, which formed the crustal rocks (gabbros and basalts) of the Ligurian ophiolites. The early detachment and sinking of garnet pyroxenite-bearing lithospheric mantle sections in the upwelling melting asthenosphere, thus, could have been responsible for both: (i) the formation of alkaline liquids; and (ii) the presence of a relict-enriched component in the source of the subsequent MORB magmatism, reflecting the garnet signature of the gabbroic intrusions and basaltic extrusions.

Evolution of the Lanzo peridotites attending the rifting of the Ligurian Tethys

As previously mentioned, lithostratigraphical studies indicate that the Lanzo peridotites are overlain by metabasites, Mn-rich metaquartzites and calcschists (i.e metamorphosed volcanites and oceanic sediments) (Lagabrielle *et al.* 1989; Pelletier & Müntener 2006), similar to other ophiolite sequences in the Alps and Apennines. This strongly suggests that the Lanzo massif was exhumed to the sea floor of the Jurassic Ligurian Tethys ocean, and was covered by basaltic volcanites and oceanic sediments.

Present knowledge of the South–Central and North Lanzo peridotite bodies indicate that: (1) they were derived from the subcontinental lithospheric mantle; (2) they underwent subsolidus exhumation at shallow levels; and (3) they were percolated by melts formed in the upwelling asthenosphere. Although the lithospheric protoliths of the different peridotite bodies record garnet-facies relict structural features and spinel-facies complete equilibrium recrystallization, prior to any melt–peridotite interaction process, the tectonic and magmatic evolutions of the South–Central and North Lanzo mantle sections were remarkably different.

The North Lanzo body exposes large areas of peridotites preserving lithospheric characteristics, which escaped spinel-facies reactive melt percolation. In fact, these peridotites were significantly recrystallized under plagioclase-facies conditions (i.e. underwent exhumation under subsolidus conditions) and, later on, were locally impregnated by aggregate MORB melts. Finally, they experienced focused migration of alkaline melts. However, the South and Central Lanzo bodies are mostly composed of peridotites profoundly modified by interaction with MORB-type fractional melts (Piccardo *et al.* 2007*a* and references therein). In fact, only sporadic, decameter–hectometer-scale remnants of the subcontinental spinel-facies lherzolite precursors are preserved, and peridotites showing significant effects of melt–peridotite interactions are dominant. Sporadically, replacive spinel harzburgite channels in Central Lanzo were used for the upwards movement of alkaline liquids (Bodinier *et al.* 1991; see discussion in Piccardo 2008). Subsequently, the South–Central Lanzo peridotite bodies were intruded by variably evolved MORB magmas, which formed gabbroic dykes ranging from less evolved olivine gabbros (Mg–Al gabbros) to more evolved oxide gabbros (Fe–Ti gabbros) (Bodinier *et al.* 1986). The gabbroic

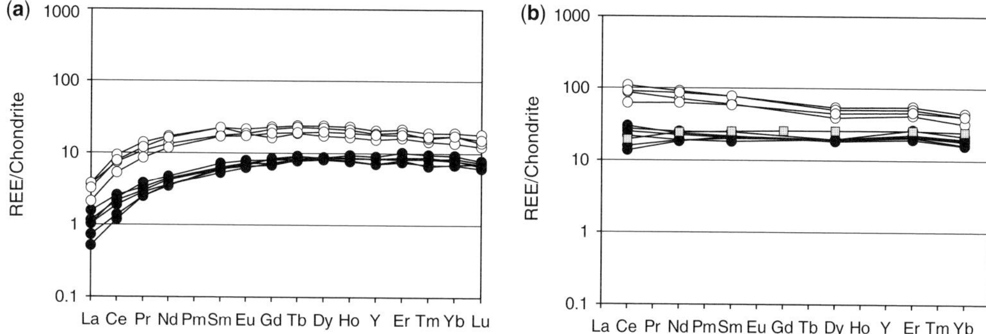

Fig. 10. (**a**) C1-normalized REE patterns of clinopyroxenes from Lanzo peridotites: (1) magmatic interstitial and megacrystic Cpx in representative replacive harzburgite–dunite channels in South Lanzo (full circles); and (2) representative impregnated plagioclase peridotites in North Lanzo (open circles). (**b**) C1-normalized REE patterns of liquids calculated in equilibrium with magmatic clinopyroxenes from: (1) representative South Lanzo replacive harzburgite–dunite channels (full circles); and (2) representative North Lanzo impregnated plagioclase peridotites (open circles).

The C1-normalized REE pattern of N-MORB (grey squares) (data from Hofmann (1988) is also reported for comparison. REE concentrations and patterns of these liquids indicate their strong similarity to aggregate MORBs.

Note that the South Lanzo liquids match well the MORB melts, whereas the North Lanzo liquids show similar REE trends but higher absolute concentrations with respect to primitive N-MORB liquids.

Simple modelling demonstrates that variable degrees of low-pressure fractional crystallization can produce the REE concentrations and patterns of liquids calculated in equilibrium with the North Lanzo Cpx starting from the South Lanzo equilibrium liquids. Accordingly, the MORB melts that impregnated the North Lanzo lherzolites can be derived by low-pressure evolution from pristine aggregate MORB melts very similar to the melts that migrated within the replacive channels of South Lanzo. Accordingly, it can be shown that the primary aggregate MORB melts underwent variable evolution by fractional crystallization during focused migration within the replacive harzburgite–dunite channels through the South Lanzo peridotite prior to infiltrating the North Lanzo peridotite.

dykes have been dated at 160 Ma by Kaczmarek *et al.* (2005), which is consistent with the inferred age of the opening of the oceanic basin (see Piccardo 2008 for a more detailed discussion).

We speculate that the lithospheric protoliths of the North and South–Central Lanzo peridotites were located at different depths in the subcontinental lithosphere during continental extension when the asthenosphere underwent decompressional

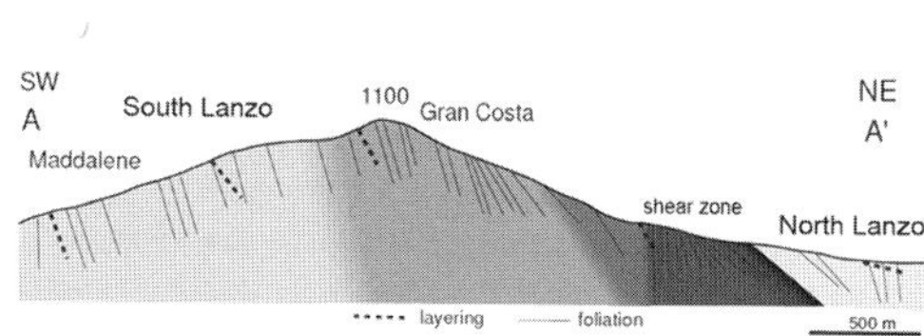

Fig. 11. Schematic topography of the shear zone cross-section (from Kaczmarek & Müntener 2008), showing that the North Lanzo body represents the hanging wall and the shear zone is completely developed in the Central Lanzo body. Increasing darkeness of grey from left to right in the Central Lanzo body (towards the contact with the North Lanzo body) corresponds to increasing deformation. Amphibole-rich mylonites are located along the contact.

melting and the asthenospheric melts infiltrated through the extending mantle lithosphere (Fig. 9).

According to this scenario, it may be inferred that early fractional melts, formed in the upwelling asthenosphere at the inception of decompressional melting, percolated and modified the deeper spinel-facies lithospheric mantle levels (i.e. the South–Central Lanzo peridotite protoliths), forming the South Lanzo reactive spinel peridotites. The shallower lithospheric mantle levels (i.e. the North Lanzo peridotite protoliths) mostly escaped this event, thus preserving their pristine fertile lherzolite compositions and pyroxenite bands. Subsequently, saturated MORB-type fractional melt increments infiltrated the already reacted and pyroxene-depleted mantle levels (i.e. the South–Central Lanzo peridotites) when they had been exhumed to plagioclase-facies conditions, thus forming the South–Central Lanzo impregnated plagioclase peridotites. The shallower mantle levels (i.e. the North Lanzo peridotites) also escaped this stage (Fig. 9b).

It can be inferred that the focused migration of aggregate MORB melts within the harzburgite–dunite channels through the South–Central Lanzo peridotite allowed these melts to pass through the deeper reactive and impregnated mantle levels (the Central–South Lanzo plagioclase peridotites) and to percolate diffusely through the shallower, more fertile mantle levels (the North Lanzo peridotites) (Fig. 9c). These melts were variably fractionated during upwelling and caused localized impregnation in the North Lanzo pyroxenite-bearing fertile lherzolites. In fact, the South Lanzo percolating liquids match well primary MORB melts, whereas the North Lanzo impregnating liquids show similar REE trends but higher absolute concentration with respect to primitive N-MORB liquids. Simple modelling confirms that variable degrees of low-pressure fractional crystallization can produce the REE concentrations and patterns of the North Lanzo melts (as calculated from the REE concentration of Cpx) starting from the concentrations of the South Lanzo primary melts (as calculated from the REE concentration of Cpx) (Fig. 10). Accordingly, it can be inferred that primary aggregate MORB melts underwent variable evolution by fractional crystallization during focused migration within the replacive harzburgite–dunite channels through the South Lanzo peridotite prior to the infiltration of the North Lanzo peridotite. Later on, the high-permeability olivine-rich channels allowed the upwards migration of alkaline liquids in both the Central and North Lanzo peridotites.

A recent structural–geochemical work (Kaczmarek & Müntener 2008) investigated the shear zone that separates the Central from the North Lanzo bodies, which are interpreted to represent, respectively, the footwall (the central part) and the hanging wall (the northern part) of an extensional system. The shear zone is mostly formed at the expense of the Central Lanzo peridotite, and deformation gradually increases in the plagioclase peridotites of the Central body towards the shear zone, from coarse-grained granular to mylonitic textures (Fig. 11). Granular rocks are characterized by Ol, Opx and Cpx porphyroclasts, and recrystallized domains where these minerals are associated with plagioclase. Mylonites are characterized by extremely stretched Opx and a fine-grained matrix composed of pyroxenes, olivine, spinel and plagioclase, while Amph-rich mylonites show a very fine-grained matrix where Ti-hornblende is associated to plagioclase (Kaczmarek & Müntener 2008). The recrystallized mineral assemblages in the fine- to ultra-fine-grained mylonite matrix suggest that the main deformation occurred on already impregnated plagioclase peridotites (i.e. later than the plagioclase-facies impregnation), and the presence of Ti-amphiboles in the recrystallized ultra-fine-grained matrix indicates fluid circulation during deformation. Kaczmarek & Müntener (2008) suggest that the mantle shear zone might act as an important permeability barrier that separates the thermal boundary layer (i.e. the Central Lanzo peridotites) from the conductively cooled mantle (i.e. the North Lanzo body). Kaczmarek & Müntener (2008) showed that the presence of the shear zone suggests rapid exhumation and indicates that the footwall of the system (i.e. the Central–South Lanzo sections) records relatively greater exhumation than the hanging wall (i.e. the North Lanzo section).

Combining available data, it can be concluded that the melt-impregnated peridotites of the Central–South Lanzo bodies were exhumed from deeper lithospheric mantle levels with respect to the North Lanzo peridotites by means of high-temperature shear zones. They were coupled at shallow mantle levels with the colder subcontinental peridotites of the North Lanzo body (Fig. 9d).

Conclusion

Available structural and petrological data indicate that the Lanzo peridotites were involved in tectonic and magmatic processes related to lithosphere extension that led to the rifting and opening of the Jurassic Ligurian Tethys basin. The North and Central–South Lanzo peridotites record different steps of the composite tectonic and magmatic evolution, which was characterized by:

- the exhumation of spinel-facies lithospheric mantle marked by the transition from spinel- to plagioclase-facies conditions, as recorded by the North Lanzo peridotites;
- the adiabatic upwelling and decompression partial melting of the underlying asthenosphere forming MORB-type single fractional melt increments that escaped aggregation, survived unmixed and migrated individually intact;
- the porous flow percolation of the single fractional melt increments through the overlying lithosphere that formed reactive depleted spinel peridotites and plagioclase-enriched impregnated peridotites, as mostly recorded by the South Lanzo peridotites;
- the coalescence of the single-melt fractions that were more efficiently mixed and completely aggregated to form MORB magmas. They passed through the South Lanzo peridotites by focused migration within replacive harzburgite–dunite channels, undergoing variable fractionation, and infiltrated through the North Lanzo peridotites.

The lithospheric protoliths of the North and South–Central Lanzo peridotites resided at different depths in the subcontinental lithosphere when the asthenosphere underwent decompressional melting during lithosphere extension and the asthenospheric melts infiltrated the mantle lithosphere. The deeper lithospheric levels (i.e. the South

Lanzo peridotites) underwent reactive melt percolation and impregnation by MORB-type fractional melts, and subsequently aggregate MORB melts passed through these peridotites migrating within focused replacive channels. The shallower lithospheric levels (i.e. the North Lanzo peridotites) were only infiltrated in places by the late aggregate MORBs. Later on, the high-permeability olivine-rich channels allowed the upwards migration of alkaline melts (as evident in the Central and North Lanzo peridotites) formed by the partial melting of garnet pyroxenite-bearing mantle sections that were initially delaminated from the deep mantle lithosphere and entrained in the upwelling melting asthenosphere.

Ongoing lithosphere extension and stretching caused progressive upwelling of the Lanzo peridotites, break-up of the continental crust and sea-floor exposure of the subcontinental lithospheric mantle. The North Lanzo peridotites, deriving from shallower mantle levels, were exhumed at more external OCT zones of the basin, close to the Adria continental margin, whereas the South Lanzo peridotites, deriving from deeper mantle levels, were exhumed and exposed at the sea floor at MIO settings of the basin.

Field, petrographical–structural and petrological–geochemical studies of Lanzo mantle peridotites allow the geodynamic evolution of the Europe–Adria extensional system leading to opening of the Jurassic Ligurian Tethys oceanic basin to be considered from a purely mantle perspective.

This work was partially supported by the Italian MIUR (Ministero dell'Istruzione, dell'Universita' e della Ricerca) (PRIN-COFIN2005: 'Lithosphere evolution induced by migration of mantle-derived melts at different geodynamic settings') and the University of Genova.

An early version of the paper was revised by G.W. Ernst and G. Ranalli. Their suggestions greatly improved the final version of the paper.

Thanks are due also to L. Guarnieri and M. Piccardo for their technical assistance.

References

Bodinier, J.-L. 1988. Geochemistry and petrogenesis of the Lanzo peridotite body, Western Alps. *Tectonophysics*, **149**, 67–88.

Bodinier, J. L. & Godard, M. 2003. *Orogenic, Ophiolitic, and Abyssal Peridotites*. Treatise on Geochemistry, **2**. Elsevier, Amsterdam, 103–170.

Bodinier, J. L., Guiroud, M., Dupuy, C. & Dostal, J. 1986. Geochemistry of basic dikes in the Lanzo massif (western Alps): petrogenetic and geodynamic implications. *Tectonophysics*, **128**, 75–95.

Bodinier, J. L., Menzies, M. A. & Thirwall, M. F. 1991. Continental to oceanic mantle transition – REE and Sr–Nd isotopic geochemistry of the Lanzo Lherzolite Massif. *In*: Menzies, M. A., Dupuy, C. & Nicolas, A. (eds) *Orogenic Lherzolites and Mantle Processes. Journal of Petrology*, Special Lherzolite Issue, 191–210.

Boudier, F. 1976. *Le massif lherzolitique de Lanzo (Alpes piemontaises). Etude Structurale et Petrologique*. These de Doctorat d'etat, University of Nantes.

Boudier, F. 1978. Structure and petrology of the Lanzo peridotite massif (Piedmont Alps). *Geological Society of America Bulletin*, **89**, 1574–1591.

Boudier, F. & Nicolas, A. 1972. Fusion partielle gabbroïque dans la lherzolite de Lanzo (Alpes piémontaises). *Schweizerische Mineralogische und Petrographische Mitteilungen*, **52**, 39–56.

Corti, G., Bonini, M., Innocenti, F., Manetti, P., Piccardo, G. B. & Ranalli, G. 2007. Experimental models of extension of continental lithosphere weakened by percolation of asthenospheric melts. *Journal of Geodynamics*, **43**, 465–483.

Hart, S. R. & Dunn, T. 1993. Experimental cpx/melt partitioning of 24 trace elements. *Contributions to Mineralogy and Petrology*, **113**, 1–8.

Hirschmann, M. M. & Stolper, E. M. 1996. A possible role for garnet pyroxenite in the origin of the 'garnet signature' in MORB. *Contributions to Mineralogy and Petrology*, **124**, 185–208.

Hirschmann, M. M., Kogiso, T., Baker, M. B. & Stolper, E. M. 2003. Alkalic magmas generated by partial melting of garnet pyroxenite. *Geology*, **31**(6), 481–484.

Hofmann, W. A. 1988. Chemical differentiation of Earth: the relationship between mantle, continental crust and oceanic crust. *Earth and Planetary Science Letters*, **90**, 297–314.

Ionov, D. A., Bodinier, J. L., Mukasa, S. B. & Zanetti, A. 2002. Mechanisms and sources of mantle metasomatism: major and trace element conditions of peridotite xenoliths from Spitzbergen in the context of numerical modelling. *Journal of Petrology*, **43**(12), 2219–2259.

Johnson, K. T. M., Dick, H. J. B. & Shimizu, N. 1990. Melting in the oceanic upper mantle: an ion microprobe study of diopsides in abyssal peridotites. *Journal of Geophysics Research*, **95**, 2661–2678.

Kaczmarek, M.-A. & Müntener, O. 2008. Juxtaposition of melt impregnation and high temperature shear zone in the upper mantle; field and petrological constraints from the Lanzo peridotite (Northern Italy). *Journal of Petrology*, **49**, 2187–2220.

Kaczmarek, M.-A., Müntener, O. & Rubatto, D. 2008. Trace element chemistry and U–Pb dating of zircons from oceanic gabbros and their relationship with whole rock composition (Lanzo, Italian Alps). *Contributions to Mineralogy and Petrology*, **155**, 295–312.

Kaczmarek, M.-A., Rubatto, D. & Müntener, O. 2005. SHRIMP U–Pb zircon dating of gabbro and granulite from the peridotite massif of Lanzo (Italy). *Geophysical Research Abstracts*, **7**, 03098.

Kelemen, P. B., Shimizu, N. & Salters, V. J. M. 1995*a*. Extraction of mid-ocean-ridge basalt from the upwelling mantle by focused flow of melt in dunite channels. *Nature*, **375**, 747–753.

Kelemen, P. B., Whitehead, J. A., Aharonov, E. & Jordahl, K. A. 1995*b*. Experiments on flow focusing in soluble porous media, with applications to melt extraction from the mantle. *Journal of Geophysical Research*, **100**, 475–496.

Lagabrielle, Y., Fudral, S. & Kienast, J.-R. 1989. La couverture océanique des ultrabasites de Lanzo (Alpes occidentales): arguments lithostratigraphiques et pétrologiques. [The oceanic cover of the Lanzo peridotite body (Western Italian Alps): lithostratigraphic and petrological evidences.] *Geodinamica Acta*, **3**, 43–55.

Marroni, M., Molli, G., Montanini, A. & Tribuzio, R. 1998. The association of continental crust rock with ophiolites in the Northern Apennine (Italy): implication for the continent-ocean transition in the Western Tethys. *Tectonophysics*, **292**, 43–66.

Montanini, A. & Tribuzio, R. 2007*a*. Petrogenesis of basalts and gabbros from an ancient continent–ocean transition (External Liguride ophiolites, Northern Italy): implications for a pyroxenite-derived component. *In*: *FIST Geoitalia 2007*, Abstract Volume, 58.

Montanini, A. & Tribuzio, R. 2007*b*. Magmatic activity during continental breakup and ocean opening recorded by the external liguride ophiolites (Northern Apennines, Italy): implications for a mixed peridotite-pyroxenite mantle source. *In*: *Epitome 2007, Geoitalia 2007*, Volume 2, 13–14.

Montanini, A., Tribuzio, R. & Anczkiewicz, R. 2006. Exhumation history of a garnet pyroxenite-bearing mantle section from a continent–ocean transition (Northern Apennine ophiolites, Italy). *Journal of Petrology*, **47**(10), 1943–1971.

Montanini, A., Tribuzio, R. & Vernia, L. 2007. Petrogenesis of basalts and gabbros from an ancient continent–ocean transition (External Liguride ophiolites, Northern Italy). *Lithos*, **101**, 453–479; doi: 10.1016/j.lithos.2007.09.007.

Müntener, O. & Piccardo, G. B. 2003. Melt migration in ophiolites: the message from Alpine–Apennine peridotites and implications for embryonic ocean basins. *In*: Dilek, Y. & Robinson, P. T. (eds) *Ophiolites in Earth History*. Geological Society, London, Special Publications, **218**, 69–89.

Müntener, O., Piccardo, G. B., Polino, R. & Zanetti, A. 2005. Revisiting the Lanzo peridotite (NW-Italy): 'asthenospherization' of ancient mantle lithosphere. *Ofioliti*, **30**(2), 111–124.

Nicolas, A. 1974. Mise en place des péridotites de Lanzo (Alpes piémontaises) Relation avec tectonique et métamorphisme alpins: conséquences géodynamiques. *Schweizerische Mineralogische und Petrographische Mitteilungen*, **54**, 449–460.

Nicolas, A. 1984. Lherzolites of the Western Alps: a structural review. *In*: Kornprobst, J. (ed.) *5th International Kimberlite Conference Proceedings*. Elsevier, Amsterdam, 333–345.

Nicolas, A. 1986. A melt extraction model based on structural studies in mantle peridotites. *Journal of Petrology*, **27**, 999–1022.

Niu, Y. 1997. Mantle melting and melt extraction processes beneath ocean ridges: evidence from abyssal peridotites. *Journal of Petrology*, **38**, 1047–1074.

Pelletier, L. & Müntener, O. 2006. High pressure metamorphism of the Lanzo peridotite and its oceanic cover, and some consequences for the Sesia-Lanzo zone (northwestern Italian Alps). *Lithos*, **90**(1), 111–130.

Petermann, M. & Hirschmann, M. M. 2003. Partial melting experiments on a MORB-like pyroxenite between 2 and 3 Gpa: constraints on the presence of pyroxenite in basalt source regions from solidus location and melting rate. *Journal of Geophysical Research*, **108**, 2125.

Piccardo, G. B. 1976. Petrologia del massiccio lherzolitico di Suvero (La Spezia). *Ofioliti*, **1**, 279–317.

Piccardo, G. B. 2003. Mantle processes during ocean formation: petrologic records in peridotites from the Alpine–Apennine ophiolites. *Episodes*, **26**(3), 193–200.

Piccardo, G. B. 2007. The Ligurian Tethys, a Jurassic ultra-slow spreading ocean. *Periodico di Mineralogia*, **76**, 67–80.

Piccardo, G. B. 2008. The Jurassic Ligurian Tethys, a fossil ultra-slow spreading ocean: the mantle perspective. *In*: Coltorti, M. & Grégoire, M. (eds) *Metasomatism in Oceanic and Continental Lithospheric Mantle*. Geological Society, London, Special Publications, **293**, 11–34.

Piccardo, G. B. & Vissers, R. L. M. 2007. The pre-oceanic evolution of the Erro-Tobbio peridotite (Voltri Massif–Ligurian Alps, Italy). *Journal of Geodynamics*, **43**, 417–449.

Piccardo, G. B., Müntener, O., Zanetti, A. & Pettke, T. 2004. Ophiolitic peridotites of the Alpine–Appenine system: mantle processes and geodynamic relevance. *International Geology Review*, **46**, 1119–1159.

Piccardo, G. B., Zanetti, A. & Müntener, O. 2007*a*. Melt/peridotite interaction in the Lanzo South peridotite: field, textural and geochemical evidence. *Lithos*, **94**(1–4), 181–209.

Piccardo, G. B., Zanetti, A., Pruzzo, A. & Padovano, M. 2007*b*. The North Lanzo peridotite body (NW-Italy): lithospheric mantle percolated by MORB and alkaline melts. *Periodico di Mineralogia*, **76**, 175–196.

Pognante, U., Rösli, U. & Toscani, L. 1985. Petrology of ultramafic and mafic rocks from the Lanzo peridotite body (western Alps). *Lithos*, **18**, 201–214.

Rampone, E., Hofmann, A. W., Piccardo, G. B., Vannucci, R., Bottazzi, P. & Ottolini, L. 1995. Petrology, mineral and isotope geochemistry of the External Liguride peridotites (Northern Apennine, Italy). *Journal of Petrology*, **36**, 81–105.

Rampone, E., Hofmann, A. W. & Raczek, I. 1998. Isotopic contrasts within the Internal Liguride ophiolite (N. Italy): the lack of a genetic mantle–crust link. *Earth and Planetary Science Letters*, **163**, 175–189.

Rampone, E., Piccardo, G. B., Vannucci, R. & Bottazzi, P. 1997. Chemistry and origin of trapped melts in ophiolitic peridotites. *Geochimica et Cosmochimica Acta*, **61**, 4557–4569.

Rampone, E., Romairone, A. & Hofmann, A. W. 2004, Contrasting bulk and mineral chemistry in depleted mantle peridotites: evidence for reactive porous flow. *Earth and Planetary Science Letters*, **218**, 491–506.

Ranalli, G., Piccardo, G. B. & Corona-Chavez, P. 2007. Softening of the subcontinental lithospheric mantle by asthenosphere melts and the continental extension/oceanic spreading transition. *Journal of Geodynamics*, **43**, 450–464.

SHAW, D. M. 1970. Trace element fractionation during anatexis. *Geochimica et Cosmochimica Acta*, **34**, 237–243.

SIMON, N. S. C. & PODLADCHIKOV, Y. Y. 2008. The effect of mantle composition on density in the extending lithosphere. *Earth and Planetary Science Letters*, **272**, 148–157; doi: 10.1016/j.epsl.2008.04.027.

VANNUCCI, R., BOTTAZZI, P., WULFF-PEDERSEN, E. & NEUMANN, E.-R. 1998. Partitioning of REE, Y, Sr, Zr and Ti between clinopyroxene and silicate melts in the mantle under La Palma (Canary Islands): implications for the nature of the metasomatic agents. *Earth and Planetary Science Letters*, **158**, 39–51.

WORKMAN, R. K. & HART, S. R. 2005. Major and trace element composition of the depleted MORB mantle (DMM). *Earth and Planetary Science Letters*, **231**, 53–72.

Composite xenoliths from Spitsbergen: evidence of the circulation of MORB-related melts within the upper mantle

MICHEL GRÉGOIRE[1]*, JUNE CHEVET[1,2] & SVEN MAALOE[3]

[1]*DTP, CNRS-UMR 5562, Observatoire Midi Pyrénées, Université Toulouse III, 14 Av. E. Belin, 31400 Toulouse, France*

[2]*GEMOC National Key Center, Macquarie University, Sydney, Australia*

[3]*Naesteboellevej 4, 5953 Tranekaer, Denmark*

**Corresponding author (e-mail: michel.gregoire@dtp.obs-mip.fr)*

Abstract: The Sverrefjell Quaternary volcano in Spitsbergen contains composite xenoliths showing lherzolite rocks cross-cut by websterite veins. These two rock types are characterized by similar major element compositions of olivines, orthopyroxenes, clinopyroxenes and spinels, as well as similar trace element composition for clinopyroxene. The clinopyroxenes of both rock types mostly display upwards convex or spoon-shaped REE (rare earth elements) patterns with a systematic enrichment in La over Ce (Ce_N/Yb_N 0.72–1.32; Sm_N/Yb_N 0.86–1.93 and La_N/Ce_N 1.27–1.93), except for one sample (SV-69) in which clinopyroxenes show a pattern characterized by low LREE compare to HREE (Ce_N/Yb_N 0.33–0.35). Metasomatic processes appear to be the most reasonable origin to form the lherzolite–websterite associations. We therefore propose that the Spitsbergen mantle has undergone at least two events: (1) a sub-alkaline (tholeiitic) metasomatism followed by (2) an alkaline metasomatic event.

This study focuses on composite xenoliths consisting of pyroxenite veins cross-cutting the peridotitic mantle from Spitsbergen. Pyroxenites and associated peridotite–pyroxenite composite xenoliths represent a minor, but significant, fraction of the mantle xenolith collections brought to the surface by alkaline magmas (e.g. Irving 1980; Grégoire *et al.* 1997, 1998; Witt-Eickschen *et al.* 1998). They are witnesses of the circulation of magmas and fluid within the upper mantle, and of the mineralogical and chemical changes associated with this circulation.

The mantle xenoliths from Spitsbergen were entrained by Quaternary alkaline basaltic magmas, whose emplacement are also connected to the activity of the Yermak hot spot, and were classified into two groups (Amundsen 1991). The first group, consisting of Cr-diopside and spinel-bearing lherzolites, corresponds to the fragments of peridotitic mantle. The second group is composed of Al-augite-rich cumulates, which represent magmatic segregates crystallized directly within the upper mantle (Amundsen 1991). Studies have been carried out on the first group to try to understand the different processes affecting the upper mantle under Svalbard. These Cr-diopside and spinel-bearing lherzolites appear to have suffered two major processes; first, an impoverishment caused by the extraction of magmas (partial melting process) and then an enrichment process during mantle metasomatic events. The presence of volatile-rich minerals (amphibole, apatite and phlogopite) in interstitial position, as well as interstitial carbonates associated with clinopyroxenes, and pockets of silicate glasses have particularly attracted the attention of previous studies (Amundsen 1991; Ionov *et al.* 1996; Ionov 1998). The studies on these different mineral phases have led to the conclusion that the upper mantle under the Svalbard has been metasomatized relatively late in its history, by magmas (fluids) displaying carbonate-rich alkaline silicate affinities. But the presence of composite xenoliths showing peridotite cross-cut by pyroxenite veins has only reported and this xenolith-type has never been studied in detail. The composite xenoliths studied in the present paper come from the Sverrefjell Quaternary volcano (Fig. 1). The study also relies on comparisons with previous studies on pristine mantle xenoliths from Spitsbergen, mostly by Ionov *et al.* (2002). The present study evidences the existence of a network of tholeiitic veins in the Spitsbergen upper mantle. This model is innovative because it offers an alternative to the classic interpretation of the patterns of rare earth elements (REE) depleted in light rare earth elements (LREE) of mantle peridotites and their constitutive clinopyroxenes in terms of partial melting residues.

Geological setting

Spitsbergen, with a surface area of 40 000 km^2, is the largest island of the Svalbard archipelago

From: COLTORTI, M., DOWNES, H., GRÉGOIRE, M. & O'REILLY, S. Y. (eds) *Petrological Evolution of the European Lithospheric Mantle*. Geological Society, London, Special Publications, **337**, 71–86.
DOI: 10.1144/SP337.4 0305-8719/10/$15.00

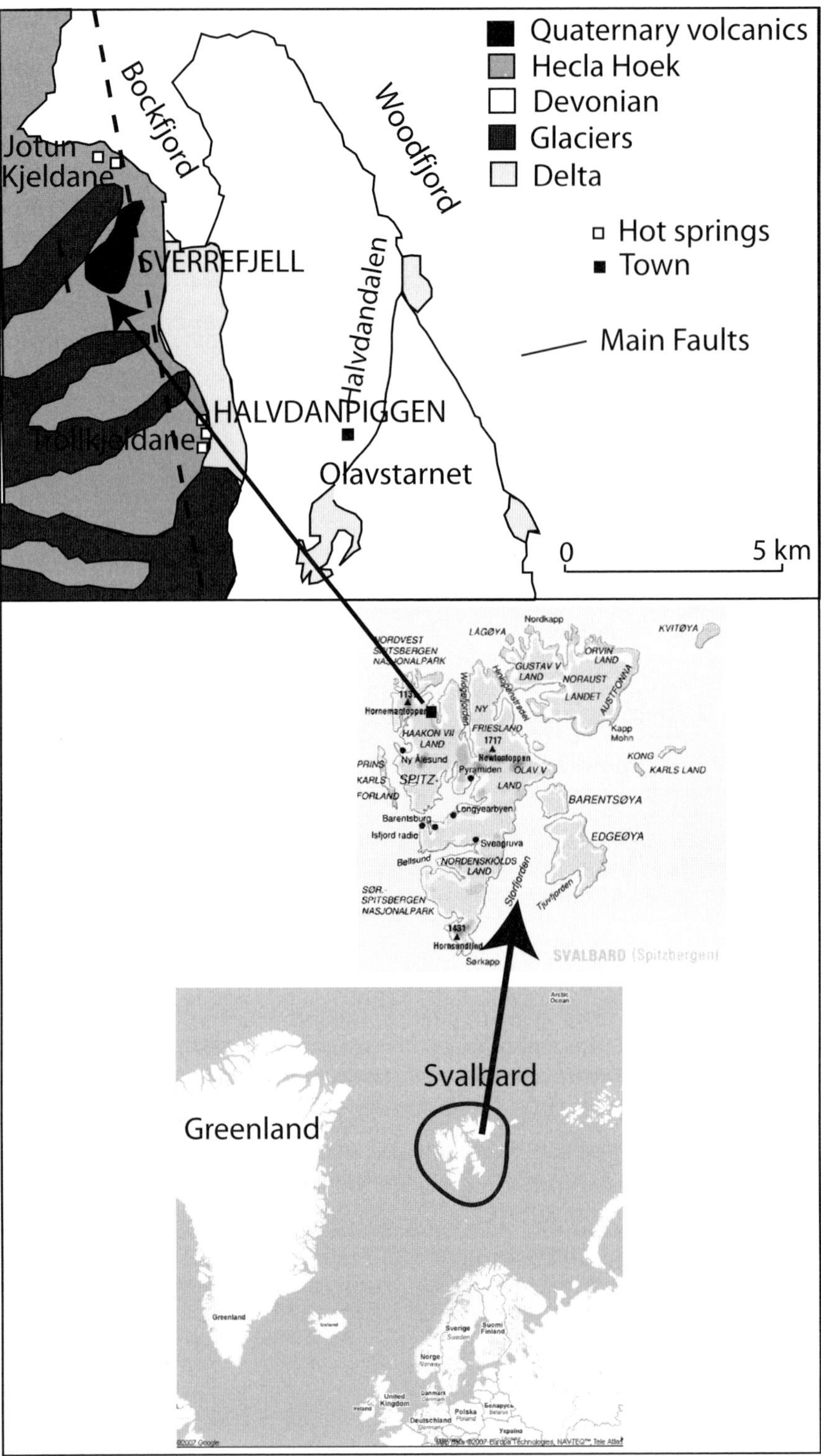

Fig. 1. Location sketch map of Spitsbergen and sampled area of this study (modified from Amundsen 1991).

(Fig. 1). This archipelago lies between northern Norway and the North Pole. Some of the northernmost examples of known mantle lherzolite xenolith localities are found here. The basement rocks consist of continental crust materials corresponding to the Proterozoic and Palaeozoic series, metamorphosed during the Caledonian orogenesis (400–450 Ma). In the north of Spitsbergen a large graben is filled by several kilometres of clastic Devonian red rocks. Some of these rocks are covered by tertiary basaltic flows (Amundsen 1991). In Bockfjord, west of the Devonian graben, there are three centres of Quaternary volcanic activity: Sverrefjell, Sigurdfjell and Halvdanpiggen. These three centres contain abundant upper-mantle and lower-crustal xenoliths (Skjelkvale *et al.* 1989). From a tectonic point of view Svalbard was a stable platform weakly subsiding from the upper Palaeozoic until the Palaeogene (Vagnes & Amundsen 1993). The early continental fragmentation in the Norwegian Greenland Sea resulted in the uplift of West Spitsbergen during the Eocene. However, the sedimentary and geomorphological data indicate that the uplift, marked by mountain peaks on Spitsbergen, is post-orogenic (post-Eocene) and preglacial (pre-Pliocene: Vagnes & Amundsen 1993). The Spitsbergen volcanism is linked to the activity of the Yermak hot spot, launched at the emplacement stage of an Iceland-type ridge, located north of Svalbard (the Nansen ridge). The magnetic anomalies indicate that this hot spot, inactive after anomaly 12, was regenerated 10 Ma ago (anomaly 5), an age coinciding with that of the Tertiary volcanism of Spitsbergen (Amundsen *et al.* 1987).

Petrography

All the samples (SV-70, SV-153-A, SV-230, SV-144, SV-69 and SV-153-B) correspond to spinel lherzolite alternating with spinel websterite veins (Fig. 2a). For this study we only had access to small pieces of sample (a maximum of 4 × 3 cm in size) required to do thin and thick sections. In sample SV-70 the vein is 1 cm wide, while in samples SV-153-A, SV-153-B and SV-230 the veins are about 0.5 cm wide. Sample SV-69 shows two veins, one of about 1 cm width and the other about 0.5 cm wide. Finally, in sample SV-144, a peridotite fragment of about 1 cm width occurs between two veins of websterite (each 1 cm wide). In all samples lherzolitic wall rocks and websteritic veins have the same textures. The types of textures range from porphyroclastic (SV-69, SV-70 and SV-153-B) to granuloblastic (SV-144, SV-153-A, SV-230). The size of porphyroclasts of olivine and pyroxenes (especially orthopyroxenes) ranges from 3 to 5 mm and from 3 to 7 mm, respectively, while undeformed neoblasts vary in size from 0.5 to 1 mm for spinels and olivines, from 0.3 to 2 mm for clinopyroxenes, and from 0.5 to 2 mm for orthopyroxenes.

All studied samples from Spitsbergen are affected by deformation, as evidenced by the numerous undulating extinctions and 'kink-bands' occurring in olivine and orthopyroxene porphyroclasts. The clinopyroxenes often contain exsolutions of orthopyroxene and vice versa. In almost all websteritic veins it is possible to observe rare poikilitic pyroxenes (orthopyroxenes and clinopyroxenes), including crystals of pyroxene, spinel and olivine (Fig. 2b). Very rarely it is possible to observe magmatic twinning in clinopyroxene.

Analytical techniques

Major and trace elements of minerals were analysed at the UMR 5563 (LMTG, Observatoire Midi-Pyrénées) of the University Paul Sabatier (Toulouse III). Major element compositions of minerals were determined with the CAMECA SX50 electron microprobe and a standard program: beam current of 20 nA and an acceleration voltage of 15 kV, 10–30 s of peak counting, 10 s of background counting, and natural and synthetic minerals as standards. Nominal concentrations were subsequently corrected using the PAP data reduction method (Pouchou & Pichoir 1984). The theoretical lower limits of detection are about 100 ppm (0.01%). The concentrations of REE and other trace elements (La, Ce, Pr, Nd, Sm, Eu, Gd, Tb, Dy, Ho, Er, Yb, Lu, Rb, Ba, Th, Sr, Zr, Ti, Y, Ni, V and Sc) of clinopyroxenes were analysed *in situ* on >120 μm-thick polished sections with a Perkin–Elmer Elan 6000 ICP-MS instrument coupled to a CETAC laser ablation module that uses a 266 nm frequency-quadrupled Nd-YAG laser. The NIST 610 and 612 glass standards were used to calibrate relative element sensitivities for the analyses. The analysis was normalized using CaO values determined by electron microprobe. The analyses were performed on intercleavage areas from the cores of clinopyroxene grains in order to obtain homogeneous results unaffected by alteration or exsolution processes. A beam diameter of 50–100 μm and a scanning rate of 20 $\mu m\ s^{-1}$ were used. Typical theoretical detection limits range from 10 to 20 ppb for REE, Ba, Rb, Th, Sr, Zr and Y; 100 ppb for Sc and V; and 2 ppm for Ti and Ni. The typical relative precision and accuracy for a laser analysis ranges from 1 to 10% (see in Dantas *et al.* 2007 for more details).

Mineral compositions

Major elements

Analyses were performed on the main mineral parageneses (so-called primary) in both peridotites and

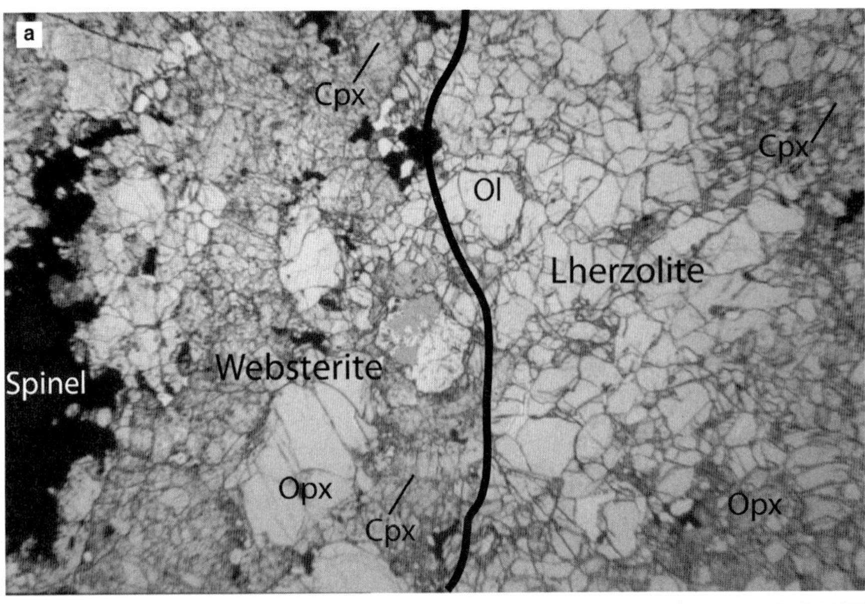

Fig. 2. (**a**) Sample SV-70, contact between the lherzolite wall rock and the websterite vein (horizontal field of view: 2 cm). (**b**) Sample SV-69: poikilitic orthopyroxene displaying inclusions of olivine and spinel in the lherzolite wall rock, a similar feature is sometimes observed in websterite veins (horizontal field of view: 1 cm). Ol, olivine; Cpx, clinopyroxene; Opx, orthopyroxene; Sp, spinel.

pyroxenites, that is, on olivine, clinopyroxene, orthopyroxene and spinel (Table 1).

In composite xenoliths from Spitsbergen the only analysed olivine in a websteritic vein comes from sample SV-69. Its chemical composition is not different from that of olivines analysed in the lherzolite. Mg# in olivines ranges from 88.6 to 90.5. CaO is less than 0.08 wt%, MnO is between 0.10 and 0.20 wt%, and NiO ranges from 0.2 to 0.5 wt%.

All analysed orthopyroxenes are enstatites characterized by a Mg# ranging from 89.5 to 91. Their TiO_2 concentrations are low, from 0.04 to 0.12 wt%, and their Al_2O_3 content ranges from 2.95 to 4.15 wt%.

All analysed clinopyroxenes are diopsides with Mg# ranging from 90.85 to 92.60 (Fig. 3). The concentrations in TiO_2, Cr_2O_3, Al_2O_3 and Na_2O range between 0.25 and 0.95 wt%, 0.3 and 1 wt%, 4.8 and 6.4 wt%, and between 1 and 1.5 wt%, respectively (Fig. 3). The concentrations of major elements in clinopyroxenes from websteritic veins and those from their hosting lherzolitic rocks are very similar. This similarity indicates that either the mineral phases forming the lherzolites and the websterites are totally re-equilibrated or they did not display any compositional difference when the veins were formed.

The spinels of composite xenoliths from Spitsbergen are all Mg–Al chromites (Fig. 4). As for the clinopyroxenes, Mg# of spinels of lherzolite wall rocks and websterite veins are very similar for almost all samples, both ranging from 76.7 to 78.5. The Mg# of spinels in sample SV-70 is slightly different from those of other spinels, that is, 64.30 in the veins and 70.54 in the wall rock. Cr# of spinel also does not display systematic and significant difference between lherzolite wall rocks and websterite veins, ranging from 5.9 to 23 (Fig. 4). Mg# and Cr# of spinel from both rock types are negatively correlated, a well-known feature in chromian spinels from mantle rocks. The TiO_2 content of all the samples ranges from 0.02 to 0.44 wt%, and in each sample this content is similar in the lherzolite wall rock and websterite vein(s). The most magnesian clinopyroxenes of the Spitsbergen composite xenoliths have Mg# similar to those of clinopyroxenes from non-composite peridotite xenoliths from Spitsbergen studied by Furnes *et al.* (1986) and Ionov *et al.* (2002). The fields of Al_2O_3, Na_2O and Cr_2O_3 compositions of clinopyroxenes are similar in our study and in those of Furnes *et al.* (1986) and Ionov *et al.* (2002), the clinopyroxenes of sample SV-69 being slightly more aluminous and those of SV-144 a little less rich in chromium (Fig. 3). The clinopyroxenes of the websterite veins of Patagonian composite mantle xenoliths (websterite–lherzolite: Dantas *et al.* 2009) are lower in MgO, but commonly higher in Na_2O and Al_2O_3 than Spitsbergen clinopyroxenes at similar Cr_2O_3 content. The clinopyroxenes of the lherzolite wall rocks of these Patagonian xenoliths display the same Mg# and Al_2O_3 content than those of Spitsbergen clinopyroxenes, but often have slightly higher Na_2O and Cr_2O_3 content (Fig. 3). The clinopyroxenes of the abyssal pyroxenite veins of the SWIR (SW Indian Ridge) (Dantas *et al.* 2007) are lower in MgO and Na_2O, but higher in Cr_2O_3 than the latter at similar Al_2O_3 content. With regard to the TiO_2 content, only the least magnesian clinopyroxenes of Spitsbergen composite xenoliths are similar to those of SWIR pyroxenites (Fig. 3), the more magnesian clinopyroxenes being richer in TiO_2. The spinels of composite samples from Spitsbergen have Cr# and Mg# similar to those of non-composite peridotite xenoliths from Spitsbergen (Ionov *et al.* 2002), except those of the composite sample SV-70 which are less magnesian. Compared to spinels of the pyroxenite veins from SWIR (Dantas *et al.* 2007), those of the composite Spitsbergen samples are more magnesian, except those of sample SV-70. They display similar Cr#, except those of sample SV-144 and those of the veins of sample SV-69 (Fig. 4).

Equilibration temperatures

The studied composite xenoliths from Spitsbergen all equilibrated in the spinel peridotite pressure–temperature P–T conditions. The upper limit of the spinel peridotite facies corresponding to the transition to the plagioclase peridotite facies lies between 0.5 and 0.7 GPa at 800 °C and between 0.7 and 0.9 GPa at 1100 °C. Its lower limit corresponding to the transition to the garnet peridotite facies lies between 1.5 and 1.6 GPa at 800 °C and between 1.6 and 1.7 GPa at 1100 °C (Herzberg 1978; Gasparik 1984; Wood & Holloway 1984). It should be noted that according to some authors an increase in the $Cr/Al + Cr$ ratio from 0.1 to 0.7 raises the pressure in the spinel peridotite field by about 0.28 GPa in the temperature range considered (Chatterjee & Terhart 1985; Webb & Wood 1986).

In our study we considered Wells (1977) and Brey & Köhler (1990) geothermometers, both based on the pair orthopyroxene–clinopyroxene. For all studied composite xenoliths the equilibrium temperatures are between 870 and 995 °C using the thermometer of Wells (1977) and between 855 and 975 °C with that of Brey & Köhler (1990). These estimated temperatures are generally lower than those reported for spinel lherzolites xenoliths in the same region by Amundsen *et al.* (1987), which are between 940 and 1170 °C. But our temperature intervals are very similar to those proposed by Ionov *et al.* (2002), that is, 840–970 °C using Wells (1977) and 885–1015 °C using Brey & Köhler (1990).

Table 1. *Representative major element compositions of the minerals of Spitsbergen composite xenoliths: Mg#: 100 × $Mg/(Mg + Fe_{total})$; Cr#: 100 × $Cr/(Cr + Al)$, both in mol%*

	Lherzolite						Websterite
Olivine	**SV-70-1**	**SV-69**	**SV-144-3**	**SV-153-A**	**SV-230**	**SV-153-B**	**SV-69**
SiO_2	40.99	41.48	41.47	41.42	41.44	42.06	41.50
TiO_2	–	–	0.01	–	0.01	–	–
Al_2O_3	–	–	–	–	–	–	–
Cr_2O_3	0.09	0.02	0.03	0.05	–	0.02	–
FeO	11.18	10.28	10.37	9.16	9.56	9.38	10.33
MnO	0.15	0.13	0.18	0.11	0.155	0.19	0.19
MgO	48.73	48.94	49.04	49.25	49.51	49.20	49.07
CaO	0.06	0.01	0.06	0.03	0.05	0.02	0.08
NiO	0.50	0.36	0.46	0.42	0.17	0.41	0.40
Total	101.71	101.23	101.61	100.44	100.90	101.27	101.58
Mg#	88.59	89.45	89.39	90.55	90.22	90.51	89.43

	Lherzolite				Websterite			
Opx	**SV-70-1**	**SV-153-A**	**SV-230**	**SV-153-B**	**SV-69**	**SV-144-3**	**SV-153-A**	**SV-153-B**
SiO_2	56.38	56.34	56.34	56.61	56.85	55.97	56.82	57.03
TiO_2	0.04	0.08	0.06	0.05	0.11	0.06	0.12	0.13
Al_2O_3	2.96	3.43	3.42	3.40	3.26	4.14	3.54	3.53
Cr_2O_3	0.38	0.32	0.36	0.38	0.26	0.13	0.04	0.36
FeO	6.86	6.03	6.01	5.96	6.65	6.79	6.29	5.96
MnO	0.18	0.10	0.13	0.11	0.20	0.17	0.19	0.15
MgO	33.31	33.80	33.72	33.77	33.19	33.13	33.48	33.91
CaO	0.63	0.63	0.60	0.58	0.46	0.61	0.65	0.59
Na_2O	–	–	–	–	–	–	–	–
K_2O	0.01	0.05	–	–	0.01	–	0.02	–
Total	100.74	100.72	100.62	100.86	100.81	100.97	101.50	101.66
Mg#	89.64	90.90	90.90	90.99	89.89	89.69	90.46	91.02

Spinel	Lherzolite SV-144-3	Lherzolite SV-70-1	Lherzolite SV-230	Lherzolite SV-153-A	Websterite SV-144-3	Websterite SV-70-1	Websterite SV-230	Websterite SV-153-A	Websterite SV-153-B	Websterite SV-69
SiO_2	–	0.14	0.05	0.06	0.04	0.07	0.05	–	0.02	0.04
TiO_2	0.02	0.44	0.14	0.1	0.02	0.25	0.14	0.10	0.14	0.05
Al_2O_3	61.92	45.45	55.47	55.08	60.11	45.92	52.48	54.10	53.98	58.90
Cr_2O_3	5.77	20.20	12.97	13.16	5.79	17.42	13.95	12.16	14.39	7.64
FeO	10.58	14.07	10.24	10.734	10.99	16.45	10.82	10.42	10.52	10.32
MnO	0.09	0.07	0.05	0.13	0.11	0.10	0.13	0.13	0.15	0.17
MgO	21.12	18.94	20.96	20.59	21.39	17.40	20.31	20.17	20.42	20.59
NiO	0.29	0.28	0.38	0.3	0.48	0.24	0.25	0.37	0.38	0.37
Total	99.96	100.00	100.47	100.33	99.27	98.35	98.39	97.66	100.18	98.23
Mg#	78.06	70.54	78.45	77.33	77.61	65.33	76.99	77.52	77.57	78.05
Cr#	5.89	22.97	13.56	13.82	6.07	20.28	15.13	13.10	15.17	8.01

Cpx	Lherzolite SV-70-1	Lherzolite SV-69	Lherzolite SV-144-3	Lherzolite SV-153-A	Lherzolite SV-230	Lherzolite SV-153-B	Websterite SV-70-1	Websterite SV-69	Websterite SV-144-3	Websterite SV-153-A	Websterite SV-230	Websterite SV-153-B
SiO_2	52.14	52.40	53.38	52.72	52.97	52.27	53.81	52.69	52.94	52.69	52.88	52.93
TiO_2	0.49	0.96	0.26	0.58	0.68	0.83	0.35	0.86	0.36	0.62	0.61	0.74
Al_2O_3	4.78	6.38	5.41	5.73	5.86	6.24	4.97	6.00	5.39	5.46	5.71	6.00
Cr_2O_3	0.83	0.48	0.38	0.79	0.86	0.98	0.95	0.52	0.30	0.73	0.91	0.88
FeO	2.69	2.35	2.77	2.49	2.46	2.43	2.83	2.34	2.78	2.35	2.23	2.35
MnO	0.03	0.06	0.03	0.12	0.12	0.10	0.09	0.09	0.11	0.03	0.05	0.15
MgO	15.50	14.99	15.88	15.81	15.56	15.44	15.82	15.20	15.75	15.63	15.59	15.50
CaO	21.46	22.16	22.35	21.86	21.38	21.28	21.49	22.10	22.16	21.72	21.35	21.44
Na_2O	1.32	1.51	1.11	1.21	1.28	1.46	1.15	1.51	1.06	1.21	1.38	1.25
K_2O	–	0.01	0.01	0.01	0.01	0.01	–	0.01	0.03	0.02	–	0.04
Total	99.24	101.41	101.55	101.38	101.15	101.11	101.47	101.30	100.93	100.42	100.72	101.28
Mg#	91.11	91.91	91.07	91.89	91.84	91.88	90.87	92.03	90.99	92.20	92.55	92.14

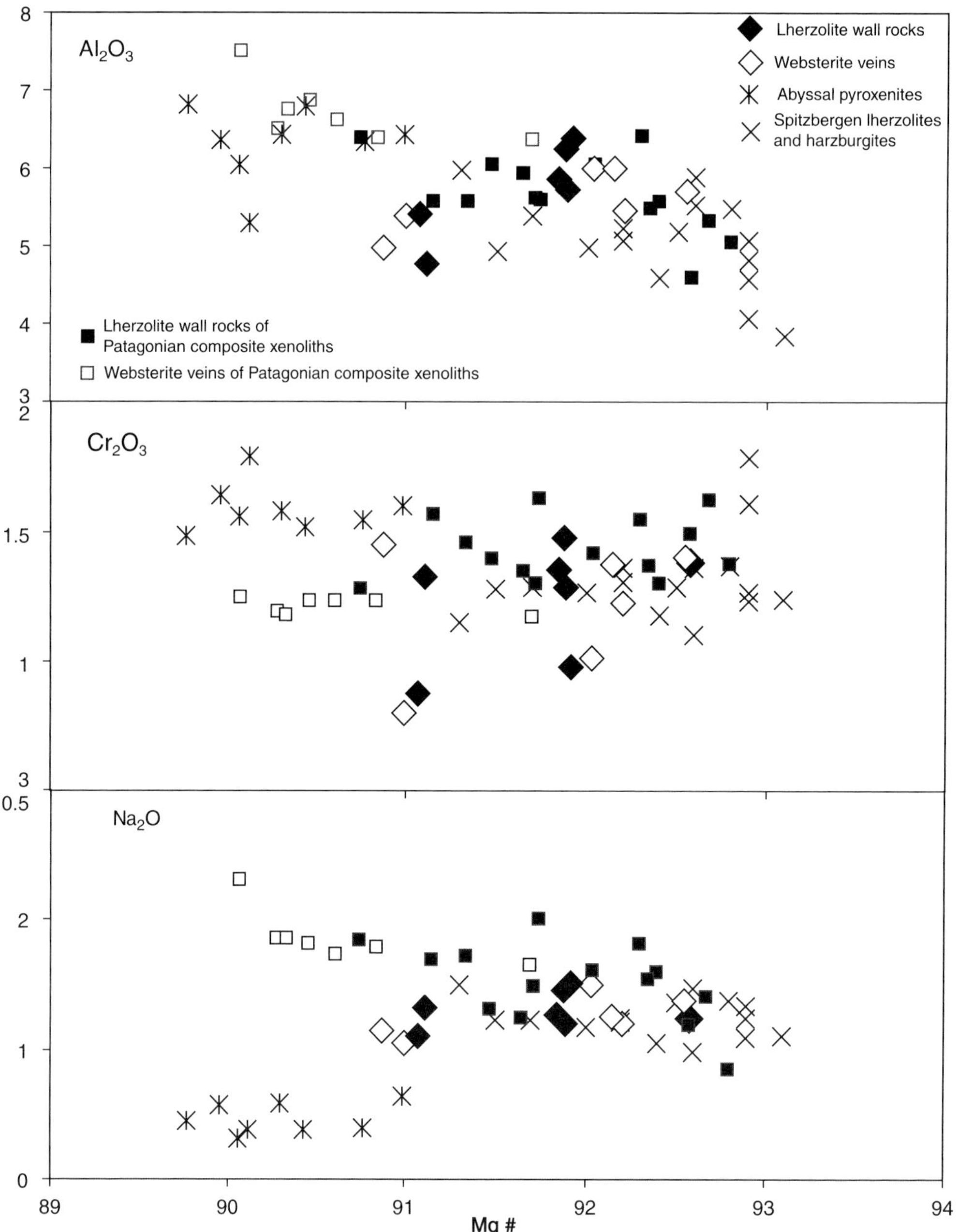

Fig. 3. Diagrams of Al_2O_3, Cr_2O_3 and Na_2O v. Mg# [$100 \times Mg/(Mg + Fe_{total})$ mol%] of clinopyroxenes of Spitsbergen composite xenoliths. Filled diamonds, lherzolite wall rocks; Empty diamonds, websterite veins; filled squares, lherzolite wall rocks of Patagonian composite xenoliths from Dantas *et al.* (2009); empty squares, websterite veins of Patagonian composite xenoliths from Dantas *et al.* (2009); crosses, Spitsbergen mantle lherzolite and harzburgite xenoliths from Ionov *et al.* (2002); double crosses: abyssal pyroxenites from SWIR from Dantas *et al.* (2007).

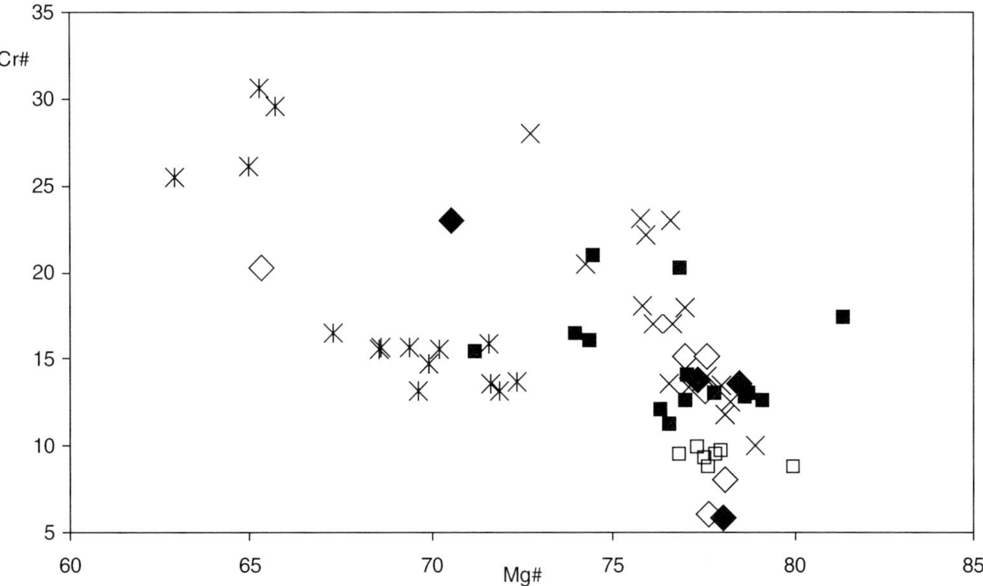

Fig. 4. Cr# [100 × Cr/(Cr + Al) mol%] v. Mg# of spinels of Spitsbergen composite xenoliths. For the key to the symbols see the legend to Figure 3.

Trace elements

For each composite xenolith sample of Spitsbergen the concentrations of trace elements were measured *in situ* in clinopyroxenes from websteritic veins and lherzolitic rocks (Table 2). The REE patterns of clinopyroxenes from veins are very similar to those of clinopyroxenes from wall rocks. The clinopyroxenes from samples SV-70, SV-144, SV-153-A, SV-153-B and SV-230 are characterized by globally upwards convex or spoon-shaped REE patterns with a systematic enrichment in La over Ce (Ce_N/Yb_N 0.72–1.32; Sm_N/Yb_N 0.86–1.93 and La_N/Ce_N 1.27–1.93: Fig. 5). All these different patterns are very close to each other in terms of REE concentration level, only those of clinopyroxenes in samples SV-70 and SV-144 being more depleted. The clinopyroxenes of sample SV-69 are characterized by LREE-depleted REE patterns (Ce_N/Yb_N being 0.33 in clinopyroxenes from the two veins and 0.35 in those of the lherzolite). The extended trace element spectra of the clinopyroxenes of samples SV-70, SV-144, SV-230, SV-153-A and SV-153-B are relatively flat, except for the negative Rb, Ba, Nb, Ta and Ti anomalies (Fig. 6). The clinopyroxenes of sample SV-69 are also characterized by negative Rb, Ba, Nb, Ta and Ti anomalies, giving them a similar shape to those of other composite xenoliths, except for the LREE that are more depleted in this sample (Fig. 6d).

The clinopyroxenes of the abyssal pyroxenites from SWIR (Dantas *et al.* 2007), such as those of our SV-69 composite xenolith, are LREE depleted. The depletion, however, is much greater in the SWIR clinopyroxenes than in the SV-69 clinopyroxenes (Fig. 5). Ionov *et al.* (2002), in their study focusing on pristine mantle peridotite xenoliths from Spitsbergen, distinguished two types of clinopyroxenes on the basis of their trace element content: type 1 (enriched in LREE compared to MREE and HREE) and type 2 (enriched in LREE and MREE compared to HREE), which do not coexist in the same sample. They explain the chemical characteristics of the two types of clinopyroxenes using a two-step model corresponding to the percolation of a single carbonate-rich silicate metasomatic agent rich in highly incompatible trace elements within a peridotitic protolith that had previously been depleted by partial melting. Volume changes in this agent, as well as differences in the distance between the peridotites and the magmatic sources of this agent (veins), would explain the differences observed between the two types of clinopyroxenes. In this model the type 1 clinopyroxenes of Ionov *et al.* (2002) occur within peridotites that were located further from the melt source and which were then circulated by lower volumes than those containing type 2, which were closer to the source.

Ionov *et al.* (2002) also estimated the two extreme bulk-rock trace element compositions of the residual peridotites before the metasomatic event. It appears that the REE patterns of the clinopyroxenes in our sample SV-69 and that of the less residual peridotite (called ‘fertile’ in Fig. 6d)

Table 2. *Averaged trace element compositions of the clinopyroxenes of Spitsbergen composite xenoliths*

Clinopyroxene	Lherzolite	Websterite	Lherzolite	Lherzolite	Websterite	Websterite	Lherzolite	Lherzolite	Websterite	Lherzolite	Websterite	Lherzolite	Websterite
Element	SV-70	SV-70	SV-125	SV-69	SV-69	SV-144	SV-144	SV-153-A	SV-153-A	SV-230	SV-230	SV-153-B	SV-153-B
No. of analyses	3	4	4	5	4	5	3	5	4	3	4	2	4
Rb	0.29	bdl	0.19	0.27	0.41	bdl	bdl	bdl	bdl	bdl	bdl	0.41	bdl
Ba	0.09	0.35	0.10	0.07	0.07	bdl	0.08	0.06	0.25	0.07	0.12	0.07	0.92
Th	0.25	0.12	0.29	0.03	0.02	0.32	0.29	0.96	0.93	0.88	0.82	0.87	0.80
U	0.28	0.21	0.29	0.02	0.04	0.09	0.10	0.29	0.32	0.28	0.28	0.25	0.32
Nb	0.05	0.03	0.50	0.02	0.03	0.16	bdl	0.06	0.06	0.06	0.01	0.05	0.08
Ta	0.02	0.03	0.05	bdl	0.01	0.06	0.04	0.05	0.06	0.06	0.03	0.06	0.05
La	3.73	2.71	2.50	0.61	0.64	2.76	2.76	6.69	6.54	6.97	6.98	6.94	6.40
Ce	6.90	5.29	5.11	3.00	3.24	4.20	4.25	9.07	8.73	9.55	9.60	10.08	9.08
Sr	98	79	60	47	50	66	66	104	111	117	105	125	117
Nd	6.15	5.16	3.61	5.86	6.57	3.19	3.26	9.60	9.97	9.30	8.83	10.45	9.39
Zr	37	31	30	51.66	49	17	17	73	64	65	61	63	63
Hf	1.11	1.28	0.99	1.61	1.77	0.41	0.41	2.00	2.10	1.56	1.82	1.51	1.60
Sm	2.05	1.82	1.51	2.68	3.02	1.24	1.23	3.18	3.33	3.24	3.02	3.55	3.25
Eu	0.82	0.70	0.65	1.07	1.21	0.53	0.51	1.33	1.29	1.19	1.17	1.16	1.22
Ti	2250	2531	2464	5629	5653	2234	2008	4181	4630	4367	3710	4543	4252
Gd	2.90	2.50	2.36	3.77	4.41	1.82	1.85	4.19	4.12	4.09	3.82	3.53	3.88
Dy	3.55	3.01	2.90	4.54	4.86	2.34	2.41	4.55	4.53	4.07	3.85	3.95	3.90
Ho	0.69	0.62	0.63	0.88	0.99	0.51	0.53	0.89	0.91	0.80	0.81	0.79	0.75
Er	1.83	1.65	1.69	2.41	2.70	1.46	1.50	2.14	2.36	2.20	2.03	2.24	2.01
Yb	1.70	1.51	1.56	2.27	2.59	1.46	1.56	2.10	2.19	1.98	1.92	2.10	1.82
Y	15	15	16	22	23	14	14	22	22	22	21	23	22
Lu	0.23	0.20	0.22	0.32	0.35	0.22	0.22	0.31	0.33	0.29	0.27	0.31	0.27

Abbreviation: bdl, below detection limit.

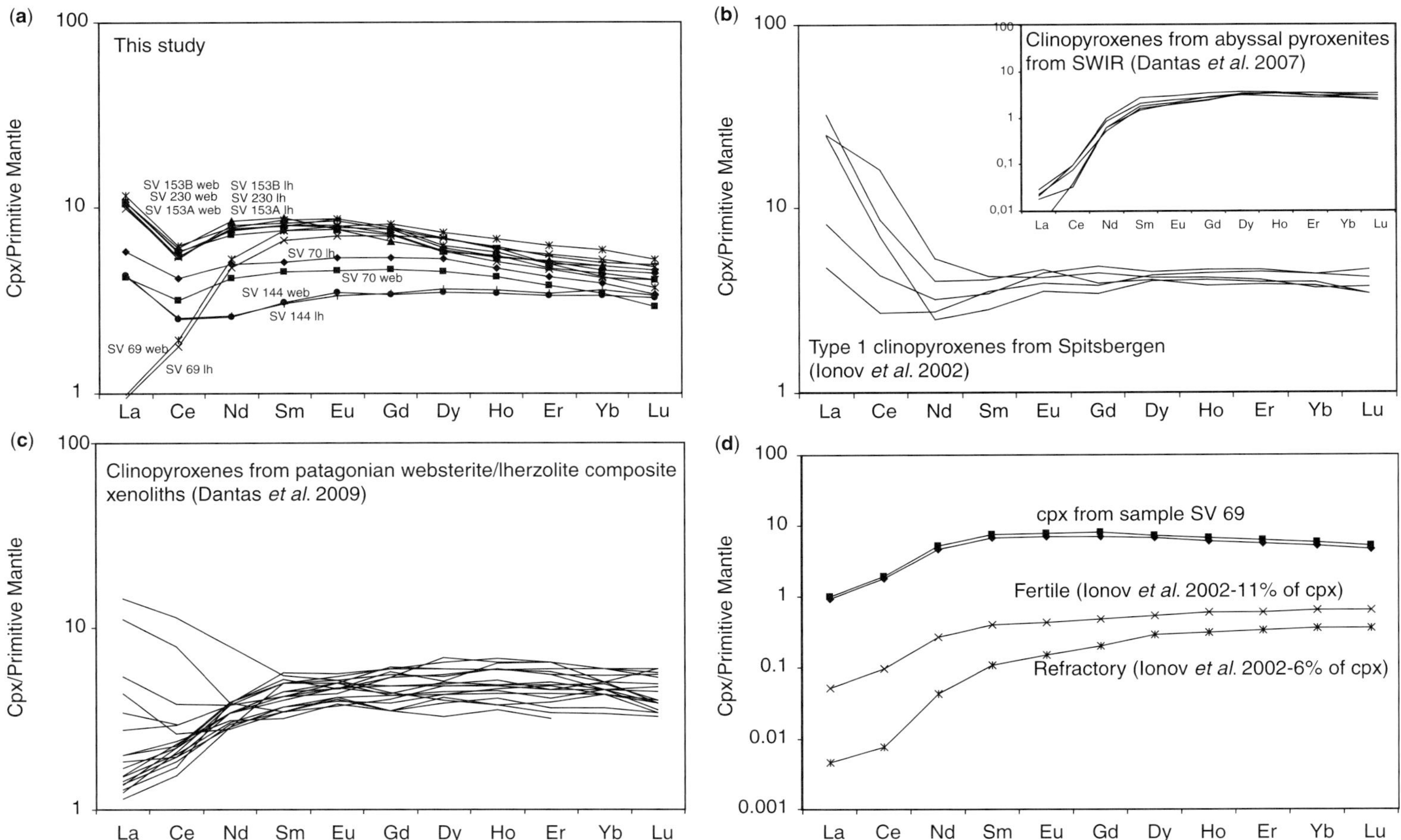

Fig. 5. Primitive mantle-normalized REE patterns of clinopyroxenes from: (**a**) this study; (**b**) Ionov *et al.* (2002) and Dantas *et al.* (2007; insert); (**c**) Dantas *et al.* (2009); and (**d**) comparison between Cpx of sample SV69 (this study) with the two extreme estimated trace element compositions of residual peridotites before the metasomatic event (see Ionov *et al.* 2002 and the text). Normalization values are from McDonough & Sun (1995).

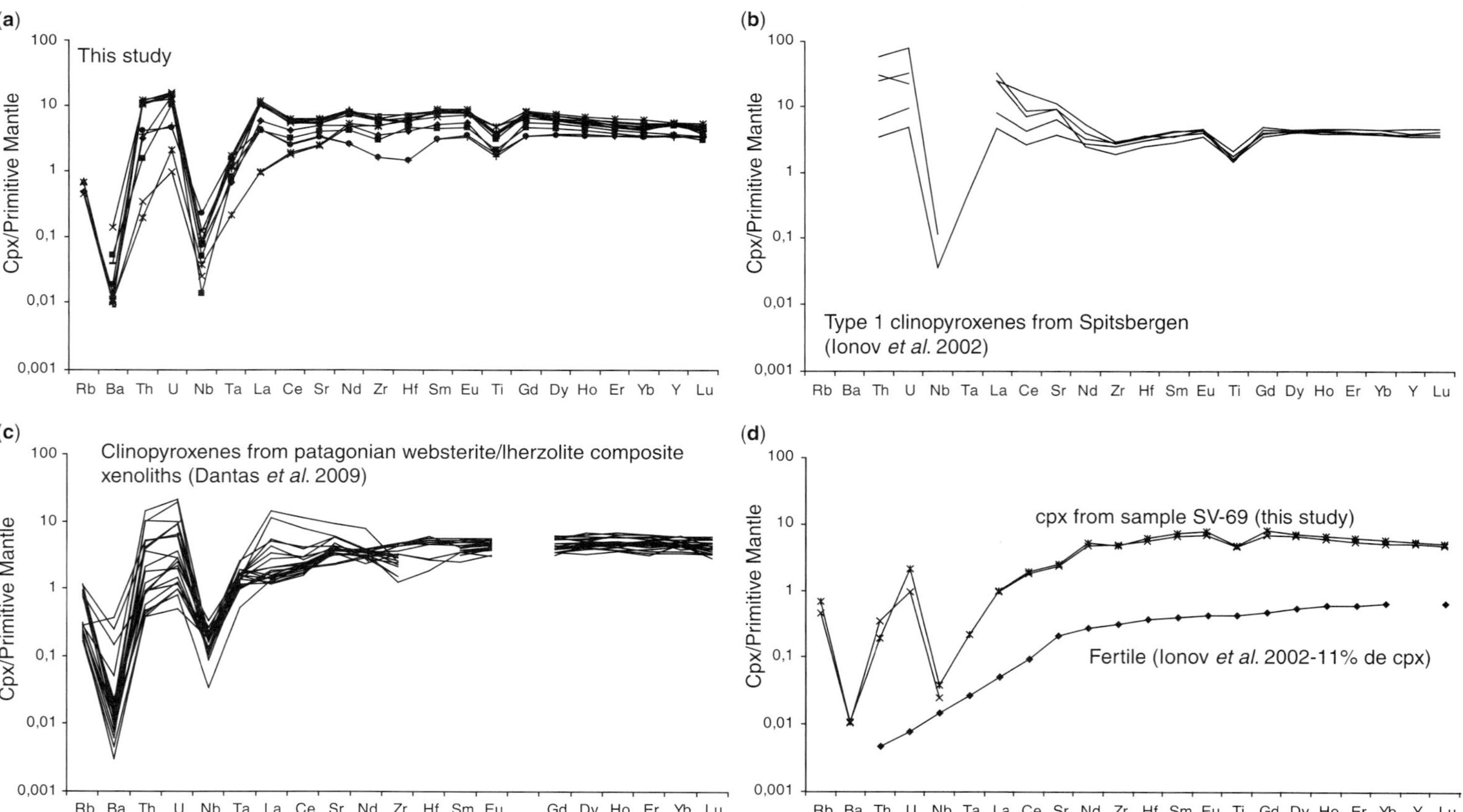

Fig. 6. Primitive mantle-normalized trace element patterns of clinopyroxenes from: (**a**) this study (the same symbols as in Fig. 5a); (**b**) Ionov *et al.* (2002); (**c**) Dantas *et al.* (2009); and (**d**) comparison between Cpx of sample SV-69 (this study) with the more fertile estimated trace element composition of residual peridotites before the metasomatic event (see Ionov *et al.* 2002 and the text). Normalization values are from McDonough & Sun (1995).

estimated by Ionov *et al.* (2002) are very similar in shape and only show different levels of concentration when estimated whole-rock content to clinopyroxene content are compared (Fig. 5). However, if we refer to the extended trace element patterns, there are clear differences between those of the SV-69 clinopyroxenes and that of the less residual peridotite of Ionov *et al.* (2002).

Finally, the REE patterns as well as the extended trace element patterns of the clinopyroxenes of the spitsbergen composite xenoliths display similarities to those of the clinopyroxenes of Ionov *et al.*'s (2002) type 1 (Figs 5 & 6). They also display similarities with those of the clinopyroxenes of the websterite–lherzolite composite mantle xenoliths from two localities in Patagonia (Cerro de los Chenques and Cerro Clark: Dantas *et al.* 2009) (Figs 5 & 6).

Discussion

It has to be noted that we worked on samples of only a few centimetres in size and all corresponding to websterite veins surrounded by only the first centimetres of the lherzolite wall rock. There is, therefore, no evidence of any modification of the percolating melt due to reactive flow through the peridotite. This is in good agreement with the fact that there is no significant difference in shape in the same sample between the trace element patterns of websteritic clinopyroxenes and those of lherzolitic clinopyroxenes. Such a feature precludes an origin of the different trace element patterns from one sample to another by fractionation/reactive flow/chromatographic column percolation of one single melt during a single percolation event.

Major and trace element compositions of the majority of studied composite samples are similar to those of the pristine mantle peridotites from Spitsbergen studied by Ionov *et al.* (2002). More specifically, most of the clinopyroxenes of composite samples have trace element compositions similar to those of type 1 clinopyroxenes of Ionov *et al.* (2002). For those type 1 xenoliths, Ionov *et al.* (2002) proposed a two-step model (see earlier). We agree with the second stage of that model and propose that most of the composite xenoliths from Spitsbergen, except sample SV-69, have suffered the same metasomatic event related to the circulation of carbonate-rich silicate melts within the upper mantle. With regard to the first step of the model, Ionov *et al.* (2002) link the depletion in the most incompatible trace element to a partial melting event, and estimate the trace element compositions of mantle peridotites after this episode of partial melting and before that of enrichment (metasomatism). The REE pattern of the clinopyroxenes from the composite websterite–lherzolite sample SV-69 is very similar in terms of shape to those estimated for the less residual peridotite protolith (Fig. 5). Sample SV-69 could, therefore, represent one of the initial theoretical premetasomatic peridotites of Ionov *et al.* (2002). However, composite sample SV-69 may not be a simple residue of partial melting because it contains veins that are tracks of the flow of melts in the mantle (e.g. Dantas *et al.* 2007). It therefore appears that another process other than partial melting should be considered to explain the LREE depleted patterns of clinopyroxenes. This LREE depletion could be caused by tholeiitic melts (MORB-like) circulating within the upper-mantle peridotites. To test this hypothesis, we calculated the REE pattern of the theoretical melt in equilibrium with the clinopyroxenes of sample SV-69 using the clinopyroxene–tholeiitic basalt partition coefficients of Fujimaki & Tatsumoto (1984). We note that the theoretical liquid in equilibrium with the SV-69 clinopyroxene displays the same REE pattern as those of some tholeiitic basalts from the North Atlantic both in terms of their shape and level of concentration (Fig. 7) (Thirwall *et al.* 1994; Fram *et al.* 1998; Brookins *et al.* 1999; Hansen & Nielsen 1999; Phillip *et al.* 2001; Peate *et al.* 2003) while those from SWIR are more LREE-depleted, as expected. Indeed, the SWIR pyroxenites are cumulates of the latest, shallowest, incremental melt fractions produced during fractional decompression melting of a N-MORB (mid-ocean ridge basalt) mantle source and not fully mixed with melt fractions produced and extracted at greater depths, whereas the Spitsbergen composite xenoliths have been circulated by MORB-like melts that are the final output from the abyssal mantle after mixing. The similarity between the theoretical liquid in equilibrium with the SV-69 clinopyroxene and some North Atlantic tholeiitic basalts allows us to propose a new first step of formation to replace Step 1 of Ionov *et al.* (2002). This first step is probably linked to the process of the opening of the North Atlantic that began about 60 Ma ago (Tegner & Duncan 1999) in the area of Svalbard. The LREE-depleted feature of the clinopyroxenes of the mantle xenoliths (this study and Ionov *et al.* 2002) might be due not simply to an episode of partial melting, but to the circulation of tholeiitic melts forming veins and reacting with the surrounding peridotites. These mantle rocks would then be metasomatized (Step 2 of Ionov *et al.* 2002) by carbonate-rich silicate melts, leading to the re-enrichment in the most incompatible trace elements of the clinopyroxenes, particularly in LREE. This second step proposed by Ionov *et al.* (2002) for the pristine mantle peridotites of Spitsbergen is reinforced by the REE patterns (but not that of sample SV-69), as well as by the

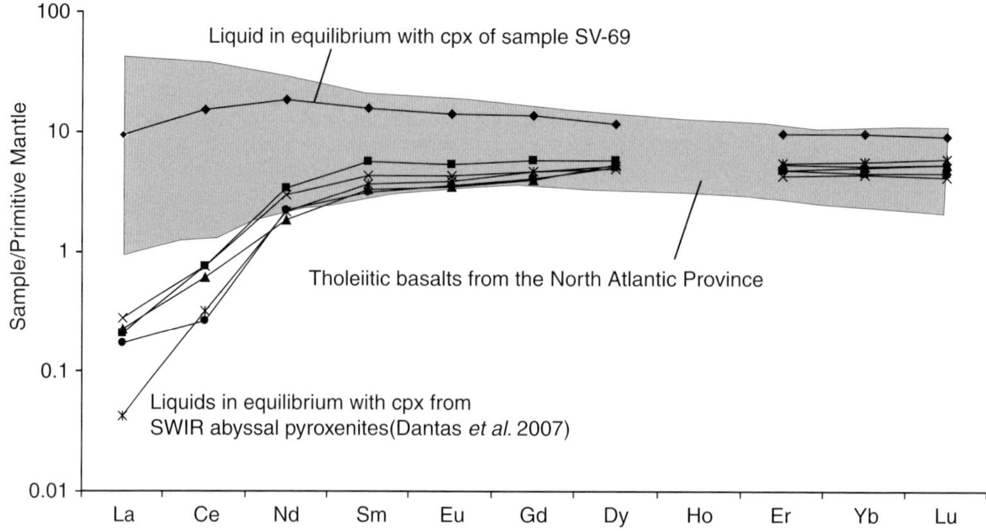

Fig. 7. Primitive mantle-normalized REE patterns of tholeiitic basalts from the North Atlantic (Thirwall *et al.* 1994; Fram *et al.* 1998; Brookins *et al.* 1999; Hansen & Nielsen 1999; Phillip *et al.* 2001; Peate *et al.* 2003), of theoretical liquid in equilibrium with the clinopyroxenes (Cpx) of the composite sample SV-69 (this study) and of theoretical liquid in equilibrium with those of the abyssal pyroxenites from SWIR (Dantas *et al.* 2007). Normalization values are from McDonough & Sun (1995).

extended trace element patterns of all the clinopyroxenes in the composite xenoliths of our study, including that of sample SV-69. Our model is very similar to the one proposed by Dantas *et al.* (2009) for a similar type of composite mantle xenolith (websterite–lherzolite) from Patagonia. Indeed, Dantas *et al.* (2009) propose that Cerro de los Chenques and Cerro Clark (Argentinian part of Patagonia) mantle composite xenoliths underwent two metasomatic events: (1) a sub-alkaline (tholeiitic) metasomatism responsible for the formation of the websterite veins followed by (2) an alkaline metasomatic event responsible for the enrichment of LREE and some other incompatible trace elements in clinopyroxenes.

The process of tholeiitic mantle metasomatism invoked in the present study is comparable with the refertilization process proposed by Piccardo (2010) for the Lanzo peridotitic massif (Italian Western Alps), where a subcontinental lithospheric mantle is progressively fertilized by MORB-like melts related to lithosphere extension that led to the rifting and opening of the Jurassic Ligurian Tethys. The Lanzo massif and Spitsbergen have undergone comparable rifting processes during their history (Vagnes & Amundsen 1993; Amundsen *et al.* 1987; Piccardo 2010), and therefore we propose that this kind of extension process should lead to significant tholeiitic (MORB-) metasomatism of the upper mantle involved in such extension zones.

Conclusions

The upper mantle beneath Spitsbergen contains mantle composite xenoliths consisting of spinel-bearing lherzolite wall rocks cross-cut by spinel-bearing olivine websterite veins. The major and trace elements of the two rock types are very similar and point to metasomatic processes as being the most reasonable origin to form the lherzolites–websterites association. In such a model websterites probably represent veins of focused melt circulation, and lherzolites the upper-mantle host rock percolated by the same melt (e.g. Bedini *et al.* 1997; Bodinier *et al.* 2008; Dantas *et al.* 2009).

Based on the petrological characteristics of the studied xenoliths and on comparisons with non-composite mantle xenoliths from Spitsbergen and Patagonia and abyssal mantle websterites we propose that the Spitsbergen mantle has undergone at least two events: (1) a sub-alkaline (tholeiitic) metasomatism leading to the formation of the websterite veins followed by (2) a carbonate-rich silicate metasomatic event. The second event corresponds to the second step of the model by Ionov *et al.* (2002). In their model they related their type 1 samples to an upper mantle located at a greater distance from the source of the metasomatic agent (veins of carbonate-rich silicate liquid) by comparison to the type 2 samples. In such a model our composite mantle rocks would also be located at a relatively greater distance from these veins. An example of such veins, which are obviously different from the tholeiitic websterite veins from the present study, may correspond to composite sample 4-90-1 of Ionov *et al.* (2002) that consisted of an amphibole wehrlite grading into a coarse-grained olivine-dominated rock that contains amphibole and phlogopite.

The present study, as well as that of Piccardo (2010), clearly demonstrate that features such as the REE-depleted patterns of clinopyroxenes that are classically related to partial melting events could, in fact, be linked to reaction processes between mantle peridotites and tholeiitic (MORB-related) melts. Future work on similar upper-mantle rocks will have to take this possibility into account, and also try to understand (unravel) the chemical reactions and change in physical properties linked to such tholeiitic metasomatic events.

H. E. F. Amundsen kindly supplied the specimens. Thin sections, electron microprobe analyses and ICP-MS analyses were performed using the facilities of the Observatoire Midi-Pyrénées, University of Toulouse III (France). We are particularly indebted to F. de Parseval and J.-F. Mena for thin-section preparation, and to P. de Parseval, F. Candaudap and R. Freydier for their help during microprobe and LA-ICP-MS data acquisition. This work was financially supported by the French Centre National de la Recherche Scientifique (CNRS). We warmly thank M. Coltorti, N. Simon and an anonymous reviewer for their constructive comments and suggestions.

References

AMUNDSEN, H. E. F. 1991. *Igneous Processes and Lithosphere Evolution: Evidence From Upper Mantle and Lower Crustal Xenoliths From Northwestern Spitsbergen and the Canary Islands.* PhD thesis, Mineralogisk-Geologisk Museum Universitet I oslo.

AMUNDSEN, H. E. F., GRIFFIN, W. L. & O'REILLY, S. Y. 1987. The lower crust and upper mantle beneath northwestern Spitsbergen: evidence from xenoliths and geophysics. *Tectonophysics*, **39**, 169–185.

BEDINI, R. M., BODINIER, J.-L., DAUTRIA, J.-M. & MORTEN, L. 1997. Evolution of LILE-enriched small melt fractions in the lithospheric mantle: a case study from the East African Rift. *Earth and Planetary Science Letters*, **153**, 67–83.

BODINIER, J. L., GARRIDO, C. J., CHANEFO, I., BRUGUIER, O. & GERVILLA, F. 2008. Origin of pyroxenite–peridotite veined mantle by refertilization reactions: evidence from the Ronda peridotite (Southern Spain). *Journal of Petrology*, **49**, 999–1025.

BREY, G. P. & KÖHLER, T. 1990. Geothermobarometry in four-phase lherzolites II. New thermobarometers, and pratical assessment of existing thermobarometers. *Journal of Petrology*, **31**, 1353–1378.

BROOKINS, D. G., REHACEK, J. & MORRIS, G. 1999. Data report: major and trace element composition, strontium, neodymium, and oxygen isotope ratios, and mineral compositions of samples. *Proceedings Ocean Drilling Program, Scientific Results*, **163**, 113–117.

CHATTERJEE, W. D. & TERHART, L. 1985. Thermodynamic calculations of peridotite phase relations in the system $MgO–Al_2O_3–SiO_2–Cr_2O_3$, with some geological applications. *Contributions to Mineralogy and Petrology*, **102**, 422–428.

DANTAS, C., CEULENEER, G., GRÉGOIRE, M., PYTHON, M., FREYDIER, R., WARREN, J. & DICK, H. B. J. 2007. Pyroxenites dredged along the south-west Indian Ridge, 9°–16°E: cumulates from incremental melt fractions produced at top of a cold melting regime. *Journal of Petrology*, **48**, 647–660.

DANTAS, C., GRÉGOIRE, M., KOESTER, E., CONCEIÇÃO, R. V. & RIECK, N. JR. 2009. The lherzolite–websterite xenolith suite from Northern Patagonia (Argentina): evidence of mantle–melt reaction processes. *Lithos*, **107**, 107–120.

FRAM, M. S., LESHER, C. E. & VOLPE, A. M. 1998. Mantle melting systematics: transition from continental to oceanic volcanism on the southeast greenland margin. *Proceedings of Ocean Drilling Program, Scientific Results*, **152**, 373–386.

FUJIMAKI, H. & TATSUMOTO, M. 1984. Lu–Hf constraints on the evolution of lunar basalts. *Journal of Geophysical Research*, **89**(B1), 445–459.

FURNES, H., PEDERSEN, R. B. & MAALOE, S. 1986. Petrology and geochemistry of spinel peridotite nodules and host basalt, Vestspitsbergen. *Norsk Geologisk Tidsskrift*, **66**, 53–68.

GASPARIK, T. 1984. Two-pyroxene thermobarometry with new experimental data in the system $CaO–MgO–Al_2O_3–SiO_2$. *Contributions to Mineralogy and Petrology*, **87**, 87–97.

GRÉGOIRE, M., COTTIN, J. Y., GIRET, A., MATTIELLI, N. & WEIS, D. 1998. The metaigneous xenoliths from Kerguelen archipelago: evidence of continent nucleation in an oceanic setting. *Contributions to Mineralogy and Petrology*, **133**, 259–283.

GRÉGOIRE, M., LORAND, J. P., COTTIN, J. Y., GIRET, A., MATTIELLI, N. & WEIS, D. 1997. Petrology of Kerguelen mantle xenoliths: evidence of a refractory oceanic mantle percolated by basaltic melts beneath the Kerguelen archipelago. *European Journal of Mineralogy*, **9**, 1085–1100.

HANSEN, H. & NIELSEN, T. F. D. 1999. Crustal contamination in paleogene east Greenland flood basalts: plumbing system evolution during continental rifting. *Chemical Geology*, **157**, 89–118.

HERZBERG, C. T. 1978. Pyroxene geothermometry and geobarometry: experimental and thermodynamic evaluation of some subsolidus phase relations involving pyroxenes in the system $CaO–MgO–Al_2O_3–SiO_2$. *Geochimica et Cosmochimica Acta*, **42**, 945–957.

IONOV, D. 1998. Trace element composition of mantle-derived carbonates and coexisting phases in peridotite xenoliths from alkali basalts. *Journal of Petrology*, **39**(11–12), 1931–1941.

IONOV, D. A., BODINIER, J. L., MUKASA, S. B. & ZANETTI, A. 2002. Mechanisms and sources of mantle metasomatism: major and trace element compositions of peridotite xenoliths from Spitsbergen in the context of numerical modelling. *Journal of Petrology*, **43**, 2219–2259.

Ionov, D. A., O'Reilly, S. Y., Kopylova, M. G. & Genshaft, Y. S. 1996. Carbonate-bearing mantle peridotite xenoliths from Spitsbergen: phase relationships, mineral compositions and trace element residence. *Contributions to Mineralogy and Petrology*, **125**, 375–392.

Irving, A. J. 1980. Petrology and geochemistry of composite ultramafic xenoliths in alkalic basalts and implications for magmatic processes within the mantle. *American Journal of Science*, **280A**, 389–426.

McDonough, W. F. & Sun, S. 1995. The composition of the Earth. *Chemical Geology*, **120**, 223–253.

Peate, D. W., Baker, J. A. et al. 2003. The prinsen of wales bjerge formation lavas, east greenland: the transition from tholeiitic to alkalic magmatism during paleogene continental breakup. *Journal of Petrology*, **44**, 279–304.

Phillip, H., Eckhardt, J.-D. & Puchelt, H. 2001. Platinum-Group Elements (PGE) in basalts of the seaward-dipping reflector sequence, SE greenland coast. *Journal of Petrology*, **42**, 407–432.

Piccardo, G. B. 2010. The Lanzo peridotite massif, Italian Western Alps: Jurassic rifting of the Ligurian Tethys. *In*: Coltorti, M., Downes, H., Grégoire, M. & O'Reilly, S. Y. (eds) *Petrological Evolution of the Lithospheric Mantle*. Geological Society, London, Special Publications, **337**, 47–69.

Pouchou, J. L. & Pichoir, F. 1984. A new model for quantitative X-ray microanalysis. Part 1: application to the analysis of homogeneous samples. *Recherche Aérospatiale*, **5**, 13–38.

Skjelkvale, B. L., Amundsen, H. E. F., O'Reilly, S. Y., Griffin, W. L. & Gjelsvik, T. 1989. A primitive alkali basaltic stratovolcano and associated eruptive centres, northwestern Spitsbergen: volcanology and tectonic significance. *Journal of Volcanology and Geothermal Research*, **37**, 1–19.

Tegner, C. & Duncan, R. A. 1999. $^{40}Ar-^{39}Ar$ Chronology for the volcanic history of the southeast greenland rifted margin. *Proceedings of the Ocean Drilling Program, Scientific Results*, **163**, 53–62.

Thirwall, M. F., Upton, B. G. J. & Jenkins, C. 1994. Interaction between continental lithosphere and the Iceland plume – Sr–Nd–Pb isotope geochemistry of tertiary basalts, NE greenland. *Journal of Petrology*, **35**, 839–879.

Vagnes, E. & Amundsen, H. E. F. 1993. Late cenozoic uplift and volcanism on Spitsbergen: caused by mantle convection? *Geology*, **21**, 251–254.

Webb, C. S. A. & Wood, B. J. 1986. Spinel–pyroxene–garnet relationships and their dependance on Cr/Al ratio. *Contributions to Mineralogy and Petrology*, **92**, 471–480.

Wells, P. R. A. 1977. Pyroxene thermometry in sample and complex system. *Contributions to Mineralogy and Petrology*, **62**, 129–139.

Witt-Eickschen, G., Kaminsky, W., Kramm, U. & Harte, B. 1998. The nature of young vein metasomatism in the lithosphere of West Eifel (Germany): geochemical and isotopic constraints from composite mantle xenoliths from the Meerfelder maar. *Journal of Petrology*, **39**, 155–185.

Wood, B. J. & Holloway, J. R. 1984. A thermodynamic model for subsolidus equilibria in the system $CaO-MgO-Al_2O_3-SiO_2$. *Geochimica et Cosmochimica Acta*, **48**, 109–124.

Insights into the origin of mantle graphite and sulphides in garnet pyroxenites from the External Liguride peridotites (Northern Apennine, Italy)

ALESSANDRA MONTANINI[1]*, RICCARDO TRIBUZIO[2,3] & DANILO BERSANI[4]

[1]*Dipartimento di Scienze della Terra, Università di Parma, via G.P. Usberti 157A, I-43100 Parma, Italy*

[2]*Dipartimento di Scienze della Terra, Università di Pavia, via Ferrata 1, I-27100 Pavia, Italy*

[3]*Centro di Studio per la Cristallochimica e per la Cristallografia, CNR, via Ferrata 1, I-27100 Pavia, Italy*

[4]*Dipartimento di Fisica, Università di Parma, Via G.P. Usberti 7A. I-43100 Parma, Italy*

**Corresponding author (e-mail: alessandra.montanini@unipr.it)*

Abstract: This paper describes a rare occurrence of graphite in non-cratonic mantle rocks. Graphite has been found in garnet clinopyroxenite layers from the External Liguride peridotites that represent slices of subcontinental lithospheric mantle exhumed at the ocean floor in Mesozoic times. The high-pressure assemblage of the pyroxenites is characterized by garnet + Al–Na-rich clinopyroxene, and testifies to an early stage of equilibration at approximately 2.8 GPa and 1100 °C. Graphite occurs as small dispersed euhedral flakes and stacks of flakes. Structural characterization by microRaman spectrometry indicates a highly ordered structure, compatible with a high-temperature mantle origin. C isotope composition of graphite has a typical mantle signature. Fe–Ni–Cu sulphides occur as accessory phases, both as blebs enclosed in silicates (E-Type) and interstitial grains (I-Type). The sulphide assemblage (Ni-free pyrrhotite, pentlandite, Cu–Fe sulphides) mainly reflects subsolidus exsolution from high-temperature Fe–Ni–Cu monosulphide solid solutions with variable Ni (up to 18 wt%) and Cu content (up to 7 wt%). The origin of E- and I-Type sulphides requires the existence of an immiscible Fe–Ni–Cu sulphide liquid, which segregated from the partial melt of the garnet pyroxenite. Graphite precipitation in the pyroxenite was presumably related to the reduction of a more oxidized carbon species interacting with the sulphide liquid as a reducing agent.

The occurrence of elemental carbon in the mantle, either as diamond or more rarely as graphite, is mostly confined to peridotite and eclogite xenoliths derived from the subcratonic lithospheric mantle (Pearson *et al.* 1994; Haggerty 1995*b*), although the reasons for this tectonic restriction remain poorly understood. The carbon source, that is, deep mantle origin v. crustal recycling, and the mechanism of elemental carbon precipitation have been the subject of a lively debate in recent decades, and a wealth of different genetic hypotheses have been proposed on the basis of diamond inclusion assemblages, stable isotope data and experimental work (Haggerty 1986, 1995*b*; Bulanova 1995; Cartigny *et al.* 1998; Navon 1999; Deines 2002; Gunn & Luth 2006; Thomassot *et al.* 2007).

In non-cratonic areas, carbon is mainly present in more oxidized forms, that is, CO_2-rich fluid inclusions in mantle minerals at low pressure and, sporadically, as carbonates (Haggerty 1995*b*; Deines 2002; Luth 2003). Graphite-bearing mantle rocks were reported for the orogenic peridotite massifs of Ronda (Davies *et al.* 1993) and Beni Bousera (Pearson *et al.* 1993), where macrocrystalline graphite occurs in some garnet pyroxenite layers as pseudomorphs after diamond, veins or disseminated flakes. Occasionally, graphite has been found in some metasomatized peridotite xenoliths (Kaeser *et al.* 2007). On the basis of ^{13}C-depleted isotopic composition, the Ronda–Beni Bousera graphite was interpreted as having originated from crustally-derived, biogenic carbon recycled into mantle through subduction (Pearson *et al.* 1993, 1994), and different mechanisms of carbon concentration were proposed to explain the various graphite occurrences (Crespo *et al.* 2006). The study of these restricted occurrences of native carbon has implications for the nature and distribution of volatiles in the non-cratonic lithospheric mantle and for the Earth volatile budget.

Sulphide minerals are common in mantle rocks (Morgan 1986; Haggerty 1995*a*), where they may

From: Coltorti, M., Downes, H., Grégoire, M. & O'Reilly, S. Y. (eds) *Petrological Evolution of the European Lithospheric Mantle*. Geological Society, London, Special Publications, **337**, 87–105.
DOI: 10.1144/SP337.5 0305-8719/10/$15.00

record partial melting and/or magmatic crystallization following immiscibility processes (Haggerty 1995*a*; Luth 2003). The investigation of sulphide minerals may provide clues about mantle processes in addition to the silicate host petrology. Sulphides are known to control the HSE (highly siderophile elements) budget of the mantle (e.g. Pattou *et al.* 1996; Alard *et al.* 2000), and sulphide metasomatism through pyroxenite partial melting has recently been invoked to explain the highly variable Os isotope systematics in the mantle sources of MORB (mid-ocean ridge basalt) (Alard *et al.* 2005; Luguet *et al.* 2008). Remarkably, only a few studies have been carried out on the petrogenesis of sulphide assemblages in pyroxenites, either in tectonically exposed ultramafic massifs (Lorand 1989*a*, *b*) or as xenoliths (De Waal & Calk 1975; Dromgoole & Pasteris 1987; Szabo & Bodnar, 1995; Guo *et al.* 1999).

We report a new occurrence of high-temperature graphite in mantle rocks from the Alpine orogenic system, i.e. the External Liguride garnet pyroxenites from the Northern Apennines (Montanini *et al.* 2006*a*). The pyroxenites occur as layers in subcontinental lithospheric mantle that represents a rare tectonic sampling of deep levels of subcontinental lithosphere exhumed at an ocean–continent transition in Mesozoic times. Although largely affected by decompression-related evolution, pristine high-pressure assemblages and textures are locally preserved. In this work the mode of occurrence and characteristics of the graphite will be illustrated along with the petrology of associated sulphides to provide constraints on the graphite origin and on the carbon provenance.

Geological and petrological framework

The External Liguride (EL) mantle bodies belong to the Northern Apennine ophiolites. These ophiolites are lithosphere remnants of the Jurassic Ligurian Tethys ocean and occur as huge blocks in sedimentary mélanges of Late Cretaceous age. They are locally associated with continental crust rocks that testify to a fossil ocean–continent transition similar to that of modern non-volcanic continental margins (Marroni *et al.* 1998). The peridotites are Ti-pargasite-bearing spinel–plagioclase lherzolites with pyroxenite layers and a fertile geochemical signature. They represent subcontinental lithospheric mantle of the former Adria–Europe system (Rampone *et al.* 1995), most probably accreted in the Proterozoic (see also Snow *et al.* 2000). Recent studies have shown that several EL peridotite bodies have been affected by refertilization by asthenospheric melts during the formation of the Ligurian Tethys in Mesozoic times (Piccardo *et al.* 2004; Piccardo 2008).

In the northern part of the External Liguride domain, the mantle rocks consist of partially serpentinized mylonitic peridotites with widespread pyroxenite layering (Montanini *et al.* 2006*a*). The pyroxenites are concordant with the foliation of the host peridotite and range in thickness from a few millimetres up to about 2 m; sharp contacts between pyroxenites and peridotites are observed. The most common layers are represented by boudinaged websterites, with a thickness of 2–10 cm, whereas metre-sized layers of garnet clinopyroxenites locally occur. These rocks, in particular, preserve an early stage of equilibration at approximately 2.8 GPa and 1100 °C. In the enclosing peridotites, the oldest recognizable texture is a spinel-facies low-strain tectonite, widely overprinted by plagioclase-facies recrystallization associated with the development of a mylonitic fabric along hectometre-size shear zones. The high-temperature decompression was subsequently overprinted by polyphase brittle deformation under decreasing temperature conditions, coupled with hydration that has been related to a brittle detachment evolution leading to continental break-up and mantle exhumation at the sea floor. The timing of the garnet- to spinel-facies decompression is unconstrained. Conversely, Lu–Hf and Sm–Nd cooling ages obtained on garnet pyroxenites (Montanini *et al.* 2006*a*) indicate that the shallow ($P < 0.9$ GPa) portion of the exhumation was related to Upper Triassic–Lower Jurassic rifting that led to continental break-up.

Analytical methods

The structural characterization of graphite was carried out by means of X-ray diffraction (XRD) and Raman spectroscopy. The XRD analyses were obtained with a Philips PW diffractometer, using Cu–Kα radiation at 40 kV and 20 mA, a step size of 0.02° (2 theta), and time per step of 1 s. The micro-Raman spectroscopic study was carried out with Jobin–Yvon Horiba LABRAM equipment equipped with an Olympus metallographic microscope. The excitation was carried out using an He–Ne Ar laser beam (632.8 nm) focused through the microscope objective. Nominal spatial resolution was around 1 μm. Position, height and width of the disorder peak, D, and order peak, O, were evaluated. In addition, second-order Raman spectra were recorded from 2500 to 3100 cm^{-1}. Graphite concentrates were obtained by manual separation of crystals from fresh surfaces, trying to avoid mechanical damage of graphite flakes that may induce an increased degree of disorder (Pasteris 1981). The graphite grains were oriented so that the laser beam was normal to (001) plane, in order to prevent any increase of D peak intensity.

The bulk stable carbon isotope ratio of graphite separates from two pyroxenites was measured at 'Activation Laboratories' (Ancaster, Ontario). Powdered sample was digested with anhydrous phosphoric acid in a Y-tube reaction vessel at 25 °C. The evolved CO_2 was cryogenically distilled from the reaction vessel into a 6 mm Pyrex tube and flame sealed. The CO_2 gas was then allowed into the ion source of a VG SIRA-10 mass spectrometer and analysed for the $^{13}C/^{12}C$ ratio. Internal standards ('*Lublin*' carbonate), periodically calibrated against NBS 19 International Standard, were run at the beginning and end of each set of samples, and used to normalize the data as well as to correct for any instrument drift. Precision and reproducibility using this technique is typically better than $\pm 0.2‰$ ($n = 10$ internal standards). All results are reported in the permil notation relative to the international PDB standard.

Mineral analyses were carried out at Dipartimento di Scienze della Terra (Università di Parma) using a JEOL-6400 electron microprobe equipped with LINK-ISIS300 energy-dispersive microanalytic system and software, which allow quantitative analyses of bulk specimens scanning the electron beam over a selected free-hand area to be obtained. Operating conditions were an accelerating voltage of 15 kV and a probe current of 0.25 nA; both minerals and synthetic compounds were used as standards.

Petrography and mineralogy of the graphite-bearing garnet pyroxenites

The pyroxenite layers include garnet websterites and Fe-rich and Mg-garnet clinopyroxenites. Graphite is restricted to the Fe-rich garnet clinopyroxenites. The enclosing peridotites and the other types of pyroxenites do not contain any vein or disseminated graphite. Field relations, petrography, mineral chemistry and tectono-metamorphic [pressure–temperature (P–T)] evolution of the graphite-bearing garnet pyroxenites have been described in detail in a previous paper (Montanini *et al.* 2006*a*). Here, the most relevant characteristics and findings will be summarized.

The garnet pyroxenite samples of this study mainly come from a thick layer (*c.* 0.8 m thick) from Rio Strega (Montanini *et al.* 2006*a*). Four samples (AM321, AM322, BAR and BA2B) were collected from different adjacent portions of the centre of the Rio Strega layer. Two samples (AM407 and AM340) were collected from the same edge of the layer, at a distance of approximately 50 cm between each other. In addition, this work considers one graphite-bearing garnet clinopyroxenite clast (E181) sampled from a polygenic breccia from Rio Parola (sample location in Montanini *et al.* 2006*a*).

The inner part of the thickest layer is made up of isotropic, coarse-grained graphite-bearing rocks, whereas the external portions display a weak foliation concordant with the enclosing mylonitic lherzolites and do not contain graphite. The high-pressure protoliths were characterized by an anhydrous bimineralic assemblage, consisting of pinkish Al-augite and red-orange garnet (Fig. 1a, b) with accessory Fe–Ni–Cu sulphides and rutile.

Primary textures and mineralogy of the garnet pyroxenites are, in places, affected by the decompressional evolution (Montanini *et al.* 2006*a*). The retrograde transformations started with a breakdown of the garnet + Al-rich clinopyroxene assemblage (Fig. 1a) into a garnet-free two-pyroxene + spinel + plagioclase assemblage. Subsequent deformation and dynamical recrystallization of the newly formed assemblage is indicated by sheared pyroxene porphyroclasts mantled by fine-grained two-pyroxene + spinel ± plagioclase aggregates. A later decompression stage is recorded by the local formation of olivine + plagioclase after two-pyroxene + spinel.

Garnet-facies clinopyroxene has a high Al_2O_3 and Na_2O content (up to 12.6 and 2.7 wt%, respectively) corresponding to high proportions of a Ca-Tschermak component (15–20 mol%) and moderate jadeite (10–16 mol%). The $Mg/(Mg + Fe^{2+})$ ratio varies in the range of 0.76–0.83 and Cr_2O_3 content is low (0.10–0.25 wt%). Similar clinopyroxene compositions were reported for Group II garnet pyroxenites from Beni Bousera (Kornprobst *et al.* 1990). Garnet composition covers the range $Prp_{53-45}Alm_{29-38}Grs_{11-20}$. The low TiO_2 (0.10–0.20 wt%) and Na_2O concentrations ($\leq$0.02 wt%) are comparable to those of the diamond-free Group II eclogites defined by McCandless & Gurney (1989). New representative analyses of primary clinopyroxene and garnet are reported in Table 1.

Results

Structural and chemical features of graphite

Graphite occurs as small dispersed euhedral flakes and stacks of flakes with grain sizes up to 2–3 mm in the inner part of the Rio Strega layer and in clast E181. Graphite distribution is very irregular. When present, its modal abundance does not exceed 0.5 vol.%. Scanning electron microscope (SEM) and optical microscope observations of graphite morphology on fresh surfaces (Fig. 1c) did not reveal any occurrence of pseudomorphs with octahedral or cubic symmetry indicative of the former occurrence of diamond, in agreement

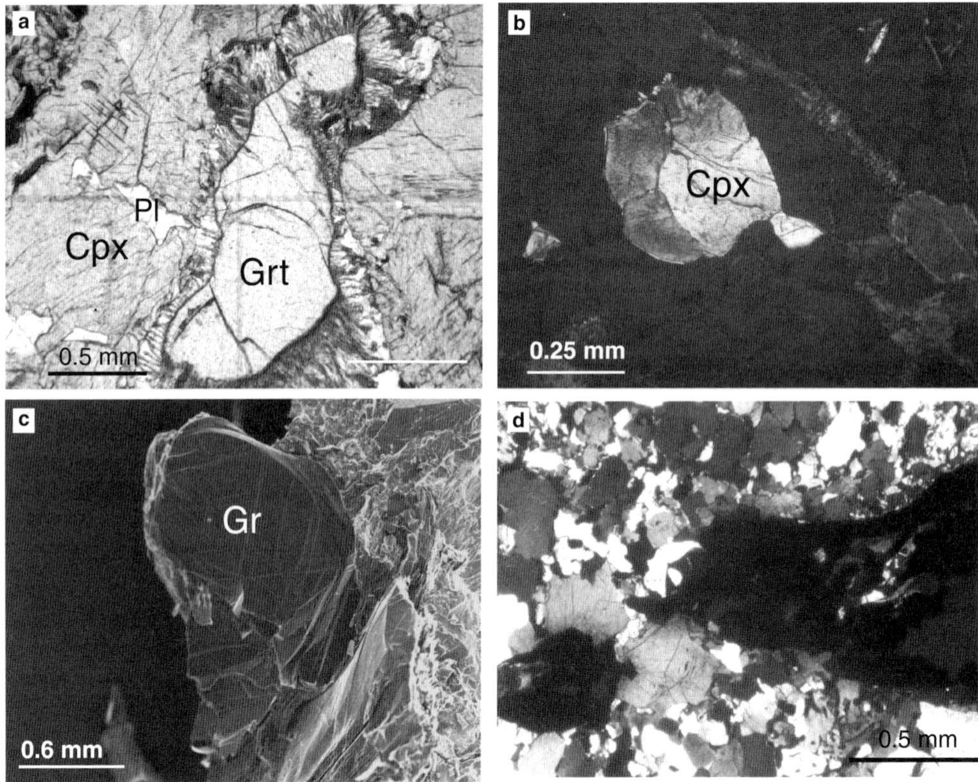

Fig. 1. Photomicrographs of garnet pyroxenite textures and graphite occurrence. Photographs were obtained with an optical microscope under plane-polarized (PPL) and crossed polarized (CPL) light and as secondary electron images (SE). (**a**) Garnet (Grt) with reaction corona formed by radial orthopyroxene–spinel–plagioclase symplectites, sample AM407 (PPL); (**b**) primary Al–Na-rich clinopyroxene (Cpx) enclosed in garnet (black), sample AM322 (CPL); (**c**) graphite flakes on a fresh surface with characteristics striations on the basal face, sample BAR (SE); (**d**) oriented graphite flakes in weakly deformed and retrogressed pyroxene + plagioclase + spinel domain (Cpx + Opx + Spl + Pl), sample AM407 (PPL).

Table 1. *Representative electron microprobe analyses of garnet and primary clinopyroxene*

Sample	AM322		BA2		AM407	M340		E181	
Phase	Grt	Cpx	Grt	Cpx	Grt	Grt	Cpx	Grt	Cpx
SiO_2	40.70	49.78	41.28	48.61	39.93	40.54	48.70	40.22	47.98
TiO_2	0.13	1.00	0.07	1.12	0.14	0.19	0.82	0.15	0.78
Al_2O_3	22.10	11.60	21.96	12.26	21.50	21.98	11.04	23.06	12.63
Cr_2O_3	0.13	0.09	0.17	0.22	0.10	0.19	0.23	0.12	0.07
FeO^T	15.24	5.05	17.23	4.88	19.39	17.04	5.85	15.70	6.43
MnO	1.48	0.12	0.24	0.15	0.59	0.40	0.06	0.30	0.21
CaO	12.96	12.00	12.22	11.92	6.35	12.65	11.17	12.72	11.31
MgO	7.14	17.67	6.77	18.60	11.41	6.64	20.36	7.81	18.92
Na_2O	bd	2.44	bd	1.71	bd	bd	1.48	bd	1.44
Total	99.88	99.72	99.94	99.47	99.41	99.63	99.71	100.08	99.77

Abbreviations: bd, below detection limit (i.e. <0.10 wt%); Grt, garnet; Cpx, clinopyroxene.

with the $P-T$ estimates reported in Montanini *et al.* (2006*a*). In the retrogressed and deformed samples, graphite is stretched along a weak plagioclase-bearing foliation (Fig. 1d).

The structural characterization of graphite by XRD shows that the basal spacing $d_{(002)}$ is close to 3.35 Å. Diffraction peaks of rhombohedral graphite were not recorded in the diffraction patterns. These features are characteristic of well-ordered graphite, in agreement with the results of Raman analyses described below.

Representative micro-Raman spectra obtained on graphite from sample AM321 are reported in Figure 2a. The first-order spectra show a narrow O peak at 1578.61 cm^{-1}, and a weak or absent disorder band D (1345–1350 cm^{-1}), with $I_D/I_O \leq 0.015$, corresponding to in-plane crystallite size $L_a \geq 2000$ Å (Wopenka & Pasteris 1993). The second-order spectra (S peak) show a resolvable doublet at about 2690 cm^{-1} (Fig. 2a). First-order Raman spectra have been measured for comparison on both graphite flakes and octahedral pseudomorphs after diamond in a garnet pyroxenite sample from the Beni Bousera massif (Fig. 2b, c). The Beni Bousera graphite similarly displays an intense O peak; the small and relatively broad peak observed at about 1331 cm^{-1} (absent in the graphite grains considered in the present study) is most probably the shifted D-defect band of relatively disordered graphite formed after diamond as a result of dispersion related to the employed He–Ne laser wavelength (see Beyssac & Chopin 2003).

Stable C isotope analyses of graphite measured in garnet pyroxenites BAR and E181 are reported in Table 2. The measured C composition ($\delta^{13}C = -4.3 \pm 0.2$, -4.5 ± 0.2‰) is close to typical $\delta^{13}C$ mantle values known for MORB–OIB (ocean island basalt) CO_2, peridotitic diamonds and carbonate from kimberlites and carbonatites (*c.* −5‰, Deines 2002; −4.5 ± 0.5‰, Thomassot *et al.* 2008 and quoted references).

Sulphide assemblage petrography and mineralogy

Sulphide minerals examined in 25 polished thin sections are very scarce, both in the thick pyroxenite layer and in the garnet pyroxenite clast E181. In most cases no more than 10 grains per thin section have been observed. Only in one sample from the edge of the layer (AM407) may their modal abundance reach approximately 1–2 vol.%.

Sulphides have been subdivided in two types on the basis of their textural occurrence, i.e. minute grains enclosed in silicates (E-Type) or larger interstitial grains (I-Type). Back-scattered SEM observations and X-ray compositional mapping show that most E-Type and I-Type sulphides are inhomogeneous polyphase grains composed mainly of Fe–Ni- and Fe-domains with minor Cu–Fe-rich domains; monomineralic grains are uncommon.

E-Type sulphides (Fig. 3a–d) are rare and small, ranging in size between 20 and 100 μm. Between five and 10 sulphide inclusions are commonly observed per thin section. They are mainly enclosed in large garnet crystals and, occasionally, in the preserved primary clinopyroxene (sample BA2B). They often have an irregular shape, show small apophyses and can be surrounded by smaller sulphide grains with rounded or vermicular shapes (Fig. 3d). The latter are occasionally associated with composite sulphide-fluid inclusions (Fig. 3e). Unfortunately, only the sulphide could be identified (as chalcopyrite) by micro-Raman analysis; the fluid had probably leaked due to the host garnet fracturing. In sample AM322, sulphide blebs with rounded (spheroidal or elongated) or polygonal shape were observed (Fig. 3b). Tiny, irregularly-shaped sulphides are locally present in the pyroxene–spinel–plagioclase symplectitic intergrowths after garnet breakdown or in the pyroxene derived from the annealing and coarsening of these symplectites. E-Type sulphides are composed of pyrrhotite (generally dominant) + pentlandite ± Cu-rich sulphide. Pentlandite (up to *c.* 40 vol.%) occurs as patches or domains with irregular shape, or more rarely, as lamellae and rims. Cu-rich sulphides (0–15–20 vol.%) are generally observed as discontinuous rims at the edge of the Fe–Ni sulphides (Fig. 3c), as commonly reported for Cu sulphides in polyphase grains (De Waal and Calk 1975; Lorand & Conquéré 1983; Szabo & Bodnar 1995; Alard *et al.* 2000; Torok *et al.* 2003; Powell & O'Reilly 2007). Monomineralic pentlandite inclusions in garnet were also found. One sulphide bleb (Fig. 3a) is a two-phase inclusion with coexisting Ni-rich (pentlandite) and Cu-rich sulphide. Trails of minute (<5 μm) Fe–Ni sulphide globules are sometimes included in garnet (Fig. 3f), but most E-Type sulphides are isolated crystals.

I-Type sulphides (Fig. 3g, h) are relatively large (*c.* 0.1–1.0 mm). They are absent in some thin sections of the central part of the layer. I-Type sulphides generally occur as anhedral lobate crystals, convex towards the core and interstitial to the silicate phases (Fig. 3g). They are composed dominantly of pyrrhotite-containing laths (Fig. 3h) or patches of pentlandite (up to *c.* 20%) and chalcopyrite/Cu-rich sulphide (0–15 vol.%) as trails of minute grains, patches or flame-like lamellae. They may have a spongy texture and include thin lamellae of magnetite or Al-rich spinel. In sample AM407, extensively retrogressed to plagioclase–spinel–pyroxene assemblage, the interstitial sulphides are rimmed by pyroxenes and/or green spinel (Fig. 3g). In samples BAR and

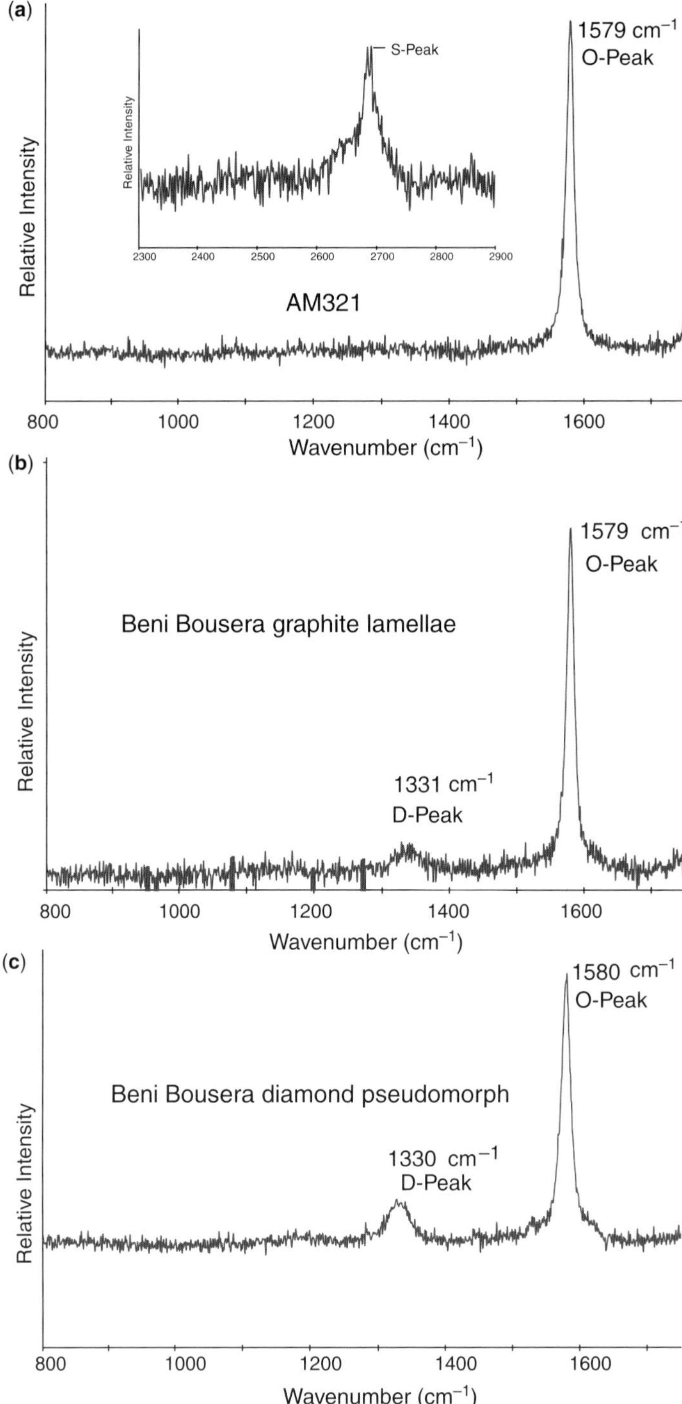

Fig. 2. Laser-Raman spectra for graphite in the garnet pyroxenite sample AM321 of this study (**a**), and comparative analyses of graphite lamellae (**b**) and pseudomorph after diamond (**c**) from a Beni Bousera pyroxenite sample.

Table 2. *Isotopic composition of graphite*

Sample	Graphite habit	$\delta^{13}C$	2s
AM322	Flake	−4.3	0.2
E181	Flake	−4.6	0.2

$\delta^{13}C = [(^{13}C/^{12}C_{sample}) / (^{13}C/^{12}C_{PDB}) - 1] \times 1000.$

AM407, the large grains are surrounded by small satellite sulphides (Fig. 3g, h) consisting of homogeneous pyrrhotite or chalcopyrite.

Fine-grained later trails of pyrrhotite in healed intergranular cracks cross-cutting the olivine-bearing domains (Fig. 3i) were observed at the edge of the thick layer, where local formation of low-temperature hydrous minerals (mainly chlorite) is observed.

Chemical composition of sulphide minerals

Point and bulk analyses of the different types of sulphides are reported in Tables 3 and 4. Following the sulphide nomenclature adopted by most authors (e.g. Dromgoole & Pasteris 1987; Szabo & Bodnar 1995; Guo *et al.* 1999), the term pyrrhotite was applied to monosulphide solid solution containing less than 5 wt% Ni with M/S [(Fe + Ni + Co + Cu)/S] ratios corresponding to those of ordered pyrrhotite (see Lorand & Conquéré 1983). As a whole, analyses of pyrrhotite, pentlandite and Cu–Fe sulphides as individual phases (Fig. 4) do not show

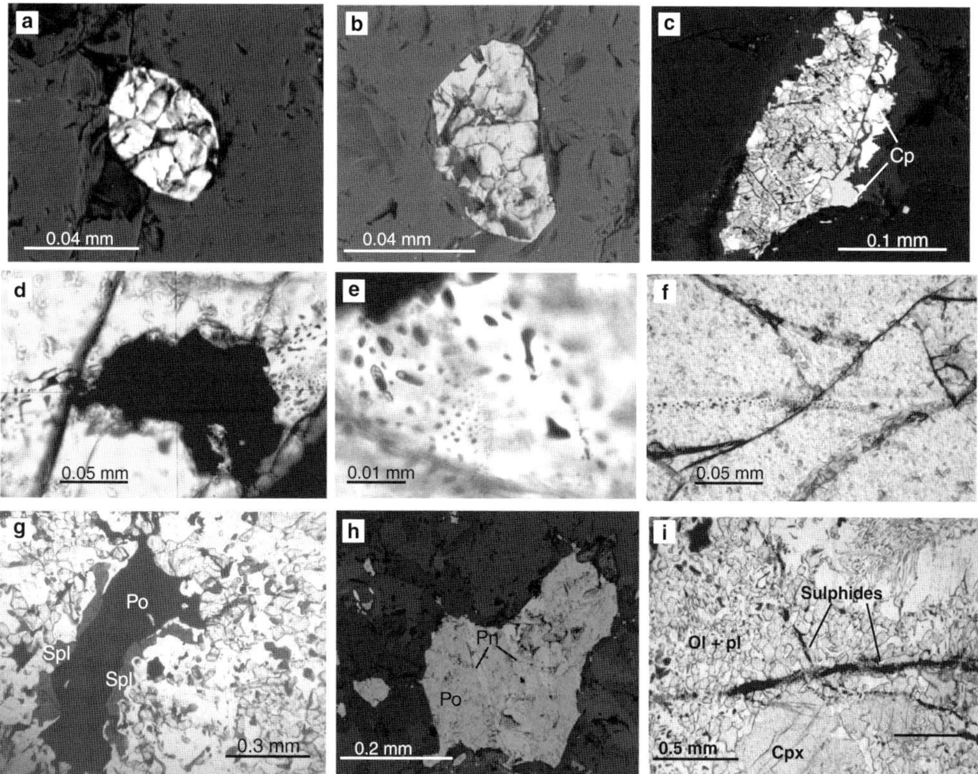

Fig. 3. Photomicrographs of sulphide textural occurrence. Photographs were obtained as back-scattered electron images (BSE) and under plane-polarized light (PPL) with optical microscope. (**a**) E-Type sulphide bleb in garnet (pentlandite + Cu–Fe sulphide, see bulk analysis 1 in Table 3), sample AM322 (BSE); (**b**) E-Type polygonal sulphide inclusion in garnet (homogeneous MSS), sample AM322 (BSE); (**c**) E-Type sulphide bleb in garnet (pyrrhotite + pentlandite with chalcopyrite rim from an original MSS), sample AM340 (BSE); (**d**) E-Type decrepitated sulphide bleb surrounded by tiny satellite sulphides and composite inclusions, sample AM322 (PPL); (**e**) enlarged portion of (d) showing composite chalcopyrite-fluid (?) inclusions (PPL); (**f**) trails of tiny Fe–Ni sulphides trapped in garnet, sample AM322 (PPL); (**g**) I-Type interstitial sulphide with green Al-rich spinel rims in retrogressed plagioclase-pyroxene-bearing domain, sample AM407 (PPL); (**h**) I-Type sulphide (pyrrhotite with pentlandite laths), sample BAR (BSE); and (**i**) trails of late sulphides cross-cutting the retrogressed Ol–plagioclase domains, sample AM407 (PPL).

Table 3. *Representative analyses of individual sulphides*

Type-1 sulphides

Sample	AM322				BA2		BAR		AM340D				E181	
Host mineral	Grt	Grt	Grt	Grt	Grt	Grt	Cpx	Cpx	Grt	Grt	Grt	Grt	Grt*	Grt*
S	33.38	35.36	33.30	39.24	37.98	32.96	39.78	38.85	40.23	41.13	33.96	34.39	38.09	31.84
Fe	34.45	27.59	28.69	60.93	60.50	31.23	60.05	54.59	59.95	24.88	29.42	31.00	61.00	30.00
Co	bd	0.87	0.71	0.06	0.24	0.49	0.74	0.60	0.63	2.24	2.03	bd	0.52	bd
Ni	0.54	37.00	37.87	0.51	0.82	35.48	0.32	6.23	0.36	32.69	35.24	0.26	0.73	bd
Cu	31.18	bd	bd	bd	bd	bd	bd	0.10	bd	bd	0.12	34.68	bd	38.65
Total	99.55	100.82	100.57	100.74	99.54	100.16	100.89	100.37	101.17	100.94	100.77	100.33	100.34	100.49
M/S ratio	1.07	1.03	1.13	0.90	0.93	1.14	0.88	0.90	0.87	0.81	1.10	1.03	0.94	1.15
Phase	ISS	pn	pn	po	po	pn	po	MSS	po	MSS	pn	cp	po	ISS

Type-2 sulphides

Sample	BAR						AM340		AM407				E181	
Host mineral	int	int	sat	int	int	int	int	int	int	int	int	sat	int	int
S	40.11	33.92	39.69	39.41	33.32	32.35	38.83	32.81	39.41	32.84	33.09	35.94	35.14	39.03
Fe	59.58	31.55	60.05	60.29	34.33	31.9	61.47	37.10	60.20	30.98	29.38	30.44	33.51	60.84
Co	0.69	1.22	0.12	0.13	1.61	bd	0.24	1.25	0.78	2.23	bd	0.15	1.31	0.53
Ni	0.43	33.91	0.74	0.78	31.90	0.36	0.79	30.40	0.13	34.00	0.10	0.25	30.76	0.42
Cu	bd	0.13	bd	bd	bd	36.04	bd	bd	bd	0.50	37.83	32.00	bd	bd
Total	100.81	100.73	100.60	100.61	101.16	100.65	101.33	101.56	100.52	100.55	100.40	98.78	100.02	100.04
M/S ratio	0.87	1.10	0.88	0.89	1.14	1.13	0.92	1.18	0.89	1.15	1.09	0.94	1.05	0.91
Phase	po	pn	po	po	pn	ISS	po	pn	po	pn	ISS	cp	pn	po

Abbreviations: Grt, garnet; Cpx, clinopyroxene; *sulphide in fine-grained Opx–Pl–Spl symplectites after garnet; bd, below detection limit (i.e. <0.10 wt%); int, interstitial sulphide; sat, satellite sulphide (see text for futher explanation); po, pyrrhotite; pn, pentlandite; cp, chalcopyrite; ISS, Cu-intermediate solid solution; MSS, monosulphide solid solution.

Table 4. *Representative analyses of bulk sulphides*

	Type-1 sulphides							Type-2 sulphides					
Sample	AM322			BA2B		AM340	E181	BAR			AM340	AM407	
Host mineral	Grt 1	Grt 2	Grt 3	Grt 4	Cpx 5	Grt 6	Grt* 7	8	9	10	11	12	13
S	34.19	36.89	38.48	35.03	31.31	37.68	36.19	36.99	36.62	34.03	36.66	36.11	39.21
Fe	31.00	50.56	58.50	47.06	31.70	27.50	56.89	53.36	52.83	52.02	58.26	51.49	55.46
Co	0.47	0.63	0.74	0.43	0.28	1.66	0.78	0.92	1.05	0.86	0.90	0.63	0.63
Ni	18.77	12.47	1.12	18.01	13.03	25.61	0.93	7.93	8.21	5.52	5.16	5.32	1.98
Cu	15.59	0.20	1.66	0.33	22.55	7.60	5.76	1.69	2.56	7.09	bd	6.79	2.73
Total	100.02	100.75	100.50	100.86	98.87	100.05	100.55	100.89	101.27	99.52	100.98	100.34	100.01
Phase	L + MSS	MSS	MSS	MSS	L	MSS	po	MSS	MSS	MSS	MSS	MSS	po

Abbreviations: Grt, garnet; Cpx, clinopyroxene; *sulphide in fine-grained Opx–Pl–Spl symplectites after garnet; bd, below detection limit (i.e. <0.10 wt%); L, liquid; MSS, monosulphide solid solution; po, pyrrhotite.

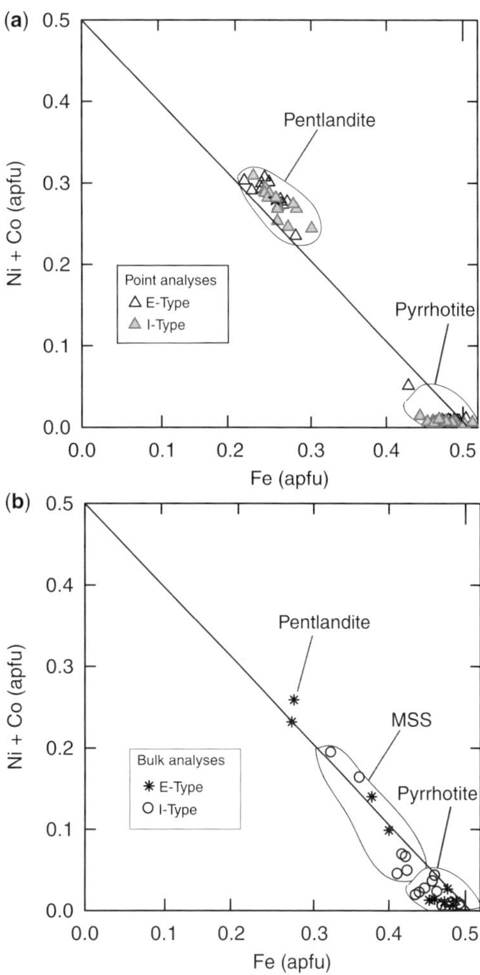

Fig. 4. Plot of Fe (at%) v. (Ni + Co) (at %) showing compositional variations of (**a**) individual and (**b**) bulk sulphide analyses in E-Type and I-Type sulphides.

any significant compositional variations between E- and I-Type sulphides. In both point and bulk sulphide analyses, Zn is below detection limit (<0.1 wt%).

Pyrrhotite in E- and I-Type sulphides has a low content of Ni (0.3–0.8 wt%), Cu (0.1–0.8 wt%) and Co (0.1–0.6 wt%). M/S ratios vary in a wide range (0.84–0.94), covering the fields of monoclinic and hexagonal pyrrhotite (Lorand & Conquéré 1983; Lorand 1989*a*). The concentrations of Ni, Cu and Co in the late trails of pyrrhotite are commonly below detection limit. Pentlandite has variable Ni content (31–38 wt%) and Ni/(Ni + Fe) molar ratios (0.48–0.57); Co content are in the range 0.5–2.3 wt%, whereas Cu is generally not detectable. The Fe–Cu sulphides include both chalcopyrite

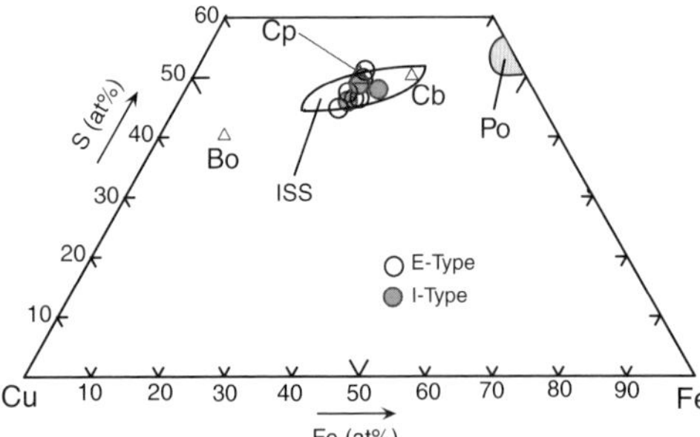

Fig. 5. Individual analyses of Cu–Fe sulphides projected in the Cu–Fe–S system at 600 °C (simplified after Cabri 1973) showing the field of intermediate solid solution (ISS) and pyrrhotite (Po). Small triangles are stoichiometrc compositions of bornite (Bo, Cu_5FeS_4), chalcopyrite (Cp, $CuFeS_2$) and cubanite (Cb, $CuFe_2S_3$).

close to ideal $CuFeS_2$ stoichiometry and compositions falling in the field of Cu-rich intermediate solid solutions (ISS according to Cabri 1973) (Fig. 5). They also have have a low Ni content (usually < 0.5 wt%, rarely up to *c.* 1 wt%). Rare Ni-rich (Ni 26–30 wt%) sulphides with low M/S ratio (*c.* 0.80) were found as irregular patches in both sulphide types.

The bulk compositions determined for polyphase E- and I-Type sulphides are metal-deficient (M/S < 1) monosulphides represented by: (i) pyrrhotite with variable Ni content (Ni 0.5–5.2 wt%) and Cu up to 4 wt%; (ii) high-temperature monosulphide solid solutions (Ni 5.5–18.0 wt%, Cu up to *c.* 7 wt%). Some E-Type sulphides have Ni- or Cu-Ni rich compositions that fall outside the field of monosulphide solid solutions at different temperatures (e.g. analyses 1 and 4–6 in Table 4, see the Discussion).

Discussion

Sulphide formation and crystallization history

Phase relations in the quaternary system Fe–Ni–Cu–S (and ternary subsystems) are well characterized from a number of experimental works (extensively reviewed in Fleet 2006, e.g. Kullerud *et al.* 1969; Craig 1973; Misra & Fleet 1973; Fleet & Pan 1994). Although the experiments were generally carried out at atmospheric pressure, the phase relations are not drastically affected by pressure and can be considered appropriate for upper-mantle sulphides. In particular, a maximum shift of phase boundaries of 50–100 °C can be conservatively assumed for the system under investigation (Brett & Bell 1969; Usselman 1975; Ryzhenco & Kennedy 1973; Guo *et al.* 1999; Tomkins *et al.* 2007). Experimentally-derived relations can be therefore applied to bulk and individual sulphide compositions analysed in the investigated rocks to infer the origin and subsolidus evolution of the sulphide assemblage, following the approach adopted in the literature for upper-mantle pyroxenites, occurring as xenoliths (De Waal & Calk 1975; Lorand & Conquéré 1983; Dromgoole & Pasteris 1987; Andersen *et al.* 1987; Guo *et al.* 1999; Lorand & Grégoire 2006) and in Alpine-type ultramafic massifs (Lorand 1989*a*, *b*).

The first crystallizing phase on the liquidus in the Fe–Ni–Cu–S system (at 1190 °C) is pyrrhotite ($Fe_{1-x}S$). With decreasing temperature, Ni enter the pyrrhotite structure forming a Fe–Ni monosulphide solid solution (MSS) $(Fe,Ni)_{1-x}S$ whose field expands towards the Ni–S join, reaching the NiS composition with further cooling between 1100 and 900 °C (Fig. 6a). Cu solubility at this stage is high, i.e. the monosulphide may contain up to about 7 wt% Cu at 1000 °C, but decreases with temperature. A Cu–Fe sulphide (sulphur-deficient intermediate solid solution, ISS) coexists with Cu-rich pyrrhotite at $T < 900$ °C, whereas chalcopyrite *sensu stricto* forms at lower temperatures (*c.* 600 °C). On further cooling, continuous solid solution between $Fe_{1-x}S$ and $Ni_{1-x}S$ is reported, with subsequent exsolution of low-temperature Ni-poor and Ni-rich monosulphide solutions (mss1 and mss2 in Fig. 6c) at approximately 200 °C. Pentlandite begins to segregate by phase separation from the monosulphide at about 600 °C and is stable also at lower temperature (Kullerud *et al.* 1969; Naldrett 1989; Etschmann *et al.* 2004), i.e. multiphase assemblages with pentlandite should be of subsolidus origin. The possible formation of pentlandite at higher temperature in the quaternary system Fe–Ni–Cu–S remains a controversial issue: according to the experiments of Sugaki & Kitakaze (1998), a high-temperature form of pentlandite may crystallize starting from 865 °C through a peritectic MSS + liquid reaction, whereas Peregoedova & Ohnenstetter (2002) found that primary pentlandite was not stable in the high-temperature association of the Fe–Ni–Cu–S system.

Recrystallization of magmatic sulphides to lower-temperature assemblages invariably occurs even for fast cooling rates (Ballhaus *et al.* 2001; Lorand & Grégoire 2006). Complex subsolidus evolution is therefore expected for mantle rocks exhumed at relatively slow rates like those investigated in this study (*c.* 1 mm a^{-1} for a cooling rate <5 °C Ma^{-1}; Montanini *et al.* 2006*a*) and information about the mantle crystallization history can be obtained only from bulk analyses.

Most bulk sulphide compositions projected in Figure 6a and b are consistent with high-temperature

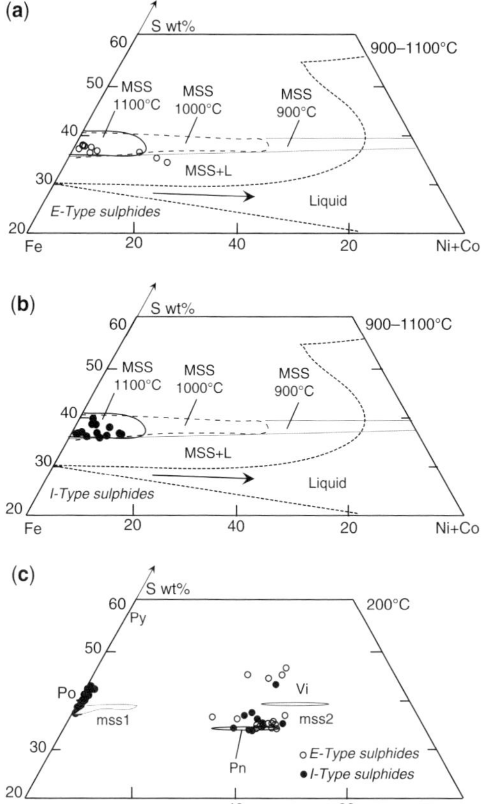

Fig. 6. Projection in the Fe-(Ni + Co)-S system of (**a**) and (**b**) bulk and (**c**) individual sulphide compositions. MSS, high-*T* monosulphide solid solution; mss1 and mss2, Ni-poor and Ni-rich low-*T* monosulphides as in Guo *et al.* (1999). Po, pyrrhotite; Pn, pentlandite; Py, pyrite; Vi, violarite. Phase relations and compositional ranges of MSS, mss1, mss2 and pentlandite from Kullerud *et al.* (1969) and Craig (1973). The arrow indicates the compositional evolution of sulphide liquid with decreasing temperature.

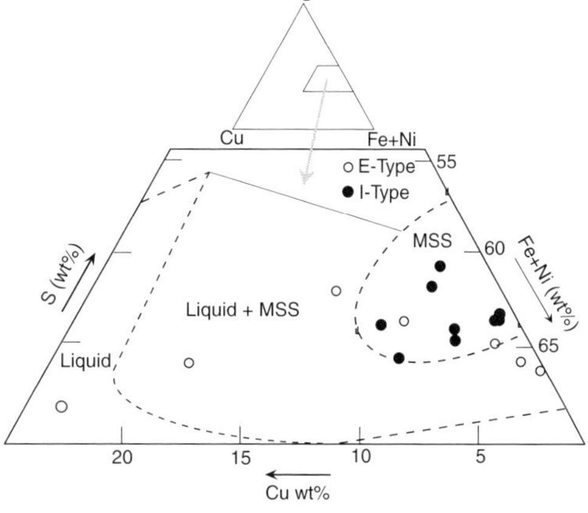

Fig. 7. Bulk sulphide analyses plotted on a (Fe + Ni)–Cu–S projection from the Fe–Ni–Cu–S system at 1000 °C (simplified phase relations after Kullerud *et al.* 1969).

(T c. 1200–1100 °C) crystallization as pyrrhotite (po) or Ni-poor monosulphide (when Ni exceeds 5 wt%) with relatively high Cu content from a sulphide melt. However, some E-Type sulphides fall outside the field of solid monosulphides for different temperatures (Fig. 6a) and most probably derive from entrapment of sulphide liquid + MSS in silicate phases. In particular, few Ni-rich compositions (analyses 4 and 6 in Table 4) plot in the two-phase liquid + MSS field in the Fe–Ni–S diagram at high temperature (Fig. 6a), whereas other inclusions (analyses 1 and 5 in Table 4) correspond to MSS (po_{ss}) coexisting with Cu–Ni-rich liquid or even pure liquid compositions (Fig. 7) and form a rough trend towards the MSS (po_{ss}) field. Sulphide blebs with comparable compositions were reported for some garnet pyroxenites from Hawaii (De Waal & Calk 1975) and in spinel peridotite xenoliths metasomatized by alkaline melts (Szabo & Bodnar 1995; Alard *et al.* 2000; Lorand & Alard 2001).

On the basis of textural and compositional characteristics, we conclude that E-Type inclusions were derived from entrapment of pyrrhotite (or MSS) ± sulphide liquid in silicates, whereas I-Type grains are the result of complete segregation of solid sulphide from an interstitial sulphide melt.

Fe/Ni ratios of bulk sulphides are controlled by those of the host–matrix silicates (Dromgoole & Pasteris 1987; Lorand 1991). The bulk sulphide compositions analysed in this study are generally consistent with those reported in the literature for the sulphide populations in mafic–pyroxenitic lithologies, which are distinct from the sulphides in peridotitic assemblage in having higher Fe and lower Ni content. In particular, Szabo & Bodnar (1995) and Guo *et al.* (1999) found that the monosulphides in peridotites have Ni abundance mainly ranging between 12 and 30 wt%, whereas Ni of sulphides from mafic cumulates and pyroxenites usually do not exceed 10 wt% (Dromgoole & Pasteris 1987; Guo *et al.* 1999; Zajacz & Szabo 2003). Similar Ni content also characterizes sulphides in eclogitic xenoliths (e.g. Meyer & Boctor 1975) and those occurring as inclusions in eclogitic-type diamonds (Deines & Harris 1995; Sobolev *et al.* 1997; Richardson *et al.* 2004).

During cooling related to exhumation, the high-temperature po_{ss}/MSS or po_{ss}/MSS + liquid mixtures recrystallized to the observed pyrrhotite + pentlandite ± Cu-rich sulphides (±Ni-rich MSS) assemblage. The Cu-free pyrrhotite should result from decreasing Cu solubility with falling temperature and re-equilibration with coexisting Cu–Fe sulphide, preferentially located at the rims of sulphide grains owing to Cu diffusion towards the sulphide edges (Torok *et al.* 2003). Preservation of Cu-rich ISS (Table 3) requires a blocking temperature of approximately 600 °C (Cabri 1973). ISS compositions could also result from electron beam overlap on submicrometric exsolutions of bornite. However, the general lack of the expected low-temperature monosulphides exsolved below approximately 300 °C (see below) may be consistent with preservation of high-temperature ISS.

Owing to the uncertainty on the revised phase relations involving pentlandite formation, it cannot be excluded that pentlandite occurring as rims around pyrrhotite might have formed at relatively low temperature through a peritectic reaction between pyrrhotite and a residual, metal-enriched sulphide melt. On the other hand, its common textural occurrence as laths or lamellae in pyrrhotite suggests an exsolution origin. Coexistence of pyrrhotite + pentlandite in interstitial sulphides of graphite-bearing rocks indicates that the graphite volume was not sufficient to buffer fO_2 (oxygen fugacity) at very reducing conditions yielding the assemblage troilite + pentlandite (Lorand 1991)

The rare patches of Ni-rich, metal-deficient sulphides cannot be considered as Ni-rich low-temperature monosulphides (mss2 of Guo *et al.* 1999) as they are enriched in sulphur, falling above the mss2 field; they could be the result of local metal loss due to alteration of pentlandite exsolutions from original mss2.

Well-preserved rounded sulphide globules are rare among E-Type sulphides, whereas irregular grains with fretted rims (often surrounded by smaller grains) and radial cracks departing from them are common. The composition of the satellite sulphides (Ni-free pyrrhotite and chalcopyrite) suggests derivation from an exsolved and re-equilibrated sulphide. Similar textures were reported for other sulphide occurrence in peridotite xenoliths (Szabo & Bodnar 1995), UHP (ultrahigh-pressure) metamorphic rocks (Hermann *et al.* 2006), and in pyroxenites and megacrysts in alkali basalts (Andersen *et al.* 1987), where they were attributed either to decrepitation (similar to the phenomenon observed for fluid inclusion in mantle rocks) or to partial melting during rapid ascent. According to Andersen *et al.* (1987), sulphide decrepitation may be due to the remobilization of a fluid (CO_2, identified by these authors in some composite inclusions) originally dissolved in the sulphide melt. The textural characteristics of the trapped sulphide blebs in garnet are therefore consistent with post-entrapment modification and structural re-arrangement during the decompressional history of the rock, although related to a slow tectonic uplift. Cracking of the host garnets during exhumation led to fractures radiating from the sulphides causing partial remobilization of the sulphides. Likewise, the arrays of tiny sulphide globules in garnet were derived from their becoming trapped into intragranular fractures of the mobilized sulphides.

The small satellite grains around I-Type interstitial sulphides, in textural equilibrium with the retrograde plagioclase-bearing assemblage formed at $T = 950–1000$ °C (Montanini *et al.* 2006*a*), are the result of subsolidus recrystallization of larger grains crystallized under high-temperature conditions (*c.* 1100 °C), coexisting with the garnet–clinopyroxene assemblage. The rims of green spinel, the spongy appearance and the occurrence of magnetite or Al-rich spinel lamellae testify for oxidation–desulfidation reactions involving the I-Type sulphides during high-temperature retrograde evolution of the pyroxenite. The later sulphide trails cross-cutting the high-temperature structures (Fig. 3i), including the olivine–plagioclase domains, probably represent a remobilization of primary I-Type sulphides related to the mantle emplacement at shallow levels (cf. Lorand 1989*a*).

Origin of the sulphide melt

The origin of E- and I-Type sulphides requires the existence of a Fe–Ni–Cu sulphide melt during the high-temperature history of the pyroxenites, in the garnet stability field. The formation of a sulphide melt is commonly attributed to immiscibility within a silicate melt; in particular, most sulphide occurrences in mantle pyroxenites are thought to have originated during magmatic crystallization of a basic melt, after reaching sulphur saturation at depth (Dromgoole & Pasteris 1987; Guo *et al.* 1999; Zajacz & Szabo 2003). A more complex scenario was proposed for the sulphides in the graphite-bearing Ronda pyroxenites (Crespo *et al.* 2006): infiltration of asthenospheric melts and partial melting of the host pyroxenite leading to a hybrid melt that reached sulphur saturation through differentiation. In addition, as the onset of partial sulphide anatexis generally occurs at lower temperatures than silicate partial melting (Szabo & Bodnar 1995; Tomkins *et al.* 2007), a sulphide melt may be present at high-temperature subsolidus conditions.

The graphite- and sulphide-bearing garnet pyroxenites of this study probably reflect a multistage petrogenetic history. The high proportions of Ca-Tschermak molecule coupled with relatively low jadeite content (Montanini *et al.* 2006*a*) are typical of clinopyroxene formed under high-pressure conditions (2.5–3.5 GPa), both as segregations from mafic melts (Thompson 1974; Eggins 1992;

Johnson 1998) or as residual phases after anhydrous partial melting of mafic rocks (Pertermann & Hirschmann 2003; Yaxley & Brey 2003). In addition, ongoing geochemical studies indicate that the garnet clinopyroxenites were subjected to a partial melting event under garnet-facies conditions (Montanini *et al.* 2006*b*).

On the basis of experimental work on partial melting of anhydrous mafic rocks (Yaxley & Green 1998; Kogiso & Hirschmann 2006; Yaxley & Sobolev 2007), relatively silica-rich (i.e. dacitic–andesitic) liquids were probably formed through low–moderate degrees of melting of the garnet pyroxenites. Solidus temperatures of garnet pyroxenites range between 1250 and 1300 °C, above the melting temperature of Fe–Ni sulphides. Therefore, pre-existing sulphides may be completely melted to form a low-viscosity, relatively mobile sulphide melt (Dobson *et al.* 2000) before the solidus temperature of the silicates is attained, although a significant pressure dependence of the solidus curve of Fe–Ni–Cu monosulphide has recently been determined by Bockrath *et al.* (2004). When the silicates starts to melt, sulphur solubility in the silicate melt (expressed as SCSS, i.e. sulphur concentration at sulphide saturation: Mavrogenes & O'Neill 1999) is a complex function of temperature, pressure and melt composition. In particular, SCSS is strongly controlled by pressure, displaying a negative exponential correlation (e.g. for a basaltic melt, SCSS is reduced by more than a half from 1.0 to 3.0 GPa: Wendlandt 1982; Mavrogenes & O'Neill 1999), whereas temperature has the opposite effect. No clear-cut fO_2 dependence was demonstrated, although a positive correlation in the interval between NNO+1 and FMQ−2 was found by several authors (Haughton *et al.* 1974; Mavrogenes & O'Neill 1999; Liu *et al.* 2007). Regarding the compositional dependence, it was inferred that sulphur solubility is mainly controlled by FeO content of the melt (Haughton *et al.* 1974; Wendlandt 1982; O'Neill & Mavrogenes 2002). Recent experiments by Liu *et al.* (2007) have stressed a more complex compositional control by the ratio of network-modifier/network-forming (and, consequently of bridging- to non-bridging oxygens) that shows a positive linear relationship with SCSS. This implies that sulphur solubility decreases markedly from basaltic to rhyolitic melts. In particular, Liu *et al.* (2007) determined SCSS values between 400 and 700 ppm at 1.0 GPa for dacitic–andesitic melts. No data are available for SCSS in silicic melts under the high-pressure conditions inferred for partial melting of the garnet pyroxenites. However, if the reduction in sulphur solubility with increasing pressure illustrated above is applicable to more silica-rich melts, a restricted sulphur solubility is expected during melting of the pyroxenite mafic protoliths, leading to early sulphur saturation and separation of small amounts of an immiscible sulphide melt, consistent with the observed trace amounts of sulphides. This melt was partly trapped in the residuum yielding minute blebs in silicates (E-Type sulphides), or crystallized as interstitial po_{ss} (I-Type sulphides). The local sulphide concentration at the margin of the garnet pyroxenite layer could be explained by preferential nucleation of sulphide melt droplets on the wall rock (as proposed by Lorand 1991 for some pyroxenite layers from Pyrenean orogenic peridotites).

Constraints on the formation of graphite

The structural features of graphite are consistent with a high temperature of formation. Raman spectra rule out that the graphite may be related to the serpentinization processes that affect the host peridotites (Pasteris 1981). Well-defined first- (O) and second-order (S) peaks, and the absence of first order D peaks, indicate a highly ordered structure compatible with a mantle origin, similar to that of graphite crystallized at high temperature in peridotite and eclogite xenoliths (Pearson *et al.* 1994). The lively debate on sources and mechanisms of native carbon precipitation in the mantle was largely devoted to the origin of diamond and, to a lesser extent, of graphite, whose occurrence in the mantle is rare (Pearson *et al.* 1994; Schulze *et al.* 1997). However, discussions about diamond origin can be generally applied to that of graphite under the appropriate P–T conditions.

The upper fO_2 limit for graphite stability at upper-mantle conditions is nominally given by the equilibrium $CO = C + \frac{1}{2}O_2$ (CCO: Luth 2003; Simakov 2006) and does not require highly reducing conditions. The CCO buffer for most geotherms lies close to or slightly below FMQ (fayalite–magnetite–quartz) (Pearson *et al.* 1994 and quoted references). Oxygen fugacity calculations for the primary garnet–clinopyoxene assemblage of the garnet pyroxenites have been performed using the method of Simakov (2006). Assuming the average P–T values determined for this assemblage ($P = 2.8$ GPa, $T = 1100$ °C: Montanini *et al.* 2006*a*), we obtained $\Delta \log fO_2^{(FMQ)} = -2.2$, that is, about two log units below the FMQ reference oxygen buffer, close to the WM (wüstibe–magnetite) buffer, thus implying fO_2 conditions compatible with graphite occurrence. Luth (1993) argued that, for a clinopyroxene–garnet silicate assemblage, graphite stability is more properly described by a reaction defining the coexistence of graphite/diamond and carbonate, namely dolomite + 2 coesite = diopside + 2C + 2O$_2$ (DCDD). At high-temperature ($\geq$1100 °C), in the diamond stability field ($P = 5.0$ GPa), DCDD is shifted approximately 1 unit below CCO (Luth 1993). If this fO_2–T correlation is extrapolated to lower-pressure conditions, the calculated fO_2 is still compatible with graphite occurrence in the garnet stability field.

Mechanisms of graphite segregation. Elemental carbon crystallization from a silicate melt is not considered as a suitable mechanism because silicate melts have low C solubility and contain relatively small amounts of transition metals (Bulanova 1995). Diamond formation was, indeed, obtained under extremely high P–T conditions only from a volatile-rich kimberlitic melt (Arima *et al.* 1993). However, recent experimental results on graphite/diamond precipitation from alkaline carbonate melts (e.g. Arima *et al.* 2002; Pal'yanov *et al.* 2002) do not seem appropriate for the system under investigation.

Graphite growth in the presence of a C–O–H fluid coming from the convecting mantle (through the reduction of CO_2 or oxidation of CH_4) was invoked as a major mechanism responsible for diamond and graphite origin in the lithospheric (subcratonic) mantle (Deines 1980; Haggerty 1986; Griffin *et al.* 2007; Malkovets *et al.* 2007). Phase equilibria involving graphite in a C–O–H fluid and mechanisms of fluid-induced precipitation of graphite were discussed by Deines (1980) and Luque *et al.* (1998). Native carbon precipitation is controlled by conditions and composition of the P–T–fO_2 fluid. The main equilibria in the C–O–H system, for high and low fO_2 conditions, respectively, are $CO_2 = C + O_2$ and $CH_4 + O_2 = C + 2H_2O$. Graphite precipitation may actually be induced in several ways, that is, isobaric cooling, hydration reactions, fO_2 variations, mixing of CO_2–CH_4 fluids and reactions catalysed by reducing agents, for example, sulphides.

Although no definite agreement exists on the variation in fO_2 variation with depth in subcontinental lithospheric mantle (Luth 2003), and the fO_2 estimates for non-cratonic mantle are mainly restricted to relatively shallow and oxidized spinel-facies peridotites (ΔFMQ between -2 and $+1$: Canil *et al.* 1990; Woodland *et al.* 1992), studies on cratonic xenoliths indicate decreasing fO_2 conditions with depth and suggest that conditions for a fluid phase dominantly composed of H_2O and CH_4 are probably reached at depths greater than 200 km (Woodland & Koch 2003; Simakov 2006; Thomassot *et al.* 2007). A CH_4 precursor fluid for the graphite of this study seems unlikely, therefore, and incompatible with the calculated depth of equilibration and fO_2 conditions.

However, a possible role for a sulphide component either as a growth medium or as a reducing agent of an oxidized carbon-bearing fluid was suggested by many authors (e.g. Marx 1972; Haggerty 1986; Bulanova 1995; Bulanova *et al.* 1998; Luque *et al.* 1998), mainly on the basis of the dominant occurrence of Fe–Ni–Cu sulphide inclusions in eclogitic diamonds. In particular, Haggerty (1986) and Bulanova (1995) speculated about graphite/diamond precipitation from a sulphide melt saturated in carbon. Haggerty (1986) also proposed that carbon crystallization may occur through the desulfidation reaction $2FeS + CO_2 = C + S_2 + 2FeO$. The occurrence of composite solid sulphide–CO_2 inclusions in pyroxenite xenoliths and pyroxene megacrysts (Andersen *et al.* 1987) suggest enhanced CO_2 solubility into a sulphide melt at depth. Recently, Gunn & Luth (2006) and Pal'yanov *et al.* (2007) have experimentally investigated carbonate interaction with a sulphide melt or solid sulphides in simplified carbonate–silicate–sulphide systems leading to the crystallization of elemental carbon from a CO_2-rich fluid supersaturated with carbon containing dissolved silicates and sulphides via reactions of decarbonation and CO_2 reduction.

In spite of the lack of detailed knowledge on the mechanisms responsible for graphite/diamond crystallization from sulphide-rich melts or sulphide-bearing CO_2-rich fluids, a genetic link between sulphides and graphite/diamond origin is supported by observations and experimental work. The occurrence of an (interstitial) sulphide melt in the garnet pyroxenites of this study could have similarly played a role in graphite formation. We propose that interaction between this melt and a oxidized carbon-bearing medium (CO_2 or carbonate) under appropriate fO_2 conditions is a viable process for the origin of the graphite. Graphite precipitation may be related to the saturation in reduced carbon reached either in the sulphide liquid or in a coexisting CO_2-bearing fluid.

The carbon source. The carbon isotope ratios can be used to provide constraints on the carbon source. The carbon isotope range of graphite in both eclogite (-14.3 to -2.8‰: Schulze *et al.* 1997 and references therein) and peridotite xenoliths (-12.3 to -3.8‰: Pearson *et al.* 1994) is similar to that of P-type (peridotitic) diamonds (-9 to -2‰: Deines 2002), whereas E-type diamonds belonging to the eclogite suite encompass a larger range of C isotope values ($\delta^{13}C$ of between -34 and -2‰). The graphite in orogenic mantle rocks from Ronda and Beni Bousera has light carbon isotope compositions ($\delta^{13}C$ -27 to -15‰: Pearson *et al.* 1991, 1993; Crespo *et al.* 2006), unlike the graphite observed in this study, which is more akin to the dominant compositions observed in eclogite xenoliths (Schulze *et al.* 1997).

Several possible carbon reservoirs, characterized by distinct isotope signatures, were proposed for the origin of mantle graphite, that is, (1) a carbon source indigenous to the mantle characterized by $\delta^{13}C$ of between −8 and −2‰, with an average value of around −5‰ (Mattey 1987; Thomassot *et al.* 2007 and quoted references); and (2) crustal carbon recycled into the mantle by subduction, namely C-bearing components released during maturation of kerogen-like, organic matter ($\delta^{13}C$ typically of between −10 and −25‰: Ohmoto 1986) or during devolatilization of marine carbonate-bearing materials ($\delta^{13}C$ −2 to +4‰: Ohmoto 1986), although not all authors favour the recycling interpretation (e.g. Deines 2002).

The C isotope composition of the graphite under investigation (Table 2) is close to the major isotopic signature for mantle carbon and incompatible with subduction-derived carbon, either from organic matter or marine carbonate recycling. The isotopic features, however, depend not only on the source of the carbon but also on the specific mechanism that causes graphite precipitation, which may be associated with significant isotopic fractionation even under high-temperature mantle conditions. In particular, graphite precipitation from a carbon species with a different oxidation state produces isotope fractionation if the graphite formed under isotopic equilibrium remains isolated from its growth medium, that is, the crystallization occurs under open-system conditions controlled by a Rayleigh-type distillation. However, formation from biogenic carbon should reflect the light composition of the precursor kerogen-like material (Pearson *et al.* 1991), making the observed C isotope values definitely incompatible with such an origin.

Fractionation would yield graphite depleted in ^{13}C isotopes (by about 3‰) relative to an original carbonate source (Bottinga 1969; Scheele & Hoefs 1992). Derivation of the relatively heavy graphite of this study from recycled marine carbonate under open-system conditions (e.g. partial degassing of CO_2 produced from a decarbonation reaction) may be hypothetically allowed (Fig. 8), if its initial $\delta^{13}C$ fell at the lower end of the marine carbonate range.

On the basis of CO_2–graphite fractionation factors (Bottinga 1969; Scheele & Hoefs 1992), it also follows that graphite would be depleted in ^{13}C with respect to a CO_2 phase. Fractionation factors ($\Delta C = \delta^{13}C_{graphite} - \delta^{13}C_{CO_2}$) vary between −5.8 and −5.1 in the temperature range 1000–1200 °C. If we assume a metasomatic CO_2-bearing fluid with MORB–OIB signature flowing through the lithosphere (Pearson *et al.* 1994; Cartigny *et al.* 1998), the first graphite to crystallize would be lighter than the observed graphite (even assuming initial $\delta^{13}C$ values spanning the entire mantle range reported above), while the residual CO_2 evolves towards heavier $\delta^{13}C$. Considering a Rayleigh-type distillation, the required $\delta^{13}C_{CO_2}$ in equilibrium with the observed graphite (+1 for T 1100 °C) may be achieved only after prolonged fractionation (Fig. 8). The absence of graphite in the whole mantle sequence except the garnet pyroxenite is inconsistent with this hypothesis, although graphite preservation could be hampered by the exhumation history (see below).

The provenance of carbon in the garnet pyroxenites cannot be univocally constrained. Nevertheless, the isotopic similarity of the graphite with carbon in MORB–OIB suggests closed-system crystallization from a carbon-bearing fluid with a mantle signature, that is, total conversion to graphite of the carbon in such a fluid within the pyroxenite. This could have been either a fluid released during partial melting of the pyroxenite protolith (carbon contained in small carbonate amounts or dissolved in silicate lattice?: see Pearson *et al.* 1991; Luth 2003) or an externally-derived fluid that interacted with the sulphide melts. The most obvious explanation for the restriction of graphite to a few garnet pyroxenites is differential carbon content of Fe-rich garnet pyroxenite and Mg-rich websterites and peridotites, that is, carbon indigenous to the garnet pyroxenite. However, if we assume that the carbon source of graphite is a metasomatic carbon-bearing fluid, it may be explained by the more reducing conditions required for the stability of elemental carbon in Mg-rich (peridotitic or websteritic) assemblages (Luth 1993) or by the difficulty

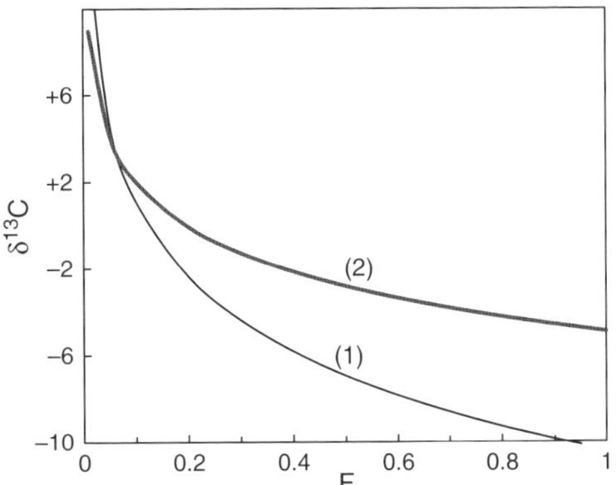

Fig. 8. Calculated variations of graphite isotope compositions achieved through a Rayleigh-type fractionation from reduction of: (1) CO_2 with initial $\delta^{13}C$ of −5‰; and (2) carbonate with intial $\delta^{13}C$ of −2‰. *F*, fraction of residual fluid; fractionation factors for CO_2–graphite and carbonate–graphite from Scheele & Hoefs (1992). See the text for further explanations.

in preserving a reduced carbon phase during exhumation at shallower lithospheric levels characterized by higher fO_2 conditions.

Conclusions

Fe-rich garnet clinopyroxenite layers in subcontinental mantle rocks from the Northern Apennines contain an uncommon occurrence of coexisting graphite and sulphides formed at high $P-T$ conditions. The pyroxenites record a complex exhumation history from deep levels of the subcontinental lithosphere (P *c.* 2.8 GPa) to the ocean floor. Graphite has structural characteristics compatible with deep-seated crystallization at mantle conditions and $^{13}\delta C$ values close to the major signature for mantle carbon. Sulphides, occurring both as inclusions in silicates (E-Type) and as interstitial grains (I-Type), are polyphase grains composed of Ni-poor pyrrhotite, pentlandite and Cu–Fe sulphides derived from exsolution and recrystallization during cooling related to slow exhumation (T 900–600 °C) of high-temperature Fe–Ni–Cu monosulphides (MSS). Bulk sulphide compositions are consistent with crystallization as MSS from a sulphide melt or, more rarely, as solidification of MSS + liquid mixtures in silicate phases. Sulphur saturation and separation of an immiscible sulphide liquid most probably occurred from a relatively silica-rich melt that originated by partial melting of the pyroxenite in the garnet stability field, followed by the trapping of small amount of the sulphide liquid in the residual rock.

A genetic link between sulphide melt and graphite origin, namely the interaction between this melt as a reducing agent and an oxidized carbon-bearing medium (CO_2 or carbonate), is suggested to explain the precipitation of graphite, in agreement with recent experiments. The source of carbon in the pyroxenite remains speculative. The isotopic composition and the restricted occurrence of graphite are consistent with closed-system crystallization from a C-bearing fluid with a mantle signature, that is, indigenous CO_2-bearing fluid released during partial melting of the pyroxenite protolith or an externally-derived metasomatic fluid of sublithospheric origin that interacted with the sulphide melts. While crustal carbon recycling from subduction-derived organic matter is precluded on the basis of heavy C isotope composition, origin from marine carbonate recycling cannot be definitely excluded if some isotopic fractionation occurred during decarbonation of a carbonate formerly occurring in the pyroxenite protolith.

This work was supported by funds of MIUR (Ministero Italiano dell'Università e della Ricerca, Progetti di Ricerca d'Interesse Nazionale). We wish to thank A. Longinelli for stimulating discussion and suggestions, and M. Beneventi for helping with mineral separation and fieldwork. L. Barchi and A. Comelli are acknowledged for SEM technical assistance and sample preparation. Constructive reviews provided by O. Alard and by an anonymous reviewer are acknowledged.

References

Alard, O., Griffin, W. L., Lorand, J. P., Jackson, S. E. & O'Reilly, S. Y. 2000. Non-chondritic distribution of the highly siderophile elements in mantle sulphides. *Nature*, **407**, 891–894.

Alard, O., Luguet, A. *et al.* 2005. *In situ* Os isotopes in abyssal peridotites bridge the isotopic gap between MORBs and their source mantle. *Nature*, **436**, 1005–1008.

Andersen, T., Griffin, W. L. & O'Reilly, S. Y. 1987. Primary sulphide melt inclusions in mantle-derived megacrysts. *Lithos*, **20**, 279–294.

Arima, M., Kozai, Y. & Akaishi, M. 2002. Diamond nucleation and growth by reduction of carbonate melts under high-pressure and high-temperature conditions. *Geology*, **30**, 691–694.

Arima, M., Nakayama, K., Akaishi, M., Yamaoka, S. & Kanda, H. 1993. Crystallisation of diamond from a silicate melt of kimberlite composition in high-pressure-temperature experiments. *Geology*, **21**, 968–970.

Ballhaus, C., Tredoux, M. & Späth, A. 2001. Phase relations in the Fe–Ni–Cu–PGE–S system at magmatic temperature and application to massive sulphide ores of the Sudbury Igneous Complex. *Journal of Petrology*, **42**, 1911–1926.

Beyssac, O. & Chopin, C. 2003. Comment on 'Diamond, former cosite and supersilicic garnet in metasedimentary rocks from the Greek Rhodop: a new ultrahigh pressure metamorphic province established' by E.D. Mposkos & Kostopolous [*Earth and Planetary Science Letters*, **192** (2001), 497–506]. *Earth and Planetary Science Letters*, **214**, 669–674.

Bockrath, C., Ballhaus, C. & Holzheid, A. 2004. Fractionation of the platinum-group elements during mantle melting. *Science*, **305**, 1951–1953.

Bottinga, Y. 1969. Calculated fractionation factors for oxygen and hydrogen isotope exchange in the system calcite–carbon dioxide–graphite–methane–hydrogen–water vapor. *Geochimica et Cosmochimica Acta*, **33**, 49–64.

Brett, R. & Bell, P. M. 1969. Melting relations in the Fe-rich portion of the system Fe–FeS at 30 kb pressure. *Earth and Planetary Science Letters*, **6**, 479–482.

Bulanova, G. P. 1995. The formation of diamond. *Journal of Geochemical Exploration*, **53**, 1–23.

Bulanova, G. P., Griffin, W. I. & Ryan, C. G. 1998. Nucleation environment of diamonds fromYakutian kimberlites. *Mineralogical Magazine*, **62**, 409–419.

Cabri, L. J. 1973. New data on phase relations in the Cu–Fe–S system. *Economic Geology*, **68**, 443–454.

Canil, D., Virgo, D. & Scarfe, C. M. 1990. Oxidation state of mantle xenoliths from British Columbia, Canada. *Contributions to Mineralogy and Petrology*, **104**, 453–462.

CARTIGNY, P., HARRIS, J. W. & JAVOY, M. 1998. Eclogitic diamond formation at Jwaneng: no room for a recycled component. *Science*, **280**, 1421–1424.

CRAIG, J. R. 1973. Pyrite-pentlandite assemblages and other low temperature relations in the Fe–Ni–S system. *American Journal of Science*, **273A**, 496–510.

CRESPO, E., LUQUE, F. J., RODAS, M., WADA, H. & GERVILLA, F. 2006. Graphite–sulfide deposits in Ronda and Beni Bousera peridotites (Spain and Morocco) and the origin of carbon in mantle-derived rocks. *Gondwana Research*, **9**, 279–290.

DAVIES, G. R., NIXON, P. H., PEARSON, D. G. & OBATA, M. 1993. Tectonic implications of graphitized diamonds from the Ronda peridotite massif, southern Spain. *Geology*, **21**, 471–474.

DEINES, P. 1980. The carbon isotopic composition of diamonds: relationship to diamond shape, color, occurrence and vapor composition. *Geochimica et Cosmochimica Acta*, **44**, 943–961.

DEINES, P. 2002. The carbon isotope geochemistry of mantle xenoliths. *Earth Science Reviews*, **58**, 247–278.

DEINES, P. & HARRIS, J. W. 1995. Sulfide inclusion chemistry and carbon isotopes of African diamonds. *Geochimica et Cosmochimica Acta*, **59**, 3173–3188.

DE WAAL, S. A. & CALK, L. C. 1975. The sulphides in the garnet pyroxenite xenoliths from Salt Lake Crater, Ohau. *Journal of Petrology*, **16**, 134–153.

DOBSON, D. P., CRICHTON, W. A. ET AL. 2000. *In situ* measurement of viscosity of liquids in the Fe-FeS system at high pressures and temperatures. *American Mineralogist*, **85**, 1838–1842.

DROMGOOLE, E. L. & PASTERIS, J. D. 1987. Interpretation of the sulfide assemblages in a suite of xenoliths from Kilbourne Hole, New Mexico. Geological Society of America, Special Paper, **215**, 25–46.

EGGINS, S. M. 1992. Petrogenesis of Hawaiian tholeiites: 1, Phase equilibria constraints. *Contributions to Mineralogy and Petrology*, **110**, 387–397.

ETSCHMANN, B., PRING, A., PUTNIS, A., GRURIC, B. A. & STUDER, A. 2004. A kinetic study of the exsolution of pentlandite $(Ni,Fe)_9S_8$ from the monosolufide solid solution (Fe,Ni)S. *American Mineralogist*, **89**, 39–50.

FLEET, M. E. 2006. Phase equilibria at high-temperatures. *In*: VAUGHAN, D. J. (ed.) *Sulfide Mineralogy and Geochemistry*. Reviews in Mineralogy and Geochemistry, **61**, 365–420.

FLEET, M. E. & PAN, Y. 1994. Fractional crystallization of anhydrous sulfide liquid in the system Fe–Ni–Cu–S, with application to magmatic sulfide deposits. *Geochimica et Cosmochimica Acta*, **58**, 3360–3377.

GARUTI, G., GORGONI, C. & SIGHINOLFI, G. P. 1984. Sulfide mineralogy and chalcophile and siderophile element abundances in the Ivrea–Verbano mantle peridotites (Western Italian Alps). *Earth and Planetary Science Letters*, **70**, 69–87.

GRIFFIN, W. L., REGE, S., ARAUJO, D., JACKSON, S. & PEARSON, N. 2007 Trace-element patterns of diamond: clues to mantle processes. *In*: *European Mantle Workshop (EMAW 2007) Abstracts*.

GUNN, S. C. & LUTH, R. W. 2006. Carbonate reduction by Fe–S–O melts at high pressure and high temperature. *American Mineralogist*, **91**, 1110–1116.

GUO, J., GRIFFIN, W. L. & O'REILLY, S. Y. 1999. Geochemistry and origin of sulfide minerals in mantle xenoliths: Qilin, Southeastern China. *Journal of Petrology*, **40**, 1125–1151.

HAGGERTY, S. E. 1986. Diamond genesis in a multiply constrained model. *Nature*, **320**, 34–38.

HAGGERTY, S. E. 1995*a*. Upper mantle mineralogy. *Journal of Geodynamics*, **20**, 331–364.

HAGGERTY, S. E. 1995*b*. A diamond trilogy: superplumes, supercontinents, and supernovae. *Science*, **285**, 851–860.

HAUGHTON, D., ROEDDER, P. L. & SKINNER, B. J. 1974. Solubility of sulphur in mafic magmas. *Economic Geology*, **69**, 451–467.

HERMANN, J., SPANDLER, C., HACK, A. & KORSAKOV, A. V. 2006. Aqueous fluids and hydrous melts in high-pressure and ultra-high pressure rocks: implications for element transfer in subduction zones. *Lithos*, **92**, 399–417.

JOHNSON, K. T. M. 1998. Experimental determination of partition coefficients for rare earth and high-field-strength elements between clinopyroxene, garnet and basaltic melt at high pressures. *Contributions to Mineralogy and Petrology*, **133**, 60–68.

KAESER, B., KALT, A. & PETTKE, T. 2007. Crystallization and breakdown of metasomatic phases in graphite-bearing peridotite xenoliths from Marsabit (Kenya). *Journal of Petrology*, **48**, 1725–1760.

KOGISO, T. & HIRSCHMANN, M. 2006. Partial melting experiments of bimineralic eclogite and the role of recycled mafic oceanic crust in the genesis of ocean island basalts. *Earth and Planetary Science Letters*, **249**, 188–199.

KORNPROBST, J., PIBOULE, M., RODEN, M. & TABIT, A. 1990. Corundum-bearing garnet clinopyroxenites at Beni Bousera (Morocco): original plagioclase-rich gabbros recrystallized at depth within the mantle? *Journal of Petrology*, **31**, 717–745.

KULLERUD, G., YUND, R. A. & MOHR, G. 1969. Phase relations in the Cu–Fe–S, Cu–Ni–S system. *In*: WILSON, H. D. B. (ed.) *Magmatic Ore Deposits*. Economic Geology Monograph, **4**, 323–343.

LIU, Y., SAMAHA, N. & BAKER, D. R. 2007. Sulfur concentration at sulfide saturation (SCSS) in magmatic silicate melts. *Geochimica et Cosmochimica Acta*, **71**, 1783–1799.

LORAND, J. P. 1989*a*. Sulfide petrology of spinel and garnet pyroxenite layers in mantle-derived spinel lherzolite massifs of Ariege and Pyrenees, France. *Journal of Petrology*, **30**, 987–1015.

LORAND, J. P. 1989*b*. The Cu–Fe–Ni sulfide component of the amphibole-rich veins from the Lherz and Freychinède spinel peridotite massifs (Northeastern Pyrenees, France): a comparison with mantle-derived megacrysts from alkali basalts. *Lithos*, **23**, 281–298.

LORAND, J. P. 1991. Sulfide petrology and sulfur geochemistry of orogenic lherzolites: a comparative study between Pyrenean bodies (France) and the Lanzo massif (Italy). *In*: MENZIES, M. A., DUPUY, C. & NICOLAS, A. (eds) *Orogenic Lherzolites and Mantle Processes*. Journal of Petrology, Special Volume, 77–95.

Lorand, J. P. & Conquéré, F. 1983. Contribution à l'étude des sulfures dans les enclaves de lherzolites à spinelle des basaltes alcalins (Massif Central et du Languedoc, France). *Bulletin de Minéralogie*, **106**, 585–606.

Lorand, J. P. & Alard, O. 2001. Platinum-group element abundances in the upper-mantle: new constraints from *in situ* and whole-rock analyses of Massif Central xenoliths (France). *Geochimica et Cosmochimica Acta*, **65**, 2789–2806.

Lorand, J. P. & Grégoire, M. 2006. Petrogenesis of base metal sulfides of some peridotites of the Kaapvaal craton (south Africa). *Contributions to Mineralogy and Petrology*, **151**, 521–538.

Luguet, A., Pearson, D. G., Nowell, G. M., Dreher, S. T., Coggon, J. A., Spetsius, Z. V. & Parman, S. W. 2008. Enriched Pt–Re–Os isotope systematics in plume lavas explained by metasomatic sulfides. *Science*, **319**, 453–456.

Luth, R. W. 1993. Diamonds, eclogites, and the oxidation state of the earth's mantle. *Science*, **261**, 66–68.

Luth, R. W. 2003. Mantle volatiles-distribution and consequences. *In*: Carlson, R. W., Holland, H. D. & Turekian, K. K. (eds) *Treatise on Geochemistry, 2. The Mantle and Core*. Pergamon, Amsterdam, 319–362.

Luque, F. J., Pasteris, J. D., Wopenka, B., Rodas, M. & Barrenechea, J. F. 1998. Natural fluid-deposited graphite: mineralogical characteristics and mechanisms of formation. *American Journal of Science*, **298**, 471–498.

Malkovets, V., Griffin, W. L., O'Reilly, S. Y. & Wood, B. J. 2007. Diamond, subcalcic garnet and mantle metasomatism: kimberlite sampling patterns define the link. *Geology*, **35**, 339–342.

Marroni, M., Molli, G., Montanini, A. & Tribuzio, R. 1998. The association of continental crust rocks with ophiolites in the Northern Apennines (Italy): implications for the continent-ocean transition in the Western Tethys. *Tectonophysics*, **292**, 43–66.

Marx, P. C. 1972. Pyrrhotite and the origin of terrestrial diamonds. *Mineralogical Magazine*, **38**, 636–638.

Mattey, D. P. 1987. Carbon isotope in the mantle. *Terra Cognita*, **7**, 31–37.

Mavrogenes, J. A. & O'Neill, H. St. C. 1999. The relative effects of pressure, temperature and oxygen fugacity on the solubility of sulfide in mafic magmas. *Geochimica et Cosmochimica Acta*, **63**, 1173–1180.

McCandless, T. E. & Gurney, J. J. 1989. Sodium in garnet and potassium in clinopyroxene: criteria for classifying mantle eclogites. *In*: Ross, J. (ed.) *Kimberlites and Related Rocks*. Geological Society of Australia, Special Publications, **2**, 827–832.

Meyer, H. O. A. & Boctor, N. Z. 1975. Sulfide-oxide minerals in eclogite from Stockdale Kimberlite, Kansas. *Contributions to Mineralogy and Petrology*, **52**, 57–68.

Misra, C. & Fleet, M. E. 1973. The chemical compositions of synthetic and natural pentlandite assemblages. *Economic Geology*, **68**, 518–539.

Montanini, A., Tribuzio, R. & Anczkiewicz, R. 2006*a*. Exhumation history of a garnet pyroxenite-bearing mantle section from a continent–ocean transition (Northern Apennine ophiolites, Italy). *Journal of Petrology*, **47**, 1943–1971.

Montanini, A., Tribuzio, R. & Thirlwall, M. 2006*b*. Garnet pyroxenite layers from the mantle peridotites of the Northern Apennine ophiolites, Italy: evidence for recycling of crustal material? *Geochimica et Cosmochimica Acta*, **70**, Supplement 1, A426.

Morgan, J. W. 1986. Ultramafic xenoliths: clues to Earth's late accretionary history. *Journal of Geophysical Research*, **91**, (B12), 12 375–12 387.

Naldrett, A. J. 1989. Magmatic sulfide deposits. *Oxford Monograph on Geology and Geophysics*, **14**, 17–38.

Navon, O. 1999. Diamond formation in the Earth's mantle. *In*: Gurney, J. J., Gurney, J. L., Pascoe, M. D. & Richardson, S. H. (eds) *The P.H. Nixon Volume, Proceedings of the VIIth International Kimberlite Conference*. Red Roof Design, Cape Town, 584–604.

O'Neill, H. St. C. & Mavrogenes, J. A. 2002. The sulfide capacity and the sulfide content at sulfide saturation of silicate melts at 1400°C and 1 bar. *Journal of Petrology*, **43**, 1049–1097.

Ohmoto, H. 1986. Stable isotope geochemistry of ore deposits. *In*: Valley, J. W., Taylor, H. P. Jr. & O'Neill, J. R. (eds) *Stable Isotopes in High Temperature Geological Processes*. Reviews in Mineralogy and Geochemistry, **16**, 491–560.

Pal'yanov, Yu. N., Borzdov, Yu. M., Bataleva, Yu. V., Sokol, A. G., Palyanova, G. A. & Kupriyanov, I. N. 2007. Reducing role of sulfides and diamond formation in the Earth's mantle. *Earth and Planetary Science Letters*, **260**, 242–256.

Pal'yanov, Yu. N., Sokol, A. G., Borzdov, Yu. M. & Khokhryakovù, A. F. 2002. Fluid-bearing alkaline carbonate melts as the medium for the formation of diamonds in the Earth's mantle: an experimental study. *Lithos*, **60**, 145–159

Pasteris, J. D. 1981. Occurrence of graphite in serpentinized olivines in kimberlite. *Geology*, **9**, 356–359.

Pattou, L., Lorand, J. P. & Gros, M. 1996. Non-chondritic platinum-group element ratios in the Earth's mantle. *Nature*, **379**, 712–715.

Pearson, D. G., Davies, G. R. & Nixon, P. H. 1993. Geochemical constraints on the petrogenesis of diamond facies pyroxenites from the Beni Bousera peridotite massif, North Morocco. *Journal of Petrology*, **34**, 125–172.

Pearson, D. G., Davies, G. R., Nixon, P. H. & Mattey, D. P. 1991. A carbon isotope study of diamond facies pyroxenites and associated rocks from the Beni Bousera peridotite, North Morocco. *In*: Menzies, M. A., Dupuy, C. & Nicolas, A. (eds) *Orogenic Lherzolites and Mantle Processes*. Journal of Petrology, Special Volume, 175–189.

Pearson, D. G., Boyd, F. R., Haggerty, S. E., Pasteris, J. D., Field, S. W., Nixon, P. H. & Pokhilenko, N. P. 1994. The characterisation and origin of graphite in cratonic lithospheric mantle: a petrological carbon isotope and Raman spectroscopic study. *Contributions to Mineralogy and Petrology*, **115**, 449–466.

Peregoedova, A. & Ohnenstetter, M. 2002. Collectors of Pt, Pd and Rh in a S-poor Fe–Ni–Cu sulfide system at 760 °C: experimental data and application to ore deposits. *Canadian Mineralogist*, **40**, 527–561.

PERTERMANN, M. & HIRSCHMANN, M. M. 2003. Anhydrous partial melting experiments on MORB-like eclogite: phase relations, phase compositions and mineral-melt partitioning of major elements at 2–3 GPa. *Journal of Petrology*, **44**, 2173–2201.

PICCARDO, G. B. 2008. The Jurassic Ligurian Tethys, a fossil ultraslow-spreading ocean: the mantle perspective. *In*: COLTORTI, M. & GRÉGOIRE, M. (eds) *Metasomatism in Oceanic and Continental Lithospheric Mantle*. Geological Society, London, Special Publications, **293**, 11–33.

PICCARDO, G. B., MÜNTENER, O., ZANETTI, A. & PETTKE, T. 2004. Ophiolitic peridotites of the Alpine–Apennine system: mantle processes and geodynamic relevance. *International Geological Review*, **46**, 1119–1159.

POWELL, W. & O'REILLY, S. 2007. Metasomatism and sulfide mobility in lithospheric mantle beneath eastern Australia: implications for mantle Re–Os chronology. *Lithos*, **94**, 132–147.

RAMPONE, E., HOFFMANN, A. W., PICCARDO, G. B., VANNUCCI, R., BOTTAZZI, P. & OTTOLINI, L. 1995. Petrology, mineral and isotope geochemistry of the External Liguride peridotites (northern Apennine, Italy). *Journal of Petrology*, **36**, 81–105.

RYZHENCO, B. & KENNEDY, G. C. 1973. The effect of pressure on the eutectic of the system Fe–FeS. *American Journal of Science*, **273**, 803–810.

RICHARDSON, S. H., SHIREY, S. B. & HARRIS, J. W. 2004. Episodic diamond genesis at Jwaneng, Botswana, implications for Kaapvaal craton evolution. *Lithos*, **77**, 143–154.

SCHEELE, N. & HOEFS, J. 1992. Carbon isotope fractionation between calcite, graphite and CO_2: an experimental study. *Contributions to Mineralogy and Petrology*, **112**, 35–45.

SCHULZE, D. J., VALLEY, J. W., VILJOEN, K. S., STIEFENHOFER, J. & SPICUZZA, M. 1997. Carbon isotope composition of graphite in mantle eclogites. *Journal of Geology*, **105**, 379–386.

SIMAKOV, S. 2006. Redox state of eclogites and peridotites from sub-cratonic upper mantle and a connection with diamond genesis. *Contributions to Mineralogy and Mineralogy*, **151**, 282–296.

SOBOLEV, N. V., KAMINSKY, F. V., GRIFFIN, W. L., YEFIMOVA, E. S., WIN, T. T., RYAN, C. G. & BOTKUNOV, A. I. 1997. Mineral inclusions in diamonds from the Sputnik kimberlite pipe, Yakutia. *Lithos*, **39**, 135–157.

SNOW, J. E., SCHMIDT, G. & RAMPONE, E. 2000. Os isotopes and highly siderophile elements (HSE) in the Ligurian ophiolites, Italy. *Earth and Planetary Science Letters*, **175**, 119–132.

SZABO, CS. & BODNAR, R. J. 1995. Chemistry and origin of mantle sulfides in spinel peridotite xenoliths from alkaline basaltic lavas, Nograd-Gomor Volcanic Field, northern Hungary and southern Slovakia. *Geochimica et Cosmochimica Acta*, **59**, 3917–3927.

SUGAKI, A. & KITAKAZE, A. 1998. High form of pentlandite and its thermal stability. *American Mineralogist*, **83**, 133–140.

THOMASSOT, E., CARTIGNY, P., HARRIS, J. W. & VILJOEN, K. S. 2007. Methane-related diamond crystallization in the Earth's mantle: stable isotope evidences from a single diamond-bearing xenolith. *Earth and Planetary Science Letters*, **257**, 362–371.

THOMPSON, R. N. 1974. Some high-pressure pyroxenes. *Mineralogical Magazine*, **39**, 768–787.

TOMKINS, A., PATTISON, D. R. M. & FROST, B. R. 2007. On the initiation of metamorphic sulfide anatexis. *Journal of Petrology*, **48**, 511–535.

TOROK, K., BALI, E., SZABÓ, C. & SZAKÁ, J. A. 2003. Sr–barite droplets associated with sulfide blebs in clinopyroxene megacrysts from basaltic tuff (Szentbékkálla, western Hungary). *Lithos*, **66**, 275–289.

USSELMAN, T. M. 1975. Experimental approach to the state of the core: Part I. The liquidus relations of the Fe-rich portion of the Fe–Ni–S system from 30–100 Kbars. *American Journal of Science*, **275**, 278–290.

WENDLANDT, R. F. 1982. Sulfide saturation of basalt and andesite melts at high pressures and temperatures. *American Mineralogist*, **67**, 877–885.

WOODLAND, A. B., KORNPROBST, J. & WOOD, B. J. 1992. Oxygen thermobarometry of orogenic lherzolite massifs. *Journal of Petrology*, **33**, 203–230.

WOODLAND, A. B. & KOCH, M. 2003. Variation in oxygen fugacity with depth in the upper mantle beneath the Kaapvaal craton, Southern Africa. *Earth and Planetary Science Letters*, **214**, 295–310.

WOPENKA, B. & PASTERIS, J. D. 1993. Structural characterization of kerogens to granulite-facies graphite: applicability of Raman microprobe spectroscopy. *American Mineralogist*, **78**, 533–557.

YAXLEY, G. M. & BREY, G. P. 2003. Phase relations of carbonate-bearing eclogite assemblages from 2.5 to 5.5 GPa: implications for petrogenesis of carbonatites. *Contributions to Mineralogy and Petrology*, **146**, 606–619.

YAXLEY, G. M. & GREEN, D. H. 1998. Reactions between eclogite and peridotite: mantle refertilisation by subduction of oceanic crust. *Schweizerische Mineralogische und Petrographische Mitteilungen*, **78**, 243–255.

YAXLEY, G. M. & SOBOLEV, A. V. 2007. High-pressure partial melting of gabbro and its role in the Hawaiian magma source. *Contributions to Mineralogy and Petrology*, **154**, 371–383.

ZAJACZ, Z. & SZABO, C. 2003. Origin of sulfide inclusions in cumulate xenoliths from Nograd–Gömör Volcanic Field, Pannonian Basin (north Hungary/south Slovakia). *Chemical Geology*, **194**, 105–117.

Mantle metasomatism by melts of HIMU piclogite components: new insights from Fe-lherzolite xenoliths (Calatrava Volcanic District, central Spain)

GIANLUCA BIANCHINI[1,2]*, LUIGI BECCALUVA[2], COSTANZA BONADIMAN[2], GEOFF M. NOWELL[3], D. GRAHAM PEARSON[3], FRANCA SIENA[2] & MARJORIE WILSON[4]

[1]*CNR – Istituto di Geoscienze e Georisorse, Via G. Moruzzi 1, 56124 Pisa, Italy*

[2]*Dipartimento di Scienze della Terra, Università di Ferrara, Via Saragat 1, 44100 Ferrara, Italy*

[3]*Department of Earth Sciences, University of Durham, South Road, Durham DH1 3LE, UK*

[4]*School of Earth and Environment, University of Leeds, Leeds LS2 9JT, UK*

**Corresponding author (e-mail: g.bianchini@igg.cnr.it)*

Abstract: Mantle xenoliths from the Calatrava Volcanic District (CLV), central Spain, are characterized by a wide compositional range that includes lherzolites (prevalent), as well as minor amounts of wehrlite, olivine (ol)-websterite and rare dunites. They generally have a bulk-rock Mg# of less than 89, lower than any primordial mantle estimates. Intra-suite variations in modal proportions are inconsistent with those predicted by melting models irrespective of the starting composition; mineral and bulk-rock variation diagrams show inconsistencies between the CLV compositions (anomalously enriched in Fe–Ti) and those predicted from the partial melting of primordial mantle material. Processes other than pure melt extraction are confirmed by the whole-rock REE (rare earth element) budget, typically characterized by LREE enrichments, with La_N/Yb_N (up to 6.7), probably related to pervasive metasomatism. CLV mantle clinopyroxenes (cpx) generally display fractionated REE patterns with upwards-convex shapes, characterized by low HREE (Tm–Lu) concentrations (typically $<6\times$ chondrite) and enrichments in middle–light REE (MREE–LREE) (Nd_N/Yb_N up to 7, La_N/Yb_N up to 5). These 'enriched' cpx compositions either result from re-equilibration of primary mantle cpx with an incoming melt, or represent cpx crystallization directly from the metasomatic agent. The latter was plausibly generated at greater depths in the presence of residual garnet (from peridotite or eclogite starting materials). Separated cpx have homogeneous $^{87}Sr/^{86}Sr$ compositions between 0.7031 and 0.7032; $^{143}Nd/^{144}Nd$ ranges from 0.51288 to 0.51295 (εNd 4.74–6.07) and $^{176}Hf/^{177}Hf$ is in the range 0.28302–0.28265 (εHf −3.6 to 9.0). Unlike mantle xenoliths and alpine-type peridotites from other Iberian occurrences, which range in composition from the depleted mantle (DM) to the enriched mantle (EM), the CLV mantle cpx approach the composition of the HIMU mantle end member, the genesis of which is generally interpreted as the result of long-term recycling of oceanic basalts/gabbros (or their eclogitic equivalent) via ancient subduction. A model is proposed for the mantle evolution under central Iberia, where sublithospheric convective instabilities – possibly triggered by the neighbouring subduction along the Betic collisional belt – could have remobilized deep domains from the mantle 'transition zone' (410–660 km), which may include relicts of older subducted slabs. Within these remobilized domains, characterized by the coexistence of peridotite and eclogite and referred to as a 'piclogite' association, the eclogites melt preferentially generating Fe–Ti rich melts characterized by a HIMU isotopic signature that infiltrates and metasomatizes the shallower lithospheric mantle.

Supplementary material: an extended dataset for calatrava xenoliths is available at http://geolsoc.org.uk/sup18410.

Most suites of mantle xenoliths and orogenic peridotites have Mg# values [molar MgO/(MgO + FeO) × 100] of more than 89 and show intra-suite variations in modal mineralogy (lherzolite–harzburgite); parallel variations in the bulk-rock chemical composition can generally be satisfactorily accounted for by partial melting processes and the extraction of basaltic magma. This is reflected in the high Mg# (>89) of the constituent silicate phases, which progressively increases in the more refractory (i.e. melt-depleted) mantle domains.

Pearson & Wittig (2008) emphasized these characteristics in mantle xenoliths from cratonic regions, which typically display very high Mg# olivine (92.6–92.8), and, to lesser extent, in peridotite xenoliths from OIB (ocean island basalt)

From: Coltorti, M., Downes, H., Grégoire, M. & O'Reilly, S. Y. (eds) *Petrological Evolution of the European Lithospheric Mantle*. Geological Society, London, Special Publications, **337**, 107–124.
DOI: 10.1144/SP337.6 0305-8719/10/$15.00

settings (where the Mg# of olivines is mostly between 91 and 92: Simon *et al.* 2008). Within off-craton localities, such as the European circum-Mediterranean region (Downes 2001), the olivine Mg# in peridotite xenoliths is 89.3–91.3 in Sardinia (Beccaluva *et al.* 2001*a*), 89.3–90.8 in the Hyblean area, Sicily (Perinelli *et al.* 2007), 89.0–89.7 in the Gharyan volcanic field (Beccaluva *et al.* 2008), 89.0–91.5 in Oranie, Tell, Algeria (Zerka *et al.* 2002), mostly between 89.0 and 92.0 in peridotite xenoliths from the Massif Central (Werling & Altherr 1997; Touron *et al.* 2008), and mostly between 89.0 and 91.5 in peridotite xenoliths from the Veneto volcanic province, where rare samples display olivine Mg# values down to 87.0 (Beccaluva *et al.* 2001*b*). Similar values are also recorded in mantle xenoliths from the Iberian peninsula, with olivine Mg# 89.4–90.9 in anhydrous peridotite xenoliths from Tallante (Beccaluva *et al.* 2004) and 89.2–91.8 in peridotite xenoliths from the Olot volcanic district (Bianchini *et al.* 2007).

In contrast, in the studied suite of mantle xenoliths from Calatrava (CLV; Fig. 1), the prevailing lherzolite lithology characteristically presents a relatively high iron content. This is reflected in the low Mg# values of both bulk rock (typically <89, down to 86.5) and the equilibrium mineral paragenesis (e.g. olivine Mg# down to 84.9), which are lower than any primordial mantle estimates (e.g. McDonough & Sun 1995 or O'Neill & Palme 1998). This suggests that the mantle beneath Calatrava may have experienced an unusual evolution, which led to significant variations in composition (and mode?) of the primary mantle paragenesis.

In this study we use a wide range of analytical data [X-ray diffraction (XRD), ICP-MS (inductively coupled mass spectrometry), EMPA (electron microprobe analysis), LAM-ICP-MS (laser ablation microprobe inductively coupled plasma mass spectrometry), TIMS (thermal ionization mass spectrometry) and MC-ICP-MS (multicollector inductively coupled plasma mass spectrometry)] in order to characterize the CLV peridotites and to constrain the nature of the metasomatic agents that modified the CLV lithospheric mantle. This, in turn, provides fresh insights into the tectono-magmatic evolution of the area, with particular regard to geochemical fluxes of mantle components between lithospheric and sublithospheric reservoirs.

Geological background

The Late Miocene–Quaternary Calatrava Volcanic Province (CLV) in central Spain, close to Ciudad Real, consists of a series of scattered volcanic centres (most of which are monogenetic) characterized by lava flows and pyroclastic deposits. This volcanic activity developed in an extensional tectonic setting in the foreland of the subduction-related Betic Cordillera (López-Ruiz *et al.* 1993). The volcanic products were erupted onto a Palaeozoic basement, and are mainly associated with NE–SW and east–west extensional and strike-slip faults (many of which appear to be reactivated Hercynian lineaments) in an area where geophysical data highlighted significant crustal thinning (down to 30 km: Bergamin & Carbo 1986).

CLV volcanism was characterized by early sporadic eruptions of leucitites (Late Miocene) followed by Pliocene–Pleistocene eruptions of alkali basalts, nephelinites and melilitites (Cebriá &

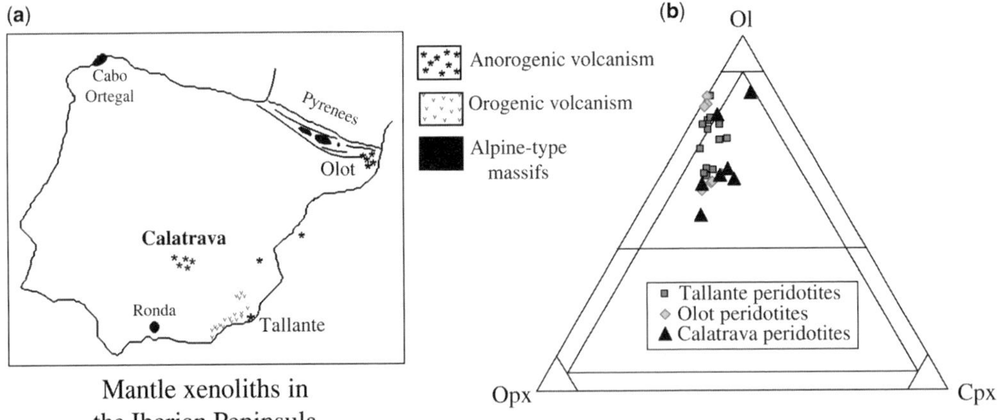

Fig. 1. (**a**) Sketch map of the Iberian Peninsula, indicating the locations of the Tertiary volcanic fields. Locations of volcanic rocks containing mantle xenoliths (Calatrava, Olot and Tallante) and Alpine-type peridotite massifs are indicated. (**b**) Modal composition of mantle xenoliths from Calatrava, Olot and Tallante in terms of olivine (ol), orthopyroxene (opx) and clinopyroxene (cpx).

López-Ruiz 1995). Recent investigations of the many diatremes and maars within the province indicate that the contemporaneous pyroclastic deposits are often carbonate-rich, suggesting that the parental magmas were CO_2-rich (Bailey *et al.* 2005).

Mantle xenoliths exhumed by the CLV volcanics have also been described by Ancochea & Nixon (1987) and Villaseca *et al.* (2010) who documented peridotite parageneses, mainly lherzolitic–wehrlitic in composition, sometimes showing distinctive accessory phases such as amphibole, phlogopite, glass and occasionally titanite (sphene).

In this study we focus our attention on a mantle xenolith population collected from the El Palo lava cone located approximately 4 km west of Ciudad Real (latitude 38°59′45″N; longitude 3°58′30″W). The samples consist of 60 xenoliths up to 5–6 cm in diameter, including Cr-diopside (green) peridotites with prevalent lherzolites and minor wehrlites, rare ol-websterite (containing black augite clinopyroxene) and dunites.

Analytical methods

The rock samples were crushed and the freshest chips were powdered in an agate mill. Major and trace elements (Ni, Co, Cr, V and Sr) were analysed on powder pellets, using a wavelength-dispersive automated Philips PW 1400 X-ray spectrometer at the Department of Earth Sciences, Ferrara University. Accuracy and precision for major elements were estimated as better than 3% for Si, Ti, Fe, Ca and K, and 7% for Mg, Al, Mn, Na and P; for trace elements (above 10 ppm) they are better than 10%. REE, Sc, Y, Zr, Hf, Nb, Th and U were analysed by ICP-MS at the Department of Earth Sciences, Ferrara University, using a VG Plasma Quad2 plus mass spectrometer. Accuracy and precision, based on replicate analyses of samples and standards, were estimated as better than 10% for all elements, well above the detection limit.

Mineral compositions were obtained at the CNR-IGG Institute of Padova using a Cameca-Camebax electron microprobe, fitted with three wavelength-dispersive spectrometers, at an accelerating voltage of 15 kV and a specimen current of 15 nA, using natural silicates and oxides as standards. Trace element analyses of clinopyroxenes were carried out at the CNR-IGG in Pavia by LAM-ICP-MS, using an Elan DRC-e mass spectrometer coupled with a Q-switched Nd:YAG laser source (Quantel Brilliant). The CaO content was used as an internal standard for both clinopyroxene and amphibole analyses. Precision and accuracy, better than 10% for concentrations at ppm level, were assessed by repeated analyses of NIST SRM 612 and BCR-2 standards.

Strontium isotopic analyses on separated clinopyroxene (*c.* 50 mg) were carried out at the School of Earth and Environment, University of Leeds. Samples were leached with hot 6 M HCl and digested with $HF–HNO_3$. Sr was separated and concentrated by conventional techniques using cation-exchange chromatographic columns, and analysed with a TIMS Finnigan-MAT TRITON mass spectrometer. The analyses of the NBS987 standard gave results analogous to the certified value, errors on the sample analyses were better than ± 0.00001 (2σ) and reproducibility, based on replicate analyses, was always better than 0.001%.

The Hf and Nd isotope compositions of clinopyroxene separates (*c.* 100 mg) were measured at the Department of Earth Sciences, Durham University. After leaching (6 M HCl, 30 min) and sample dissolution, pre-concentration was performed using a two-column procedure that employed a 5 ml cation separation as the first step, using 1 N HF–1 N HCl to elute Hf and 6 N HCl to elute Nd, followed by a mixed sulphuric acid–H_2O_2 anion column for final purification of the Hf (Dowall *et al.* 2003). Measurements of samples and standards (JMC-475 and J&M) were made on a ThermoFinnigan Neptune plasma ionization multicollector mass spectrometer, with precision and accuracy better than 0.001% (Pearson & Nowell 2004, 2005). During the analytical sessions, the average value of the J&M Nd standard was 0.511097 ± 0.000012 (2σ, $n = 12$), the average value of the JMC-475 Hf standard was 0.282143 ± 0.000008 (2σ, $n = 8$) and errors on the sample analyses were generally better than ± 0.00002 (2σ).

Petrography and mineral chemistry of lherzolite xenoliths from Calatrava

The CLV lherzolite xenoliths considered in this study are typically characterized by a four-phase mineral assemblage including olivine (ol), orthopyroxene (opx), clinopyroxene (cpx) and spinel (sp). As shown in microphotographs of Figure 2, CLV lherzolites exhibit protogranular textures overprinted by widespread pyrometamorphic reaction domains (grain boundaries and patches characterized by subgraining: Pike & Schwarzman 1976). The latter consist of spongy clinopyroxene, and reaction rims/patches around orthopyroxenes and spinel with ‘secondary’ microcrysts of olivine, clinopyroxene, feldspar, and rarely phlogopite, apatite and carbonate, as well as interstitial glassy blebs. These reaction textures, recorded in different xenolith suites from various localities of Calatrava (Humphreys *et al.* 2008*a*; Villaseca *et al.* 2010), do not reflect interaction with the host basalts and were plausibly formed at mantle depths before

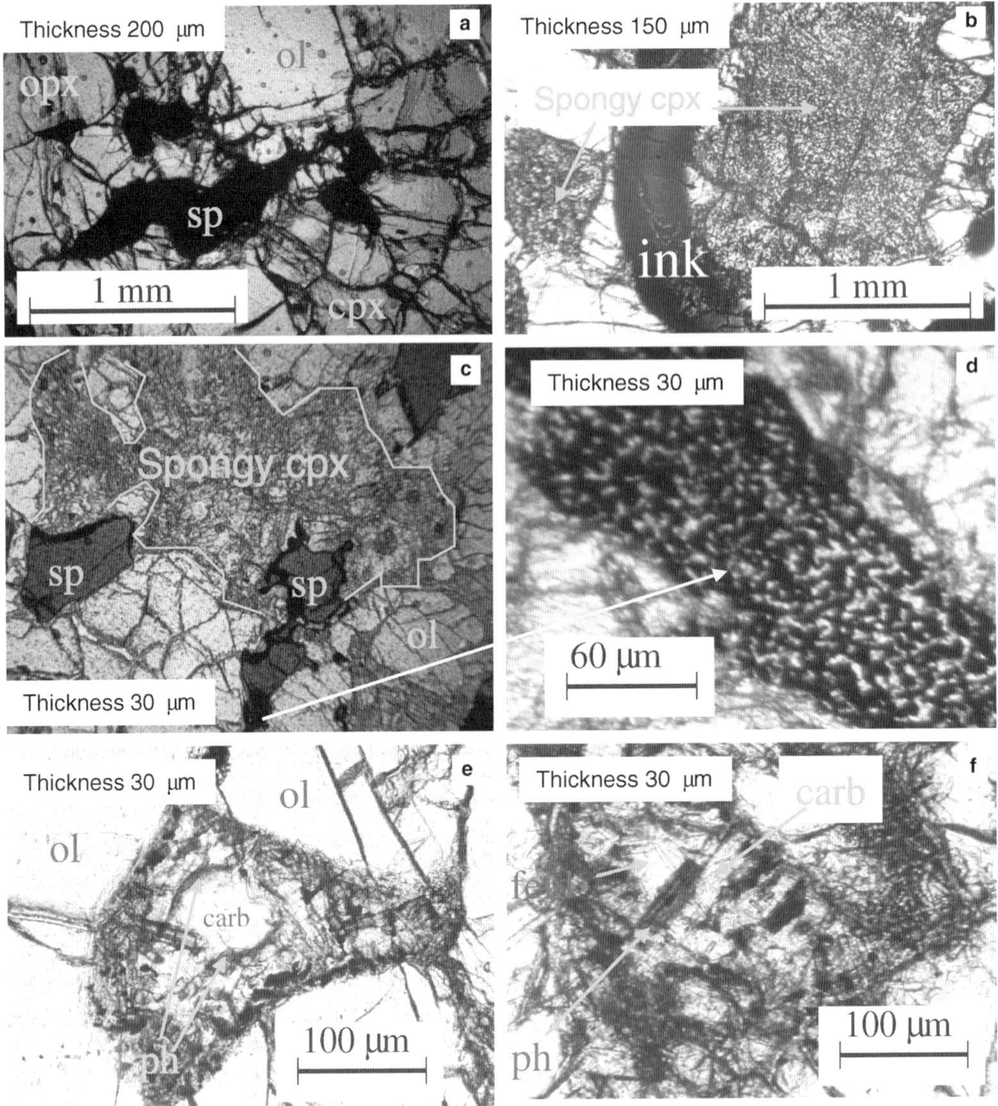

Fig. 2. Petrographical features of mantle xenoliths from Calatrava. (**a**) Typical protogranular texture of a well-equilibrated sample; (**b**) typical spongy clinopyroxene; (**c**) spongy clinopyroxene and spinels characterized by brown core and black rims; (**d**) magnification of a spinel rim consisting of a fine aggregate of vermicular Cr–Fe–Ti-rich spinel and feldspar; and (**e**) and (**f**) minute patches containing carbonate, feldspar and phlogopite. Abbreviations: ol, olivine; opx, orthopyroxene; cpx, clinopyroxene; sp, spinel; feld, feldspar; carb, carbonate; ph, phlogopite.

the exhumation of the xenoliths (Siena *et al.* 1991; Coltorti *et al.* 1999, 2000; Beccaluva *et al.* 2001*b*, 2008; Delpech *et al.* 2004; Bonadiman *et al.* 2005).

Electron microprobe analyses of the constituent phases in the CLV lherzolites are reported in Table 1 and in the supplementary data available on the Society's website. Primary olivine (ol) consists of large kink-banded crystals (up to 1 mm) with Fo content ranging from 86.4 to 89.6. Lower Fo content in olivine (down to Fo 83.6) are recorded in olivine microcrysts in reaction patches. Anomalous low Fo values have also been recorded in mantle xenoliths from other Calatrava occurrences (Humphreys *et al.* 2008*a*; Villaseca *et al.* 2010).

Table 1. *Average composition of (**a**) olivines, (**b**) orthopyroxenes, (**c**) clinopyroxenes and (**d**) spinels from the Calatrava lherzolite xenoliths. Mg# is expressed as* $[Mg/(Mg + Fe^{2+})] \times 100$ *and Cr# as* $[Cr/(Cr + Al)] \times 100$ *from recalculated mineral analyses. A more extended dataset, including the averaged analyses, is available as Supplementary Material (see p. 107)*

Sample	CLV8	CLV11	CLV12	CLV15a	CLV16	CLV18b	CLV54b	CLV58
(**a**) Olivines	(2)	(4)	(4)	(6)	(3)	(4)	(2)	(2)
SiO_2	40.52	39.80	40.36	41.74	40.57	41.02	40.38	40.81
TiO_2	0.01	0.03	0.03	0.01	0.00	0.01	0.02	0.01
Al_2O_3	0.00	0.01	0.01	0.00	0.01	0.02	0.02	0.03
FeO_{tot}	11.51	14.24	11.39	10.08	12.91	10.49	10.22	11.34
MnO	0.12	0.23	0.22	0.18	0.19	0.11	0.19	0.14
MgO	47.30	45.00	46.98	48.69	46.21	48.35	48.71	47.72
CaO	0.08	0.15	0.11	0.03	0.09	0.13	0.08	0.11
Na_20	0.02	0.01	0.01	0.01	0.01	0.01	0.00	0.00
NiO	n.a.	0.15	0.23	0.32	n.a.	n.a.	0.29	0.31
Total	99.56	99.62	99.32	101.06	99.99	100.14	99.89	100.47
Mg#	88.0	84.9	88.0	89.6	86.5	89.1	89.5	88.2
(**b**) Orthopyroxenes		(3)	(2)	(2)	(3)	(2)	(2)	(2)
SiO_2		53.40	54.22	56.82	54.16	54.29	53.80	54.57
TiO_2		0.36	0.26	0.08	0.18	0.09	0.17	0.15
Al_2O_3		5.24	5.32	3.61	5.10	5.63	5.56	4.77
FeO_{tot}		8.73	7.72	6.41	8.23	6.58	6.37	7.47
MnO		0.17	0.20	0.18	0.17	0.09	0.14	0.21
MgO		29.93	31.33	33.19	30.72	31.65	31.68	31.74
CaO		1.11	1.04	0.52	1.03	1.09	1.03	1.11
Na_2O		0.07	0.09	0.01	0.07	0.13	0.13	0.09
Cr_2O_3		0.36	0.43	0.25	0.40	0.43	0.47	0.65
Total		99.37	100.61	101.06	100.06	99.98	99.33	100.76
Mg#		85.9	88.0	89.7	86.9	89.6	90.0	88.6
(**c**) Clinopyroxenes	(5)	(4)	(7)	(5)	(3)	(6)	(4)	(5)
SiO_2	51.19	51.92	49.34	53.23	50.64	51.68	51.09	51.33
TiO_2	0.57	0.86	2.19	0.38	0.57	0.30	0.92	0.43
Al_2O_3	5.00	2.71	5.18	5.06	6.73	6.28	3.26	6.36
FeO_{tot}	3.46	3.96	4.46	2.78	4.32	3.44	4.07	4.00
MnO	0.10	0.10	0.10	0.10	0.12	0.11	0.08	0.10
MgO	15.83	16.37	14.34	15.56	14.96	15.89	15.65	15.70
CaO	21.94	22.02	22.00	21.46	19.52	20.28	23.09	19.94
Na_2O	0.51	0.54	0.59	1.25	1.24	0.95	0.46	1.23
Cr_2O_3	0.95	0.96	1.07	0.71	0.80	0.80	0.96	0.99
Total	99.55	99.46	99.26	100.53	98.90	99.73	99.58	100.08
Mg#	89.50	88.90	85.10	90.80	86.60	89.30	88.50	88.60
(**d**) Spinels	(2)	(1)	(2)		(2)	(2)	(2)	(2)
TiO_2	0.36	0.76	0.53		0.36	0.14	0.19	0.32
Al_2O_3	48.68	52.00	51.00		49.11	54.57	54.12	44.61
FeO_{tot}	14.00	15.18	15.60		16.96	12.79	11.87	15.68
MnO	0.07	0.11	0.12		0.16	0.09	0.08	0.14
MgO	18.14	17.96	18.30		18.01	20.12	20.16	18.08
CaO	0.30	0.00	0.04		0.02	0.00	0.01	0.09
Cr_2O_3	12.42	11.61	11.25		12.76	9.95	10.50	17.69
NiO	n.a.	0.16	0.31		n.a.	n.a.	0.30	0.25
Total	93.96	97.78	97.14		97.38	97.67	97.22	96.87
Cr#	14.6	13.0	12.9		14.8	10.9	11.5	21.0
Mg#	75.9	71.9	74.2		73.3	79.5	80.4	75.6

The number of average analyses is reported in parentheses.
Abbreviation: n.a., not analysed.

The opx composition varies in the following ranges: Wo 1.0–2.2, En 83.7–88.5, Fs 10.2–14.1, with Mg# 85.9–90. The cpx composition varies in the following ranges: Wo 44.3–48.4, En 43.8–49.0, Fs 5.0–7.9, with Mg# 85.1–90.1; TiO_2 wt% is 0.3–0.6 in texturally equilibrated clinopyroxenes unaffected by reaction, but increases in reacted cpx (especially those of samples CLV11, CLV12 and CLV54b). The application of the two-pyroxene geothermometers of Brey & Köhler (1990) using the compositions of large equilibrated crystals from samples less affected by secondary reaction textures (CLV16, CLV18b and CLV58) gives nominal equilibration temperatures in the range 1040–1100 °C, which appear slightly higher than those recorded in comparable lherzolites from other Spanish mantle xenoliths occurrences (Beccaluva *et al.* 2004; Bianchini *et al.* 2007).

Spinels typically show holly-leaf or lobate shapes and are generally characterized by brown cores with Mg# 71.9–80.4 and Cr/(Cr + Al) 0.11–0.21; black rims have lower Mg# (down to 40.0) and higher Cr/(Cr + Al) (up to 0.5), and gradually transform into a fine aggregate of vermicular Cr–Fe–Ti-rich spinel and feldspar.

Feldspar occurs only as a reaction product in the fine-grained reaction patches, varying in composition between andesine-labradorite (An 44–55, Ab 43–54) and variably potassic alkali feldspar (Ab 47–91, Or 7–52). This wide compositional spectrum encompasses the complete range of feldspar compositions in continental peridotite xenoliths (e.g. Ionov *et al.* 1999; Grégoire *et al.* 2000; Beccaluva *et al.* 2001*b*, 2008; Delpech *et al.* 2004; Bonadiman *et al.* 2005), and cannot be interpreted as an infiltration effect of the host lava (a plagioclase-bearing basanite).

Some carbonate-bearing patches – consisting of minute (<400 μm) ocelli-like textures that contain an intergrowth of calcite, feldspar, Ti-rich phlogopite and apatite – may be primary, in the sense that they appear unrelated to surface weathering or hydrothermal alteration. Similar carbonate patches in xenoliths from Calatrava have also been recorded by Humphreys *et al.* (2008*a*).

As noted above, and illustrated in Figures 3 and 4, the mineral compositions in Calatrava lherzolites are distinctly different from those of other Spanish occurrences (i.e. Tallante: Beccaluva *et al.* 2004; Olot: Bianchini *et al.* 2007), as well as from those of other European peridotite xenolith populations (Downes & Coltorti 2008). FeO enrichment in the pyroxenes is correlated with TiO_2, suggesting that the causative agent that lowered the Mg# of the CLV lherzolites was enriched in both Fe and Ti (Fig. 4).

It has to be noted that new analyses of other mantle xenoliths collected in the Calatrava volcanic field, from the volcanic centres of Cerro Pelado and El Aprisco, seem to confirm Fe and Ti enrichment trends of the constituting mineral phases; these enrichments also characterize the composition of glass films and veins (see xenoliths from Cerro Pelado: Villaseca *et al.* 2010).

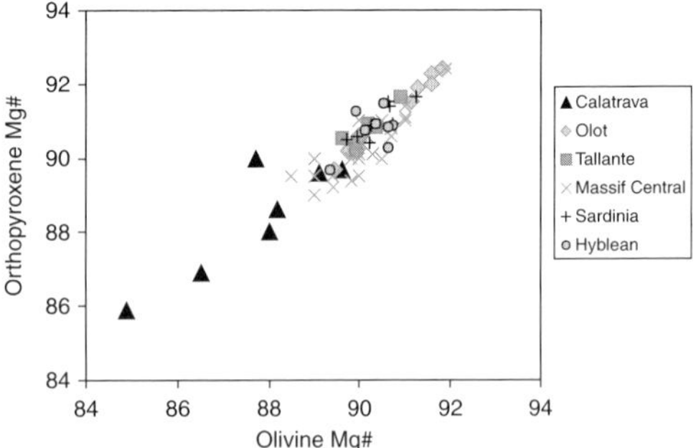

Fig. 3. Mg# of orthopyroxene v. Mg# of olivine. Mg# is expressed as $[Mg/(Mg + Fe^{2+})] \times 100$ from recalculated mineral analyses (a.f.u. on the basis of six and four oxygens, respectively). Representative compositions of Calatrava lherzolite xenoliths are compared with those recorded in mantle xenoliths from Olot (Bianchini *et al.* 2007), Tallante (Beccaluva *et al.* 2004) Massif Central (Werling & Altherr 1997; Touron *et al.* 2008), Sardinia (Beccaluva *et al.* 2001*a*) and the Hyblean area (Perinelli *et al.* 2008). (a.f.u., atomic formula units.)

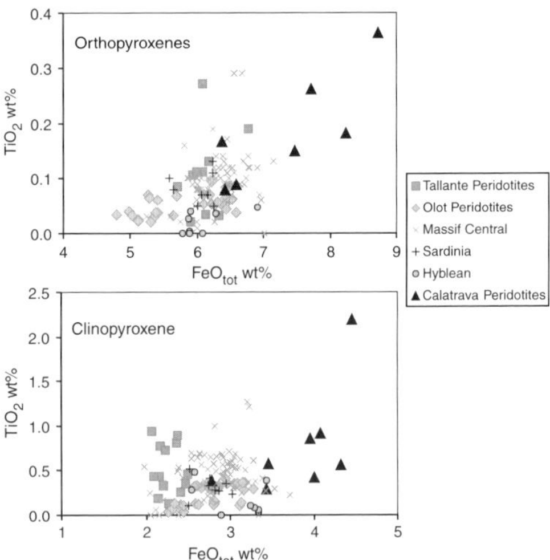

Fig. 4. TiO_2 v. FeO_{tot} content of orthopyroxene and clinopyroxene in Calatrava lherzolite xenoliths compared with those in mantle xenoliths from Olot (Bianchini *et al.* 2007), Tallante (Beccaluva *et al.* 2004) Massif Central (Werling & Altherr 1997; Touron *et al.* 2008), Sardinia (Beccaluva *et al.* 2001*a*) and the Hyblean area (Perinelli *et al.* 2008).

Whole-rock geochemistry of lherzolite xenoliths from Calatrava

Bulk-rock major and trace element compositions are reported in Table 2. Major element mass-balance calculations between the mineral phases and the whole-rock compositions indicate that in the investigated lherzolites, olivine varies between 49 and 84%, orthopyroxene between 6 and 35% and clinopyroxene between 5 and 18%.

At first sight, the modal proportions resemble those expected for residues produced by variable degrees of melt extraction from a fertile mantle source. CaO and Al_2O_3 variations as a function of MgO are comparable with those observed in other mantle xenolith suites from the Mediterranean area, and closely match theoretical residual compositions (batch and fractional melting, between 10 and 20 kbar: Niu 1997). However, significant inconsistencies between the Calatrava bulk-rock compositions and those predicted on the basis of model partial melting trends are highlighted by the anomalously high FeO_{tot} and TiO_2 content, suggesting that melt extraction was not the only process to

Table 2. *Major (wt%) and trace element (ppm) whole-rock analyses of representative lherzolite xenoliths from Calatrava. Modal proportions (%) of the major phases are calculated by mass balance based upon the major element bulk-rock and mineral chemistry. Mg# is expressed as* $[Mg/(Mg + Fe_{tot})] \times 100$

Sample	CLV8	CLV9a	CLV16	CLV18b	CLV49d	CLV54b	CLV58
Texture	Protogranular texture						
SiO_2	45.04	46.40	45.08	45.72	42.62	45.33	43.71
TiO_2	0.24	0.17	0.28	0.11	0.12	0.11	0.09
Al_2O_3	2.68	4.24	2.45	2.80	1.03	3.03	1.21
FeO	9.84	8.94	10.58	8.58	10.26	8.43	9.90
MnO	0.15	0.13	0.14	0.13	0.14	0.13	0.14
MgO	37.92	36.50	38.13	39.69	43.49	40.02	43.56
CaO	4.11	3.34	3.27	2.96	2.30	2.89	1.38
Na2O	0.01	0.16	0.00	0.00	0.00	0.00	0.00
K_2O	0.01	0.11	0.05	0.01	0.04	0.00	0.01
P_2O_5	0.00	0.01	0.01	0.00	0.00	0.05	0.00
Mg#	87.3	87.9	86.5	89.2	88.3	89.4	88.7
Ni	1887	1815	1532	2069	2225	2049	2200
Co	124	98	115	108	122	107	121
Cr	3132	4222	2118	2542	1374	3199	1737
V	89	93	87	70	42	64	41
Sr	17	38	78	14	37	21	10
Sc	31.1	30.4		32.4	14.1	20.8	12.1
Y	4.35	4.09	2.60	4.89	1.28	2.88	1.22
Zr	18.1	11.8	11.5	9.68	7.69	5.20	5.26
Hf	0.54	0.26	0.41	0.22	0.25	0.14	0.15
Nb	1.08	1.92	1.74	0.57	1.67	1.50	0.89
La	2.37	1.92	1.26	1.15	0.93	1.38	0.48
Ce	4.65	4.09	3.16	2.88	2.09	2.08	1.31
Pr	0.72	0.47	0.47	0.45	0.26	0.22	0.17
Nd	3.30	1.86	2.27	2.09	1.25	0.85	0.78
Sm	0.81	0.50	0.61	0.60	0.26	0.25	0.22
Eu	0.27	0.19	0.20	0.19	0.09	0.09	0.07
Gd	0.76	0.56	0.66	0.61	0.29	0.33	0.21
Tb	0.14	0.10	0.10	0.12	0.04	0.06	0.03
Dy	0.70	0.57	0.54	0.67	0.23	0.42	0.23
Ho	0.14	0.14	0.10	0.15	0.04	0.10	0.05
Er	0.37	0.41	0.27	0.44	0.12	0.30	0.13
Tm	0.05	0.06	0.04	0.07	0.02	0.04	0.02
Yb	0.31	0.38	0.21	0.41	0.10	0.29	0.13
Lu	0.04	0.06	0.03	0.07	0.02	0.05	0.02
Th	0.03	0.14	0.11	0.04	0.11	0.06	0.05
U	0.02	0.06	0.15	0.03	0.10	0.08	0.02
ol	59	49	62	61	84	57	78
opx	22	35	22	25	6	30	17
cpx	18	15	15	14	10	11	5
sp	1	1	1	1	2	2	0

affect the mantle beneath the Calatrava volcanic district (Fig. 5).

The positive correlation between FeO_{tot} and TiO_2 in our CLV lherzolites implies an interaction with metasomatic agents significantly enriched in both elements, precluding a relation with alkaline magmas similar to the host basalts (see the following discussion).

Processes other than pure melt extraction are confirmed by the whole-rock incompatible element budget, characterized by positively fractionated rare earth element (REE) patterns and overall LREE enrichment. Chondrite (Ch)-normalized REE patterns (Fig. 6) show that heavy (H)REE concentrations in most samples range between 1.8× and 2.7× chondrite. The two lherzolite samples with subchondritic HREE content (CLV49d and CLV58) are characterized by a minimum cpx abundance (down to 5%). However, the observed HREE variation is not simply attributable to the depletion degree by extraction of basic melts, as there is no precise correlation between modal cpx abundance and HREE content. Most samples display a LREE enrichment (La_N/Yb_N up to 6.7) exceeding that recorded in comparable lherzolites from other Spanish mantle xenolith occurrences (Beccaluva *et al.* 2004; Bianchini *et al.* 2007), with the exception of sample CLV18b which shows a less fractionated REE pattern with HREE = 2.7× chondrite and $La_N/Yb_N = 2.0$. This LREE (and Fe–Ti) enrichment clearly indicates that metasomatizing agents widely affected the uppermost lithospheric mantle beneath Calatrava.

Pyroxene trace element composition of mantle xenoliths from Calatrava

Clinopyroxene and orthopyroxene trace element compositions, reported in Table 3 (and in the

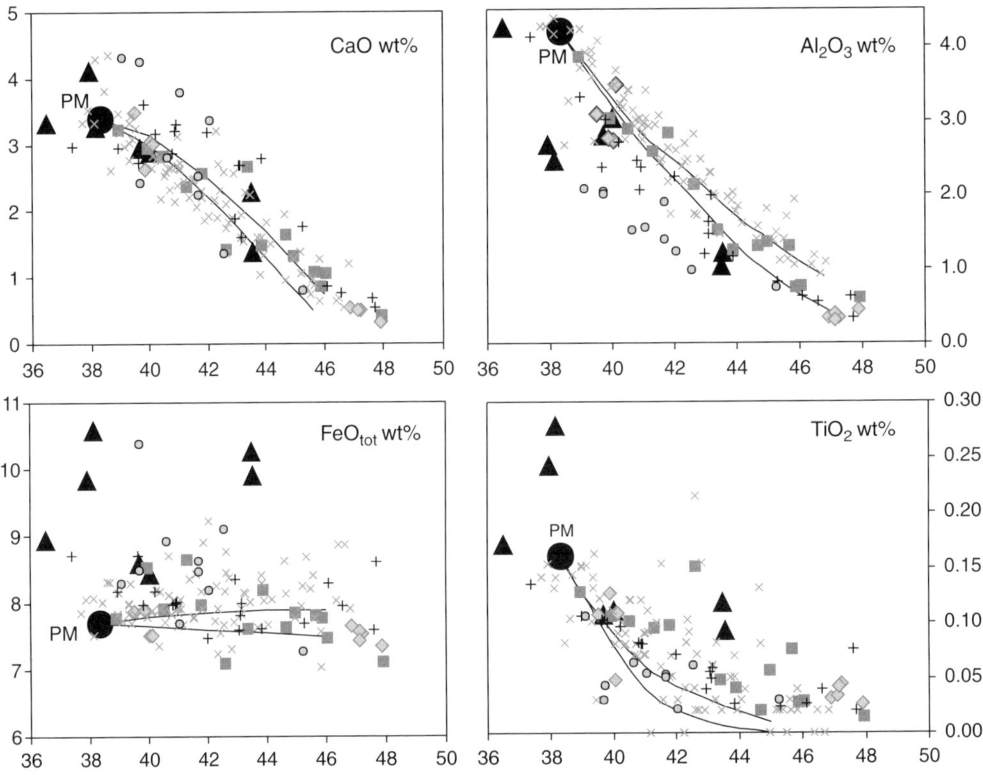

Fig. 5. Bulk-rock variations of Al_2O_3, CaO, FeO_{tot} and TiO_2 as a function of MgO. Compositions of Calatrava Lherzolites (black triangles) are compared with those of peridotite xenolith suites from Tallante (grey squares; Beccaluva *et al.* 2004), Olot (grey diamond; Bianchini *et al.* 2007), Massif Central (grey crosses; Lenoir *et al.* 2000), Sardinia (black crosses; Beccaluva *et al.* 2001*a*) and the Hyblean area (grey circles; Sapienza & Scribano 2000; Perinelli *et al.* 2008). Envelopes of theoretical residual compositions after batch and fractional melting (at 10 and 20 kbar) starting from a model fertile lherzolite (PM) are reported after Niu (1997).

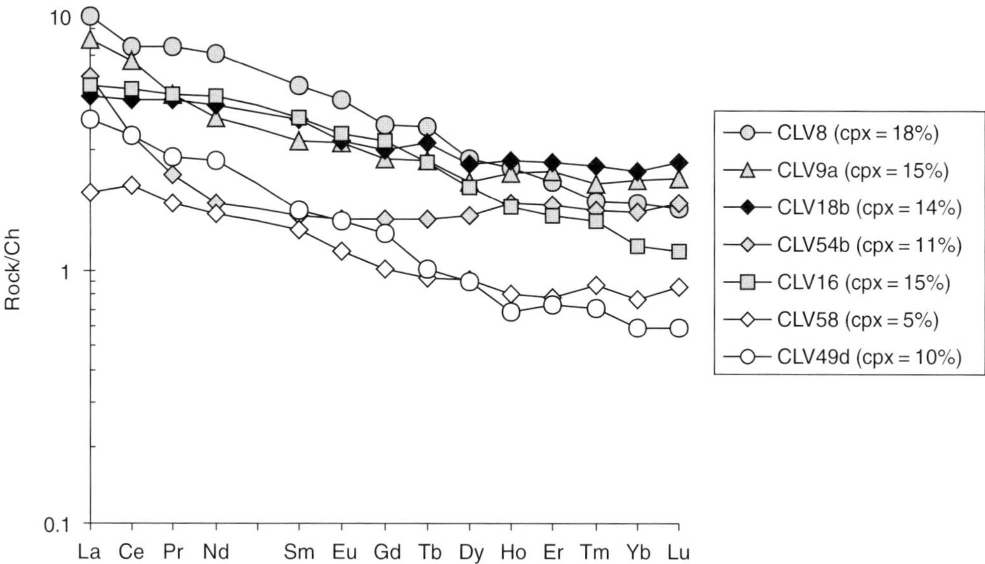

Fig. 6. Chondrite (Ch)-normalized bulk-rock REE patterns of Calatrava lherzolites. Normalizing factors from McDonough & Sun (1995).

Table 3. *Average trace elements composition of clinopyroxenes and orthopyroxene from the Calatrava lherzolite xenoliths. A more extended dataset, including the averaged analyses, is available as Supplementary Material (see p. 107)*

Sample	CLV8	CLV15a	CLV16	CLV16	CLV18b	CLV54b	CLV58	CLV58
Mineral	cpx	cpx	cpx	opx	cpx	cpx	cpx	opx
No. of spots*	3	6	4	1	3	2	4	2
Rb	0.83	0.04	<0.03	<0.03	0.36	0.21	<0.03	<0.03
Sr	59.5	89.9	107	0.39	65.4	118	95.7	0.84
Y	8.97	6.23	12.4	1.07	15.9	8.03	10.7	0.95
Zr	29.1	20.1	48.3	4.21	16.5	15.7	31.8	3.56
Nb	0.65	0.79	0.60	0.03	1.97	0.99	0.73	0.04
La	1.66	2.24	3.69	0.02	4.26	3.54	3.32	<0.02
Ce	6.71	7.86	13.7	0.06	8.96	9.97	12.0	0.07
Pr	1.23	1.37	2.21	0.02	0.98	1.62	1.89	0.01
Nd	6.63	7.20	11.5	0.06	4.13	7.71	9.17	0.14
Sm	2.02	2.05	3.15	0.08	1.55	2.03	2.42	0.08
Eu	0.79	0.65	1.15	0.03	0.51	0.69	0.91	0.02
Gd	2.05	2.03	3.01	0.11	2.34	2.29	2.35	0.09
Tb	0.33	0.27	0.51	0.01	0.41	0.31	0.36	0.01
Dy	2.21	1.53	2.81	0.12	2.48	1.53	2.12	0.10
Ho	0.43	0.24	0.52	0.02	0.51	0.27	0.43	0.04
Er	0.98	0.54	1.24	0.16	1.59	0.79	1.00	0.17
Tm	0.12	0.08	0.15	0.02	0.25	0.10	0.16	0.03
Yb	0.65	0.51	0.80	0.31	1.82	0.56	0.96	0.24
Lu	0.08	0.05	0.13	0.03	0.24	0.09	0.13	0.04
Hf	1.66	0.89	1.72	0.11	0.64	0.64	1.04	0.10
Ta	0.07	0.07	0.11	0.01	0.21	0.07	0.09	<0.01
Th	0.03	0.07	0.12	0.01	0.36	0.11	0.07	<0.01
U	0.03	0.02	0.27	0.01	0.11	0.34	0.08	0.01

*No. of spots, number of averaged analyses.

supplementary data available on the Society's website), are plotted as chondrite-normalized REE patterns in Figure 7. CLV18b, the sample with the least fractionated bulk-rock REE pattern, shows a flat cpx REE pattern (*c.* 10× chondrite) only characterized by a slight enrichment of La and Ce. Clinopyroxenes in the remaining samples display fractionated REE patterns with upwards-convex shapes, low HREE concentrations (2×–6× chondrite) and uniform enrichments in MREE–LREE (Nd_N/Yb_N up to 6, La_N/Yb_N up to 5).

Mantle clinopyroxenes from another xenolith occurrence of the Calatrava volcanic field (Cerro Pelado) show REE profiles characterized by similar upwards-convex shape (with the peak located at Nd) combined with low HREE content (Lu_N down to 9× Ch: Villaseca *et al.* 2010).

Orthopyroxene (analysed only in samples CLV16 and CLV58) presents negatively fractionated REE patterns with HREE approaching the chondrite value and LREE down to <0.1× chondrite, as shown by mantle orthopyroxene from the Calatrava occurrence of Cerro Pelado that also displays a severe LREE depletion (Villaseca *et al.* 2010).

The distinctive REE patterns of most cpx are not consistent with those of primitive mantle domains that are invariably characterized by flat, unfractionated REE patterns with abundances of approximately 10×–12× Ch (Johnson *et al.* 1990; O'Neill & Palme 1998). Their positively fractionated REE patterns also preclude a residual origin after partial melting, as in this case cpx should be characterized by MREE/HREE<1 and LREE/MREE ≪ 1 (e.g. Bodinier & Godard 2003).

Similar cpx REE profiles have not been recorded in comparable lherzolites from other Spanish occurrences (Beccaluva *et al.* 2004; Shimizu *et al.* 2004; Bianchini *et al.* 2007) and resemble only those observed in metasomatic (neo-crystallized) cpx from the Olot harzburgites.

In general, similar REE patterns are relatively rare in cpx from off-craton spinel lherzolite xenoliths (Bonadiman *et al.* 2005), and resemble those observed in metasomatized clinopyroxenes from cratonic garnet-bearing peridotite xenoliths (e.g. Menzies *et al.* 1987; Pearson *et al.* 2003). The observed 'enriched' cpx may result from the re-equilibrium of pristine mantle cpx with incoming metasomatic melts and/or direct crystallization from these melts (Simon *et al.* 2003).

The highly fractionated HREE patterns of these cpx also suggest that the metasomatic agents were formed in the presence of residual garnet from a peridotite and/or an eclogite source.

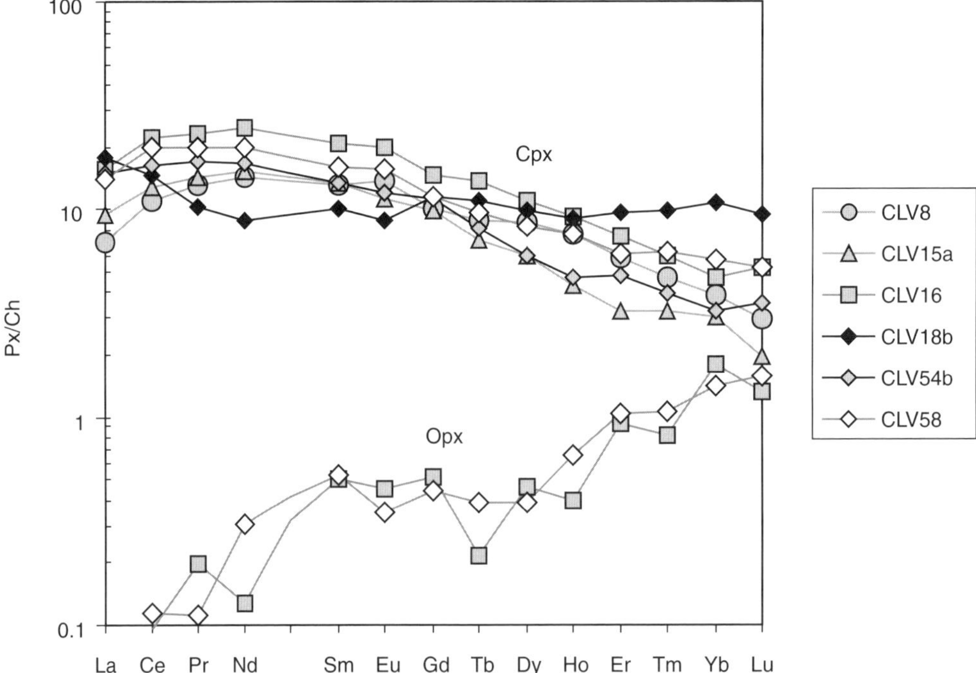

Fig. 7. Representative chondrite-normalized REE patterns of clinopyroxene and orthopyroxene from Calatrava lherzolites. Normalizing factors after McDonough & Sun (1995).

Isotopic characteristics of mantle xenoliths from Calatrava

Strontium isotopic compositions have been determined for clinopyroxenes separated from six distinct lherzolites, and reveal a homogeneous $^{87}Sr/^{86}Sr$ composition between 0.70312 and 0.70328 (Table 4). Consequently, there is no co-variation of Sr isotope compositions with trace element systematics.

Clinopyroxenes in samples CLV8, CLV16 and CLV18b have also been analysed for Nd and Hf isotopes: $^{143}Nd/^{144}Nd$ ranges from 0.51288 to 0.51295 (εNd 4.74–6.07) and $^{176}Hf/^{177}Hf$ from 0.28302 to 0.28265 (εHf −3.57 to 9.41). $^{143}Nd/^{144}Nd$ appears positively correlated with the Sm/Nd ratio, and the least metasomatized sample CLV18b (cpx Nd_N/Yb_N 0.7–0.9) has the highest $^{143}Nd/^{144}Nd$ value.

In contrast, the Lu–Hf system is characterized by a negative correlation between Lu/Hf and $^{176}Hf/^{177}Hf$, totally precluding a residual origin for the clinopyroxene within the Calatrava peridotite xenoliths. Instead, this relationship indicates control of Hf (and probably Nd) isotopes via mixing of a pristine mantle with a metasomatic agent. The lack of Sr isotope variation probably indicates that the relative level of Sr in the metasomatic component was so elevated that it dominated the Sr isotope composition of the clinopyroxene.

The $^{176}Hf/^{177}Hf$ isotope composition of the metasomatized samples CLV8 and CLV16 ($^{176}Hf/^{177}Hf$ 0.28300–0.28304) is more radiogenic than that of the unmetasomatized sample CLV18b. This metasomatic signature may approach that of the causative agents, supporting the idea that the metasomatic melts were formed by partial melting of a garnet-bearing source (high Lu/Hf ratio), possibly including eclogite components.

In the Sr–Nd, Sr–Hf and Nd–Hf isotope diagrams of Figure 8 it is notable that – in contrast to mantle xenoliths from other Iberian occurrences (Beccaluva *et al.* 2004; Bianchini *et al.* 2007) where compositions range from depleted mantle (DM) to enriched mantle (EM) – the Calatrava mantle clinopyroxenes are a very close approximation of the notional HIMU mantle end member.

Table 4. *Sr–Nd–Hf isotope composition of clinopyroxene separates from the Calatrava lherzolite xenoliths*

	$^{87}Sr/^{86}Sr$	$^{143}Nd/^{144}Nd$	$^{176}Hf/^{177}Hf$
CLV8 cpx	0.703135	0.512904	0.283035
CLV15a cpx	0.703246		
CLV16 cpx	0.703276	0.512881	0.282998
CLV18b cpx	0.703146	0.512949	0.282668
CLV54b cpx	0.703123		
CLV58 cpx	0.703153		

This is also supported by Sr–Nd isotopic data carried out on whole-rock peridotite xenoliths from other localities of the Calatrava volcanic field (Villaseca *et al.* 2010). In particular, peridotite samples from Cerro Pelado display a remarkable analogy, being characterized by $^{87}Sr/^{86}Sr$ 0.70320–0.70333 and $^{143}Nd/^{144}Nd$ 0.51292–0.51296 (only one outlier has $^{87}Sr/^{86}Sr$ 0.70368 and $^{143}Nd/^{144}Nd$ 0.51278).

The dominance of this mantle component has been recognized on the basis of Sr–Nd–Pb isotopic data for the Plio-Quaternary CLV lavas (Cebriá & López-Ruiz 1995), and can be interpreted as the isotopic fingerprint of the metasomatic agents that pervasively affected the CLV mantle section.

The HIMU-type isotopic signature characterizes cpx variably affected by metasomatic reaction, irrespective of the degree of textural equilibrium. This suggests a homogeneous isotopic composition for the metasomatic agents over time.

Discussion

The major element compositions of both bulk rocks and constituent minerals in the CLV lherzolites coherently indicate that the local lithospheric mantle is characterized by anomalous low Mg#. The occurrence of anhydrous Fe-rich peridotites has previously been recorded in mantle xenolith suites, although they usually represent a subordinate lithotype with respect to the entire population. As far as we know, anhydrous Fe-rich lherzolites seem to be predominant only in two other mantle xenolith populations: lherzolite–wehrlite suites from the Tok volcanic field in SE Siberia (Ionov *et al.* 2005) and from Oki-Dogo Island, SW Japan (Abe *et al.* 2003).

Ionov *et al.* (2005) explained the Tok Fe-rich lherzolites and wehrlites as the result of large-scale mantle refertilization induced by the reactive percolation (at a high melt/matrix ratio) of silica-undersaturated melts through refractory mantle peridotites. A similar conclusion has also been reached for deformed dunitic peridotites recovered from the Kimberley area, South Africa, in which olivines, characterized by low Mg# (88–86), possibly reflect a significant refertilization of the pristine mantle composition (Boyd *et al.* 1983).

Interpretation of the CLV lherzolites as portions of pristine mantle unaffected by melting processes is not supported by their major/trace element budget, which conflicts with models of a mantle evolution simply based on melt extraction (Niu 2004). Instead, their geochemical characteristics,

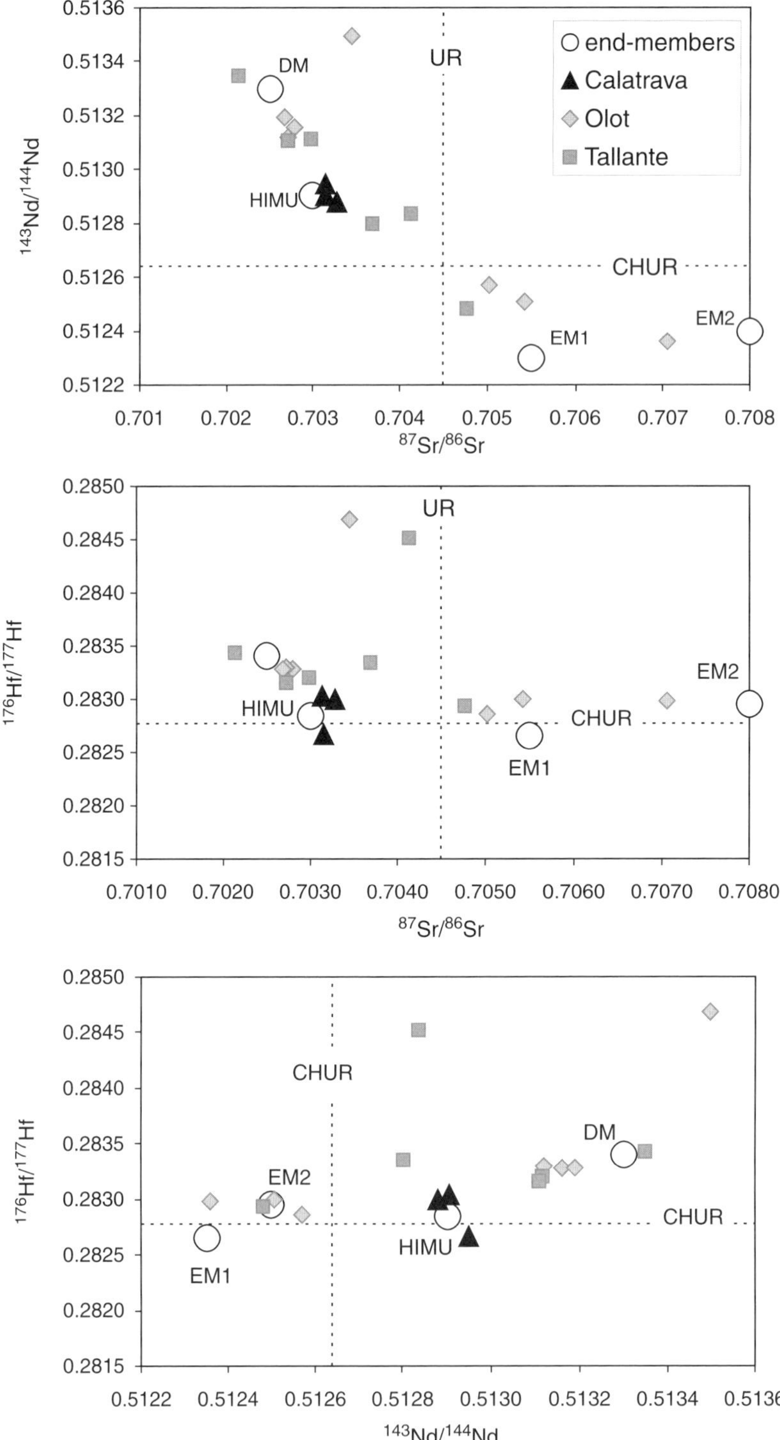

Fig. 8. Sr–Nd–Hf isotopic variations for clinopyroxene separates from Calatrava lherzolite xenoliths, compared with those from Olot (Bianchini *et al.* 2007) and Tallante (anhydrous peridotite group: Beccaluva *et al.* 2004, and authors' unpublished data). Isotopic mantle end members (DM, HIMU, EMI and EMII) are after Zindler & Hart (1986) and Salters & White (1998).

specifically the REE patterns of both bulk rocks and cpx and the unusual Fe–Ti enrichment, suggest interaction between a peridotite matrix and an incoming metasomatizing melt generated from deeper mantle portions in the presence of residual garnet, including eclogitic domains.

This hypothesis is evaluated in Figure 9, which reports the FeO–TiO_2 bulk-rock composition of the investigated Calatrava lherzolites, together with the composition of other xenolith suites from the Mediterranean region. Figure 9 also shows the 'eclogite line' connecting the minimum and maximum FeO–TiO_2 pairs of bimineralic eclogites from orogenic massifs and xenoliths (Pearson *et al.* 2003 and references therein; Jacob 2004 and references therein; Appleyard *et al.* 2007). Calculations based on major element data have been carried out using a simple mixing equation between the least metasomatized CLV lherzolites (CLV18b) and a hypothetical liquid derived by 20% melting of a theoretical bimineralic MORB-type eclogite (Pertermann & Hirschmann 2003). To simulate eclogite melting we applied the one dimensional steady-state major element equation of Asimow & Stolper (1999), and the mineral–melt element partitioning coefficients of Pertermann & Hirschmann (2002, 2003) and Pertermann *et al.* (2004). These theoretical eclogite-derived melts are characterized by a basaltic–andesitic composition (TiO_2 2.7–3.3 wt%, FeO_{tot} 17.3–18.0 wt% and SiO_2 up to 53.3 wt%) and favourably compare with those experimentally determined at 3.0 GPa (Pertermann & Hirschmann 2003). The subsequent interaction of these eclogite-derived melts with the Calatrava mantle has been evaluated using the Asimow & Stolper (1999) major element equation for a simple, one-dimensional equilibrium flow column, and mineral–melt partitioning coefficients from the Geochemical Earth Reference Model (GERM) website (http://earthref.org/GERM/). Numbers reported in Figure 9 represent the calculated amount of melt that plausibly metasomatized the Calatrava mantle. This model is able to reproduce the positive TiO_2–FeO trend of the Calatrava

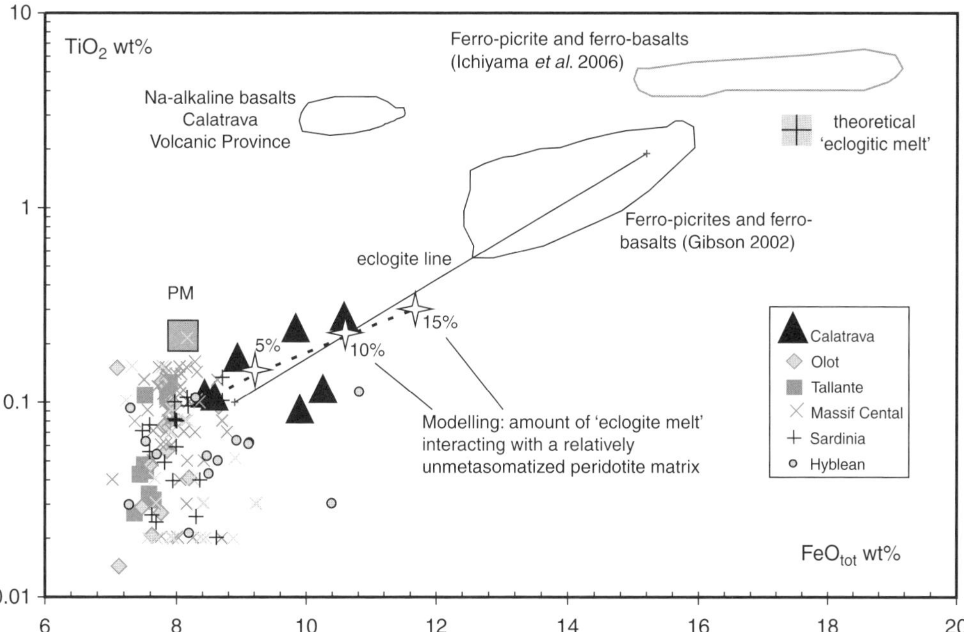

Fig. 9. Bulk rock TiO_2 v. FeO_{tot} (wt%) comparing the compositions of peridotites from Calatrava, Olot (Bianchini *et al.* 2007), Tallante (Beccaluva *et al.* 2004), Massif Central (Lenoir *et al.* 2000), Hyblean area (Sapienza & Scribano 2000; Perinelli *et al.* 2008) and the compositional range of Calatrava alkali basalts (López-Ruiz *et al.* 1993). The primordial mantle (PM) composition is from McDonough & Sun (1995). The 'eclogite line' indicates the minimum and maximum FeO_{tot}–TiO_2 pairs included in a large dataset of bimineralic eclogites recorded from orogenic massifs and xenoliths (Pearson *et al.* 2003 and references therein; Jacob 2004 and references therein; Appleyard *et al.* 2007). Also shown are the results of a mixing model between the unmetasomatized Calatrava lherzolite (CLV18b) and a hypothetical melt (large crossed grey square) derived by 20% melting of a bimineralic eclogite (see the text for further details). In this framework the Calatrava peridotites could have been modified by interaction with an eclogite-derived melt (up to 10–12%), similar in composition to natural ferro-picrites and ferro-basalts (Gibson 2002; Ichiyama *et al.* 2006).

lherzolites (Fig. 9), which requires a maximum of 10–12% of the melt infiltrating and reacting with mantle domains relatively unmetasomatized such as the CLV18b lherzolite.

The incompatible element compositions of Calatrava protogranular clinopyroxenes provide further constraints on the characteristics of the metasomatic agents. Melt compositions in equilibrium with the clinopyroxene cores were calculated using clinopyroxene–eclogite melt partition coefficients (Pertermann & Hirschmann 2002; Pertermann *et al.* 2004) experimentally determined for high-pressure Al-rich clinopyroxene. The calculated melts, reported in Figure 10, show positively fractionated REE patterns and Sr/Y ratios compatible with eclogite-derived melts (Fig. 10) (Pertermann *et al.* 2004; Gao *et al.* 2007).

The isotopic fingerprint of the CLV lherzolites (and associated lavas) with its HIMU affinity is in agreement with the above model, given that the HIMU mantle end member is classically interpreted as the result of long-term recycling of oceanic basalts/gabbros (or their eclogitic equivalents) within the mantle via ancient subduction (Weaver 1991; Carlson 1995; Hofmann 1997; Stracke *et al.* 2005). The time-integrated genesis of this mantle component seems to preclude a direct connection with the Cenozoic subduction processes along the Betic Cordillera, and Wilson & Downes (1991) suggested 400–500 Ma as a proper timespan required to maturate similar isotopic compositions.

Significantly, HIMU-like Sr–Nd–Pb isotopic compositions have been recorded in some garnet

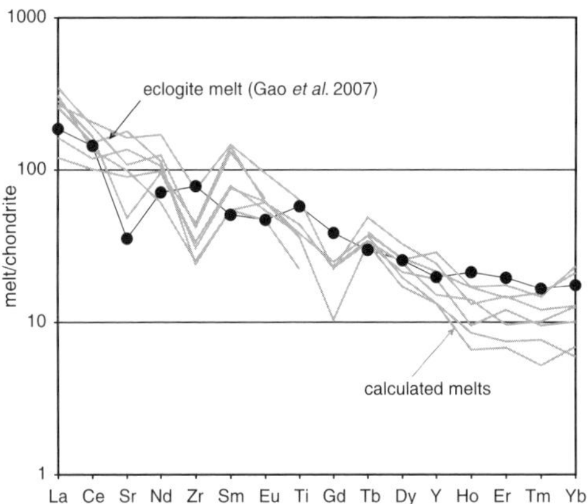

Fig. 10. Chondrite-normalized trace element compositions of melts calculated to be in equilibrium with cores of protogranular clinopyroxenes from Calatrava lherzolites. The theoretical melts favourably compare with the eclogite-derived melt of Gao *et al.* (2007). Normalization values from McDonough & Sun (1995).

pyroxenites/eclogites from the Cabo Ortegal complex in NW Spain, which has been interpreted as an exhumed Palaeozoic mantle wedge (Santos *et al.* 2002 and references therein). In addition, Hf isotopic compositions approaching those of the Calatrava mantle rocks have been observed in some ultrahigh-pressure eclogite rocks of Palaeozoic age (Kylander-Clark *et al.* 2007).

The age of the proposed eclogitic component is unlikely to be much older because eclogites and pyroxenites would develop very diverse isotopic signatures over timescales that cover billions of years (Pearson & Nowell 2004; Jacob *et al.* 2005).

This may imply that, under the CLV district, sub-lithospheric convective instabilities – possibly triggered by the neighbouring Cenozoic subduction along the Betic collisional belt – could have remobilized deep-mantle domains from the mantle 'transition zone' (410–660 km) that possibly include eclogitic rocks, that is, relicts of older (Palaeozoic?) subducted slabs.

Conclusions

The importance of eclogites and pyroxenites as a contributing source to magma generation within the convecting mantle has recently gained recognition (Anderson 2006; Sobolev *et al.* 2007), and experimental studies in various carbonated eclogite systems (Hammouda 2003; Dasgupta *et al.* 2004; Yaxley & Brey 2004) suggest that such eclogite domains would preferentially melt with respect to the surrounding peridotite, potentially generating a wide range of magma compositions characterized by comparatively lower values of Mg# (relative to typical peridotite melts: Ichiyama *et al.* 2006; Kogiso & Hirschmann 2006).

Significantly, intimate associations of Fe-rich peridotite and eclogite/garnet pyroxenite have been observed in Palaeozoic European peridotite massifs (Medaris *et al.* 2005), including that of Cabo Ortegal, northern Spain (Santos *et al.* 2002). These associations could be considered actual examples of the theoretical 'mingled' mantle composition defined as piclogite by Anderson (1989 and references therein).

Convective instabilities generated as a dynamic response to subduction have recently been referred to as 'splash plumes' (Davies & Bunge 2006). Such mantle instabilities could displace material from the top of the mantle 'transition zone' (410–660 km depth), which may represent a graveyard for subducted slabs of oceanic lithosphere, and remobilize foundered blocks of eclogite that which deform and mix with the surrounding peridotite as they rise (Farnetani & Samuel 2005).

Within this intimate association of eclogite boudins in a peridotite matrix, referred to as a 'piclogite' association, the eclogites will preferentially melt (as they have a lower solidus temperature than the surrounding peridotite), generating melts characterized by various silica-saturation, Fe–Ti enrichments and HIMU-like Sr–Nd–Pb isotopic signatures. These melts would then infiltrate and metasomatize shallower mantle domains.

The high CO_2 content typical of the CLV volcanic products, reflected in diatremes and maars characterized by carbonate-rich pyroclastic deposits (Bailey *et al.* 2005; Humphreys *et al.* 2008*b*), the observed presence of carbonate-bearing patches in the investigated xenoliths and the recurrent presence of wehrlites in the xenolith suites from Calatrava (Ancochea & Nixon 1987; Bailey *et al.* 2005; Humphreys *et al.* 2008*a*, *b*; Villaseca *et al.* 2010), are consistent with the proposed model given that subducting oceanic lithosphere can drag carbonated pelagic sediments down into the deep mantle.

In conclusion, we speculate that the geochemical and isotopic characteristics of the CLV lherzolites might provide clues for understanding the HIMU-like fingerprint that is ubiquitous throughout the anorogenic volcanic districts of the central–western Mediterranean area (Wilson & Downes 1991; Beccaluva *et al.* 1998, 2005, 2007; Wilson & Bianchini 1999; Wilson & Patterson 2001; Wilson 2007; Lustrino & Wilson 2007; Bianchini *et al.* 2008).

The authors gratefully acknowledge the constructive criticism of S. Arai and of an anonymous referee, as well as the fruitful discussion with M. Coltorti, M. Grégoire and H. Downes that greatly improved the manuscript. Particular thanks are extended to R. Carampin (EMPA), R. Tassinari (XRF and ICP-MS) and A. Zanetti (LAM-ICP-MS) for their analytical assistance.

References

Abe, N., Takami, M. & Arai, S. 2003. Petrological feature of spinel lherzolite xenolith from Oki-Dogo Island: an implication for variety of the upper mantle peridotite beneath southwestern Japan. *The Island Arc*, **12**, 219–232.

Ancochea, E. & Nixon, P. H. 1987. Xenoliths in the Iberian Peninsula. *In*: Nixon, P. H. (ed.) *Mantle Xenoliths*. Wiley, Chichester, 119–124.

Anderson, D. L. 1989. *Theory of the Earth*. Blackwell Scientific, Boston, MA.

Anderson, D. L. 2006. Speculations on the nature and cause of mantle heterogeneity. *Tectonophysics*, **416**, 7–22.

Appleyard, C. M., Bell, D. R. & Le Roex, A. P. 2007. Petrology and geochemistry of eclogite xenoliths from the Rietfontein kimberlite, Northern Cape, South Africa. *Contributions to Mineralogy and Petrology*, **154**, 309–333.

Asimow, P. D. & Stolper, E. M. 1999. Steady-state mantle–melt interactions in one dimension: 1. Equilibrium transport and melt focusing. *Journal of Petrology*, **40**, 475–494.

Bailey, K., Garson, M., Kearns, S. & Velasco, A. P. 2005. Carbonate volcanism in Calatrava, central Spain: a report on the initial findings. *Mineralogical Magazine*, **69**, 907–915.

Beccaluva, L., Bianchini, G., Bonadiman, C., Coltorti, M., Macciotta, G., Siena, F. & Vaccaro, C. 2005. Within-plate Cenozoic volcanism and lithospheric mantle evolution in the western-central Mediterranean area. *In*: Finetti, I. (ed.) *Crop Project – Deep Seismic Exploration of the Central Mediterranean Italy*. Elsevier, Amsterdam, 641–664.

Beccaluva, L., Bianchini, G., Bonadiman, C., Siena, F. & Vaccaro, C. 2004. Coexisting anorogenic and subduction-related metasomatism in mantle xenoliths from the Betic Cordillera (southern Spain). *Lithos*, **75**, 67–87.

Beccaluva, L., Bianchini, G., Coltorti, M., Perkins, W. T., Siena, F., Vaccaro, C. & Wilson, M. 2001*a*. Multistage evolution of the European lithospheric mantle: new evidence from Sardinian peridotite xenoliths. *Contributions to Mineralogy and Petrology*, **142**, 284–297.

Beccaluva, L., Bonadiman, C., Coltorti, M., Salvini, L. & Siena, F. 2001*b*. Depletion events, nature of metasomatizing agent and timing of enrichment processes in lithospheric mantle xenoliths from Veneto Volcanic Province. *Journal of Petrology*, **42**, 173–187.

Beccaluva, L., Bianchini, G., Ellam, R. M., Marzola, M., Oun, K. M., Siena, F. & Stuart, F. M. 2008. The role of HIMU metasomatic components in the African lithospheric mantle: petrological evidence from the Gharyan peridotite xenoliths, NW Libya. *In*: Coltorti, M. & Grégoire, M. (eds) *Metasomatism in Oceanic and Continental Lithospheric Mantle*. Geological Society, London, Special Publications, **293**, 253–277.

Beccaluva, L., Bianchini, G. et al. 2007. Intraplate lithospheric and sublithospheric components in the Adriatic domain: nephelinite to tholeiite magma generation in the Paleogene Veneto Volcanic Province, Southern Alps. *In*: Beccaluva, L., Bianchini, G. & Wilson, M. (eds) *Cenozoic Volcanism in the Mediterranean Area*. Geological Society of America, Special Papers, **418**, 131–152.

Beccaluva, L., Siena, F. et al. 1998. Nephelinitic to tholeiitic magma generation in a transtensional tectonic setting: an integrated model for the Iblean volcanism, Sicily. *Journal of Petrology*, **39**, 1547–1576.

Bergamin, J. F. & Carbo, A. 1986. Discusion de modelos para la corteza y manto superior en la zona sur de l area centroiberica, basados en anomalias gravimetricas. *Estudios Geológicos*, **42**, 143–146.

Bianchini, G., Beccaluva, L., Bonadiman, C., Nowell, G., Pearson, G., Siena, F. & Wilson, M. 2007. Evidence of diverse depletion and metasomatic events in harzburgite-lherzolite mantle xenoliths from the Iberian plate (Olot, NE Spain): implications for lithosphere accretionary processes. *Lithos*, **94**, 25–45.

Bianchini, G., Beccaluva, L. & Siena, F. 2008. Post-collisional and intraplate Cenozoic volcanism in the rifted Apennines/Adriatic domain. *Lithos*, **101**, 125–140.

Bodinier, J.-L. & Godard, M. 2003. Orogenic, ophiolitic, and abyssal peridotites. *In*: Carlson, R. W. (ed.) *The Mantle and Core. Treatise on Geochemistry*, **2**, 103–170.

Bonadiman, C., Beccaluva, L., Coltorti, M. & Siena, F. 2005. Kimberlite-like metasomatism and 'garnet signature' in spinel-peridotite xenoliths from Sal, Cape Verde Archipelago: relics of a subcontinental mantle domain within the Atlantic Oceanic lithosphere? *Journal of Petrology*, **46**, 2465–2493.

Boyd, F. R., Jones, R. A. & Nixon, P. H. 1983. Mantle metasomatism: the Kimberley dunites. *Carnegie Institution of Washington Yearbook*, **82**, 330–336.

Brey, G. P. & Köhler, T. P. 1990. Geothermobarometry in four phases lherzolites II. New thermobarometers and practical assessment of existing thermobarometers. *Journal of Petrology*, **31**, 1353–1378.

Carlson, R. W. 1995. Isotopic inferences on the chemical structure of the mantle. *Journal of Geodynamics*, **20**, 365–386.

Cebriá, J.-M. & López-Ruiz, J. 1995. Alkali basalts and leucitites in an extensional intracontinental plate setting: the late Cenozoic Calatrava Volcanic Province (central Spain). *Lithos*, **35**, 27–46.

Coltorti, M., Beccaluva, L., Bonadiman, C., Salvini, L. & Siena, F. 2000. Glasses in mantle xenoliths as geochemical indicators of metasomatic agents. *Earth and Planetary Science Letters*, **183**, 303–320.

Coltorti, M., Bonadiman, C., Hinton, R. W., Siena, F. & Upton, B. G. J. 1999. Carbonatite metasomatism of the oceanic upper mantle: evidence from clinopyroxenes and glasses in ultramafic xenoliths of Grande Comore, Indian Ocean. *Journal of Petrology*, **40**, 133–165.

Dasgupta, R., Hirschmann, M. M. & Withers, A. C. 2004. Deep global cycling of carbon constrained by the solidus of anhydrous, carbonated eclogite under upper mantle conditions. *Earth and Planetary Science Letters*, **227**, 73–85.

Davies, J. H. & Bunge, H.-P. 2006. Are splash plumes the origin of minor hotspots? *Geology*, **34**, 349–352.

Delpech, G., Grégoire, M., O'Reilly, S. Y., Cottin, J. Y., Moine, B., Michon, G. & Giret, A. 2004. Feldspar from carbonate-rich silicate metasomatism in the shallow oceanic mantle under Kerguelen Islands (South Indian Ocean). *Lithos*, **75**, 209–237.

Dowall, D. P., Nowell, G. M. & Pearson, D. G. 2003. A 2-column procedure for the pre-concentration of Sr, Nd and Hf for isotopic analysis by plasma ionisation multi-collector mass spectrometry. *In*: Holland, J. G. & Tanner, S. D. (eds) *Plasma Source Mass Spectrometry: Applications and Emerging Technologies*. Royal Society of Chemistry, Cambridge, 321–337.

Downes, H. 2001. Formation and modification of the shallow sub-continental lithospheric mantle: a review of geochemical evidence from ultramafic xenolith suites and tectonically emplaced ultramafic massifs of Western and Central Europe. *Journal of Petrology*, **42**, 233–250.

Downes, H. & Coltorti, M. 2008. Mantle Xenoliths in Space and Time in Europe. *In*: *EGU Abstracts Volume EGU General Assembly*, Vienna, Austria, 13–18 April.

Farnetani, C. G. & Samuel, H. 2005. Beyond the thermal plume paradigm. *Geophysical Research Letters*, **32**, L07311.

Gao, J., John, T., Klemd, R. & Xiong, X. 2007. Mobilization of Ti–Nb–Ta during subduction: evidence from rutile-bearing dehydration segregations and veins hosted in eclogite, Tianshan, NW China. *Geochimica et Cosmochimica Acta*, **71**, 4974–4996.

Gibson, S. A. 2002. Major element heterogeneity in Archean to Recent mantle plume starting-heads. *Earth and Planetary Science Letters*, **195**, 59–74.

Grégoire, M., Lorand, J. P., O'Reilly, S.Y. & Cottin, J. Y. 2000. Armalcolite-bearing, Ti-rich metasomatic assemblages in harzburgitic xenoliths from the Kerguelen Islands: implications for the oceanic mantle budget of high-field strength elements. *Geochimica et Cosmochimica Acta*, **64**, 673–694.

Hammouda, T. 2003. High-pressure melting of carbonated eclogite and experimental constraints on carbon recycling and storage in the mantle. *Earth and Planetary Science Letters*, **214**, 357–368.

Hofmann, A. W. 1997. Mantle geochemistry: the message from oceanic volcanism. *Nature*, **385**, 219–229.

Humphreys, E. R., Bailey, K., Hawkesworth, C. J. & Wall, F. 2008*a*. Mantle xenoliths from the Calatrava Volcanic Province, Spain – evidence for carbonatite–silicate interaction in the upper mantle. *Eos, Transactions of the American Geophysical Union*, **89**(53), Abstract #V43F-2200.

Humphreys, E. R., Bailey, K., Wall, F., Hawkesworth, C. J. & Kearns, S. 2008*b*. Highly heterogeneous mantle sampled by rapidly erupted carbonate volcanism. *In*: *9th International Kimberlite Conference Extended Abstract*, No. 9IKC-A-00255.

Ichiyama, Y., Ishiwatari, A., Hirahara, Y. & Shuto, K. 2006. Geochemical and isotopic constraints on the genesis of the Permian ferropicritic rocks from the Mino–Tamba belt, SW Japan. *Lithos*, **89**(1–2), 47–65.

Ionov, D. A., Chanefo, I. & Bodinier, J.-L. 2005. Origin of Fe-rich lherzolites and wehrlites from Tok, SE Siberia by reactive melt percolation in refractory mantle peridotites. *Contributions to Mineralogy and Petrology*, **150**, 335–353.

Ionov, D. A., Grégoire, M. & Prikhod'ko, V. S. 1999. Feldspar–Ti-oxide metasomatism in off-cratonic continental and oceanic upper mantle. *Earth and Planetary Science Letters*, **165**, 37–44.

Jacob, D. E. 2004. Nature and origin of eclogite xenoliths from kimberlites. *Lithos*, **77**, 295–316.

Jacob, D. E., Bizimis, M. & Salters, V. J. M. 2005. Lu–Hf and geochemical systematics of recycled ancient oceanic crust: evidence from Roberts Victor eclogites. *Contributions to Mineralogy and Petrology*, **148**, 707–720.

Johnson, K. T. M., Dick, H. J. B. & Shimizu, N. 1990. Melting in the oceanic upper mantle: an ion microprobe study of diopsides in abyssal peridotites. *Journal of Geophysical Research*, **95**, 2661–2678.

KOGISO, T. & HIRSCHMANN, M. M. 2006. Partial melting experiments of bimineralic eclogite and the role of recycled mafic oceanic crust in the genesis of ocean island basalts. *Earth and Planetary Science Letters*, **249**, 188–199.

KYLANDER-CLARK, A. R. C., HACKER, B. R., JOHNSON, C. M., BEARD, B. L., MAHLEN, N. J. & LAPEN, T. J. 2007. Coupled Lu–Hf and Sm–Nd geochronology constrains prograde and exhumation histories of high- and ultrahigh-pressure eclogites from western Norway. *Chemical Geology*, **242**, 137–154.

ICHIYAMA, Y., ISHIWATARI, A., HIRAHARA, Y. & SHUTO, K. 2006. Geochemical and isotopic constraints on the genesis of the Permian ferropicritic rocks from the Mino–Tamba belt, SW Japan. *Lithos*, **89**, 47–65.

LENOIR, X., GARRIDO, C. J., BODINIER, J.-L. & DAUTRIA, J.-M. 2000. Contrasting lithospheric mantle domains beneath the Massif Central (France) revealed by geochemistry of peridotite xenoliths. *Earth and Planetary Science Letters*, **181**, 359–375.

LÓPEZ-RUIZ, J., CEBRIÁ, J. M., DOBLAS, M., OYARZUN, M., HOYOS, M. & MARTÍN, C. 1993. The late Cenozoic alkaline volcanism of the Central Iberian Peninsula (Calatrava Volcanic Province, Spain): intra-plate volcanism related to extensional tectonics. *Journal of the Geological Society, London*, **150**, 915–922.

LUSTRINO, M. & WILSON, M. 2007. The circum-Mediterranean anorogenic Cenozoic igneous province. *Earth Science Reviews*, **81**, 1–65.

MCDONOUGH, W. F. & SUN, S. 1995. The composition of the Earth. *Chemical Geology*, **120**, 223–253.

MEDARIS, G. JR, WANG, H., JELÍNEK, E., MIHALJEVIČ, M. & JAKEŠ, P. 2005. Characteristics and origins of diverse Variscan peridotites in the Gföhl Nappe, Bohemian Massif, Czech Republic. *Lithos*, **82**, 1–23.

MENZIES, M. A., ROGERS, N., TINDLE, A. & HAWKESWORTH, C. J. 1987. Metasomatic and enrichment processes in lithospheric peridotites, an effect of asthenosphere–lithosphere interaction. *In*: MENZIES, M. A. & HAWKESWORTH, C. J. (eds) *Mantle Metasomatism*. Academic Press, London, 313–361.

NIU, Y. 1997. Mantle melting and melt extraction processes beneath ocean ridges: evidence from abyssal peridotites. *Journal of Petrology*, **38**, 1047–1074.

NIU, Y. 2004. Bulk-rock major and trace element compositions of abyssal peridotites: implications for mantle melting, melt extraction and post-melting processes beneath mid-ocean ridges. *Journal of Petrology*, **45**, 2423–2458.

O'NEILL, H. ST. C. & PALME, H. 1998. Composition of the silicate Earth: implications for accretion and core formation. *In*: JACKSON, I. (ed.) *The Earth's Mantle: Structure, Composition, and Evolution*. Cambridge University Press, Cambridge, 3–126.

PEARSON, D. G. & NOWELL, G. M. 2004. Re–Os and Lu–Hf isotope constraints on the origin and age of pyroxenites from the Beni Bousera peridotite massif: implications for mixed peridotite–pyroxenite melting models. *Journal of Petrology*, **54**, 439–455.

PEARSON, D. G. & NOWELL, G. M. 2005. Accuracy and precision in plasma ionisation multi-collector mass spectrometry: constraints from neodymium and hafnium isotope measurements. *In*: HOLLAND, J. G. & BANDURA, D. R. (eds) *Plasma Source Mass Spectrometry. Current Trends and Future Developments*. Royal Society of Chemistry, Cambridge, 284–314.

PEARSON, D. G. & WITTIG, N. 2008. Formation of Archaean continental lithosphere and its diamonds: the root of the problem. *Journal of the geological Society, London*, **165**, 895–914.

PEARSON, D. G., CANIL, D. & SHIREY, S. B. 2003. Mantle samples included in volcanic rocks: xenoliths and diamonds. *In*: CARLSON, R. W. (ed.) *The Mantle and Core. Treatise on Geochemistry*, **2**, 171–275.

PERINELLI, C., SAPIENZA, G. T., ARMIENTI, P. & MORTEN, L. 2008. Metasomatism of the upper mantle beneath the Hyblean Plateau (Sicily): evidence from pyroxenes and glass in peridotite xenoliths. *In*: COLTORTI, M. & GRÉGOIRE, M. (eds) *Metasomatism in Oceanic and Continental Lithospheric Mantle*. Geological Society, London, Special Publications, **293**, 197–221.

PERTERMANN, M. & HIRSCHMANN, M. M. 2002. Trace-element partitioning between vacancy-rich eclogitic clinopyroxene and silicate melt. *American Mineralogist*, **87**, 1365–1376.

PERTERMANN, M. & HIRSCHMANN, M. M. 2003. Anhydrous partial melting experiments on MORB-like eclogite: phase relations, phase compositions and mineral–melt partitioning of major elements at 2–3 GPa. *Journal of Petrology*, **44**, 2173–2201.

PERTERMANN, M., HIRSCHMANN, M. M., HAMETNER, K., GÜNTHER, D. & SCHMIDT, M. W. 2004. Experimental determination of trace element partitioning between garnet and silica-rich liquid during anhydrous partial melting of MORB-like eclogite. *Geochemistry Geophysics Geosystems*, **5**, Q05A01.

PIKE, J. E. N. & SCHWARZMAN, E. C. 1976. Classification of textures in ultramafic xenoliths. *Journal of Geology*, **85**, 49–61.

SALTERS, V. J. M. & WHITE, W. M. 1998. Hf isotope constraints on mantle evolution. *Chemical Geology*, **145**, 447–460.

SANTOS, J. F., SCHÄRER, U., GIL IBARGUCHI, J. I. & GIRARDEAU, J. 2002. Genesis of pyroxenite-rich peridotite at Cabo Ortegal (NW Spain): geochemical and Pb–Sr–Nd isotope data. *Journal of Petrology*, **43**, 17–43.

SAPIENZA, G. & SCRIBANO, V. 2000. Distribution and representative whole-rock chemistry of deep-seated xenoliths from the Iblean Plateau, South-Eastern Sicily, Italy. *Periodico di Mineralogia*, **69**, 185–204.

SHIMIZU, Y., ARAI, S., MORISHITA, T., YURIMOTO, H. & GERVILLA, F. 2004. Petrochemical characteristics of felsic veins in mantle xenoliths from Tallante (SE Spain): an insight into activity of silicic melt within the mantle wedge. *Transactions of the Royal Society of Edinburgh: Earth and Environmental Science*, **95**, 265–276.

SIENA, F., BECCALUVA, L., COLTORTI, M., MARCHESI, M. & MORRA, V. 1991. Ridge to hot-spot evolution of the Atlantic lithospheric mantle: evidence from Lanzarote peridotite xenoliths (Canary Islands). *In*: MENZIES, M. A., DUPUY, C. & NICOLAS, A. (eds) *Orogenic Lherzolites and Mantle Processes, Journal of Petrology*, Special Volume, 271–290.

SIMON, N. S. C., IRVINE, G. J., DAVIES, G. R., PEARSON, D. G. & CARLSON, R. W. 2003. The origin of garnet and clinopyroxene in depleted Kaapvaal peridotites. *Lithos*, **71**, 289–322.

SIMON, N. S. C., NEUMANN, E.-R., BONADIMAN, C., COLTORTI, M., DELPECH, G., GREGOIRE, M. & WIDOM, E. 2008. Ultra-refractory domains in the oceanic mantle lithosphere sampled as mantle xenoliths at ocean islands. *Journal of Petrology*, **49**, 1223–1251.

SOBOLEV, A. V., HOFMANN, A. W. *ET AL.* 2007. The amount of recycled crust in sources of mantle-derived melts. *Science*, **316**, 412–417.

STRACKE, A., HOFMANN, A. W. & HART, S. R. 2005. Fozo, HIMU and the rest of the mantle zoo. *Geochemistry, Geophysics, Geosystems*, **6**, 1–20.

TOURON, S., RENAC, C., O'REILLY, S. Y., COTTIN, J.-Y. & GRIFFIN, W. L. 2008. Characterization of the metasomatic agent in mantle xenoliths from Devès, Massif Central (France) using coupled *in situ* trace-element and O, Sr and Nd isotopic compositions. *In*: COLTORTI, M. & GRÉGOIRE, M. (eds) *Metasomatism in Oceanic and Continental Lithospheric Mantle*. Geological Society, London, Special Publications, **293**, 177–196.

VILLASECA, C., ANCOCHEA, E., OREJANA, D. & JEFFRIES, T. E. 2010. Composition and evolution of the lithospheric mantle in central Spain: inferences from peridotite xenoliths from the Cenozoic Calatrava volcanic field. *In*: COLTORTI, M., DOWNES, H., GRÉGOIRE, M. & O'REILLY, S. Y. (eds) *Petrological Evolution of the European Lithospheric Mantle*. Geological Society, London, Special Publications, **337**, 125–151.

WEAVER, B. L. 1991. The origin of ocean island basalt end-member compositions: trace element and isotopic constraints. *Earth and Planetary Science Letters*, **104**, 381–397.

WERLING, F. & ALTHERR, R. 1997. Thermal evolution of the lithosphere beneath the French Massif Central as deduced from geothermobarometry on mantle xenoliths. *Tectonophysics*, **275**, 119–141.

WILSON, M. 2007. Tertiary-Quaternary magmatism in Europe: how has it influenced or been influenced by the evolution of the lithosphere? *In*: *EMAW 2007, Workshop on the Petrological Evolution of the European Lithospheric Mantle: Archean to Present Day*, 29–31 August 2007, Ferrara, Italy.

WILSON, M. & BIANCHINI, G. 1999. Tertiary–Quaternary magmatism within the Mediterranean and surrounding regions. *In*: DURAND, B., JOLIVET, L., HORVÁTH, F. & SÉRANNE, M. (eds) *The Mediterranean Basin: Tertiary Extension within the Alpine Orogen*. Geological Society, London, Special Publications, **156**, 141–168.

WILSON, M. & DOWNES, H. 1991. Tertiary–Quaternary extension-related alkaline magmatism in Western and Central Europe. *Journal of Petrology*, **32**, 811–849.

WILSON, M. & PATTERSON, R. 2001. Intraplate magmatism related to short-wavelength convective instabilities in the upper mantle: Evidence from the Tertiary–Quaternary volcanic province of Western and Central Europe. *In*: ERNST, R. E. & BUCHAN, K. L. (eds) *Mantle Plumes: Their Identification Through Time*. Geological Society of America, Special Paper, **352**, 37–58.

YAXLEY, G. M. & BREY, G. P. 2004. Phase relations of carbonate-bearing eclogite assemblages from 2.5 to 5.5 GPa: implications for petrogenesis of carbonatites. *Contributions to Mineralogy and Petrology*, **146**, 606–619.

ZERKA, M., COTTIN, J.-Y., GRÉGOIRE, M., LORAND, J.-P., MEGARTSI, M. & MIDOUN, M. 2002. Les xénolites ultramafiques du volcanisme alcalin quaternaire d'Oranie (Tell, Algérie occidentale), témoins d'une lithosphère cisaillée et enrichie. [Ultramafic xenoliths from Quaternary alkali volcanism from Oranie (Tell, western Algeria): witnesses of a sheared and enriched lithosphere]. *Comptes Rendus Geoscience*, **334**, 387–394.

ZINDLER, A. & HART, S. R. 1986. Chemical geodynamics. *Annual Review of Earth and Planetary Sciences*, **14**, 493–571.

Composition and evolution of the lithospheric mantle in central Spain: inferences from peridotite xenoliths from the Cenozoic Calatrava volcanic field

C. VILLASECA[1]*, E. ANCOCHEA[1], D. OREJANA[1] & T. E. JEFFRIES[2]

[1]*Departamento de Petrología y Geoquímica, UCM-CSIC, Facultad de Geología, c/Jose Antonio Novais, 2, 28040 Madrid, Spain*

[2]*Department of Mineralogy, Natural History Museum, London SW7 5BD, UK*

**Corresponding author (e-mail: granito@geo.ucm.es)*

Abstract: Spinel lherzolite xenoliths from the Cenozoic Calatrava volcanic field provide a sampling of the lithospheric mantle of central Spain. The xenoliths are estimated to originate from depths of 35–50 km. Trace element content of clinopyroxene and Cr-number in spinel indicate low degrees of partial melting ($\leq 5\%$) of the xenoliths. Although a major element whole-rock model suggests wider degrees of melting, the Calatrava peridotite chemistry indicates a moderately fertile mantle beneath central Spain. Calatrava peridotite xenoliths bear evidence for interaction with two different metasomatic agents. The enrichment in LREE (light rare earth element), Th, U and Pb, and the negative anomalies in Nb–Ta in clinopyroxene and amphibole from xenoliths of El Aprisco, indicate that the metasomatic agent was probably a subduction-related melt, whereas the enrichment in MREE in clinopyroxene from xenoliths of the Cerro Pelado centre suggests an alkaline melt similar to the host undersaturated magmas. These metasomatic agents are also consistent with the chemistry of interstitial glasses found in xenoliths of the two volcanic centres. Differences in metasomatism but also in mantle composition is supported by Sr–Nd whole-rock data, which show a more radiogenic nature for Sr isotopes of samples from the El Aprisco centre ($^{87}Sr/^{86}Sr$ ratios of 0.7035–0.7044 instead of 0.7032–0.7037 for samples from Cerro Pelado). The timing of the subduction-related metasomatic stage is unconstrained, although the Calatrava intraplate volcanism intrudes an old Variscan lithospheric section reworked during the converging plate system affecting SE Iberia in the Tertiary. The presence of wehrlite types within the Calatrava peridotite xenoliths is here interpreted as a reaction of host lherzolites with silica-undersaturated silicate melts that could be related to the Calatrava alkaline magmatism. The Sr–Nd isotopic composition of Calatrava peridotites plot within the European asthenospheric reservoir (EAR) mantle, these values represent more enriched signatures than those found in the other Spanish Cenozoic alkaline province of Olot.

Studies of ultramafic xenoliths exhumed by Cenozoic volcanic activity have provided substantial information regarding the nature of the subcontinental lithospheric mantle (e.g. Nixon 1987; Downes 2001). In the Iberian peninsula three main Cenozoic volcanic fields have provided significant mantle-derived xenolith suites since studies from the last century: SE Spain (Ossan 1889), Olot (San Miguel de la Cámara 1936) and Calatrava (Ancochea & Nixon 1987) (Fig. 1). Scarce ultramafic xenoliths have also been described in the Cofrentes volcanic area (Ancochea & Nixon 1987; Seghedi *et al.* 2002), and mantle-derived xenoliths have been found in Upper Permian subvolcanic dykes of the Spanish Central System, although they represent mafic–ultramafic cumulates instead of real mantle peridotitic fragments (Orejana *et al.* 2006; Villaseca *et al.* 2007; Orejana & Villaseca 2008).

In this work we study the chemical composition of the Calatrava sample suite, including major and trace elements for the constituent minerals, and major, trace elements and Nd and Sr isotopes for whole rocks. As for many spinel lherzolite xenolith suites, our data indicate the decoupling of chemical features caused by melt extraction during partial melting and subsequent metasomatism. This study, together with that of Bianchini *et al.* (2010), are the first attempts to characterize the subcontinental mantle beneath central Spain.

Geological setting

The Calatrava volcanic field comprises more than 200 volcanic centres in an area of around 5500 km^2 (Ancochea 1982). The volcanic field is exclusively formed by monogenetic edifices,

From: Coltorti, M., Downes, H., Grégoire, M. & O'Reilly, S. Y. (eds) *Petrological Evolution of the European Lithospheric Mantle*. Geological Society, London, Special Publications, **337**, 125–151.
DOI: 10.1144/SP337.7 0305-8719/10/$15.00

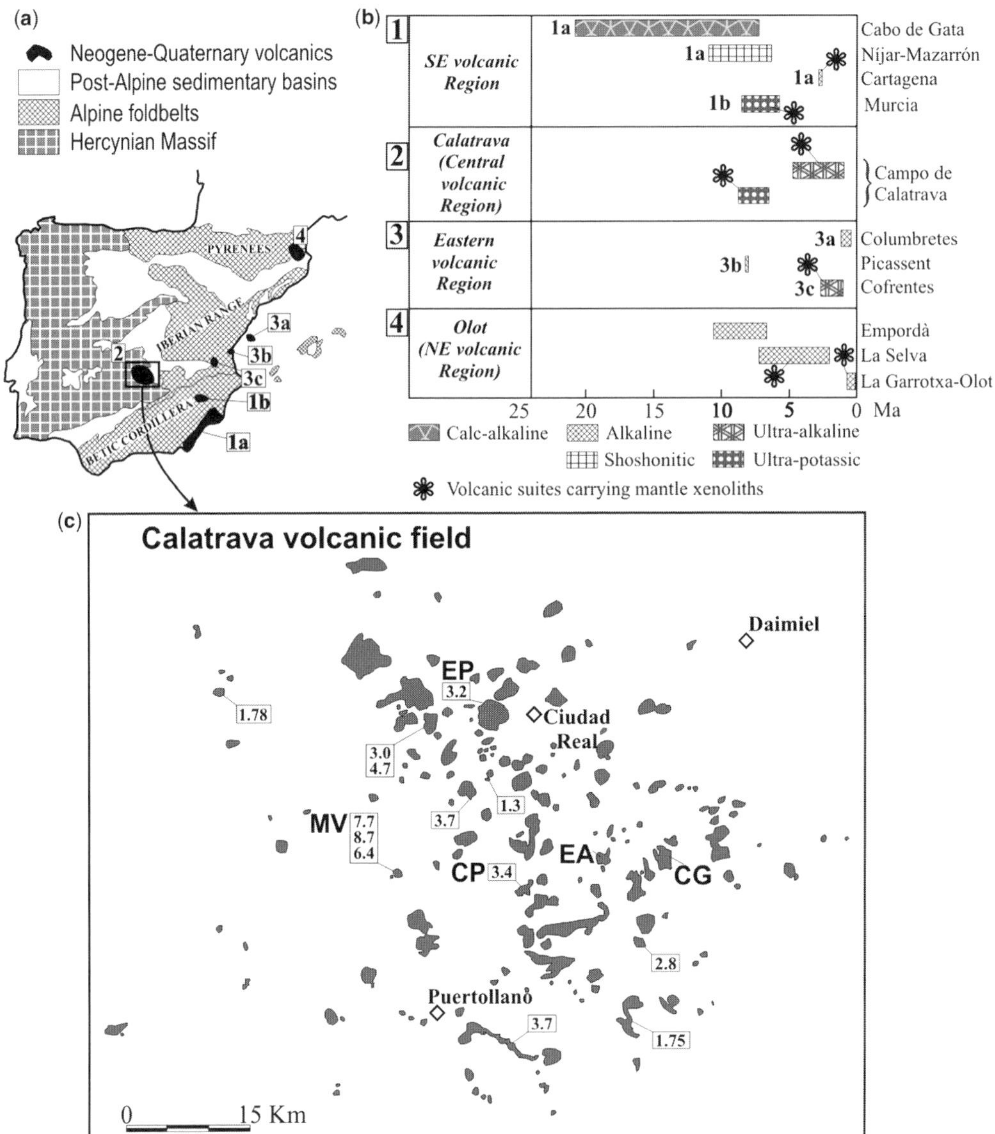

Fig. 1. (**a**) Sketch map of the Iberian Peninsula showing the location of the Cenozoic volcanic fields. (**b**) Geochronological table of the different magmatic suites showing those that carry lherzolite xenoliths. (**c**) Calatrava volcanic field and location of the volcanic centres mentioned in this study: EA, El Aprisco; EP, El Palo (studied by Bianchini *et al.* 2010); CP, Cerro Pelado; CG, Cerro Gordo; MV, Morrón de Villamayor. Numbers inside squares refer to geochronological data (Ancochea 2004).

suggestive of small and short-lived shallow magma chambers. Most edifices are strombolian cones but more than 50 hydromagmatic tuff rings or maars have been also described. Both types of volcanic edifices yield peridotite xenoliths. A first minor ultrapotassic event at around 8.7–6.4 Ma was followed by the eruption of alkaline basalts, basanites, and olivine nephelinites and melilitites from 3.7 to around circa 0.7 Ma (Ancochea 2004). Xenoliths studied in this work have been sampled from the pyroclastic deposits of two undersaturated alkaline basaltic volcanoes: the Cerro Pelado centre (38°48′32″N, 3°54′56″W), an olivine nephelinite scoria cone, and in El Aprisco (38°50′05″N; 3°50′00″W), an olivine melilitite maar. Smaller lherzolite xenoliths appear in other volcanic centres, including that of the early ultrapotassic event (El Morrón de Villamayor, Cerro Gordo, El Palo, etc.) (Fig. 1), but they have not been sampled for this study.

The similarity in normalized trace element patterns and the Sr–Nd–Pb isotopic homogeneity of the primary alkaline and ultralkaline basic magmas suggest that the Calatrava basaltic suites are derived from different degrees of partial melting of an enriched, but relatively homogeneous mantle source (Ancochea 1982; Cebriá & López-Ruiz 1995). The isotopic composition suggests a HIMU-like reservoir similar to that defined for the European asthenospheric mantle (Granet *et al.* 1995). An enriched asthenospheric reservoir (EAR) is in agreement with the suggested presence of garnet and phlogopite within mantle sources based on trace element geochemistry of Calatrava volcanics (e.g. Ancochea 1982; López-Ruiz *et al.* 2002).

The Calatrava volcanic field is a typical intracontinental zone within the Neogene–Present western and central European province. Similar to the NE Spanish volcanic region (also called the Olot volcanic field), this area is located in the limit of small late Cenozoic sedimentary basins trangressive over Variscan terranes. Volcanic vent distribution does not follow a clear spatial pattern and their geodynamical setting is controversial, with theories invoking a gigantic megafault system affecting the western Mediterranean European block (López-Ruiz *et al.* 2002) or volcanic clustering related to asthenospheric mantle upwelling (hot-spot or diapir instabilities) in a pre-rifting stage (Ancochea 1982; López-Ruiz *et al.* 1993).

Analytical methods

The major element mineral composition has been analysed at the Centro de Microscopía Electrónica 'Luis Bru' (Complutense University of Madrid) using a Jeol JXA-8900 M electron microprobe with four wavelength-dispersive spectrometers. Analytical conditions were an accelerating voltage of 15 kV and an electron beam current of 20 nA, with a beam diameter of 5 μm. Elements were counted for 10 s on the peak and 5 s on each background position. The measurement of Ca in olivine was repeated in some crystals with high content of this element (samples 65290 and 72674) using EMP conditions of 20 kV, 50 nA with counting times of 60 s. To reduce alkali loss during glass analysis, we lowered the beam current to 10 nA and the beam was defocused to 10 μm. Corrections were made using an on-line ZAF method. Detection limits are 0.02 wt% for Al, Na, K and P, 0.03 wt% for Ti, Fe, Mn, Mg, Ni and Cr, and 0.04 wt% for Si.

We have determined the *in situ* concentrations of 30 trace elements (REE, Ba, Rb, Sr, Th, U, Nb, Ta, Pb, Zr, Hf, Y, Sc, V, Co, Zn and Cr) in clinopyroxene (cpx), orthopyroxene (opx), olivine (ol) and amphibole (amph) on >130 μm-thick polished sections using laser ablation (LA-ICP-MS) at the Natural History Museum of London using an Agilent 7500CS ICP-MS coupled to a New Wave UP213 laser source (213 nm frequency-quadrupled Nd–YAG laser). The counting time for one analysis was typically 90 s (40 s measuring gas blank to establish the background and 50 s for the remainder of the analysis). The diameter of the laser beam was around 50 μm. The NIST 612 glass standard was used to calibrate relative element sensitivities for the analyses of the silicate minerals. Each analysis was normalized to Ca or Si (Al for spinel) using concentrations determined by electron microprobe. Detection limits for each element were in the range of 0.01–0.06 ppm, except for Sc and Cr (0.11 and 0.73 ppm, respectively).

Eleven spinel peridotite xenoliths from two volcanic centres (El Aprisco and Cerro Pelado) were used in this investigation. The whole-rock major and trace element composition was analysed at ACTLABS. The samples were melted using $LiBO_2$ and dissolved with HNO_3. The solutions were analysed by inductively coupled plasma atomic emission spectrometry (ICP-AES) for major elements, whereas trace elements were determined by ICP mass spectrometry (ICP-MS). Uncertainties in major elements are bracketed between 1 and 3%, except for MnO (5–10%) and P_2O_5 (>10%). The precision of ICP-MS analyses at low concentration levels was evaluated from repeated analyses of the international standards BR, DR-N, UB-N, AN-G and GH. The precision for Rb, Sr, Zr, Y, V, Hf and most of the REE were in the range 1–5%, whereas they range from 5 to 10% for the rest of trace elements, including Tm. Some samples had concentrations of certain elements below detection limits (K_2O 0.01%; Rb 1; Zr 1; Nb 0.2; Tb 0.01; Ho 0.01; Tm 0.005; Lu 0.002; Hf 0.1; Ta 0.01; Th 0.05; U 0.01). More information on the procedure, precision and accuracy of ACTLABS ICP-MS analyses is available at www.actlabs.com.

Sr–Nd isotopic analyses were carried out at the CAI de Geocronología y Geoquímica Isotópica of the Complutense University of Madrid, using an automated VG Sector 54 multicollector thermal ionization mass spectrometer with data acquired in multidynamic mode. Isotopic ratios of Sr and Nd were measured on a subset of whole-rock powders. The analytical procedures used in this laboratory have been described elsewhere (Reyes *et al.* 1997). Repeated analysis of NBS 987 gave $^{87}Sr/^{86}Sr = 0.710249 \pm 30$ (2σ, $n = 15$) and for the JM Nd standard the $^{143}Nd/^{144}Nd = 0.511809 \pm 20$ (2σ, $n = 13$). The 2σ error on the ε(Nd) calculation is ± 0.4.

Petrography and mineral chemistry

The studied Calatrava mantle xenoliths are rounded medium-size samples (from 5 to 45 cm in diameter) that show no evidence of alteration or host basalt infiltration. Xenoliths equilibrated in the spinel peridotite stability field and display a wide modal variation from lherzolite to minor wehrlite types. Modal composition was determined by mass-balance calculations from the main minerals and the major element compositions of the whole rocks, using the least-squares inversion method of Albarède (1995). Within the 11 analysed rock samples, 10 are lherzolites and only one is a wehrlite (sample 72674) (Fig. 2). Mantle xenoliths from the El Aprisco centre tend to have orthopyroxene more abundant than clinopyroxene, whereas those from Cerro Pelado are more clinopyroxene-rich, even wehrlitic in composition (Fig. 2). The lherzolitic–harzburgitic mantle xenoliths from Olot (Bianchini *et al.* 2007; Galán *et al.* 2008) have been plotted for comparison, and are similar in modal composition to those from SE Spain (Tallante: Beccaluva *et al.* 2004) (not shown). Calatrava lherzolites are richer in clinopyroxene and poorer in olivine than other Spanish mantle xenolith suites (Fig. 2). Scarce phlogopite-rich clinopyroxenites (glimmerite varieties) have been found at Cerro Pelado (Ancochea & Nixon 1987) but they have not been sampled for this study.

Although in accessory amounts, the studied peridotite xenoliths usually have interstitial volatile-rich phases indicative of modal metasomatism: amphibole in samples from the El Aprisco centre, and phlogopite from those of the Cerro Pelado maar. Only one xenolith is an anhydrous lherzolite (sample 72690 from El Aprisco). The wehrlite 72674 shows trace amounts of phlogopite included in clinopyroxene. Although peridotite xenoliths from Cerro Pelado with both hydrous minerals, amphibole and phlogopite, have been described previously (Ancochea & Nixon 1987) we did not find this type.

Most peridotite xenoliths have a coarse-grained texture of protogranular aspect, defined by a grain size greater than 2 mm and commonly equigranular (Fig. 3a). Some porphyroclastic textures or more

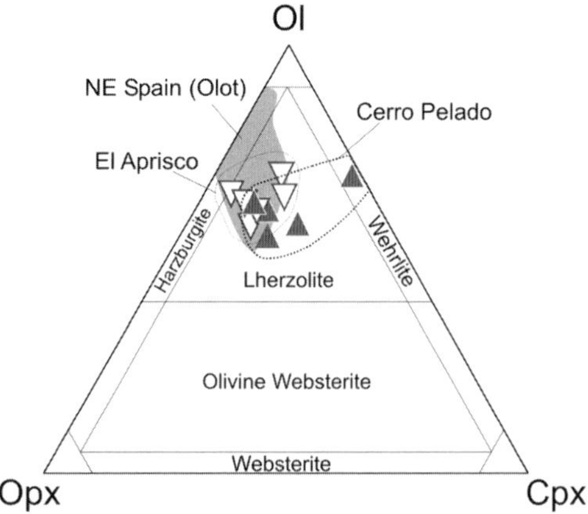

Fig. 2. Modal proportions of the studied Calatrava mantle xenoliths calculated using the mass-balance method of Albarède (1995). The modal compositional field from mantle xenoliths from NE Spain (Olot) is taken from Bianchini *et al.* (2007) and Galán *et al.* (2008).

inequigranular varieties also appear, in which olivine or orthopyroxene are commonly the porphyroclasts. Grain boundaries are usually curvilinear defining mosaic or triple-junction textures. No phase layering, foliation or lamination have been found.

Olivine crystals may have different sizes even in a single sample. Some fine-grained interstitial crystals, spatially related to spinel microaggregates or spinel coronas as described below, have been considered to be of second generation. Both orthopyroxene and clinopyroxene show mutual lamellae, and commonly a second superimposed spinel lamellae. Spinel occurs as discrete, dispersed interstitial grains that usually show some fine-grained polycrystalline coronas of amphibole (only in the El Aprisco outcrop) (Fig. 3b–d). Some scarce spinel–pyroxene–amphibole symplectite has been also observed, but the clearly hydrous character of the corona rejects the possibility of a reaction between pre-existing garnet and matrix olivine, as suggested in other spinel symplectitic lherzolite xenoliths (Ackerman *et al.* 2007) (Fig. 3c). Some spinel grains have sieve textures defined by a partial corona of a new fine-grained spinel-2 associated to vesicular glass and second-generation olivine (Fig. 3f). These textural features have been interpreted as re-equilibrations or reactions with a

Fig. 3. Photomicrographs of representative Calatrava mantle xenoliths. *Lherzolite 72691* (El Aprisco): (**a**) equigranular texture; (**b**) interstitial amphibole around major peridotite minerals (spinel, clinopyroxene, orthopyroxene) (BSE image); and (**c**) symplectitic intergrowths of spinel-2, amphibole, clinopyroxene-2 (an amph–cpx intergrowth is shown on the left) within orthopyroxene (BSE image). *Lherzolite 72689* (El Aprisco) showing: (**d**) amphibole aureoles around spinel; and (**e**) a complex reaction zone showing the breakdown of spinel to clinopyroxene-2, olivine-2 and vesicular

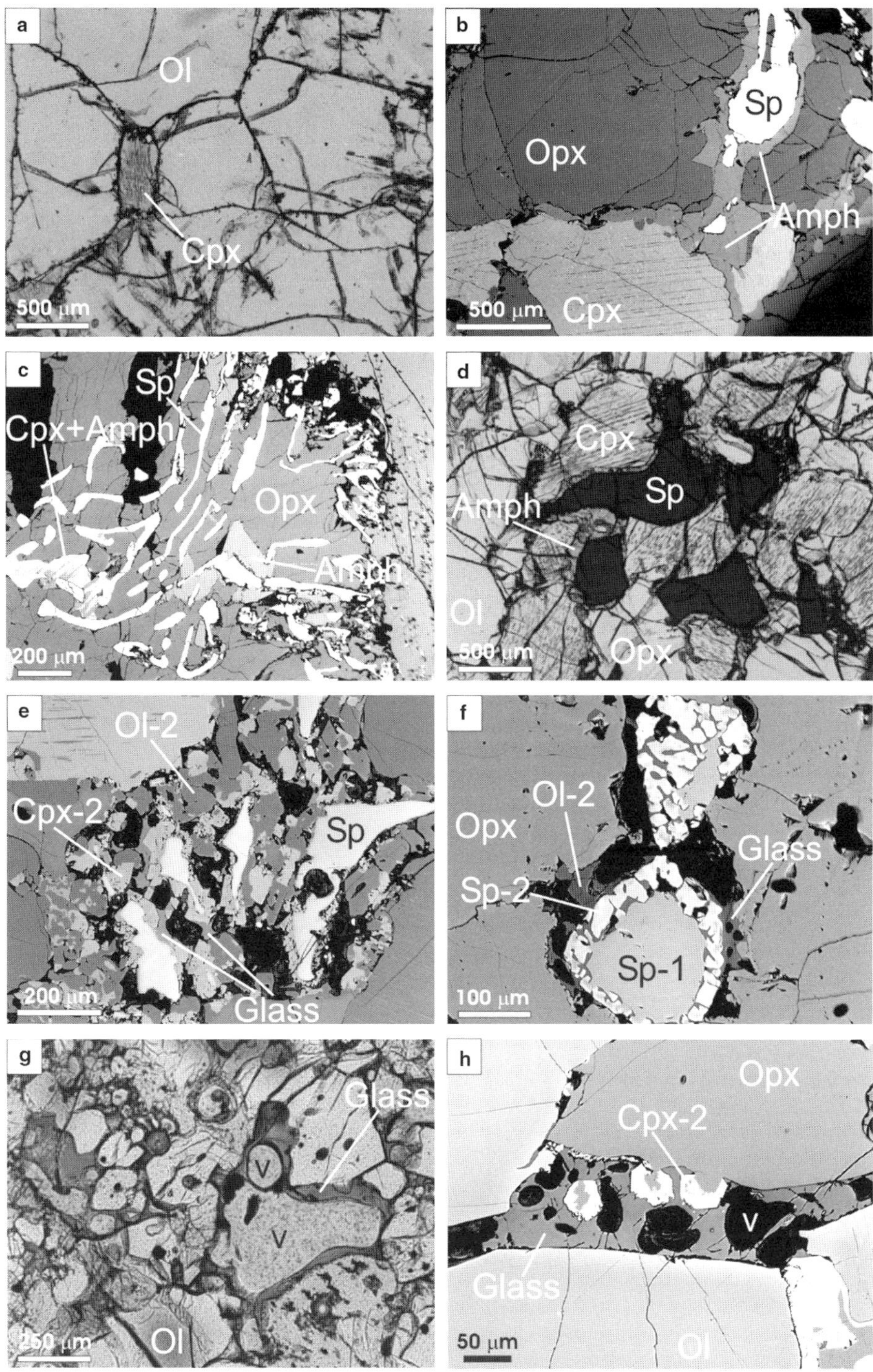

Fig. 3. (*Continued*) glass (BSE image). *Lherzolite 65290* (Cerro Pelado): (**f**) reaction zones around primary spinel (Sp-1) composed of cellular spinel-2, microgranular olivine-2 and vesicular glass (BSE image). *Wehrlite 72674* (Cerro Pelado): (**g**) highly vesicular interstitial glass (BSE image); and (**h**) vesicular glass vein with associated cpx-2 crystals (BSE image). Ol, olivine; Opx, orthopyroxene; Cpx, clinopyroxene; Sp, spinel; Amph, amphibole; v, vesicle.

percolating intergranular melt (Shaw & Dingwell 2008). In sample 65290 it is possible to see a complex corona of symplectitic spinel-2 and interstitial glass rimming primary spinel (Fig. 3f). Acicular pentlandite, interstitial to olivine, is very scarce in some lherzolites.

Accessory amounts of hydrous minerals (amphibole or phlogopite) are present in most of the Calatrava lherzolite xenoliths (except in anhydrous lherzolite 72690). They are mostly intergranular phases forming small anhedral crystals. Amphibole in most lherzolites forms coronas around spinel grains or intergrowths with clinopyroxene around spinel symplectites (Fig. 3b–d), in textures similar to those described in other peridotite xenolith suites (e.g. Coltorti *et al.* 2004, 2007*b*).

Some xenoliths also show interstitial brown glass with small vesicle or bubble-like structures (Fig. 3e–h). Moreover, some interstitial glasses appear as part of a complex reaction zone that involves many of the primary lherzolite minerals, but especially spinel, which is surrounded either by symplectitic intergrowths of new spinel-2 and olivine-2 with interstitial glass (Fig. 3f) or by a microaggregate of newly formed cpx–ol–glass (Fig. 3e).

The wehrlite 72674 is porphyroclastic in texture and the two pyroxenes do not show lamellar exsolution, as is typical in the lherzolite types. Wehrlite clinopyroxene shows a marked poikilitic texture with multiple glass, apatite, phlogopite and fluid-rich inclusions. Olivine grains show locally deformation twins and most crystals have smooth curvilinear boundaries. Only trace amounts of spinel (as microinclusions in olivine and clinopyroxene) appear in this sample. The wehrlite also shows interstitial vesicular glass, which is mainly concentrated in the fine-grained section of the inequigranular texture, defining some interconnected veining (Fig. 3g, h).

Major element mineral composition

The Calatrava hydrous mantle xenoliths consist of variable proportions of magnesian olivine, orthopyroxene and clinopyroxene, aluminous spinel, and accessory amounts of calcic amphibole or phlogopite, the compositions of which are summarized in Table 1. All minerals are unzoned and homogeneous within a single crystal.

The Mg-number for olivine mostly ranges from 89.2 to 91.5, although neoformed varieties (oliv-2) and olivine from wehrlite 72674 show lower Mg-numbers (88.4 and 84.5–86.0, respectively). Olivine-2 in lherzolites also show slightly higher CaO and lower NiO content than Mg-rich olivine (Table 1), features typical of metasomatism (e.g. Coltorti *et al.* 1999; Ionov *et al.* 2005). Olivine in the wehrlite shows the highest TiO_2 content (up to 0.09 wt%), and high CaO and low NiO content (Table 1).

Orthopyroxene has a similar range of Mg-numbers than olivine, mostly from 88.7 to 92.3 (Table 1), but it shows a wider range in content of Cr_2O_3 (0.13–0.52), Al_2O_3 (3.00–5.74) and CaO (0.12–1.09), always in a common range for abyssal peridotites (Bonatti & Michael 1989). Correlatively to olivine, orthopyroxene of wehrlite 72674 shows lower Mg-numbers (85.0–86.6), and higher content of TiO_2 (0.22–0.41), CaO (0.94–3.86) and Na_2O (0.12–0.25), than orthopyroxene from associated lherzolites (Fig. 4a). Lherzolite 72691 with the lowest averaged Al content of orthopyroxene (and the highest Mg-Cr values) could be the most depleted peridotite of the xenolith suite (Fig. 4a).

Clinopyroxene also shows a wide range of Mg-numbers (87.6–92.7) with cpx-2 (neocrystals related to intergranular reaction zones) having low values (88.2–90.0) but variable Al–Ti–Cr content. For example, the cpx-2 analysis 71 in lherzolite 72689 (related to a spinel reaction zone with glass) shows the highest Al–Cr–Ti content of the whole analysed clinopyroxene population (Fig. 4b). Clinopyroxene from wehrlite 72674 is also Fe–Ti-enriched, as are the other minerals of this xenolith (Fig. 4b, c) (Table 1). Lherzolite 72691 shows the lowest Al clinopyroxene, reinforcing the idea of being the most depleted peridotite xenolith.

Primary spinel has low Cr-numbers (from 8.3 to 10.8) and a narrow Mg-number range (0.75–0.78) (Table 1). TiO_2 content is generally low, ranging from 0.01 to 0.16 wt%. Spinel from the most depleted lherzolites (samples 72691 and 55570) shows a wider range of composition, with relatively higher Cr values and lower Al content than spinel from associated lherzolites (Fig. 5). In lherzolite 55570 cores of large spinel crystals show the highest Al and Mg content (and the lowest Cr values) compared with rims or interstitial rods, which are associated with intergranular amphibole. In fact, the smallest interstitial spinel crystals in lherzolite 55570 and the symplectitic spinel from lherzolite 72691 show the highest Cr-numbers (13.0–17.7) and the lowest Al_2O_3 contents (51–55 wt%) of the Calatrava lherzolites. These contents are similar to those of the sieved sp-2 of lherzolites 65290 and 65298, which also show high TiO_2 content (up to 0.44 wt%), suggestive of a reaction with a Ti-rich metasomatic agent (Perinelli *et al.* 2008*b*) (Fig. 5). Residual spinel micrograins preserved as inclusions in major minerals of wehrlite 72674 show the highest Ti and Cr (TiO_2 up to 3.0 wt% and Cr-numbers in the range of 52.4–54.1), and the lowest Al–Mg content, of the studied peridotite xenoliths (Table 1). Owing to

Table 1. *Major element composition of representative minerals from the Calatrava mantle xenoliths*

Representative olivines

	El Aprisco						Cerro Pelado			
Sample	55569 A39	55570 A18	72688 A142	72689 A82	72690 A38	72691 A120	65290 A23	65298 A2	65298 21 (oliv-2)	72674 A95
SiO_2	41.33	40.56	41.05	39.97	41.20	40.63	40.58	40.67	40.69	40.03
TiO_2	bdl	bdl	bdl	bdl	0.01	bdl	0.02	0.01	0.01	0.03
Al_2O_3	bdl	bdl	0.02	bdl	bdl	bdl	0.03	bdl	bdl	bdl
FeO	9.63	9.18	9.88	9.53	9.31	8.66	10.40	9.29	11.36	14.85
MnO	0.05	0.18	0.20	0.18	0.06	0.13	0.16	0.07	0.13	0.12
MgO	49.60	49.54	49.61	49.85	49.74	50.68	49.01	50.43	48.71	45.53
CaO	0.03	0.04	0.09	0.04	0.02	0.05	0.06	0.09	0.13	0.09
Na_2O	bdl	0.02	0.01	bdl	0.02	0.01	bdl	0.01	bdl	bdl
K_2O	0.01	bdl	bdl	bdl	bdl	bdl	bdl	bdl	bdl	bdl
NiO	0.41	0.47	0.22	0.42	0.33	0.42	0.21	0.36	0.12	0.20
Cr_2O_3	bdl	0.08	0.01	0.07	0.02	0.03	0.07	0.03	0.04	bdl
Total	101.06	99.98	101.09	99.99	100.71	100.58	100.49	100.92	101.15	100.91
XMg	90.19	90.59	89.97	90.31	90.50	91.25	89.36	90.62	88.44	84.47

Representative orthopyroxenes

	El Aprisco						Cerro Pelado			
Sample	55569 A36	55570 A32	72688 A59	72689 A76	72690 A41	72691 A65	65290 A27	65298 A23	65298 A3	72674 A110
SiO_2	55.08	54.47	55.38	54.97	54.73	54.71	54.19	54.24	53.88	53.64
TiO_2	0.15	0.12	0.03	0.06	0.04	0.04	0.18	0.07	0.08	0.37
Al_2O_3	4.72	5.00	3.49	3.35	5.42	4.08	5.74	5.49	5.39	3.94
FeO	6.26	5.77	6.54	6.41	5.80	5.45	6.30	6.11	6.24	8.46
MnO	0.14	0.14	0.15	0.15	0.12	0.13	0.14	0.15	0.13	0.15
MgO	32.80	32.56	33.39	33.94	32.34	33.66	31.53	32.96	32.81	28.54
CaO	0.99	1.10	0.54	0.43	1.37	0.62	1.04	0.88	0.80	3.86
Na_2O	0.11	0.11	0.02	0.03	0.13	0.06	0.15	0.12	0.15	0.25
K_2O	bdl	bdl	0.02	bdl	bdl	bdl	bdl	bdl	bdl	bdl
NiO	0.08	0.14	bdl	0.08	0.08	0.15	0.04	0.18	0.14	bdl
Cr_2O_3	0.33	0.53	0.18	0.23	0.34	0.38	0.40	0.35	0.33	0.44
Total	100.68	99.93	99.74	99.66	100.38	99.29	99.70	100.55	99.95	99.64
XMg	90.31	90.94	90.11	90.38	88.94	91.66	89.92	90.59	90.35	85.78

(*Continued*)

Table 1. *Continued*

Representative clinopyroxenes

	El Aprisco							Cerro Pelado				
Sample	55569 A30	55570 A30	72688 A57	72689 71 (Cpx-2)	72689 A75	72690 A42	72691 A66	65290 A31	65298 22 (Cpx-2)	65298 A4	72674 101 (Cpx-2)	72674 A96
SiO_2	52.29	51.75	51.89	49.86	51.05	52.37	52.57	51.19	53.03	51.36	53.97	49.49
TiO_2	0.46	0.23	0.42	0.80	0.41	0.33	0.01	0.54	0.51	0.52	0.62	1.39
Al_2O_3	6.90	5.72	4.87	8.11	6.07	7.61	4.44	7.37	4.43	7.11	3.12	6.70
FeO	2.60	2.87	2.86	3.08	2.73	2.46	2.33	3.46	4.21	3.31	6.11	4.80
MnO	0.05	0.06	0.14	0.02	0.10	bdl	0.05	0.09	0.17	0.06	0.16	bdl
MgO	14.94	15.81	15.66	15.65	15.35	14.44	16.64	15.72	17.72	15.66	17.90	14.39
CaO	19.69	21.15	22.26	18.80	21.58	20.55	21.29	18.90	18.17	19.09	16.57	20.95
Na_2O	1.62	1.35	0.87	1.19	0.98	1.64	0.89	1.60	1.21	1.61	1.04	1.04
K_2O	bdl	bdl	bdl	bdl	bdl	0.01	bdl	0.01	bdl	bdl	0.03	bdl
NiO	0.04	0.08	0.03	0.01	0.03	0.07	0.05	bdl	bdl	0.01	0.01	0.01
Cr_2O_3	0.79	0.83	0.54	1.50	0.70	0.94	0.61	0.74	0.97	0.66	0.63	0.59
Total	99.36	99.85	99.55	99.03	98.99	100.42	98.88	99.63	100.41	99.39	100.14	99.34
XMg	91.10	90.79	90.72	90.06	90.98	91.56	92.65	88.98	88.20	89.33	83.94	84.20
XCr	0.07	0.09	0.06	0.08	0.07	0.07	0.10	0.09	0.13	0.06	0.12	0.09

Representative spinels

	El Aprisco							Cerro Pelado				
Sample	55569 A40	55570 A16	72688 A144	72689 A90	72690 A55	72691 121 sympl.	72691 56 sympl.	65290 A29	65290 A30 (Sp-2)	65298 A5 (Sp-2)	65298 A1	72674 A13
SiO_2	0.07	0.02	0.13	0.07	0.06	0.05	0.04	0.08	0.11	0.11	0.08	0.11
TiO_2	0.13	0.10	0.10	0.06	0.10	0.01	bdl	0.19	0.41	0.38	0.14	2.49
Al_2O_3	56.51	52.23	55.96	57.78	58.40	53.85	51.03	56.97	53.15	51.64	56.83	20.76
Cr_2O_3	10.17	15.73	9.61	8.84	10.15	13.10	16.32	10.17	13.11	14.46	9.31	35.80
FeO	11.97	12.09	13.23	12.00	10.57	10.90	11.73	12.55	11.99	11.10	11.15	26.37
MnO	0.04	0.10	0.15	0.11	0.041	0.10	0.07	0.02	0.11	0.13	0.10	0.19
NiO	0.34	0.33	0.29	0.48	0.39	0.40	0.25	0.32	0.34	0.28	0.34	0.11
MgO	20.41	19.65	19.81	20.49	21.12	20.82	19.90	20.23	20.62	20.93	21.56	13.23
CaO	0.03	0.03	0.09	0.01	bdl	0.03	0.06	0.01	0.02	0.03	0.02	0.18
Na_2O	0.04	0.01	0.04	0.01	0.01	bdl	bdl	bdl	bdl	bdl	0.01	bdl
K_2O	bdl	bdl	bdl	bdl	bdl	bdl	bdl	0.01	bdl	bdl	0.02	bdl
Total	100.00	100.28	99.74	100.12	100.84	99.54	99.65	100.78	100.13	99.36	99.93	99.22
XCr	10.78	16.81	10.33	9.30	10.43	14.05	17.66	10.71	14.21	15.83	9.90	53.64
XMg	0.75	0.74	0.73	0.75	0.78	0.77	0.75	0.74	0.76	0.77	0.78	0.48

Representative amphiboles and phlogopites

	Amphibole						Phlogopite		
Sample	55569 A27	55570 A35	72689 A85	72691 A55	72688 A51	72688 A68	65298 A13	65298 A19	72674 A116
SiO_2	43.83	42.65	42.79	43.63	42.19	42.59	37.64	38.13	35.98
TiO_2	1.48	0.58	0.73	0.42	1.33	1.33	3.10	3.12	6.76
Al_2O_3	14.25	15.43	15.05	14.97	15.18	15.09	17.90	18.18	15.70
FeO	3.72	3.82	4.04	3.72	4.40	4.16	4.17	3.88	8.08
MnO	0.04	0.14	0.05	0.09	0.12	0.04	0.05	bdl	0.09
MgO	17.22	17.86	18.09	18.64	17.67	17.54	21.35	20.54	17.66
CaO	11.48	10.74	11.17	10.83	11.76	11.69	0.02	0.05	0.10
Na_2O	3.51	3.75	3.59	3.62	3.50	3.41	0.79	0.76	0.81
K_2O	0.02	0.20	bdl	0.11	0.06	0.07	9.15	9.37	8.91
NiO	0.25	0.10	0.19	0.08	0.07	0.04	bdl	bdl	bdl
Cr_2O_3	0.57	1.39	0.72	1.34	0.73	0.77	0.78	0.81	0.16
Total	95.54	95.17	95.50	96.02	96.20	95.92	94.93	94.83	94.25
XMg	89.18	89.29	88.87	89.93	87.75	88.27	90.14	90.43	79.58
XCr	2.64	5.71	3.10	5.66	3.14	3.30	2.85	2.90	0.69

Interstitial glasses

	El Aprisco		Cerro Pelado					
Sample	72689 A22	72689 A24	65290 A34	65290 A37	72674 A7	72674 A1	72674 A5	72674 A109
SiO_2	57.73	56.79	55.66	54.60	53.29	53.68	58.17	59.74
TiO_2	0.84	0.80	1.02	1.08	1.92	2.36	1.49	1.78
Al_2O_3	22.22	21.87	21.51	21.63	18.87	18.99	16.52	16.69
FeO	2.01	2.25	2.35	1.86	5.17	4.63	3.75	3.43
MnO	bdl	0.03	0.03	0.03	0.08	0.12	0.02	0.07
MgO	3.09	3.48	3.37	2.74	2.25	1.89	1.87	1.82
CaO	6.70	7.07	4.77	4.81	4.19	3.90	2.37	2.30
Na_2O	5.54	5.02	4.36	5.07	6.26	6.06	5.19	3.34
K_2O	0.06	0.03	4.33	4.87	5.27	5.18	5.44	5.17
P_2O_5	0.11	0.08	0.23	0.20	0.35	0.29	0.24	0.20
Cr_2O_3	0.06	0.12	0.37	0.37	0.07	0.04	0.03	0.03
Total	98.42	97.34	98.00	97.24	97.77	97.18	95.10	94.56
XMg	0.73	0.73	0.72	0.72	0.44	0.42	0.47	0.49

Abbreviation: bdl, below detection limits.

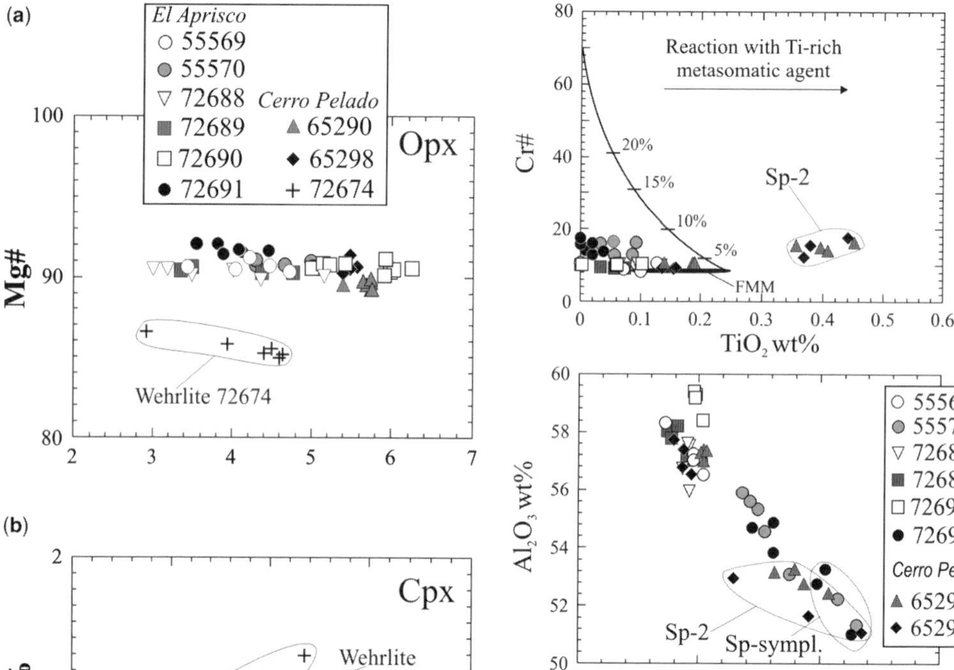

Fig. 5. Composition of spinel from the Calatrava peridotite xenoliths. The trend of partial melting of a fertile MORB mantle (FMM) is shown for comparison. The line of reaction with a metasomatic agent (or a partial melt) is taken from Perinelli *et al.* (2008*b*) and Shaw & Dingwell (2008).

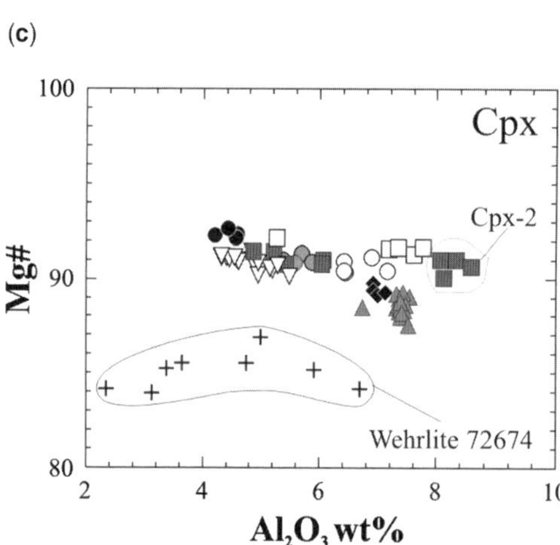

Fig. 4. Major element compositions of pyroxenes in Calatrava peridotite xenoliths. Mg# is the Mg-number. The grey arrow in (**b**) shows clinopyroxene compositional variation with a progressive increase in peridotite partial melting (Uysal *et al.* 2007).

this extreme composition they have not been plotted in Figure 5.

The low Cr-number of primary spinels from the Calatrava lherzolite xenoliths combined with the averaged Mg-number of the coexisting olivine make them plot in the OSMA (olivine–spinel mantle array) closer to fertile mantle compositions (Arai 1992) than other lherzolite xenolith suites from Iberia (Fig. 6). The Cr-spinel of wehrlite 72674 plots outside the OSMA, further to the right of the Olot harzburgite field (Fig. 6).

Amphibole is a typical interstitial mineral locally surrounding primary spinel in the lherzolite samples from the El Aprisco maar. Its composition is mainly pargasite. Mg-numbers range from 86.9 to 91.2, and are positively correlated with Cr_2O_3 content (0.7–1.4 wt%) but inversely with TiO_2 (1.48–0.69 wt%) and CaO (11.9–11.0 wt%) content. Amphibole in lherzolites 72691 and 55570 shows the highest Mg- and Cr-numbers (89.3–91.2 and 4.8–5.7, respectively), and the lowest TiO_2 and CaO content (0.3–0.6 and 10.4–10.8 wt%,

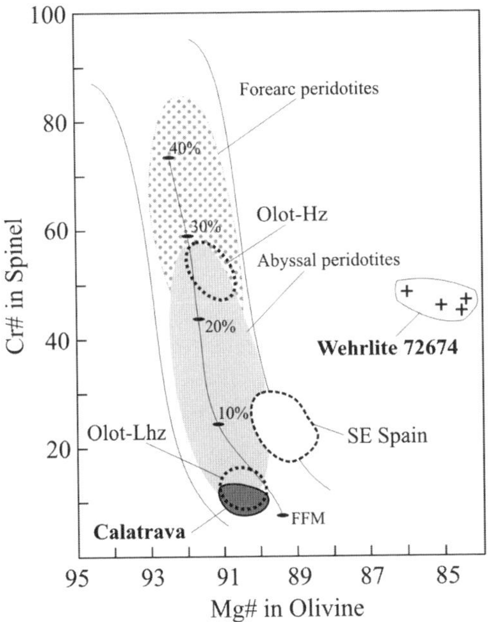

Fig. 6. Cr-number (100×Cr/(Cr + Al) in spinel v. Mg-number in coexisting olivine for Iberian mantle xenoliths. Mineral data from the Olot volcanic field (Lhz, lherzolites; Hz, harzburgites) are taken from Bianchini *et al.* (2007) and Galán *et al.* (2008). Data from the SE Spain volcanic field (mainly from the Tallante volcanic centre) are from Beccaluva *et al.* (2004) and Shimizu *et al.* (2004). OSMA–olivine–spinel mantle array (Arai 1994). FMM: fertile MORB mantle and partial melting trends are from Arai (1992). Fields of forearc and abyssal peridotites are from Pearce *et al.* (2000).

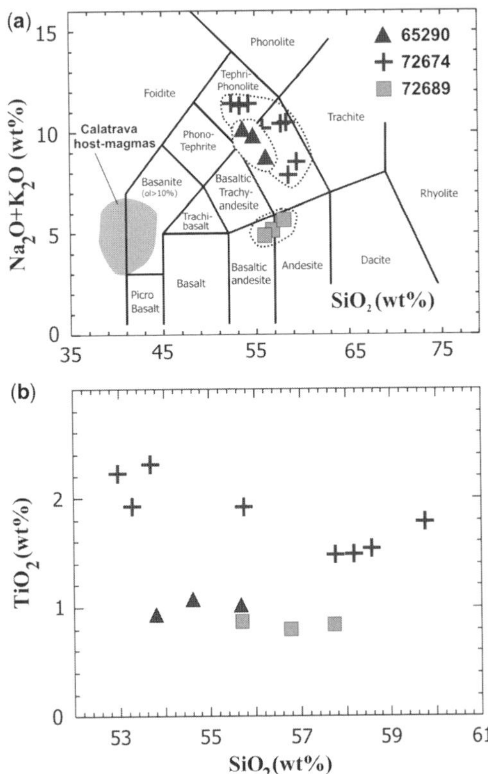

Fig. 7. Chemistry of glasses from Calatrava peridotite xenoliths. (**a**) The total alkali–silica (TAS) diagram of Le Bas *et al.* (1986) for glasses in Calatrava mantle xenoliths. The grey field for Calatrava volcanic rocks is taken from Ancochea (2004). (**b**) TiO_2–silica (in wt%) diagram.

respectively). The low-K content of Calatrava amphiboles (<0.11 wt%) is remarkable compared to other mantle xenolith suites (Coltorti *et al.* 2007*a*), except those from some Olot lherzolites (OLT-21 and OLT-23 samples: Bianchini *et al.* 2007).

Phlogopite is only found in mantle xenoliths from the Cerro Pelado centre with Mg-number (90.2–90.4) similar to coexisting clinopyroxene but with higher Cr-number (2.8–2.9) (Table 1). Phlogopite included in clinopyroxene from wehrlite 72674 is markedly poorer in Cr and Mg values (0.7 and 79.6, respectively) but richer in TiO_2 content (6.8 wt%), similar in composition to phlogopites from associated glimmerite xenoliths (anal. 6 in table 32 of Ancochea & Nixon 1987).

Glass veins and interstitial glasses have been observed in some Calatrava mantle xenoliths. In xenoliths from the Cerro Pelado centre the glasses are alkaline in composition and they plot mostly in the field of trachy-andesite in the total alkalis–silica (TAS) diagram (Fig. 7). However, interstitial glass from lherzolite 72689 from the El Aprisco volcano is subalkaline in composition owing to its low K_2O content (Table 1), and plots mostly in the field of basaltic andesite in the TAS diagram. The Calatrava glasses are compositionally heterogeneous, and thus, the high TiO_2 and FeO (and low Al_2O_3 and MgO) content shown by glasses from wehrlite 72674 (1.8–2.3 and 3.3–5.8 wt%, respectively) (Table 1) are quite remarkable (Fig. 7).

Pressure–temperature P–T *estimates*

The lithospheric thickness in central Spain seems to be almost constant over a large area at around 110 km, while the depth to the Moho is around 32 ± 2 km (Fernàndez *et al.* 1998). The depth of xenolith extraction is constrained to less than 70 km (*c.* 20 kbar) by the absence of garnet in the lherzolite suite. Although for spinel peridotites there are no precise geobarometers (e.g. Pearson *et al.* 2005; Ackerman *et al.* 2007), we have used

Table 2. P–T *estimates on the Calatrava mantle xenoliths*

	T (°C)			P (kbar)
	T_{WB}	T_{BM}	T_{BK}	P_{NU}
El Aprisco xenoliths				
55569 cpx30opx36	1093	1185	1052	14.6
55570 cpx30opx32	1116	1220	1073	9.7
72688 cpx57opx59	1003	1036	866	6.4
72689 cpx75opx76	1054	1109	938	8.9
(cpx-2)71opx76	1182	1345	1163	13.9
72690 cpx42opx41	1079	1062	980	15.0
72691 cpx66opx65	1115	1274	994	7.6
Cerro Pelado xenoliths				
65290 cpx31opx27	1149	1382	1112	13.4
65298 cpx4opx3	1165	1321	1123	13.8
(cpx-2)22opx23	1255	1420	1209	7.0
72674 cpx96opx108	1056	1050	986	7.6
(cpx-2)101opx110	1243	1500	1207	7.0

the estimations of Nimis & Ulmer (1998) for clinopyroxene in order to approximate the depth of extraction of the lherzolite xenoliths (Table 2). Pressure estimates range from 14–15 kbar, for weakly metasomatized lherzolites (e.g. anydrous lherzolite 72690, Table 2), to 7–8 kbar for the most depleted lherzolites (72691) or wehrlite varieties (72674). These low-pressure values are unrealistic for mantle depths but suggest that xenoliths could have been trapped from a shallow lithospheric mantle close to Moho depths. The second generation of clinopyroxene, related either to glass veining in lherzolite 72689 (cpx 71, Table 2) or to micro-aggregates around spinel in lherzolite 65298 (cpx 22, Table 2), yields P estimates mostly overlapping those from primary clinopyroxene (13.9–7.0 kbar). These unrealistic low-pressure estimates are also in disagreement with the absence of replacement of spinel by plagioclase in the studied samples.

Temperatures were calculated using three calibrations of the two-pyroxene geothermometer (Wood & Banno 1973; Bertrand & Mercier 1985; Brey & Köhler 1990). In general there is a good agreement between the different methods used, although Bertrand & Mercier's (1985) temperatures are slightly higher than the two other estimates (Table 2). Most lherzolites yield T estimates in the range of 1000–1165 °C, whereas neoformed clinopyroxene give a still higher temperature range of 1180–1255 °C. We have to bear in mind that most of the neoformed clinopyroxene is related to interstitial glass (e.g. cpx 71 in lherzolite 72689 and cpx 101 in wehrlite 72674, Table 2). Thus, geothermometry suggests some kind of reheating at depth to explain the thermal increase (*c.* 100 °C). The xenolith reaction with high-temperature magmas (e.g. silica-undersaturated alkaline magmas have liquidus temperatures of around 1250–1300 °C: Perinelli *et al.* 2008*a*) may explain the higher temperature of the second-generation clinopyroxene in the Calatrava peridotites.

Trace element mineral composition

Trace element analyses of representative main minerals from two lherzolites are reported in Table 3. REE primitive mantle-normalized patterns for clinopyroxene of amphibole-lherzolite (72688 from the El Aprisco centre) are LREE-enriched (La_N/Sm_N = 10.4–13.7) and almost flat in HREE (Fig. 8). The coexisting orthopyroxene displays a similar trend in some LREE patterns (although showing an inflection to Nd instead of Sm as in cpx patterns) but are positively fractionated in HREE (Fig. 8). Olivine shows a similar REE pattern to orthopyroxene, although more spiky and less HREE enriched (not shown, Table 3). Similarly, multitrace element patterns normalized to primitive mantle for clinopyroxene show a marked Th–U enrichment, and Nb–Ta and, to a lesser extent, Zr–Ti negative anomalies (Fig. 8). Coexisting orthopyroxene and, less markedly, olivine show similar trace element patterns.

Amphibole REE patterns perfectly mimic those of the coexisting clinopyroxene (Fig. 8). Incompatible element patterns are also similar to those of clinopyroxene but with higher LILE (large ion lithophile elements) and Ti content. The high Th–U content of amphibole and associated clinopyroxene suggests an enrichment process related to melt instead of a volatile-rich fluid as the metasomatizing agent (see also Coltorti *et al.* 2007*b*). This marked Th–U positive anomaly highlights the prominent Nb–Ta negative anomaly of the El Aprisco

Table 3. *Trace element composition of minerals from the Calatrava lherzolite xenoliths*

	72688 (El Aprisco)						65290 (Cerro Pelado)		
	ol ($n = 4$)	cpx ($n = 12$)	cpx-2 ($n = 3$)	opx ($n = 4$)	amph ($n = 21$)	sp ($n = 2$)	ol ($n = 9$)	cpx ($n = 24$)	opx ($n = 12$)
Sc	6.783	77.70	76.43	23.88	66.59	0.279	6.542	62.83	27.89
V	16.62	202	199	86.83	331	351	5.470	187	98.93
Cr	457	3803	4117	1840	4568	47750	128	4235	2868
Co	116	18.70	19.00	54.35	32.87	187	139	19.44	57.45
Zn	37.88	14.60	15.26	28.38	11.46	885	78.99	11.70	46.71
Rb	0.059	0.048	0.051	0.053	3.377	0.084	0.064	0.048	0.052
Sr	0.180	172	168	0.935	529	0.054	0.060	146	0.803
Y	0.214	18.63	18.467	1.137	33.629	0.023	0.113	24.26	2.320
Zr	0.138	11.36	11.30	0.632	12.13	0.074	0.227	119	12.94
Nb	0.017	0.014	0.015	0.014	0.154	0.094	0.025	1.118	0.082
Ba	0.281	2.862	0.189	0.142	845	0.084	0.142	0.089	0.086
Hf	0.068	0.758	0.717	0.052	0.798	0.076	0.039	3.585	0.332
Ta	0.012	0.011	0.009	0.014	0.010	0.025	0.012	0.472	0.023
Pb	0.050	1.311	1.357	0.057	5.543	0.056	0.045	0.051	0.037
Th	0.080	4.115	4.020	0.253	5.356	0.028	0.014	0.108	0.012
U	0.018	1.203	1.183	0.052	1.467	0.016	0.009	0.015	0.007
La	0.047	27.60	27.10	0.195	40.63	0.021	0.013	5.399	0.024
Ce	0.044	31.31	30.57	0.218	44.92	0.017	0.013	17.18	0.136
Pr	0.011	2.278	2.110	0.018	3.276	0.008	0.006	3.448	0.034
Nd	0.084	6.235	5.527	0.107	9.175	0.104	0.090	19.53	0.244
Sm	0.062	1.458	1.313	0.071	2.220	0.060	0.058	5.391	0.139
Eu	0.019	0.556	0.558	0.024	0.909	0.018	0.015	1.637	0.046
Gd	0.068	2.378	2.460	0.104	4.124	0.107	0.062	5.575	0.201
Tb	0.009	0.433	0.447	0.014	0.746	0.013	0.007	0.814	0.040
Dy	0.052	3.391	3.220	0.132	5.710	0.059	0.049	5.101	0.343
Ho	0.018	0.729	0.730	0.038	1.277	0.015	0.011	0.967	0.090
Er	0.050	2.250	2.093	0.159	3.794	0.051	0.048	2.542	0.320
Tm	0.013	0.319	0.308	0.028	0.552	0.011	0.010	0.355	0.054
Yb	0.120	2.198	2.080	0.302	3.666	0.075	0.067	2.148	0.446
Lu	0.018	0.308	0.294	0.065	0.545	0.013	0.012	0.312	0.075
Sum LREE	0.25	68.88	66.62	0.61	100.22	0.21	0.18	50.95	0.58
Sum HREE	0.35	12.01	11.63	0.84	20.42	0.35	0.27	17.81	1.57
Sum REE	0.62	81.44	78.80	1.47	121.55	0.58	0.46	70.40	2.19

Analysed phases are primary crystals except those marked as type-2 (cpx-2 of 72688 sample).

amphibole that is higher than those described in peridotite xenoliths, with amphiboles generated in the mantle wedge above subduction zones (supra-subduction amphiboles of Coltorti *et al.* 2007*a*).

Clinopyroxene in lherzolite 65290 from the Cerro Pelado maar displays completely different REE and trace element patterns to the previous lherzolite. It shows a peculiar upwards-convex shape of LREE patterns (with the peak located at Nd) combined with low HREE content (Lu_N down to $9 \times Ch$). Incompatible element-normalized patterns show a positive fractionation up to the REE where the pattern becomes almost flat, except for a small Ti negative anomaly (Fig. 8). Orthopyroxene (and olivine) does not mimic the REE patterns of coexisting clinopyroxene, instead showing a marked positive fractionation in the REE profiles. Incompatible element patterns are more spiky than those of the coexisting clinopyroxene but fluctuating close to $0.1-1 \times Ch$ values (Fig. 8). A positive Zr–Hf anomaly in the Cerro Pelado orthopyroxene contrasts with the small negative one in El Aprisco.

Similar LREE–Th–U-enriched clinopyroxenes to those of lherzolite 72688 have been described in the Olot lherzolite xenolith suite (Bianchini *et al.* 2007; Galán *et al.* 2008), but the upwards-convex LREE shape has been only described in Olot harzburgites, although displaying a more severe HREE–Zr–Ti depletion (Bianchini *et al.* 2007). Fe-rich lherzolites from the northermost El Palo volcano (Bianchini *et al.* 2010) show clinopyroxene REE patterns similar to those from the Cerro Pelado lherzolite 65290, although with lower total REE content and less prominent convex shape at the MREE. Moreover, clinopyroxene from the Cerro Pelado lherzolite also shows a slightly

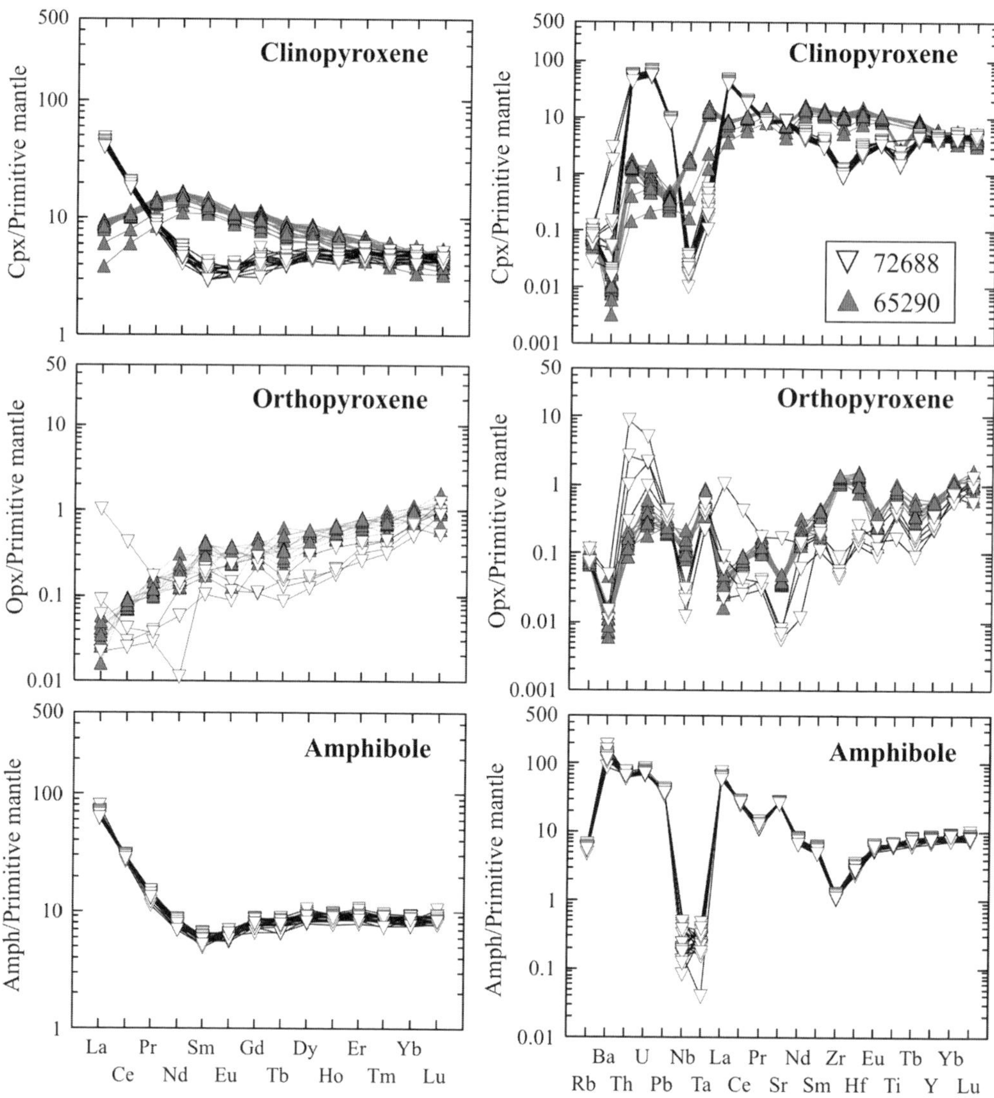

Fig. 8. Primordial mantle-normalized REE and trace element patterns for minerals in the Calatrava peridotite xenoliths. Normalizing values from McDonough & Sun (1995).

higher Zr–Hf–Ta–(Nb)–Y content than those from El Palo. The trace element content of clinopyroxenes from the SE Spain peridotite xenoliths (Beccaluva *et al.* 2004) are completely different to those of Iberian alkaline volcanic fields.

Whole-rock composition

Major and trace elements

Whole-rock major and trace element composition of Calatrava mantle xenoliths are reported in Table 4. They are close to primordial mantle composition or moderately depleted, but their lower Ni, Cr and MgO content when compared to other Spanish mantle xenolith suites (Fig. 9) suggests a more fertile lithospheric mantle composition beneath central Spain. In most samples the Al_2O_3, CaO, FeO and TiO_2 content exibits negative correlations with MgO (Fig. 9), which is a common feature in mantle xenolith suites (e.g. Downes 2001; Ackerman *et al.* 2007; Bianchini *et al.* 2007). This is attributed to depletion during partial melting of the lithospheric mantle (Niu 1997). The most clinopyroxene-rich samples of Calatrava xenoliths (including wehrlite 72674) define a change in the slope of this compositional trend or an opposite evolution (e.g. positive Al_2O_3 and SiO_2 correlations with MgO) (Fig. 9). The wehrlite sample (72674) shows the most extreme composition of this trend (the highest FeO and TiO_2, and the lowest SiO_2, Al_2O_3 and Cr–Ni, content).

Most of the studied Calatrava lherzolites have high Mg-numbers (0.91–0.92), which are close to primordial mantle estimates (Table 4). However, lower Mg-numbers were determined for wehrlite (0.87, Table 4) and the Ti–Fe-rich lherzolite 65294 (0.89, Table 4). Lower Mg values have been found in other Calatrava lherzolites (see Bianchini *et al.* 2010), which also show a higher Fe-rich character than our studied xenoliths (Fig. 9).

Whole-rock REE patterns are varied in Calatrava xenoliths (Fig. 10). Lherzolites from the El Aprisco centre show a wide compositional trend varying from LREE-enriched steep patterns (72691 sample), through the downwards-convex shape of LREE patterns (72688 sample), identical to those of its corresponding clinopyroxene (see also Fig. 8), towards the slightly negative REE patterns (although slightly upwards-convex in the LREE) (sample 55569) (Fig. 10a). The two samples with lower LREE content are either anhydrous (72690) or contain very minor modal amounts of amphibole (55569). They have similar trace element patterns to the least refractory Olot lherzolites (Beccaluva *et al.* 2004).

Xenoliths from the Cerro Pelado centre show a more homogeneous shape of REE patterns from a LREE-enriched (e.g. wehrlite 72674) to a slightly upwards-convex LREE pattern, similar to its corresponding clinopyroxene (sample 65290) (Fig. 10b). As for the major elements, lherzolites from the El Palo centre (Bianchini *et al.* 2010) show REE patterns more similar to those of the Cerro Pelado lherzolites. Most of the LREE-enriched patterns are attributed to cryptic metasomatism, their variability related both to the nature of the metasomatic agent and the efficiency of the metasomatic process (e.g. Bianchini *et al.* 2007).

In addition to LREE, studied Calatrava xenoliths are also enriched in LILE, Th and U (Fig. 10), although there is some scattering in the data owing to their low trace element concentrations, sometimes below analytical detection limits (Table 4). By contrast, the less LREE-enriched samples show the lowest LILE, P, Th and U content. Most lherzolite xenoliths from the El Aprisco centre display Nb–Ta (Zr–Hf and Ti) negative anomalies in trace element patterns (Fig. 10c).

Sr–Nd isotopes

Isotopic data for the Calatrava lherzolites define two overlapping compositional fields, depending on the sampled volcanic centre. El Aprisco lherzolites have slightly more radiogenic Sr ($^{87}Sr/^{86}Sr$ ratios 0.7035–0.7044) than the Cerro Pelado samples ($^{87}Sr/^{86}Sr$ ratios 0.7032–0.7037) (Table 5) (Fig. 11). Lherzolite xenoliths from this latter volcanic centre show a composition more similar to the host magmas than xenoliths from El Aprisco (Fig. 11). The isotopic field of the Cerro Pelado lherzolites includes the compositional field defined by peridotite xenoliths from the El Palo volcano (Bianchini *et al.* 2010). The whole lherzolite isotopic field defines a wider compositional field than that of their host magmas. This suggests a slightly more enriched character of the Calatrava xenoliths when compared to lithospheric–asthenospheric mantle sources of the host alkaline ultrabasic volcanic magmas. This enriched isotopic signature is remarkable when compared to mantle xenoliths from the other Iberian Cenozoic intraplate alkaline volcanic field (Olot province), which plot towards the depleted mantle composition (Fig. 11) (Bianchini *et al.* 2007). Sr–Nd data from lherzolite xenoliths from the Tallante centre (SE Spain) are highly heterogeneous, plotting along the whole mantle array (not shown) (Beccaluva *et al.* 2004).

Calatrava mantle xenoliths have geochemical affinities to the HIMU-source of ocean island basalts (OIB) and to the EAR, recently called the common mantle reservoir (CMR) of the Circum-Mediterranean anorogenic Cenozoic igneous (CiMACI) province (Lustrino & Wilson 2007). The Sr–Nd isotopic composition of the Calatrava xenoliths shows many similarities with mantle xenoliths from northern domains of the French Massif Central (Downes *et al.* 2003) or those from the Rhön region of Germany (Witt-Eickschen & Kramm 1997).

Discussion

Melting and depletion of the mantle sources of the xenoliths

Clinopyroxene is the main host for HREE in the Calatrava spinel peridotites and, consequently, the degree of partial melting can be estimated by its trace element content (Norman 1998). Although the dataset is limited to two samples, Ti and Na show positive correlation with HREE in clinopyroxene and, thus, the use of Norman's modelling is appropriate. Modelling indicates low degrees of melting for both lherzolite xenoliths, irrespective of the type of melting (batch or fractional), yielding less than 5% of partial melting (Fig. 12). As proposed by several authors (Frey *et al.* 1985; Takazawa *et al.* 2000) for other suites, the major element abundances of the studied Calatrava peridotites could also be used to roughly estimate degrees

Table 4. *Major (wt%) and trace element (ppm) whole-rock analyses of the Calatrava mantle xenoliths*

	El Aprisco						Cerro Pelado				
Sample	55569	55570	72688	72689	72690	72691	58498	65290	65294	65298	72674
SiO_2	43.62	44.72	43.76	44.36	44.04	44.2	44.52	44.51	43.94	44.67	42.35
TiO_2	0.10	0.04	0.12	0.11	0.10	0.06	0.06	0.20	0.43	0.13	0.36
Al_2O_3	2.92	1.86	3.47	4.01	3.73	2.81	2.93	3.76	3.03	3.32	1.84
Fe_2O_3	8.8	8.52	9.11	9.05	9	8.63	8.27	8.75	10.51	8.6	12.99
MnO	0.13	0.13	0.13	0.13	0.13	0.13	0.13	0.13	0.15	0.13	0.16
MgO	41.35	43.46	40.16	40.01	40.05	42.16	41.93	37.89	37.02	41.01	36.92
CaO	2.51	1.29	3.75	2.96	2.49	2.03	2.34	3.28	3.85	3.12	5.25
Na_2O	0.23	0.35	0.28	0.29	0.32	0.49	0.36	0.36	0.32	0.3	0.32
K_2O	bdl	0.13	0.04	0.05	bdl	0.05	0.04	0.07	0.21	0.03	0.1
P_2O_5	0.02	0.07	0.02	0.01	0.01	0.02	0.02	0.03	0.05	0.03	0.07
Total	99.69	100.57	100.85	100.98	99.88	100.58	100.59	98.98	99.51	101.33	100.36
Mg#	0.92	0.92	0.91	0.91	0.91	0.92	0.92	0.91	0.89	0.92	0.87
Ba	16	41	41	23	10	74	38	21	35	10	25
Rb	bdl	bdl	bdl	bdl	bdl	bdl	bdl	bdl	3	bdl	bdl
Sr	5	58	45	14	3	60	49	28	41	26	43
Th	bdl	0.52	0.8	0.6	0.16	1.29	0.45	bdl	0.33	0.15	0.27
U	bdl	0.16	0.29	0.18	0.11	0.22	0.12	bdl	bdl	bdl	bdl
Zr	bdl	bdl	bdl	bdl	6	bdl	bdl	19	15	bdl	14
Nb	bdl	5.2	bdl	bdl	bdl	6.8	bdl	1.4	2.8	1	3.4
Y	2.6	1.5	3.7	2.8	2.7	4.1	2.6	4.1	4	3.4	3.2
Sc	14	10	16	14	12	13	14	16	13	15	15
V	59	22	74	54	50	36	50	68	79	62	73

Co	98	107	101	89	100	107	100	92	110	97	119
Cr	1940	2180	2510	1870	2310	2680	2730	2610	2440	2230	1340
Ni	1730	1990	1730	1570	1710	1900	1790	1520	1790	1680	1000
Ta	bdl	0.13	bdl	bdl	bdl	bdl	bdl	0.13	0.22	bdl	0.17
Hf	bdl	bdl	bdl	bdl	0.2	bdl	bdl	bdl	0.5	bdl	0.5
La	0.21	4.66	5.37	2.03	0.34	5.74	3.31	1.55	3.85	2.07	3.47
Ce	0.4	8.29	7.13	2.59	0.65	11.8	9.09	4.31	7.09	4.03	6.54
Pr	0.06	0.81	0.44	0.17	0.09	1.37	1.16	0.62	0.95	0.45	0.88
Nd	0.4	2.48	1.33	0.78	0.51	5.13	3.7	2.9	4.07	1.74	3.85
Sm	0.17	0.46	0.29	0.22	0.18	1.15	0.74	0.83	1.01	0.42	1.01
Eu	0.07	0.14	0.12	0.10	0.09	0.34	0.23	0.29	0.34	0.15	0.34
Gd	0.3	0.37	0.41	0.3	0.27	0.9	0.52	0.8	0.96	0.44	0.88
Tb	bdl	bdl	bdl	bdl	bdl	0.14	bdl	0.13	0.15	bdl	0.13
Dy	0.39	0.25	0.58	0.44	0.42	0.74	0.42	0.76	0.82	0.52	0.68
Ho	bdl	bdl	0.13	bdl	bdl	0.14	bdl	0.14	0.15	0.11	0.11
Er	0.27	0.14	0.41	0.3	0.3	0.39	0.25	0.4	0.39	0.36	0.29
Tm	bdl	bdl	0.06	bdl	bdl	0.06	bdl	0.06	0.05	0.06	bdl
Yb	0.27	0.13	0.39	0.31	0.33	0.37	0.24	0.39	0.31	0.37	0.2
Lu	0.05	bdl	0.06	0.05	0.05	0.05	0.04	0.06	0.05	0.06	bdl
Ol	67.9	64.5	62.5	55.6	59.8	61.7	62.6	54.4	57.5	60.7	69.3
Cpx	13.8	6.0	16.7	13.2	12.6	9.5	11.4	17.7	22.7	14.9	28.4
Opx	15.8	28.5	17.4	27.5	24.8	26.3	24.7	25.6	17.9	22.9	1.5
Sp	2.5	1.0	3.4	3.7	2.9	2.6	1.3	2.3	2.0	1.5	0.1
F*	13.1	20.9	8	7.9	8.3	17.3	16.8	1.5	–	13.3	–

*Partial melting degrees estimated using Takazawa *et al.* (2000) method.

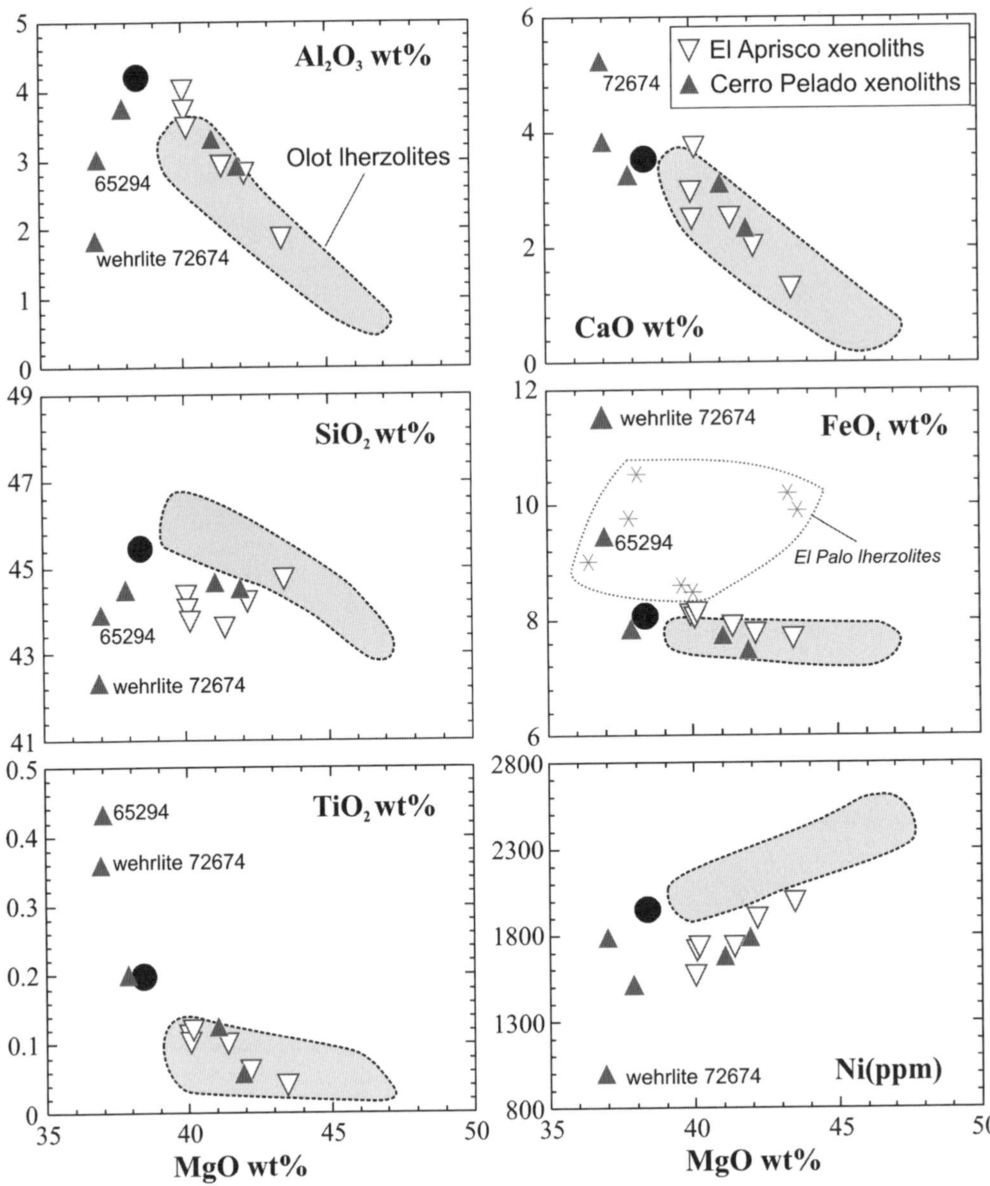

Fig. 9. Variations in whole-rock major oxides with MgO (wt%). Data of lherzolite xenoliths from NE Spain (Olot) (grey field) (Bianchini *et al.* 2007; Galán *et al.* 2008: excepting sample BB.12.030) are shown for comparison. The data for Tallante xenoliths overlap the Olot field (see also Bianchini *et al.* 2007). Data from peridotite xenoliths from the other Calatrava centre (El Palo: Bianchini *et al.* 2010) is only included in the MgO v. FeO_t plot, as they mostly overlap with our data in other diagrams. The composition of primitive mantle (McDonough & Sun 1995) is represented by a filled circle.

of melting of a compositionally homogeneous source in a melting residue model. Using batch melting modelling with a $K_{d\,FeO/MgO}^{olivine}$ of 0.3 and assuming the primitive mantle values of McDonough & Sun (1995) as source composition, we find that the series of peridotites reflect residues from approximately 2 to 21% melting (lherzolites 65290 and 55570, respectively; Table 4). Inferred degrees of melting for lherzolites 72688 and 65290 estimated using both methods, whole-rock major element composition and clinopyroxene trace element content, yield similar results (although

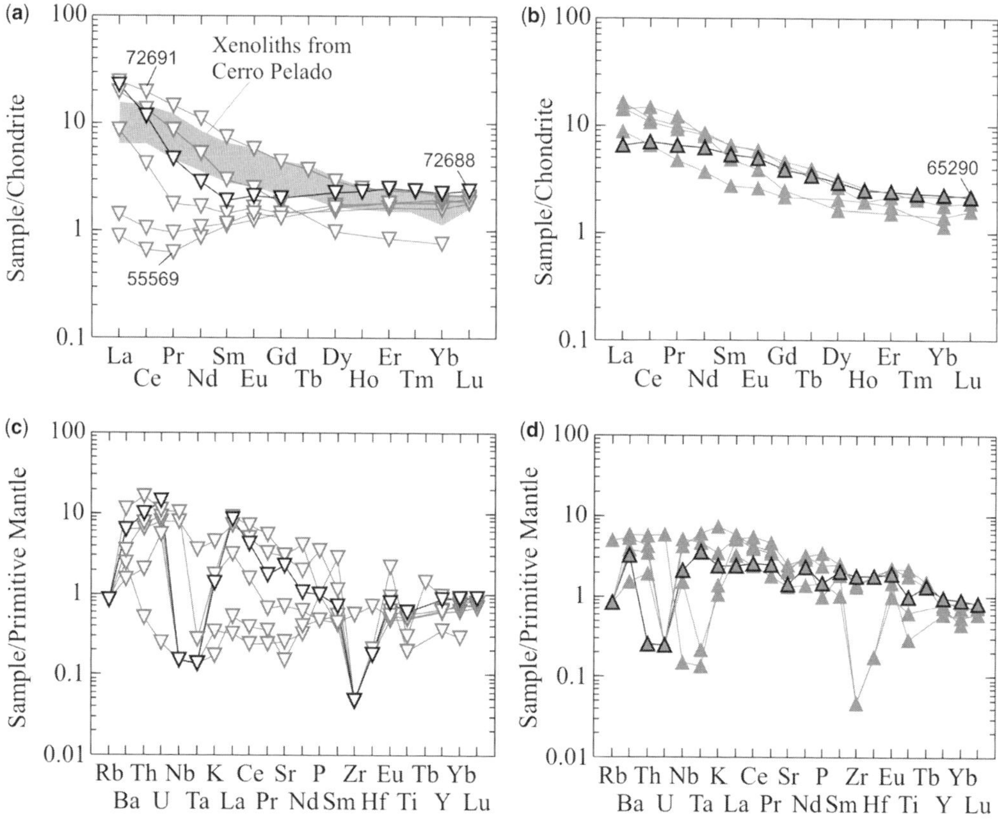

Fig. 10. Chondrite-normalized REE patterns and primordial mantle-normalized trace element patterns for whole rocks of the Calatrava peridotite xenoliths. (**a**), (**c**) REE and trace element patterns of lherzolite xenoliths from the El Aprisco centre. (**b**), (**d**) REE and trace element patterns of mantle xenoliths from the Cerro Pelado centre. Normalizing values from McDonough & Sun (1995).

Table 5. *Whole-rock (Sr–Nd) isotopic composition of the Calatrava mantle xenoliths*

	$^{87}Sr/^{86}Sr$	εSr	$^{143}Nd/^{144}Nd$	εNd
El Aprisco				
55569	0.704350 ± 06	−2.1	0.512976 ± 63	6.6
55570	0.703515 ± 06	−14.0	0.512847 ± 16	4.1
72688	0.703686 ± 05	−11.6	0.513073 ± 22	8.5
72689	0.703768 ± 06	−10.4	0.513051 ± 43	8.1
72690	0.704251 ± 05	−3.5		
72691	0.703675 ± 05	−11.7	0.512804 ± 03	3.2
Cerro Pelado				
58498	0.703675 ± 06	−11.7	0.512777 ± 03	2.7
65290	0.703200 ± 06	−18.5	0.512957 ± 17	6.2
65294	0.703198 ± 06	−18.5	0.512950 ± 06	6.1
65298	0.703331 ± 05	−16.6	0.512922 ± 05	5.5
72674	0.703219 ± 05	−18.2	0.512917 ± 05	5.4

Uncertainties for the $^{87}Sr/^{86}Sr$ and $^{143}Nd/^{144}Nd$ ratios are 2σ (mean) errors in the last two digits.

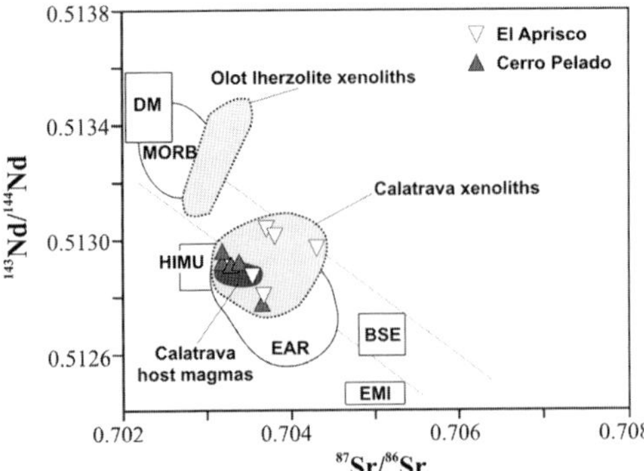

Fig. 11. $^{143}Nd/^{144}Nd$ v. $^{87}Sr/^{86}Sr$ diagram for the Calatrava mantle xenoliths compared to the composition of the host volcanic magmas (López-Ruiz *et al.* 2002). The compositional field of NE Spain (Olot) lherzolite xenoliths (Bianchini *et al.* 2007) is also reported. Mantle end members (DM, MORB, HIMU, BSE and EMI) are from Zindler & Hart (1986). EAR after Downes *et al.* (2003).

the whole-rock method gives higher F values). The samples wehrlite 72674 and lherzolite 65294 have not been used in this model as they have a more fertile composition than primitive mantle estimates. Nevertheless, there is no evidence for such a high partial melting degree in the Calatrava xenoliths. The absence of harzburgites in the studied xenoliths contrasts with the common presence of this rock type in other mantle xenolith suites that have suffered degrees of melting higher than 10% (e.g. Olot mantle xenoliths: Galán *et al.* 2008). These F values are also higher than those estimated by the major element chemistry of minerals

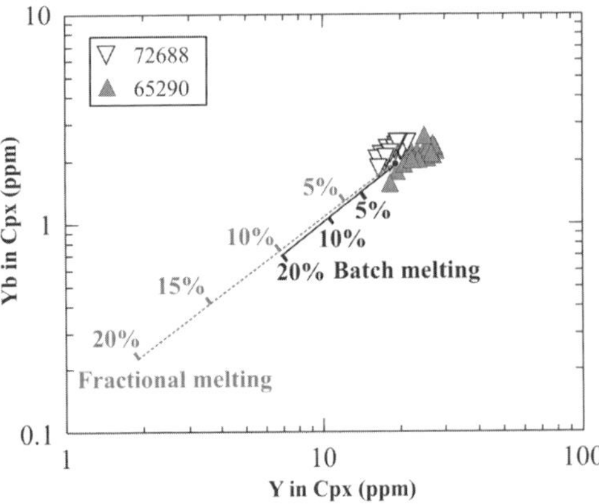

Fig. 12. Partial melting results for Y and Yb in clinopyroxene, following the model of Norman (1998). Both fractional and batch melting suggest low degrees of melting (<5%) for the analysed lherzolites. See the text for further explanation.

(specially the low Cr-numbers of primary spinel; see Fig. 6), suggesting that with the current dataset most of the Calatrava peridotite xenoliths are slightly–moderately depleted and their chemistry is close to fertile mantle composition.

The moderate degree of melting of the Calatrava peridotites contrast with data from the others Iberian mantle xenolith suites. The Olot lherzolites yield modelled melting fractions up to 17% (Bianchini *et al.* 2007), being markedly higher when involving harzburgite types (up to 30–40% of melting: Bianchini *et al.* 2007; Galán *et al.* 2008). In SE Spain, the presence of harzburgite and opx-rich lherzolite xenoliths also suggests a large degree of partial melting of lithospheric mantle (Beccaluva *et al.* 2004; Shimizu *et al.* 2004), although detailed partial melting models have not been applied. The high Cr-numbers of spinel from peridotite xenoliths from SE Spain volcanics is also in accordance with a wider and higher degree of partial melting, and a more residual nature of those mantle xenoliths (Fig. 6). The lower Cr, Ni and MgO content of the Calatrava xenoliths (Table 4 and Fig. 9) is in agreement with a more fertile character of the lithospheric mantle beneath central Spain.

A discussion on the possibility that the xenoliths may have a similar composition to the sources for the alkaline Calatrava volcanic field is beyond the scope of this paper. We note, however, that the relatively fertile character of most xenoliths and the Sr–Nd isotopic signatures of them suggest a genetical relationship with the associated volcanism. Cebriá & López-Ruiz (1995) suggested that the Calatrava volcanic rocks constitute a suite generated by variable degrees of partial melting ($F = 5–17\%$) of a phlogopite (amphibole)-bearing enriched mantle source with 1.2–10 times the primitive mantle values for incompatible elements. Most of the studied peridotite xenoliths show this compositional range, especially the phlogopite-bearing xenoliths from the Cerro Pelado centre (Fig. 10). Nevertheless, the HREE depleted character of most of the Calatrava volcanic rocks is indicative of garnet-bearing sources (Cebriá & López-Ruiz 1995; López-Ruiz *et al.* 2002) and, therefore, they may have a deeper mantle derivation.

Origin of the glasses

The presence of glass veins in mantle xenoliths may be the result of several different formation mechanisms: (1) decompression melting during transport; (2) host magma infiltration; (3) reaction between a percolating melt and host peridotite at mantle depths; and (4) involvement of previously formed metasomatic phases (e.g. amphibole, phlogopite) during partial melting in the mantle (Yaxley *et al.*

1997; Yaxley & Kamenetsky 1999; Coltorti *et al.* 2000). The contrasting geochemical composition of analysed interstitial glasses in the Calatrava xenoliths with ultrabasic host nephelinite–melilitite magmas (Fig. 7) suggests that host magma infiltration could not explain the origin of these low melt fractions.

Most of the glasses in the Calatrava xenoliths contain microcrysts of clinopyroxene, olivine and spinel (Fig. 3e–h). The Al^{VI}/Al^{IV} ratios close to 1 in this newly formed clinopyroxene (cpx-2, Table 1) indicate high-pressure crystallization (Aoki & Kushiro 1968). As discussed above, the chemistry of cpx-2 related to glasses leads to similar geobarometric estimates as for primary clinopyroxene (Table 2). Thus, the formation of glasses might have occurred at mantle depths, and these metasomatically enriched mantle fragments were later accidentally entrained as xenoliths in the Calatrava alkaline magmas. The vesicular aspect of the intergranular glasses and veins (Fig. 3e–h), typical of volcanic–subvolcanic emplacement levels, suggests that volatile exsolution after xenolith entrainment has occurred.

The variability observed in glass composition between different Calatrava peridotite xenoliths implies variations in the metasomatic agent and in the peridotite minerals involved in partial melting processes. Glasses from the Cerro Pelado xenoliths show a great similarity in chemical composition with experimental studies on the low of degree melting of peridotites, which usually yield trachyandesitic glass compositions (Draper & Green 1997; Perinelli *et al.* 2008*a*). The glass composition is used as a good geochemical indicator of the metasomatic agent (Coltorti *et al.* 2000). The major element composition of glasses from the Cerro Pelado xenoliths are indicative of a Na-alkali silicate metasomatism owing to its high $TiO_2 + K_2O$ content (in the range of 5.3–7.2) (Table 1). Moreover, the involvement of an alkaline metasomatism in the Cerro Pelado xenoliths is also consistent with the trace element geochemistry of their primary clinopyroxenes, as discussed later. The melts were partially modified by crystallization of a secondary assemblage of clinopyroxene + olivine + spinel or by reaction with primary peridotite minerals. The origin of cellular or sieve textures in peridotite minerals (e.g. spinel, Fig. 3f) is consistent with incongruent dissolution, in which the part of the crystal in direct contact with an interstitial melt is first dissolved, followed by nucleation and growth of the new phase. Strong contrasts in Ti and Cr content between primary spinel and the new cellular spinel-2 is consistent with the presence of melt (Shaw & Dingwell 2008). However, the involvement of previous metasomatic hydrous phases in the melting reaction is not precluded, as suggested in other cases (Yaxley *et al.* 1997; Yaxley & Kamenetsky 1999), but this deserves better constrained mineral data in more detailed studies. Glass in wehrlite 72674 forms some veining, indicating melt mobility through grain boundaries of the peridotite matrix. However, an interaction of peridotite with its host melt has not been observed.

Interstitial glass in lherzolite 72689 from El Aprisco is markedly subalkaline (Fig. 7). Its chemical composition, characterized by extremely low K (and Ti) content (Table 1), is similar to glasses originated during carbonatitic metasomatism (Coltorti *et al.* 2000) or by infiltration of subduction-related carbonate–silicate melts (Demény *et al.* 2004). Nevertheless, the Qtz–Hy-normative composition of this glass, combined with its slightly peraluminous character (Table 1), precludes a carbonate-rich percolant agent. SiO_2-oversaturated glasses with low Ti–P–K content in peridotite xenoliths are usually interpreted as having originated by amphibole breakdown (Chazot *et al.* 1996; Ban *et al.* 2005) or by infiltration of subduction-derived silicate melts–aqueous fluids (Ishimaru & Arai 2009) or by a combination of both processes. Glass in hydrous lherzolite 72689 appears in a complex reaction zone involving the primary lherzolite minerals and producing a second crystal generation of clinopyroxene, olivine and spinel (Fig. 3e). These low-K glasses are uncommon in Iberian mantle xenoliths but they are described in some other xenolith suites (Chazot *et al.* 1996; Ismail *et al.* 2008). Locally, the melting has involved the former amphibole present in spinel aureoles. This is supported by the similar Na_2O/K_2O ratios displayed by amphibole and glass composition in lherzolite xenoliths from El Aprisco (glasses: $Na_2O/K_2O = 80–180$; amphiboles: $Na_2O/K_2O = 45–700$), and the tendency for the subalkaline composition of glasses generated in partially melted natural amphibole-bearing ultramafic rocks (Perinelli *et al.* 2008*a*). Moreover, the composition of the interstitial glass in lherzolite 72689 is very similar to that of amphibole-derived melts from lherzolites from West Eifel (Ban *et al.* 2005), including typical low abundances of P_2O_5 and TiO_2, although El Aprisco glasses have a lower K_2O content, in accordance with the extremely low-K character of the amphibole in this sample. The absence of phlogopite in the El Aprisco xenoliths would also explain the low Ti and K content of analysed interstitial glasses of lherzolite 72689.

Metasomatism of the xenoliths

Evidence of metasomatism in the Calatrava peridotites is provided by: (i) a strong enrichment in LREE, Th and U in clinopyroxene; (ii) the occurrence of metasomatic hydrous minerals (amphibole, phlogopite); and (iii) the presence of intergranular glasses. Most of the studied xenoliths display evidence of modal metasomatism, with the development of reaction zones and aureoles at the expense of spinel (except for the lherzolite sample 72690 from El Aprisco). The involvement of a previous silicate metasomatism is supported by the trace element chemistry of primary clinopyroxenes, as discussed further below.

The presence of LREE-rich clinopyroxenes with negative anomalies in Nb, Ta and Zr (but no Hf) has been also described in some spinel lherzolites from Olot (sample Olt8: Bianchini *et al.* 2007), the French Massif Central (Downes *et al.* 2003; Touron *et al.* 2008) and Eifel (Witt-Eickschen & Kramm 1977) within the CiMACI province. Some of these xenoliths lack volatile-rich phases, so the clinopyroxene composition is a primary mantle feature related to the infiltration of LREE–Th–U-enriched silicate melts (some of them of probable subalkaline affinity: Touron *et al.* 2008). Clinopyroxenes from Calatrava mantle xenoliths have a high Ti/Eu and moderate $(La/Yb)_N$ (Fig. 13), as well as low $(Nd/Hf)_N$ (<2.3) and $(Gd/Ti)_N$ (<4.8), which are geochemical indicators that the metasomatic medium was essentially a silicate melt rather than a carbonate-rich melt fraction (Coltorti *et al.* 1999; Xu *et al.* 2003; Ismail *et al.* 2008).

In terms of REE (and HFSE) content, the metasomatic amphibole perfectly mimics the coexisting clinopyroxene (sample 72688) and shows LREE enrichment at a comparable HREE content (10 × Ch). The same trace element mimicry is also displayed by the small neoformed type-2 clinopyroxene which shows similar REE patterns

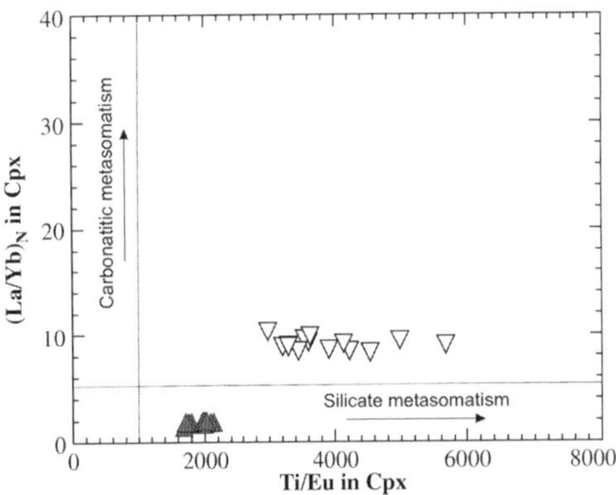

Fig. 13. $(La/Yb)_N$ v. Ti/Eu in clinopyroxenes from Calatrava lherzolite xenoliths. (La/Yb) ratios are normalized to chondrite values from McDonough & Sun (1995).

to the primary crystals. Trace element content might be controlled by intermineral distribution coefficients more than by extrinsic parameters (P, T, fO_2), and K_d^{REE} for clinopyroxenes–amphiboles is fairly close to 1 (e.g. Bianchini *et al.* 2007; Galán *et al.* 2008). Geochemical features of amphiboles from mantle xenoliths are good indicators of the metasomatic agents involved. The low Ti–Nb–Zr content of amphiboles from the Calatrava xenoliths (and their correlative high Zr/Nb and Ti/Nb ratios, >25 and >50 000, respectively) are chemical features indicative of suprasubduction amphiboles (S-Amph types of Coltorti *et al.* 2007*a*), and they might record subduction-related metasomatic components, even in a multistage metasomatic history. Glass chemistry in hydrous lherzolite 72689 from El Aprisco also suggests the involvement of a subduction-related metasomatism (see the previous subsection).

The estimated composition of melts in equilibrium with clinopyroxene and amphibole from lherzolite 72688 points to the involvement of a metasomatic agent with marked Nb–Ta negative anomalies and flat HREE patterns, very different to the incompatible trace element patterns shown by the host ultrabasic alkaline melts (Fig. 14). The Nb–Ta trough (and prominent Th–U–Pb positive anomalies) in all of the primary minerals of the studied Calatrava mantle xenolith 72688 (clinopyroxene, orthopyroxene and olivine) suggests that an important metasomatic event occurred before xenolith entrapment within the lithospheric mantle. Other Calatrava mantle xenoliths also show Nb–Ta negative (and Th–U positive) anomalies in whole-rock composition (mainly from the El Aprisco centre), which suggests that this kind of metasomatism was common in the lithospheric mantle of central Spain. Most geochemical agents with prominent Th–U positive and Nb–Ta negative anomalies are related to continental lithosphere, suggesting that this chemical imprint could be inherited from some kind of subduction-derived metasomatism. Contamination by Variscan continental crust in mantle sources has recently been proposed, by Prelevic *et al.* (2008), as the origin for the nearby Cenozoic lamproites from SE Spain. However, no significant contamination by continental crust in the Calatrava peridotites is recorded by the isotopic data.

We have also estimated the composition of melts in equilibrium with clinopyroxene from lherzolite 65290 (from the Cerro Pelado centre) (Fig. 14a). The trace element content of the inferred melts are comparable to the undersaturated magma found in the area, suggesting a strong link between metasomatism and the magmatism of the Calatrava volcanic field in this case. This is in agreement with the less radiogenic Sr isotope composition of this

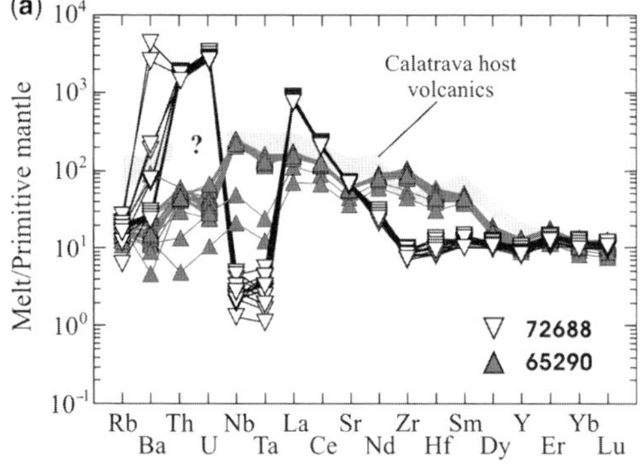

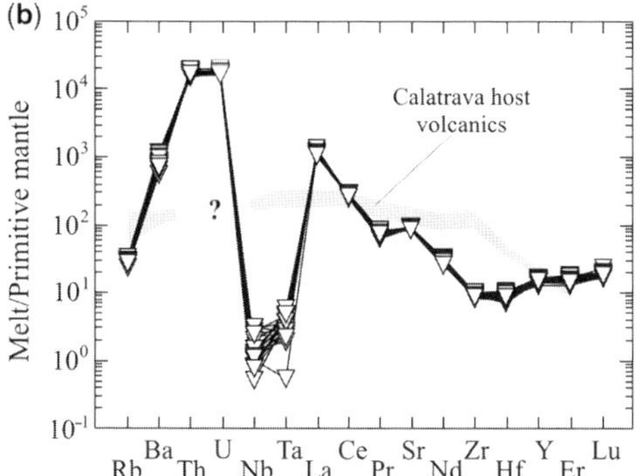

Fig. 14. Trace element composition of melts in equilibrium with (**a**) clinopyroxenes and (**b**) amphiboles from Calatrava lherzolites. (a) Data calculated using cpx–basaltic melt partition coefficients (Hart & Dunn 1993), except for Rb (Foley *et al.* 1996) and Ta (Adam & Green 2003). (b) Data calculated using amph–melt partition coefficients of La Tourrette *et al.* (1995). The average composition of Calatrava host volcanics is from Cebriá & López-Ruiz (1995). Normalizing values for primitive mantle from McDonough & Sun (1995).

xenolith compared to those from El Aprisco, and the coincidence in isotopic composition with the host basaltic magmas (Table 5 and Fig. 11). It is also in agreement with the composition of associated interstitial glasses, which involves a Na-rich alkaline melt. Thus, different silicate metasomatic agents are involved in the genesis of the wide range of Calatrava peridotite xenoliths.

Cenozoic Calatrava volcanics record a HIMU component in their lithospheric–asthenospheric mantle sources based on their Sr–Nd–Pb isotopic composition (López-Ruiz *et al.* 2002). The Sr–Nd isotopic composition of mantle xenoliths from central Spain also plot close to the HIMU compositional field (Fig. 11). Nevertheless, studied mantle xenoliths reflect a previous subduction-related metasomatic imprint. Isotopic composition of mantle xenoliths from central Spain could be the result of the mixing between a more depleted mantle source (asthenospheric component) and a deeply recycled enriched component, as recently suggested in some French Massif Central mantle xenoliths (Touron *et al.* 2008), making arguable the involvement of a HIMU component. In this sense, mantle xenoliths from the northernmost El Palo volcano indicate a long-term recycling of oceanic subducted components (Bianchini *et al.* 2010) that is not recorded in our studied lherzolite xenolith suite. Thus, additional sampling and new isotope data (Pb, O, etc.) are required to solve this problem.

On the origin of Calatrava wehrlite xenoliths

The presence of ultramafic rock types, which are not likely to be partial melting residues from fertile mantle in peridotite xenolith suites, poses the problem of their origin and significance. Wehrlites are common in many lherzolite and harzburgite xenoliths suites. Their origin has been discussed in terms of cumulates from mafic melts trapped in mantle depths or by reaction of residual wall-rock peridotites with migrating magmatic liquids, either carbonate-rich or subsaturated Fe–Ti-rich silicate melts (e.g. Hauri *et al.* 1993; Coltorti *et al.* 1999; Ionov *et al.* 2005; Beard *et al.* 2007). Wehrlite 75674, from the Calatrava xenoliths, has no cumulate textures, contains some olivine grains showing strain and deformation twins, and the majority of crystals have smooth curvilinear boundaries. Moreover, this wehrlite shows similar geochemical signatures to those of lherzolite 65294 and the other Cerro Pelado xenoliths (Figs 9–11). Thus, our petrographical observations and whole-rock composition rule out a cumulate origin for the Calatrava wehrlite studied.

The similar Al content in the orthopyroxenes from the lherzolites and wehrlites (Fig. 4a) suggest that the original rock was lherzolitic. Moreover, major element trends shown by the Cerro Pelado xenoliths are similar to those described for the Fe-rich lherzolite–wehrlite (LW) series formed by reactive melt percolation in associated lherzolites (Ionov *et al.* 2005). The overlapping of trace element patterns (wehrlite and Cerro Pelado lherzolites have almost identical REE patterns, Fig. 10b) also indicates that wehrlite might have been originated by the transformation of associated slightly residual lherzolites, by percolation of metasomatic melts. Accordingly, associated lherzolites must be the protolith of the wehrlite. The lack of a marked LREE enrichment, in addition to a strong Ti–Fe enrichment of the wehrlite, suggest that the metasomatic agent was not a carbonate-rich silicate melt (Hauri *et al.* 1993; Coltorti *et al.* 1999). This is also in agreement with the lack of carbonatite metasomatism in the Calatrava peridotites, together with the absence of carbonate-rich magmatism in central Spain.

We suggest that a metasomatic agent had reacted with the primary lherzolite orthopyroxene to produce clinopyroxene and to displace its modal composition towards wehrlite. If this is correct, then the melt must have been undersaturated in silica and relatively rich in Ca, K, Fe and Ti, which would have increased these components in the wehrlite (Fig. 9). In this respect, if we use the whole-rock Mg# and Cpx/Opx modelling of Ionov *et al.* (2005) our Calatrava samples 75674 and 65294 match a model involving a reaction of host peridotites with evolved silica-undersaturated silicate melts (Mg numbers = 0.6–0.7) at low melt/rock ratios (*c.* 2 for *R* parameter = 0.2) (fig. 14 of Ionov *et al.* 2005).

Addressing possible origins for such percolating alkaline silicate melts is complicated. The mineral composition of olivine, clinopyroxene and the scarce spinel of the studied wehrlite xenolith are similar to those obtained in experimental work on metasomatism induced by alkaline magmas of nephelinite composition with upper-mantle peridotites (Perinelli *et al.* 2008*a*). Calatrava volcanism involved silica-undersaturated alkaline melts of either potassic or sodic composition (e.g. Ancochea 1982; López-Ruiz *et al.* 2002), and ultrapotassic magmatism occurred prior to these more sodic alkaline events that carried the studied mantle xenoliths. The similarity in Sr–Nd isotopic ratios of the wehrlite and its host volcanic rocks suggests that the Calatrava alkaline magmatism could have been involved in the genesis of the wehrlites. Furthemore, as explained above, a similar alkaline metasomatic imprint has been described for the associated lherzolite xenoliths.

Conclusions

The mineral chemistry and whole-rock composition of the Calatrava lherzolite xenoliths is typical of peridotites that have undergone small–moderate degrees of partial melting, mostly less than 10%, although the more varied xenolith population described in the literature suggests a more heterogenous nature for this lithospheric section. Superimposed on this moderately fertile character there is a complex metasomatism of uncertain timing produced by silicate melts. One metasomatic agent (the El Aprisco lherzolite suite) could be genetically associated with subduction-derived components. Another metasomatic agent recorded in the studied peridotite xenoliths (the Cerro Pelado lherzolite suite) is an alkaline silicate melt producing a geochemical signature similar to that of the undersaturated alkaline Calatrava magmatism. Metasomatism related to tholeiitic melts or subducted oceanic crust, as sugested for the Calatrava El Palo lherzolite suite (Bianchini *et al.* 2010), is not recorded in the studied peridotites. This reveals the wide range in upper-mantle heterogeneous composition that can be found beneath a small volcanic province.

The presence of wehrlite types within the studied Calatrava peridotite xenoliths is interpreted as a reaction of host lherzolites with silica-undersaturated silicate melts. The lack of petrographical evidence, the overlapping of trace element content and the Sr–Nd isotopic ratios of wehrlite and associated lherzolites rule out a cumulate origin. The strong Fe–Ti enrichment of the wehrlite indicates that the metasomatic agent was not a carbonate-rich silicate melt. The development of wehrlite types by the reactive percolation of alkaline magmas similar to those that generate the Calatrava volcanic field, but at higher melt/rock ratios than associated lherzolites, must be studied in more detail in future work. Indeed, the possibility of the existence of Fe-rich lherzolite–wehrlite series rocks in the underlying Calatrava upper mantle is insinuated by the present data (see also Bianchini *et al.* 2010).

Most of the metasomatically newly formed minerals mimic the trace element content of the primary phases and, thus, the main metasomatism recorded by studied peridotite xenoliths is likely to be related to 'primary' metasomatism at mantle sources rather than during magma transport.

The proposed sequence of events for the studied Calatrava mantle xenoliths could be as follows.

(1) Cryptic and modal (e.g. amphibole with suprachondritic Ti/Nb and Zr/Nb ratios) metasomatism by subduction-related components. The timing of this metasomatism is unconstrained, although a Tertiary subduction, as suggested in other Cenozoic volcanic fields (Piromallo *et al.* 2008; Bianchini *et al.* 2010), could not be discounted with the current available data.
(2) Localized alkaline metasomatism shortly before magma entrainment as evidenced by some lherzolite xenoliths from Cerro Pelado (sample 65290). This could evolve to widespread replacement towards wehrlite types (sample 72674) at higher rates of melt percolation. This silica-undersaturated silicate metasomatic agent could be associated with early stages of the Cenozoic alkaline magmatism in central Spain.
(3) Partial melting occurred in some xenoliths at mantle depths and related to similar metasomatic agents that have enriched the xenoliths previously. The participation in the melting of the previously metasomatically introduced

phases (e.g. amphibole, phlogopite) is not precluded with the current data.

(4) Erosion of previously metasomatized lithospheric mantle at spinel-facies conditions, at around 35–50 km depth. Rapid decompression would promote the vesiculation and quenching of previously formed interstitial glass, together with localized intergranular recrystallization around spinel and the other peridotite minerals.

Finally, the Sr–Nd isotopic composition of the Calatrava mantle xenoliths plot within the EAR or CMR upwelling mantle beneath Europe; these values represent more enriched signatures than those found in the other Spanish Cenozoic alkaline province (Olot) but show many similarities with mantle xenoliths from the northern domains of the French Massif Central and those from the Rhön region in Germany.

We thank A. F. Larios and J. G. del Tánago for their assistance with the electron microprobe analyses in the CAI of Microscopía Electrónica (UCM). Also J. M. Fuenlabrada Pérez and J. A. Hernández Jiménez from the CAI of Geocronología y Geoquímica (UCM) for their help in analysing samples by TIMS. H. Downes is greatly thanked for her careful English revision and for inviting us to collaborate in this monographic volume. Detailed revision by V. Cvetkovic, T. Ntaflos and M. Coltorti greatly improved the quality of the manuscript. This work was supported by grants CGL-2006-03414 and CGL-2008-05952 of the Ministerio de Educación y Ciencia of Spain. This research also received support from the SYNTHESYS project GB-TAF-2768 funded by the European Community Research Infrastructure Action under the FP6 'Structuring the European Research Area' Programme.

References

Ackerman, L., Mahlen, N., Jelínek, E., Medaris, G. Jr, Ulrych, J., Strnad, L. & Mihaljevic, M. 2007. Geochemistry and evolution of subcontinental lithospheric mantle in Central Europe: evidence from peridotite xenoliths of the Kozákov volcano, Czech Republic. *Journal of Petrology*, **48**, 2235–2260.

Adam, J. & Green, T. H. 2003. The influence of pressure, mineral composition and water on trace element partitioning between clinopyroxene, amphibole and basanitic melts. *European Journal of Mineralogy*, **15**, 831–841.

Albarède, F. 1995. *Introduction to Geochemical Modelling*. Cambridge University Press, Cambridge.

Ancochea, E. 1982. *Evolución espacial y temporal del volcanismo reciente de España Central*. PhD thesis, Complutense University, Madrid.

Ancochea, E. 2004. La región volcánica del Campo de Calatrava. *In*: Vera, J. A. (ed.) *Geología de España*. SGE-IGME, Madrid, 676–677.

Ancochea, E. & Nixon, P. H. 1987. Xenoliths in the Iberian peninsula. *In*: Nixon, P. H. (ed.) *Mantle Xenoliths*. Wiley, Chichester, 119–124.

Aoki, K. & Kushiro, I. 1968. Some clinopyroxenes from ultramafic inclusions in Dreiser Eiher, Eifel. *Contributions to Mineralogy and Petrology*, **18**, 326–337.

Arai, S. 1992. Chemistry of chromian spinel in volcanic rocks as a potential guide to magma chemistry. *Mineralogical Magazine*, **56**, 173–184.

Arai, S. 1994. Characterization of spinel peridotites by olivine–spinel compositional relationships: review and interpretation. *Chemical Geology*, **113**, 191–204.

Ban, M., Witt-Eickschen, G., Klein, M. & Seck, H. A. 2005. The origin of glasses in hydrous mantle xenoliths from the West Eifel, germany: incongruent break down of amphibole. *Contributions to Mineralogy and Petrology*, **148**, 511–523.

Beard, A. D., Downes, H., Mason, P. R. D. & Vetri, V. R. 2007. Depletion and enrichment processes in the lithospheric mantle beneath the Kola Peninsula (Russia): evidence from spinel lherzolite and wehrlite xenoliths. *Lithos*, **94**, 1–24.

Beccaluva, L., Bianchini, G., Bonadiman, C., Siena, F. & Vaccaro, C. 2004. Coexisting anorogenic and subduction-related metasomatism in mantle xenoliths from the Betic Cordillera (southern Spain). *Lithos*, **75**, 67–87.

Bertrand, P. & Mercier, J. C. 1985. The mutual solubility of coexisting ortho- and clinopyroxene: toward an absolute geothermometer for the natural system. *Earth and Planetary Science Letters*, **76**, 109–122.

Bianchini, G., Beccaluva, L., Bonadiman, C., Nowell, G., Pearson, G., Siena, F. & Wilson, M. 2007. Evidence of diverse depletion and metasomatic events in harzburgite–lherzolite mantle xenoliths from the Iberian plate (Olot, NE Spain): implications for lithosphere accretionary processes. *Lithos*, **94**, 25–45.

Bianchini, G., Beccaluva, L., Bonadiman, C., Nowell, G. M., Pearson, D. G., Siena, F. & Wilson, M. 2010. Mantle metasomatism by melts of HIMU piclogite components: new insights from Fe-lherzolite xenoliths (Calatrava Volcanic District, central Spain). *In*: Coltorti, M., Downes, H., Grégoire, M. & O'Reilly, S. (eds) *Petrological Evolution of the European Lithospheric Mantle*. Geological Society, London, Special Publications, **337**, 107–124.

Bonatti, E. & Michael, P. J. 1989. Mantle peridotites from continental rifts to ocean basins to subduction zones. *Earth Planetary Science Letters*, **91**, 297–311.

Brey, G. P. & Köhler, T. 1990. Geothermobarometry in four-phase lherzolites; II, New thermobarometers, and practical assessment of existing thermobarometers. *Journal of Petrology*, **31**, 1353–1378.

Cebriá, J. M. & López-Ruiz, J. 1995. Alkali basalts and leucitites in an extensional intracontinental plate setting: the late Cenozoic Calatrava Volcanic Province (central Spain). *Lithos*, **35**, 27–46.

Chazot, G., Menzies, M. & Harte, B. 1996. Silicate glasses in spinel lherzolite from Yemen: origin and chemical composition. *Chemical Geology*, **134**, 159–179.

Coltorti, M., Beccaluva, L., Bonadiman, C., Faccini, B., Ntaflos, T. & Siena, F. 2004. Amphibole genesis via metasomatic reaction with clinopyroxene in mantle xenoliths from Victoria Land, Antarctica. *Lithos*, **75**, 115–139.

Coltorti, M., Beccaluva, L., Bonadiman, C., Salvini, L. & Siena, F. 2000. Glasses in mantle xenoliths as geochemical indicators of metasomatic agents. *Earth Planetary Science Letters*, **183**, 303–320.

Coltorti, M., Bonadiman, C., Faccini, B., Grégoire, M., O'Reilly, S. Y. & Powell, W. 2007*a*. Amphiboles from suprasubduction and intraplate lithosperic mantle. *Lithos*, **99**, 68–74.

Coltorti, M., Bonadiman, C., Faccini, B., Ntaflos, T. & Siena, F. 2007*b*. Slab melt and intraplate metasomatism in Kapfenstein mantle xenoliths (Styrian Basin, Austria). *Lithos*, **94**, 66–89.

Coltorti, M., Bonadiman, C., Hinton, R. W., Siena, F. & Upton, B. G. J. 1999. Carbonatite metasomatism of the oceanic upper mantle: evidence from clinopyroxenes and glasses in ultramafic xenoliths of Grande Comore, Indian Ocean. *Journal of Petrology*, **40**, 133–165.

Demény, A., Vennemann, T. W. et al. 2004. Trace element and C–O–Sr–Nd isotope evidence for subduction-related carbonate–silicate melts in mantle xenoliths (Pannonian Basin, Hungary). *Lithos*, **75**, 89–113.

Downes, H. 2001. Formation and modification of the shallow sub-continental lithospheric mantle: a review of geochemical evidence from ultramafic xenolith suites and tectonically emplaced ultramafic massifs of western and central Europe. *Journal of Petrology*, **42**, 233–250.

Downes, H., Reichow, M. K., Mason, P. R. D., Beard, A. D. & Thirlwall, M. F. 2003. Mantle domains in the lithosphere beneath the French Massif Central: trace element and isotopic evidence from mantle clinopyroxenes. *Chemical Geology*, **200**, 71–87.

Draper, D. S. & Green, T. H. 1997. *P–T* phase realtions of silicic, alkaline, aluminous mantle-xenoliths glasses under anhydrous and C–O–H fluid saturated conditions. *Journal of Petrology*, **38**, 1187–1224.

Fernàndez, M., Marzán, I., Correia, A. & Ramalho, E. 1998. Heat flow, heat production, and lithospheric thermal regimen in the Iberian Peninsula. *Tectonophysics*, **291**, 29–53.

Foley, S. F., Jackson, S. E., Fryer, B. J., Greenough, J. D. & Jenner, G. A. 1996. Trace element partition coefficients for clinopyroxene and phlogopite in an alkaline lamprophyre from Newfounland by LA-ICP-MS. *Geochimica et Cosmochimica Acta*, **60**, 629–638.

Frey, F. A., Suen, C. J. & Stockman, H. W. 1985. The Ronda high temperature peridotite: geochemical and petrogenesis. *Geochimica et Cosmochimica Acta*, **49**, 2469–2491.

Galán, G., Oliveras, V. & Paterson, B. A. 2008. Types of metasomatism in mantle xenoliths enclosed in Neogene–Qaternary alkaline mafic lavas from Catalonia (NE Spain). *In*: Coltorti, M. & Grégoire, M. (eds) *Metasomatism in Oceanic and Continental Lithospheric Mantle*. Geological Society, London, Special Publications, **293**, 121–153.

Granet, M., Wilson, M. & Achauer, U. 1995. Imaging a mantle plume beneath the French Massif Central. *Earth Planetary Science Letters*, **136**, 281–296.

Hart, S. R. & Dunn, T. 1993. Experimental cpx/melt partitioning of 24 trace element. *Contributions to Mineralogy and Petrology*, **113**, 1–8.

Hauri, E. H., Shimizu, N., Dieu, J. J. & Hart, S. R. 1993. Evidence for hotspot-related carbonatite metasomatism in the oceanic upper mantle. *Nature*, **365**, 221–227.

Ishimaru, S. & Arai, S. 2009. Highly silicic glasses in peridotite xenoliths from Avacha volcano, Kamchatka arc; implications for melting and metasomatism within sub-arc mantle. *Lithos*, **107**, 93–106.

Ismail, M., Delpech, G., Cottin, J. Y., Grégoire, M., Moine, B. N. & Bilal, A. 2008. Petrological and geochemical constraints on the composition of the lithospheric mantle beneath the Syrian rift, northern part of the Arabian plate. *In*: Coltorti, M. & Grégoire, M. (eds) *Metasomatism in Oceanic and Continental Lithospheric Mantle*. Geological Society, London, Special Publications, **293**, 223–251.

Ionov, D. A., Chanefo, I. & Bodinier, J.-L. 2005. Origin of Fe-rich lherzolites and wehrlites from Tok, SE Siberia by reactive melt percolation in refractory mantle peridotites. *Contributions to Mineralogy and Petrology*, **150**, 335–353.

La Tourrette, T., Hervig, R. L. & Holloway, J. R. 1995. Trace element partitioning between amphibole, phlogopite, and basanite melt. *Earth Planetary Science Letters*, **135**, 13–30.

Le Bas, M. J., Le Maitre, R. W., Streckeisen, A. & Zanetin, B. A. 1986. Chemical classification of volcanic rocks based in the total alkali-silica diagram. *Journal of Petrology*, **27**, 745–750.

López-Ruiz, J., Cebriá, J. M. & Doblas, M. 2002. Cenozoic volcanism I: the Iberian peninsula. *In*: Gibbons, W. & Moreno, T. (eds) *The Geology of Spain*. Geological Society, London, 417–438.

López-Ruiz, J., Cebriá, J. M., Doblas, M., Oyarzun, R., Hoyos, M. & Martín, C. 1993. The late Cenozoic alkaline volcanism of the central Iberian Peninsula (Calatrava Volcanic Province, Spain): intraplate volcanism realted to extensional tectonics. *Journal of the Geological Society, London*, **150**, 915–922.

Lustrino, M. & Wilson, M. 2007. The circum-Mediterranean anorogenic Cenozoic igneous province. *Earth Science Reviews*, **81**, 1–65.

McDonough, W. F. & Sun, S. 1995. The composition of the Earth. *Chemical Geology*, **120**, 223–253.

Nimis, P. & Ulmer, P. 1998. Clinopyroxene geobarometry of magmatic rocks Part 1: an expanded structural geobarometer for anhydrous and hydrous, basic and ultrabasic systems. *Contributions to Mineralogy and Petrology*, **133**, 122–135.

Niu, Y. 1997. Mantle melting and melt extraction processes beneath ocena ridges: evidence from abyssal peridotites. *Journal of Petrology*, **38**, 1047–1074.

Nixon, P. H. 1987. *Mantle Xenoliths*. Wiley, Chichester.

Norman, M. D. 1998. Melting and metasomatism in the continental lithosphere: laser ablation ICPMS anlysis of minerals in spinel lherzolites from eastern Australia. *Contributions to Mineralogy and Petrology*, **130**, 240–255.

Orejana, D., Villaseca, C. & Paterson, B. A. 2006. Geochemistry of pyroxenitic and hornblenditic xenoliths in alkaline lamprophyres from the Spanish Central System. *Lithos*, **86**, 167–196.

Orejana, D. & Villaseca, C. 2008. Heterogeneous metasomatism in cumulate xenoliths from the Spanish Central System: implications for percolative fractional crystallization of lamprophyric melts.

In: Coltorti, M. & Grégoire, M. (eds) *Metasomatism in Oceanic and Continental Lithospheric Mantle*. Geological Society, London, Special Publications, **293**, 101–120.

Ossan, A. 1889. Beitrage zur kenntniss der eruptivegesteine des Capo de Gata (Almería). *Zeitschrift der Deutschen Geologischen Gesellschaft*, **51**, 306–311.

Pearce, J. A., Barker, P. F., Edwards, S. J., Parkinson, I. J. & Leat, P. T. 2000. Geochemistry and tectonic significance of peridotites from the South Sandwich arc-basin systems, south Atlantic. *Contributions to Mineralogy and Petrology*, **139**, 36–53.

Pearson, D. G., Canil, D. & Shirey, S. B. 2005. Mantle samples included in volcanic rocks: xenoliths and diamonds. *In*: Carlson, R. W. (ed.) *The Mantle and Core. Treatise on Geochemistry*, **2**, 171–275.

Perinelli, C., Orlando, A., Conte, A. M., Armienti, P., Borrini, D., Faccini, B. & Misiti, V. 2008*a*. Metasomatism induced by alkaline magma in the upper mantle of northern Victoria Land (Antarctica): an experimental approach. *In*: Coltorti, M. & Grégoire, M. (eds) *Metasomatism in Oceanic and Continental Lithospheric Mantle*. Geological Society, London, Special Publications, **293**, 279–302.

Perinelli, C., Sapienza, G. T., Armienti, P. & Morten, L. 2008*b*. Metasomatism of the upper mantle beneath the Hyblean Plateau (Sicily): evidence from pyroxenes and glass in peridotite xenoliths. *In*: Coltorti, M. & Grégoire, M. (eds) *Metasomatism in Oceanic and Continental Lithospheric Mantle*. Geological Society, London, Special Publications, **293**, 197–221.

Piromallo, C., Gasperini, D., Macera, P. & Faccenna, C. 2008. A late-Cretaceous contamination episode of the European–Mediterranean mantle. *Earth and Planetary Science Letters*, **268**, 15–27.

Prelevic, D., Foley, S. F., Romer, R. & Conticelli, S. 2008. Mediterranean tertiary lamproites derived from multiple source components in postcollisional geodynamics. *Geochimica et Cosmochima Acta*, **72**, 2125–2156.

Reyes, J., Villaseca, C., Barbero, L., Quejido, A. J. & Santos, J. F. 1997. Descripción de un método de separación de Rb, Sr, Sm y Nd en rocas silicatadas para estudios isotópicos. *Congreso Ibérico de Geoquímica*, **I**, 46–55.

San Miguel de la Cámara, M. 1936. *Estudio de las rocas eruptivas de España*. Memoria Academia de Ciencias de Madrid (Ciencias Naturales), **6**.

Seghedi, I., Brändle, J. L., Szakács, A., Ancochea, E. & Vaselli, O. 2002. El manto litosférico en el sureste de España: Aportaciones de los xenolitos englobados en rocas alcalinas del Mioceno-Plioceno. *Geogaceta*, **32**, 27–30.

Shaw, C. S. J. & Dingwell, D. B. 2008. Experimental peridotite–melt reaction at one atmosphere: a textural and chemical study. *Contributions to Mineralogy and Petrology*, **155**, 199–214.

Shimizu, Y., Arai, S., Morishita, T., Yurimoto, H. & Gervilla, F. 2004. Petrochemical characteristics of felsic veins in mantle xenoliths from Tallante (SE Spain): an insight into activity of silicic melt within the mantle wedge. *Transactions of the Royal Society of Edinburgh: Earth Sciences*, **95**, 265–276.

Takazawa, E., Frey, F. A., Shimizu, N. & Obata, M. 2000. Whole rock compositional variations in an upper mantle peridotite (Horman, Hokkaido, Japan): Are they consistent with a partial melting process? *Geochimica et Cosmochimica Acta*, **64**, 695–716.

Touron, S., Renac, C., O'Reilly, S. Y., Cottin, J. Y. & Griffin, W. L. 2008. Characterization of the metasomatic agent in mantle xenoliths from Devès, Massif Central (France) using coupled *in situ* trace-element and O, Sr and Nd isotopic compositions. *In*: Coltorti, M. & Grégoire, M. (eds) *Metasomatism in Oceanic and Continental Lithospheric Mantle*. Geological Society, London, Special Publications, **293**, 177–196.

Uysal, I., Kaliwoda, M., Karsli, O., Tarkian, M., Sadiklar, M. B. & Ottley, C. J. 2007. Compositional variations as a result of partial melting and melt–peridotite interaction in an upper mantle section from the Ortaca area, SW Turkey. *Canadian Mineralogist*, **45**, 1471–1493.

Villaseca, C., Orejana, D., Paterson, B. A., Billstrom, K. & Pérez-Soba, C. 2007. Metaluminous pyroxene-bearing granulite xenoliths from the lower continental crust in central Spain: their role in the genesis of Hercynian I-type granites. *European Journal of Mineralogy*, **19**, 463–477.

Witt-Eickschen, G. & Kramm, U. 1997. Mantle upwelling and metasomatism beneath central Europe: geochemical and isotopic constraints from mantle xenoliths from the Rhön (Germany). *Journal of Petrology*, **38**, 479–493.

Wood, B. J. & Banno, S. 1973. Garnet–orthopyroxene and Orthopyroxene–clinopyroxene relationships in simple and complex systems. *Contributions to Mineralogy and Petrology*, **42**, 109–124.

Xu, X., O'Reilly, S. Y., Griffin, W. L. & Zhou, X. 2003. Enrichment of upper mantle peridotite: petrological, trace element and isotopic evidence in xenoliths from SE China. *Chemical Geology*, **198**, 163–188.

Yaxley, G. M. & Kamenetsky, V. 1999. Insitu origin for glass in mantle xenoliths from southeastern Australia: insight from trace element compositions of glasses and metasomatic phases. *Earth and Planetary Science Letters*, **172**, 97–109.

Yaxley, G. M., Kamenetsky, V., Green, D. H. & Fallon, T. J. 1997. Glasses in mantle xenoliths from western Victoria, Australia, and their relevance to mantle processes. *Earth and Planetary Science Letters*, **148**, 433–446.

Zindler, A. & Hart, S. R. 1986. Chemical geodynamics. *Annual Review of Earth and Planetary Sciences*, **14**, 493–571.

Geochemical and Sr–Nd isotopic characteristics and pressure–temperature estimates of mantle xenoliths from the French Massif Central: evidence for melting and multiple metasomatism by silicate-rich carbonatite and asthenospheric melts

M. YOSHIKAWA[1]*, T. KAWAMOTO[1], T. SHIBATA[1] & J. YAMAMOTO[1,2]

[1]*Institute for Geothermal Sciences, Graduate School of Science, Kyoto University, Beppu 874-0903, Japan*

[2]*Present address: Woods Hole Oceanographic Institution, Woods Hole, MA 02543, USA*

**Corresponding author (e-mail: masako@bep.vgs.kyoto-u.ac.jp)*

Abstract: Ultramafic xenoliths from Mont Briançon, Ray Pic and Puy Beaunit in the French Massif Central show variable mineral compositions that indicate a residual origin after various degrees of partial melting of a fertile peridotite. Furthermore, trace element and Sr–Nd isotopic variations of clinopyroxenes indicate mixing processes between depleted mantle and enriched components such as asthenospheric melt and silicate carbonatite melt. Pyroxene geothermometer and CO_2 geobarometer estimates are 860–1060 °C at 0.92–1.10 GPa for Mont Briançon, 930–980 °C at 0.89–1.04 GPa for Ray Pic and 840–940 °C at 0.59–0.71 GPa for Puy Beaunit. From south to north, the xenoliths show the following trends: (1) deeper to shallower origin; (2) more depleted mineral compositions, suggesting higher degrees of partial melting; and (3) more enriched isotopes and trace elements, indicating a mixing process with a silicate-rich carbonatite melt characterized by high H_2O and K_2O, possibly during Variscan subduction.

Cenozoic volcanism in the French Massif Central brought to the surface many peridotite and pyroxenite xenoliths that have been extensively studied in terms of petrography, trace elements and isotope chemistry. Coisy & Nicolas (1978) reported a dominance of porphyroclastic and mosaic (deformed) peridotite xenoliths in the centre of the Massif Central, and coarse-grained (weakly deformed or undeformed) xenoliths in marginal regions. Based on these observations, the authors suggested a diapiric rise of the upper mantle, causing crustal thinning in this region. Downes & Dupuy (1987) reported that the undeformed xenoliths had Sr–Nd isotopic ratios and rare earth element (REE) chondrite-normalized patterns similar to those of the source of mid-ocean ridge basalts (MORB), whereas the deformed xenoliths showed enriched signatures.

Based on petrographical observations, and the whole-rock major and trace element compositions of xenoliths, Lenoir *et al.* (2000) divided the subcontinental lithospheric mantle (SCLM) beneath the Massif Central into two distinct lithospheric domains, north and south of latitude 45°30′N. The authors suggested that the compositional differences between the two domains could not be explained solely by the upwelling of a Cenozoic plume; instead, they proposed that the differences might indicate the existence of distinct lithospheric blocks assembled during the Variscan orogeny. These two different lithospheric domains can be distinguished based on seismic anisotropy (Babuska *et al.* 2002). Downes *et al.* (2003) reported a difference in trace element and isotope compositional data between clinopyroxenes of xenoliths from the northern and southern domains.

Previous studies have described the relationships among the deformation textures, equilibration temperature and geochemistry of xenoliths from the southern domain (at Ray Pic and Borée: Zangana *et al.* 1997; Xu *et al.* 1998). These studies ascribed the differences in equilibrium temperature to the depth of xenolith residence; therefore, variations in geochemical and isotopic compositions occur not only between the north and south domains, but also as a function of depth.

The Puy Beaunit xenoliths from the northern domain show secondary protogranular–equigranular textures with spherical spinel, and locally show a transition to charnockitic paragenesis with increasing silica content (Mercier & Nicolas 1975). Clinopyroxene of the Puy Beaunit xenoliths is characterized by high Sr and low Nd isotopic compositions compared with those of other peridotite xenoliths from the French Massif Central (e.g. Downes & Dupuy 1987). This isotopic character is

From: Coltorti, M., Downes, H., Grégoire, M. & O'Reilly, S. Y. (eds) *Petrological Evolution of the European Lithospheric Mantle*. Geological Society, London, Special Publications, **337**, 153–175.
DOI: 10.1144/SP337.8 0305-8719/10/$15.00

similar to that of granulite xenoliths derived from the lower crust beneath the French Massif Central (e.g. Downes & Dupuy 1987). The isotopic and textural signatures of the Puy Beaunit xenoliths are possibly the result of mechanical mixing (e.g. Mercier & Nicolas 1975; Downes & Dupuy 1987) associated with the intrusion of mantle lithologies into the lower crust or the accretion of such lithologies.

To confirm the existence of relationships between depth of residence and textural and geochemical variations within the xenoliths, we present new data on the major element compositions of minerals and glass, density of CO_2 fluid inclusions, trace element compositions of clinopyroxene and phlogopite, and Sr–Nd isotopic compositions of clinopyroxene within peridotite and pyroxenite xenoliths from Mont Briançon, Ray Pic and Puy Beaunit. We also discuss the pressure–temperature ($P-T$) conditions and chemical characteristics of metasomatic agents, and the isotopic composition of the uppermost mantle beneath the Massif Central.

Geological background

The Massif Central forms part of the European Variscan belt (Fig. 1), which is dominated by metamorphic and granitic rocks. This basement formed during the Variscan orogeny as a result of the collision between the European and African continental plates (Matte 1986). The Massif Central is also the largest Cenozoic magmatic province in the Western European Rift System. From the Palaeocene to the Holocene, rifting and associated graben development occurred in association with basaltic magmatism. Michon & Merle (2001) divided the volcanism into three magmatic phases: the pre-rift, rift-related and major magmatic events. During the pre-rift event, scattered volcanoes located along the grabens effused magmas ranging in composition from melilitite to alkali basalt. During the rift-related event, alkalic–nephelinitic volcanism was limited to the northern part of the Massif Central and areas close to the grabens. The major magmatic events occurred from 15 Ma to the Recent, and involved extensive subalkaline and alkaline volcanism. Most of the peridotite and pyroxenite xenoliths from the Massif Central are found within these magmas.

Texture of peridotite xenoliths

We analysed the geochemistry and isotopic compositions of six lherzolite xenoliths, one harzburgite xenolith and one pyroxenite xenolith collected from Mont Briançon (BR), Ray Pic (BU) and Puy Beaunit (BE, Fig. 1). Thin slabs (*c.* 100 μm) were made with samples and then doubly polished. Modal analysis (1500 points in each sample) revealed that samples BR1, BR3, BR4, BU4, BU5 and BE4 are spinel lherzolites, whereas BE11 is a spinel harzburgite and BE1 is an olivine websterite (Table 1). The peridotite xenoliths from Mont Briançon and Ray Pic show a primary protogranular texture (Fig. 2a, b), while the peridotite xenoliths from Puy Beaunit show a secondary protogranular texture (Fig. 2c) according to the nomenclature of Mercier & Nicolas (1975). CO_2 fluid inclusions are abundant in most samples, and pyroxene generally contains more fluid inclusions than does olivine. The fluid inclusions occur in intracrystalline positions along planes and are of various types (e.g. negative crystal and small equant shapes: see Roedder 1984; Sapienza *et al.* 2005).

The Mont Briançon and Ray Pic xenoliths contain coarse (up to 6 mm) kink-banded olivine and slightly smaller orthopyroxene (opx; up to 5 mm) grains showing curvilinear grain boundaries. With the exception of BR1 xenolith, lamellae of opx and tiny rectangular inclusions are commonly observed in clinopyroxene (cpx) grains (up to 3 mm in size). Spinel (up to 2 mm in size) generally has a vermicular shape and is typically in contact with pyroxene (Fig. 2c). The BR1 xenolith displays holly-leaf–subhedral rounded spinel (up to 1 mm in size). The BR4 and BU5 xenoliths contain fewer CO_2 fluid inclusions than do the other samples.

In samples BE1 and BE4, both olivine and opx are up to 3 mm in size while cpx is smaller (up to 2 mm). Olivine shows kink-bands, and spinel occurs as round interstitial grains (0.1–0.7 mm) in sample BE4. Clinopyroxene grains of sample BE11 have spongy rims and contain tiny inclusions of spinel. Sample BE4 contains conspicuous patches of phlogopite associated with glass, olivine and spinel (Fig. 2d). These patches are observed along a vein and are generally in contact with pyroxene grains (Fig. 2c). Phlogopite is rounded in shape, and is surrounded by a glass pool that contains small crystals of olivine and spinel (Fig. 2d). Wilson & Downes (1991) reported glass patches associated with amphibole and/or phlogopite in xenoliths from Puy Beaunit, Massif Central.

Sample preparation and analytical method

Major element compositions of minerals were measured from mounted and polished separates in epoxy resin using an energy dispersive X-ray spectrometer (EDS) attached to a scanning electron microscope housed at the Institute of Geothermal

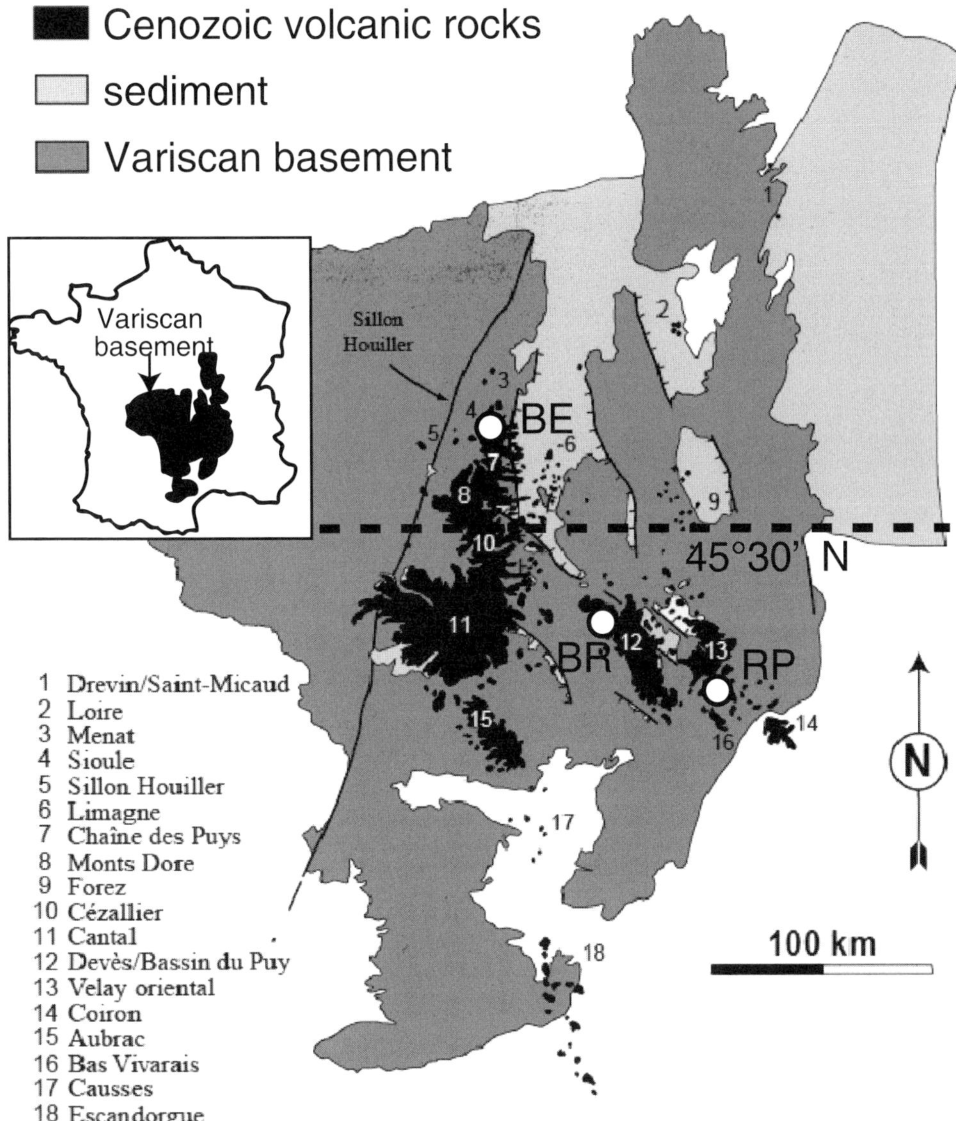

Fig. 1. Geological map of the French Massif Central (Michon & Merle 2001) with sample localities (BR, Mont Briançon; RP, Ray Pic; BE, Puy Beaunit). The boundary (dashed line) between the northern and southern lithospheric domains (Lenoir *et al.* 2000) is also shown.

Sciences (I.G.S.), Kyoto University, in Beppu, Japan. The analysed area of minerals and glass was over a few μm^2 or $15 \times 20\ \mu m^2$ using a 0.5 nA Faraday cup current and an analysis time of 200 s. Na loss from silicic glass was minimized by analysing a larger area ($>10 \times 10\ \mu m^2$: Nakajima & Arima 1998; Varela *et al.* 1999). Analysis of minerals and glasses using thin (*c.* 100 μm) slabs fails to produce a total of 100% oxides (70–97% depending on thin slabs), although it gives a total of 100% for standard analyses. We did not find any explanation, and only guess that this failure is possibly caused by a poor electric connection in the aggregate samples.

Trace element compositions of cpx were analysed *in situ* on thin polished slabs (*c.* 100 μm) using a 266 nm UV Nd YAG laser (CETAC LSX 200) connected to a ThermoFisher X2 quadruple

Table 1. *Modal compositions (%) of the French Massif xenoliths*

	Mont Briançon			Ray Pic		Puy Beaunit		
	BR1	BR3	BR4	BU4	BU5	BE1	BE4	BE11
ol	61.4	60.4	56.3	55.7	69.9	10.3	57.3	67.5
opx	30.0	24.3	23.2	23.0	15.8	39.9	27.6	26.8
cpx	5.1	13.4	11.3	15.8	10.3	44.9	8.7	4.6
sp	3.5	1.9	9.2	5.5	4.0	3.3	4.4	1.1
phl	–	–	–	–	–	–	1.1	–
glass	–	–	–	–	–	–	0.9	–
alter	–	–	–	–	–	1.6	–	–

Abbreviations: ol, olivine; opx, orthopyroxene; cpx, clinopyroxene; sp, spinel; phl, phlogopite; alter, altered part.

inductively coupled plasma mass spectrometer (Q-ICP-MS) system housed at the I.G.S. at Beppu. The count time for each spot was 120 s, and the size of the laser was 50, 100 or 150 μm, according to the concentration of the analysed element. The laser frequency and energy were 6 Hz and 7.6 mJ cm^{-2}, respectively. Helium was used as a carrier gas to enhance the transport efficiency of the ablated material. The helium carrier gas was mixed with argon gas as a make-up gas before entering the ICP. The operating parameters for Q-ICP-MS analyses were as follows; nebulizer gas flow 0.81 l min^{-1}; auxiliary gas flow 1.2 l min^{-1}; plasma gas flow 1.4 l min^{-1}; and ICP RF power 1400 W. The NIST610 glass standard was used to calibrate relative element sensitivities for the analyses and our analytical protocol consisted of analysing the standard glass at the

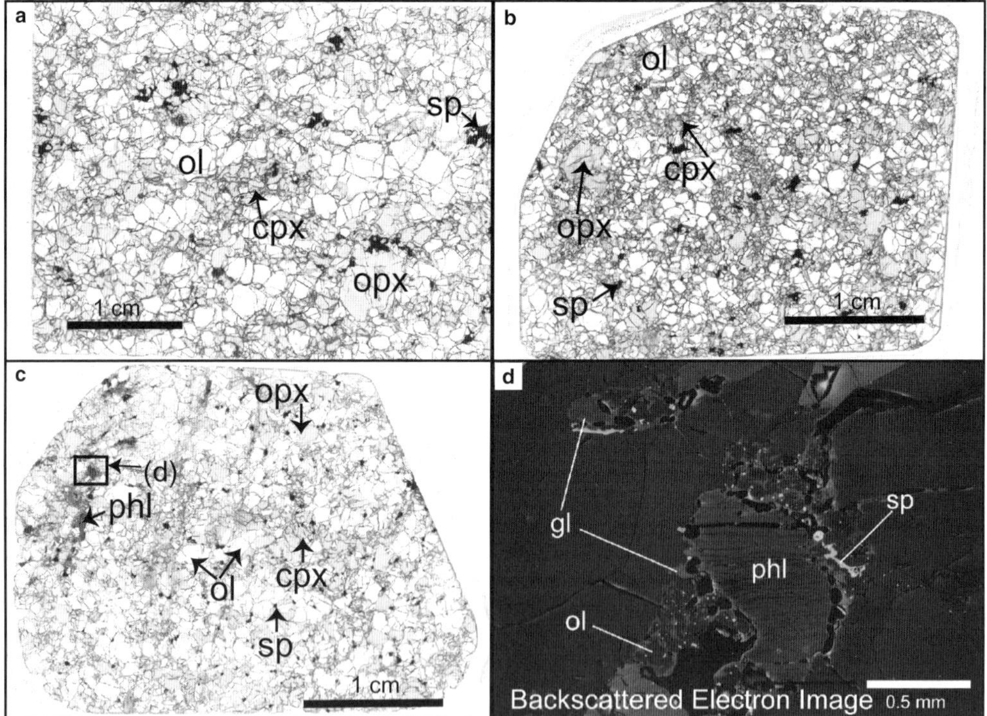

Fig. 2. Photomicrograph of samples with plane-polarized light: (**a**) primary protogranular texture (sample BU5); (**b**) primary protogranular texture (sample BR1); and (**c**) secondary protogranular texture (sample BE4). The backscattered electron image in (**d**) is the glass vein within sample BE4, and the location is shown in (c). ol, olivine; opx, orthopyroxene; cpx, clinopyroxene; sp, spinel; phl, phlogopite; gl, glass.

beginning, middle and end of each 3-h measurement. We analysed Si as an internal reference isotope for data reduction. The NIST612 glass was analysed as an unknown, and typical analytical reproducibility and accuracy (RSD %) for elements analysed by laser-ICP-MS was 5–10% (but 15–20% for Rb and Gd).

To determine the depths of origin of the xenoliths we applied the micro-Raman densimeter following the procedure described by Yamamoto *et al.* (2002). The xenoliths contained a large number of fluid inclusions of almost pure CO_2; indeed, the obtained Raman spectra showed no trace of components other than CO_2. CO_2 fluid inclusions with negative crystal shape retain the internal pressure of fluid inclusions. If the density of the CO_2 is known, extrapolation of the isochore to the equilibrium temperature estimated using a mineral geothermometer enables estimation of the pressure of CO_2 at the time when the fluid inclusions were thermally equilibrated with the surrounding host mineral. For micro-Raman analyses, we selected CO_2 fluid inclusions with negative crystal shapes, located away from the crystal edge. Unpolarized Raman spectra of CO_2 fluid inclusions were collected using a Raman microscope (Kaiser HoloLab 5000 system) consisting of a 532 nm YAG laser (17 mW at each sample position), holographic transmission gratings and a 2048-pixel CCD detector over the spectral range from 291 to 4573 cm^{-1}, housed at the I.G.S., Kyoto University, in Beppu. An excitation laser beam was focused on spots 1 or 2 μm in diameter using a ×100 or ×50 objective lens, respectively. The accumulation time was typically 40 s. Raman spectra of CO_2 have two main peaks: v_+ (1388 cm^{-1}) and v_- (1285 cm^{-1}). The split (Δ) of the Fermi diad increases with increasing pressure (density) of CO_2, as indicated by the polynomial equation that describes the relationship between split and density in the density range between 0.1 and 1.24 g cm^{-3} (Yamamoto & Kagi 2006). Before undertaking measurements of unknown samples each day, we measured the Δ value of a standard fluid inclusion ($\Delta = 104.686$ cm^{-1}) preserved in an olivine crystal from Ennokentiev, Far East Russia, the value of which was determined at the same time as the polynomial equation of Δ, and the density was determined by Yamamoto *et al.* (2002). We therefore employed this Δvalue as a standard, using it to normalize the data.

Clinopyroxene grains were separated by hand-picking under a binocular microscope from crushed and sieved xenolith samples. The separated cpx was then cleaned with distilled 6 M HCl at 80 °C and rinsed three times with Milli-Q water. The final cpx separates were selected by careful hand-picking before being washed with distilled 6 M HCl and rinsed three times with Milli-Q water in a clean room. The acid-washed cpx was decomposed by $HF + HClO_4$, as described by Yokoyama *et al.* (1999). Details of the employed analytical procedures, including chemical separation and mass spectrometry, can be found in Yoshikawa & Nakamura (1993) and Shibata & Yoshikawa (2004). Mass spectrometry was performed using a ThermoFinnigan MAT 262 instrument housed at the I.G.S., Kyoto University, at Beppu. Normalizing factors used to correct the isotopic fractionation of Sr and Nd were $^{86}Sr/^{88}Sr = 0.1194$ and $^{146}Nd/^{144}Nd = 0.7219$, respectively. Standard solution values were $^{87}Sr/^{86}Sr = 0.710279 \pm 28$ (2σ) for NIST 987 and $^{143}Nd/^{144}Nd = 0.511851 \pm 13$ (2σ) for La Jolla. Total procedural blanks for Rb, Sr, Sm and Nd were <30, <100, <10 and <10 pg, respectively. Analytical reproducibility of Rb, Sr, Sm and Nd abundances by isotope dilution method was better than 2% (2σ).

Major element compositions of minerals

Southern domain (Mont Briançon and Ray Pic)

The forsterite contents (Fo) of analysed olivines (Fo_{89-90}), and the Mg-number (molar 100 × Mg/(Mg + Fe)) of opx (89–91) and cpx (90–92) are consistent with previously determined values (Fo_{89-92}, Mg-number of opx 89–91; Mg-number of cpx 88–93: Werling & Altherr 1997; Zangana *et al.* 1997; Touron *et al.* 2008) (Table 2). The Al_2O_3 content of cpx ranges from 5.4 to 7.1 wt%, that is, within the range of previously reported values for peridotite xenoliths from the southern domain (2.4–7.2 wt%: Werling & Altherr 1997; Zangana *et al.* 1997; Touron *et al.* 2008) (Fig. 3a). Spinel Cr-number (molar 100 × Cr/(Cr + Al)) and Mg-number range from 8 to 25 and from 78 to 83, respectively. These two values show a negative correlation (Fig. 3b) consistent with the nature of compositional variations produced by partial melting and melt extraction (e.g. Dick & Bullen 1984; Hellebrand *et al.* 2001).

Northern domain (Puy Beaunit)

The Fo content of olivines (Fo_{90}), and Mg-number of opx (90–91) and cpx (91–94) from peridotites, are within the range of values reported previously for the northern domain (Fo_{90-92}; Mg-number of opx 89–92; Mg-number of cpx 89–94: Werling & Altherr 1997; Downes *et al.* 2003), and are similar to values determined from xenoliths in the southern domain. The Al_2O_3 content of cpx from peridotites

Table 2. *Major element compositions of minerals and glass vein*

Sample	BR1	BR3	BR4	BU4	BU5	BE1	BE4	BE4	BE11
Rock type	L	L	L	L	L	ol.WEB	L	L	H
Mineral	ol	ol	ol	ol	ol	ol	ol	ol in gl	ol
SiO_2	40.89	41.07	40.63	41.12	41.06	40.79	41.18	39.68	41.28
TiO_2	0.07	0.11	0.07	b.d.l.	b.d.l.	b.d.l.	0.09	0.04	0.08
Al_2O_3	0.25	0.10	0.14	0.26	0.23	0.25	0.09	0.01	0.13
FeO^T	9.71	9.88	10.08	9.56	10.83	11.95	9.57	8.23	9.34
MnO	0.19	0.18	0.11	0.28	0.19	0.20	0.24	0.07	0.08
MgO	48.48	48.91	48.10	49.05	48.23	47.00	48.76	48.55	49.14
CaO	0.09	0.06	0.05	0.05	0.03	0.05	0.03	0.18	0.12
NiO	0.42	0.48	0.48	0.56	0.54	0.32	0.41	0.38	0.42
Total	100.10	100.79	99.66	100.88	101.11	100.56	100.37	97.14	100.59
Fo	89.9	89.8	89.5	89.7	88.8	87.5	90.1	91.3	90.4
Sample	BR1	BR3	BR4	BU4	BU5	BE1	BE4		BE11
Rock type	L	L	L	L	L	ol.WEB	L		H
Mineral	opx	opx	opx	opx	opx	opx	opx		opx
SiO_2	54.95	56.04	55.60	55.21	55.51	56.46	56.63		57.21
TiO_2	0.14	0.08	0.08	0.06	0.18	0.04	0.10		0.14
Al_2O_3	3.93	3.58	3.80	3.91	4.02	2.65	2.48		1.34
Cr_2O_3	0.56	0.42	0.26	0.29	0.28	0.33	0.45		0.31
FeO^T	6.02	6.20	6.86	6.96	6.94	7.92	6.57		6.08
MnO	0.28	0.10	0.10	0.12	0.25	0.04	0.19		0.23
MgO	32.44	33.24	32.73	32.59	32.51	32.61	33.48		34.27
CaO	0.93	0.49	0.52	0.59	0.56	0.75	0.48		0.39
Na_2O	0.36	0.35	0.30	0.33	0.28	0.30	0.17		0.16
Total	99.61	100.50	100.25	100.06	100.53	101.10	100.55		100.13
Mg-number	90.6	90.5	89.5	89.4	89.3	88	90.1		90.9

Sample	BR1	BR3	BR4	BU4	BU5	BE1	BE4	BE11	BE4	BE4
Rock type	L	L	L	L	L	ol.WEB	L	H	L	L
Mineral	cpx	cpx	cpx	cpx	cpx	cpx	cpx	cpx	phl	glass
SiO_2	52.04	52.31	51.92	52.18	51.68	52.98	53.32	53.02	37.75	58.86
TiO_2	0.59	0.36	0.70	0.53	0.84	0.08	0.13	0.09	1.21	1.12
Al_2O_3	5.35	6.06	6.91	5.98	7.05	3.41	3.70	2.92	16.86	18.94
Cr_2O_3	1.33	1.29	0.75	0.85	0.95	0.64	1.09	0.99	1.31	0.06
FeO^T	2.73	2.28	2.50	2.98	2.84	2.58	1.82	2.78	4.29	2.77
MnO	0.02	0.08	0.11	0.09	0.00	0.03	0.18	0.22	b.d.l.	b.d.l.
MgO	15.82	14.45	14.57	15.14	14.43	15.82	15.72	16.66	21.03	2.72
CaO	19.96	20.56	20.63	20.74	19.85	22.42	21.95	21.89	0.05	6.91
Na_2O	1.39	1.79	1.80	1.33	1.78	0.67	1.10	0.68	0.25	1.05
K_2O	b.d.l.	b.d.l.	b.d.l.	b.d.l.	b.d.l.	b.d.l.	b.d.l.	b.d.l.	10.11	9.01
NiO	0.03	b.d.l.	b.d.l.	b.d.l.	b.d.l.	b.d.l.	b.d.l.	b.d.l.	0.07	0.30
Total	99.26	99.18	99.89	99.82	99.42	98.63	99.01	99.25	92.93	101.74
Mg-number	91.2	91.9	91.2	90.1	90.1	91.6	93.9	91.4	89.7	63.6

Sample	BR1	BR3	BR4	BU4	BU5	BE4	BE4	BE11
Rock type	L	L	L	L	L	L	L	H
Mineral	sp	sp	sp	sp	sp	sp	sp in gl	sp
TiO_2	0.33	0.11	0.19	0.15	0.13	0.03	0.75	0.13
Al_2O_3	46.16	54.87	59.18	57.97	60.67	43.04	29.63	31.46
Cr_2O_3	22.50	14.17	8.54	10.10	7.97	24.45	32.07	37.57
FeO^T	9.43	7.70	8.21	9.32	9.44	10.01	9.72	9.20
Fe_2O_3	0.00	0.47	1.00	0.88	0.00	0.61	2.84	0.55
MnO	0.08	0.17	0.03	0.17	0.18	0.20	b.d.l.	b.d.l.
MgO	18.82	21.36	21.50	20.74	20.93	18.07	16.24	17.51
NiO	0.18	0.34	0.58	0.47	0.52	0.56	0.56	0.71
Total	97.50	99.19	99.23	99.80	99.84	96.97	91.81	97.13
Mg-number	78.1	83.2	82.4	79.9	79.8	76.3	74.9	77.2
Cr-number	24.6	14.8	8.8	10.5	8.1	27.6	42.1	44.5

Abbreviations: L, spinel lherzolite; H, spinel harzburgite; ol.WEB, olivine websterite; ol, olivine; opx, orthopyroxene; cpx, clinopyroxene; sp, spinel; gl, glass; phl, phlogopite; total Fe as FeO^T; Fo, forsterite content of olivine; b.d.l., below detection limit; Mg-number = molar ($100 \times Mg/(Mg + Fe^{2+})$); Cr-number = molar ($100 \times Cr/(Cr + Al)$).
FeO and Fe_2O_3 calculated assuming a stoichiometric spinel.

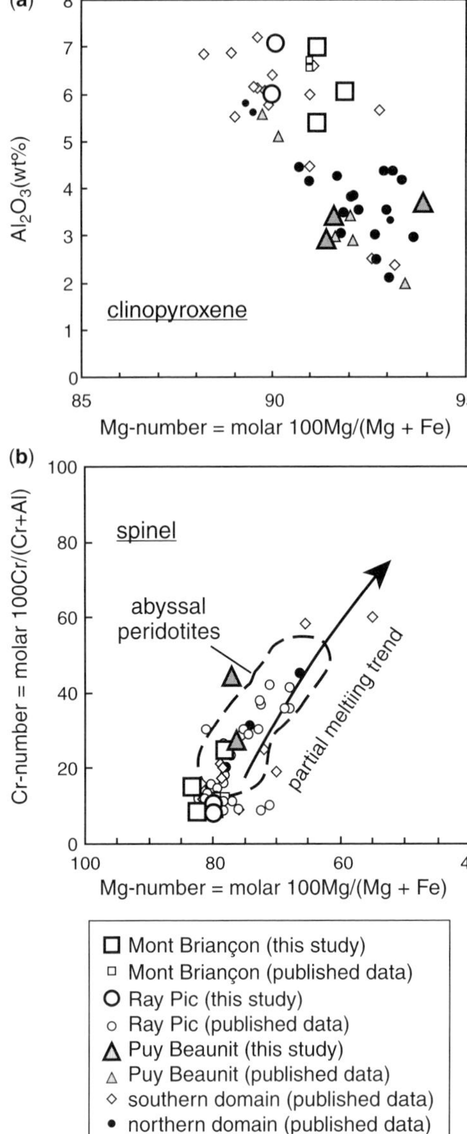

Fig. 3. (**a**) Al_2O_3 (wt%) v. Mg-number (molar $100 \times Mg/(Mg + Fe)$) of clinopyroxene. (**b**) Cr-number (molar $100 \times Cr/(Cr + Al)$) v. Mg-number of spinel. Data from the southern and northern domain xenoliths (Werling & Altherr 1997; Zangana *et al.* 1997; Touron *et al.* 2008), Mont Briançon (Touron *et al.* 2008) and Puy Beaunit (Downes *et al.* 2003) are also shown.

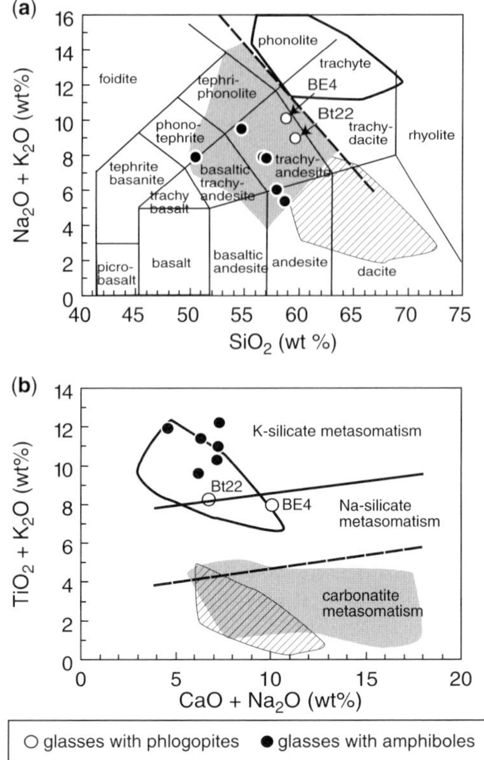

Fig. 4. (**a**) Total alkalis v. SiO_2; and (**b**) $TiO_2 + K_2O$ v. $CaO + Na_2O$ diagrams for glasses with hydrous minerals in the Massif Central xenoliths. Data, except for the glass within sample BE4, are from Wilson & Downes (1991). Grey, empty and oblique hatched fields are the composition ranges of glasses observed in worldwide peridotite xenoliths related to carbonatite, K-silicate and supra-subduction setting metasomatism, respectively (Coltorti *et al.* 2000, 2007*a*). Dashed and solid lines are the boundaries for classifications of metasomatic agents between carbonatite and other metasomatism, and between K- and Na-silicate metasomatism, respectively (Coltorti *et al.* 2000).

and the pyroxenite vary from 2.9 to 3.7 wt%, that is, within the range of values reported previously for northern peridotite xenoliths (1.9–5.8 wt%: Werling & Altherr 1997; Downes *et al.* 2003) (Fig. 3a). The BE1 pyroxenite records the lowest Fo content (Fo_{88}) among the Puy Beaunit xenoliths.

Phlogopite-glass vein

Olivines within the phlogopite-glass vein of sample BE4 have higher Fo values (Fo_{91}) and higher CaO concentrations (0.18 wt%) than olivine within the host lherzolite (Fo_{90}, 0.03 wt% CaO). Similar characteristics have been reported for olivine in glassy patches from peridotite xenoliths (e.g. Frey & Green 1974; Delpech *et al.* 2004). Glass within the phlogopite-bearing vein of sample BE4 (Fig. 2c, d) is a trachyandesitic in composition, according to data plotted on a total alkalis–silica (TAS) diagram (Fig. 4a and Table 2). This glass and the previously reported glass associated with phlogopite from a Puy Beunit xenolith

(Bt 22: Wilson & Downes 1991) plot into the field of carbonatite- and K-silicate-metasomatized glasses (as defined by Coltorti *et al.* 2000, 2007*a*) in a TAS diagram (Fig. 4a) and $TiO_2 + K_2O$ v. $CaO + Na_2O$ diagram (Fig. 4b). These observations suggest that the Puy Beaunit glasses are genetically related to silicate-rich carbonatite melt, although it is uncertain that the glass was formed directly from such melt.

Based on mass-balance calculations using the major element compositions of glass and minerals of sample BE4, we obtained the following reaction in the SiO_2–TiO_2–Al_2O_3–Cr_2O_3–MgO–FeO–CaO–Na_2O–K_2O system:

$$135\,\text{ol} + 81\,\text{opx} + 31\,\text{cpx} + 14\,\text{sp} + 89\,\text{phl} = 234\text{ secondary ol} + 16\text{ secondary sp} + 100\,\text{gl}$$

(ol, olivine; phl, phlogopite; sp, spinel; gl, glass). We therefore propose that the destabilization of phlogopite and primary pyroxenes resulted in the formation of glass, secondary olivine and spinel.

The BE4 glass is characterized by a low Na_2O concentration (1.05 wt%, Table 2), much lower than values reported previously for interstitial glass from peridotite xenoliths with or without phlogopite and/or amphibole, although the total alkali composition of the BE4 glass (10.06 wt%) is similar to that of the Bt22 glass (9.22 wt%; Fig. 4 and Wilson & Downes 1991).

P–T conditions

The palaeogeothermal structure of regions has been discussed previously based on the equilibrium temperature and pressure obtained from ultramafic xenoliths (e.g. O'Reilly & Griffin 1985; Choi *et al.* 2005). Many geothermometers are based on the composition of coexisting phases in ultramafic xenoliths (e.g. Wood & Banno 1973; Wells 1977; Bertrand & Mercier 1985; Brey & Köhler 1990; Köhler & Brey 1990). In the present study we calculated equilibrium temperatures based on the chemical compositions of Ca-rich and Ca-poor pyroxenes, as described by Wells (1977) (T_{Wells}) and Brey & Köhler (1990) (T_{BK}). T_{BK} is calculated assuming a pressure of 1.0 GPa. The temperatures calculated for the ultramafic xenoliths are as follows: 870–1020 °C (T_{Wells}) and 860–1060 °C (T_{BK}) for Mont Briançon; 930 °C (T_{Wells}) and 960–980 °C (T_{BK}) for Ray Pic; and 840–900 °C (T_{Wells}) and 840–940 °C (T_{BK}) for Puy Beaunit (Table 3).

Working in the French Massif Central, Werling & Altherr (1997) estimated *P–T* conditions by applying the two-pyroxene thermometer (Brey & Köhler 1990) and the Ca-in-olivine barometer (Köhler & Brey 1990) to ultramafic xenoliths from many localities: temperatures range from 700 to 1280 °C and pressures from 0.68 to 1.80 GPa. Use of the Ca-in-olivine geobarometer, however, involves three limitations: (1) the extrapolation of high-*P* (garnet stability field) experimental data to spinel-lherzolite and the insensitivity to pressure of Ca in olivine in the *P–T* stability field of spinel peridotite (O'Reilly *et al.* 1997); (2) the difficulty in obtaining high-precision analyses of Ca in olivine (Lee *et al.* 1996); and (3) the fact that the reliability of the barometer depends on the state of equilibrium in each sample because of the high diffusion velocity of Ca in olivine relative to that in pyroxene (Werling & Altherr 1997). Depths at which ultramafic xenoliths were sampled by the host magma have previously been estimated based on the internal pressure of CO_2 fluid inclusions using microthermometry (e.g. Roedder 1983, 1984). Although this method is precise, it is difficult to make a reliable visual observation of the homogenization of biphase CO_2 fluid in small inclusions of <5 μm in diameter (Yamamoto & Kagi 2008). In the present samples most of the CO_2 fluid inclusions are around 5 μm or less in diameter; therefore, we applied Raman spectroscopic barometry as proposed by Yamamoto *et al.* (2002, 2007). The uncertainty involved in density estimates was less than 0.1%, an indication of only a negligible effect on the estimated pressures and depths. The pressures calculated using T_{Wells} and T_{BK} are similar (Table 3), with values ranging from 0.92 to 1.10 GPa for Mont Briançon, 0.89–1.04 GPa for Ray Pic and 0.59–0.71 GPa for Puy Beaunit. Depths were calculated based on a lithosphere density of 2.85 g cm^{-3}. The calculated depth ranges for Mont Briançon, Ray Pic and Puy Beaunit xenoliths are 33.3–39.8, 32.1–37.6 and 21.2–25.6 km, respectively (Table 3).

Most of the xenoliths analysed in this study were equilibrated within the spinel stability field (Fig. 5). One Puy Beaunit xenolith (BE11) seems to have been equilibrated in the stability field of plagioclase peridotite; however, it lacks modal plagioclase. Its pressure was determined based on the density of CO_2 fluid inclusions in olivine, which yield a lower value than that obtained from similar inclusions in pyroxenes. This difference is attributed to rheological differences between minerals (Yamamoto *et al.* 2002, 2007, 2008; Yamamoto & Kagi 2008). Decrepitation of CO_2 fluid inclusions in peridotite xenoliths (Roedder 1984; Sapienza *et al.* 2005) or CO_2 flushing from lithospheric mantle to lower crust (Santosh & Omori 2008) have been suggested as possible mechanisms of reducing CO_2 fluid density. It is possible to identify the partial loss of CO_2 by decrepitation based on observations of thin slabs

Table 3. *Densities obtained by the micro-Raman spectroscopic analyses and equilibrium temperature, and pressures and depths calculated using the equilibrium temperatures*

Sample	Density ($g\ cm^{-3}$)	T_{Wells} (°C)	P_{Wells} (GPa)	+/−	Depth (km)	+/−	T_{BK} (°C)	P_{BK} (GPa)	+/−	Depth (km)	+/−	Host mineral
Mont Briançon												
BR1	1.15	1015	1.07	0.02	38.5	0.9	1060	1.10	0.02	39.8	0.9	opx/cpx
BR3	1.19	870	1.06	0.02	38.2	0.8	864	1.05	0.02	38.1	0.8	cpx
BR4	1.13	885	0.92	0.03	33.3	1.0	912	0.94	0.03	34.0	1.0	opx
Ray Pic												
BU4	1.10	933	0.89	0.03	32.1	1.1	961	0.91	0.03	32.8	1.1	cpx
BU5	1.15	934	1.01	0.02	36.3	0.8	981	1.04	0.02	37.6	0.8	cpx
Puy Beaunit												
BE1	1.02	902	0.71	0.03	25.6	1.0	848	0.67	0.03	24.3	1.0	opx
BE4	1.01	879	0.68	0.04	24.4	1.4	838	0.65	0.04	23.5	1.4	opx
BE11	0.97	840	0.59	0.06	21.2	2.1	941	0.65	0.06	23.3	2.1	ol

T_{Wells} and T_{BK} are equilibrim temperatures calculated by Wells (1977) and Brey & Köhler (1990), respectively.
P_{Wells} and P_{BK} are pressures calculated using T_{Wells} and T_{BK}, respectively.
Abbreviations: ol, olivine; opx, orthopyroxene; cpx, clinopyroxene.

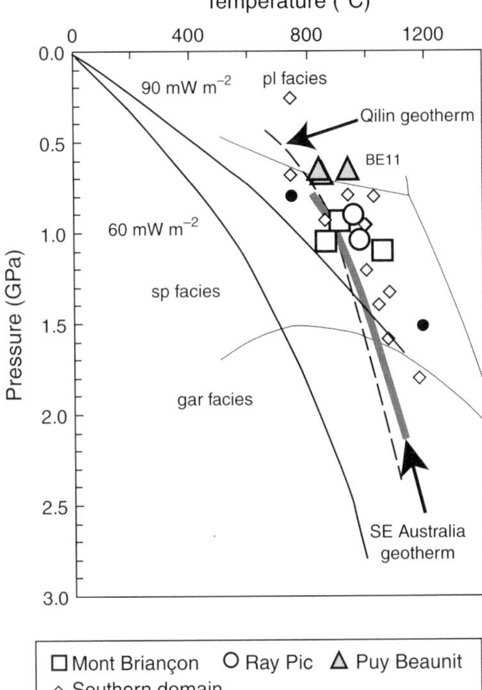

Fig. 5. Pressure v. temperature diagram estimated using the micro-Raman barometer of Yamamoto & Kagi (2006) and the two-pyroxene thermometer of Brey & Köhner (1990) compared with previously published data (Werling & Altherr 1997). The model of conductive continental geotherms with surface heat flows of 60 and 90 mW m^{-2} are from Pollack & Chapman (1977). Geotherms for Qilin, eastern China from Xu *et al.* (1996) and for SE Australia from O'Reilly & Griffin (1985) are also shown. Facies boundaries for peridotite in the Na–Ca–Al–Mg–Al–Si system are after Gasparik (1987).

(e.g. Yamamoto *et al.* 2002). Any evidence of disturbance of CO_2 fluid density in the mantle by a geological event such as CO_2 flushing would be completely erased within several days, as fluid inclusions rapidly stretch in response to changing pressure and temperature (Yamamoto *et al.* 2002, 2007, 2008).

The estimated temperatures are higher than those predicted by the conductive model at the depth of the Moho discontinuity (as proposed by Werling & Altherr 1997) and similar to the geotherm obtained from other continental peridotite xenoliths (Fig. 5). Such elevated geotherms have previously been explained by heat transport associated with lithospheric thinning and the intrusion of dykes into the crust (e.g. O'Reilly & Griffin 1985).

Trace element compositions of clinopyroxene

Southern domain (Mont Briançon and Ray Pic)

The analysed cpx from southern xenoliths (except BR1) is generally depleted in light REEs (LREEs) such as La, Ce and Nd, and depleted in large ion lithophile elements (LILEs) such as Ba and Sr relative to heavy REEs (HREEs) such as Yb and Lu ($La_N/Yb_N = 0.1–0.3$, where the subscript N indicates normalization to the primary mantle values provided by McDonough & Sun 1995) (Fig. 6a and Table 4). The HREE contents are within a limited range (e.g. $Yb_N = 4–5$: Table 4). Similar results have been reported for xenoliths from Mont Briançon and Ray Pic ($La_N/Yb_N = 0.06–0.6$, $Yb_N = 4–6$; Fig. 6a) by De Vries (2007) and Touron *et al.* (2008). Trace element patterns of residual cpx after simple partial melting are expected to be depleted in LREEs and LILEs because of their highly incompatibility character. In addition, HREE content of cpx in residual peridotite decreases with increasing degree of partial melting (e.g. Johnson *et al.* 1990; Hellebrand *et al.* 2001). The trace element characteristics of the southern domain xenoliths suggest that these xenoliths were residues left after the extraction of melts formed by low degrees of partial melting (Touron *et al.* 2008). The cpx within the BR1 xenolith shows an enriched pattern ($La_N/Yb_N = 4$), and is depleted in Rb, Ba and Nb. Its Yb content ($Yb_N = 2$; Table 4) is lower than that of cpx from the other Mont Briançon xenoliths, and its trace element pattern is similar to that of the amphibole and phlogopite-bearing lherzolite ML40 ($La_N/Yb_N = 4$, $Yb_N = 3$) from Marais de Limagne, as reported by Touron *et al.* (2008) (Fig. 6a). These features cannot be explained by simple partial melting and melt extraction; instead, they indicate that these cpx were affected by a LILEs and LREEs enrichment process, probably related to mantle metasomatism (e.g. Frey & Green 1974; Yoshikawa & Nakamura 2000).

Northern domain (Puy Beaunit)

The cpx of Puy Beaunit xenoliths show extensive LREE- and LILE-enriched patterns (e.g. $La_N/Yb_N = 2–11$), with depletion of Zr and Hf relative to low HREE content ($Yb_N < 3$; Fig. 6b). The BE1 and BE4 cpx show positive anomalies of Pb and Sr (Fig. 6a). High Sr content of cpx from Puy Beaunit xenoliths has also been reported by Downes *et al.* (2003). These features are different from those of cpx from the BR1 xenolith, indicating contrasting metasomatic agents. Depletion of Ba

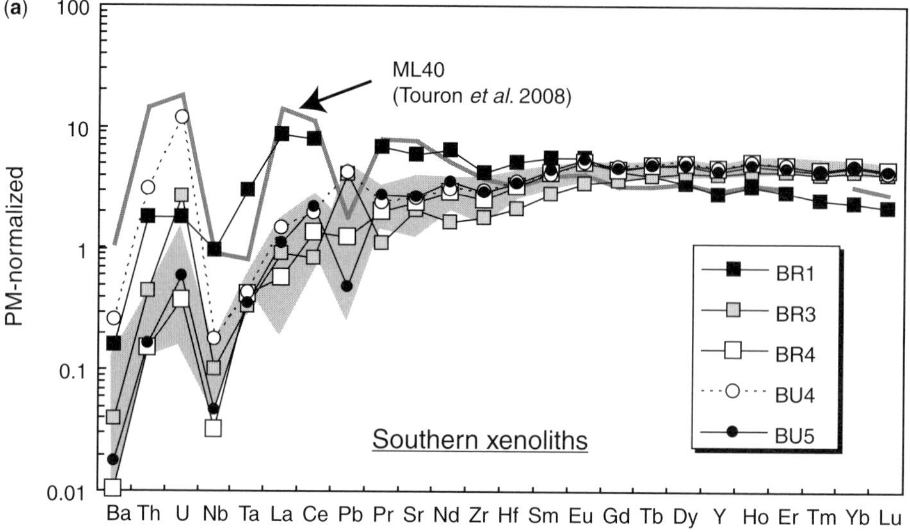

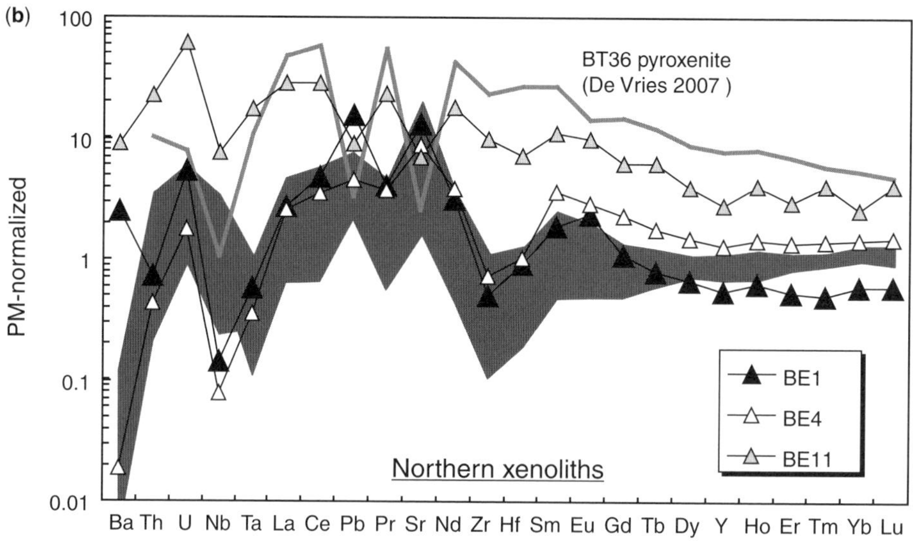

Fig. 6. Primitive mantle-normalized trace element patterns of clinopyroxene from (**a**) Mont Briançon and Ray Pic xenoliths in the southern domain, and from (**b**) Puy Beaunit xenoliths in the northern domain. The normalized values are from McDonough & Sun (1995). The right-hand grey field of (a) is a range of LREE- and LILE-depleted clinopyroxene from the southern domain xenoliths (De Vries 2007; Touron *et al.* 2008). The dark grey field of (b) is a range of clinopyroxene from Puy Beaunit xenoliths with Pb positive anomalies (De Vries 2007).

in BE4 cpx probably reflects its strong affinity for phlogopite (e.g. Green 1994). Clinopyroxenes of BE1 and BE4 have trace element patterns similar to those of cpx with positive Pb anomalies (De Vries 2007) (grey field in Fig. 6b) from the Puy Beaunit xenoliths. The BE11 cpx shows negative Sr and Pb anomalies, as also reported for BT36 pyroxenite from Puy Beaunit (De Vries 2007).

Sr–Nd isotopic compositions of clinopyroxene

Southern domain (Mont Briançon and Ray Pic)

The $^{87}Sr/^{86}Sr$ and $^{143}Nd/^{144}Nd$ ratios for cpx of the southern domain xenoliths range from 0.70217 to 0.70324 and from 0.51293 to 0.51355, respectively

Table 4. *Trace element compositions of clinopyroxenes and phlogopite in the French Massif Central xenoliths (ppm) analysed by laser ICP-MS*

	Mont Briançon			Ray Pic		Puy Beaunit			
	BR1	BR3	BR4	BU4	BU5	BE1	BE4		BE11
	cpx	cpx	cpx	cpx	cpx	cpx	cpx	phl	cpx
Rb	0.16	0.07	b.d.l.	0.79	0.11	0.40	0.35	352	3.16
Sr	118	40.8	47.9	51.7	52.9	264	173	82.1	139
Y	12.1	17.5	19.7	19.5	18.6	2.36	5.52	0.22	12.0
Zr	44.7	18.9	26.9	31.5	30.8	5.36	7.60	4.60	104
Nb	0.62	0.07	0.02	0.12	0.03	0.09	0.05	8.18	5.06
Ba	1.04	0.26	0.07	1.68	0.12	16.4	0.12	958	60.4
La	5.53	0.59	0.37	0.95	0.72	1.74	1.65	0.20	18.4
Ce	13.2	1.40	2.29	3.26	3.72	8.10	5.95	0.29	47.9
Pr	1.72	0.28	0.51	0.61	0.71	1.02	0.94	0.19	5.85
Nd	8.12	2.05	3.65	3.89	4.41	3.96	4.82	0.20	22.8
Sm	2.25	1.16	1.74	1.68	1.80	0.76	1.48	0.24	4.50
Eu	0.86	0.54	0.78	0.80	0.84	0.37	0.44	0.19	1.53
Gd	2.34	1.97	2.45	2.55	2.48	0.59	1.23	0.19	3.40
Tb	0.40	0.39	0.48	0.49	0.48	0.08	0.17	0.18	0.61
Dy	2.36	2.84	3.37	3.48	3.29	0.45	1.01	0.26	2.64
Ho	0.49	0.65	0.76	0.77	0.73	0.09	0.21	0.19	0.61
Er	1.28	1.90	2.13	2.11	2.04	0.23	0.61	0.24	1.29
Tm	0.17	0.28	0.30	0.30	0.29	0.03	0.10	0.15	0.28
Yb	1.06	1.91	2.12	2.08	2.06	0.26	0.65	0.19	1.12
Lu	0.15	0.28	0.30	0.28	0.29	0.04	0.10	0.16	0.27
Hf	1.46	0.60	0.93	1.01	0.98	0.26	0.29	0.23	2.05
Ta	0.11	0.01	0.02	0.02	0.01	0.02	0.01	0.66	0.65
Pb	b.d.l.	0.61	0.19	0.64	0.07	2.33	0.68	2.20	1.34
Th	0.14	0.03	0.01	0.25	0.01	0.06	0.03	0.22	1.79
U	0.04	0.05	0.01	0.24	0.01	0.11	0.04	0.30	1.27

Abbreviation: b.d.l., below detection limit.

(Table 5 and Fig. 7). These data (except those for BR3) are generally within the range of values reported previously for cpx from this region ($^{87}Sr/^{86}Sr = 0.70195–0.70375$, $^{143}Nd/^{144}Nd = 0.51285–0.51347$: Downes & Dupuy 1987; Zangana *et al.* 1997; Touron *et al.* 2008). In contrast, cpx of sample BR3 from Mont Briançon has a higher $^{143}Nd/^{144}Nd$ ratio than those reported previously for mantle cpx from the southern domain. This high Nd isotopic ratio may reflect the time-integrated evolution of a residual cpx. The BR1 cpx, which shows an enriched trace element

Table 5. *Sr and Nd isotopic data of clinopyroxenes*

	Rb ($\mu g\ g^{-1}$)	Sr ($\mu g\ g^{-1}$)	$^{87}Sr/^{86}Sr$	$2\sigma_m$	Sm ($\mu g\ g^{-1}$)	Nd ($\mu g\ g^{-1}$)	$^{147}Sm/^{144}Nd$	$^{143}Nd/^{144}Nd$	$2\sigma_m$
Mont Briançon									
BR1	0.001	128	0.703243	±09	2.57	9.09	0.1707	0.512928	±07
BR3	0.001	45.6	0.702673	±11	1.08	2.12	0.3083	0.513549	±30
BR4	0.0003	47.4	0.702173	±08	1.71	3.65	0.2826	0.513245	±14
Ray Pic									
BU5	0.005	52.3	0.703007	±10	1.90	4.54	0.2530	0.512986	±12
Puy Beaunit									
BE1	0.068	228	0.704173	±10	0.856	4.45	0.1163	0.512578	±16
BE4	0.007	155	0.705677	±10	1.11	4.56	0.1471	0.512306	±11
BE11	0.605	111	0.704297	±09	5.05	26.2	0.1166	0.512763	±11

Abundances of Rb, Sr, Sm and Nd were determined using the isotope dilution method.

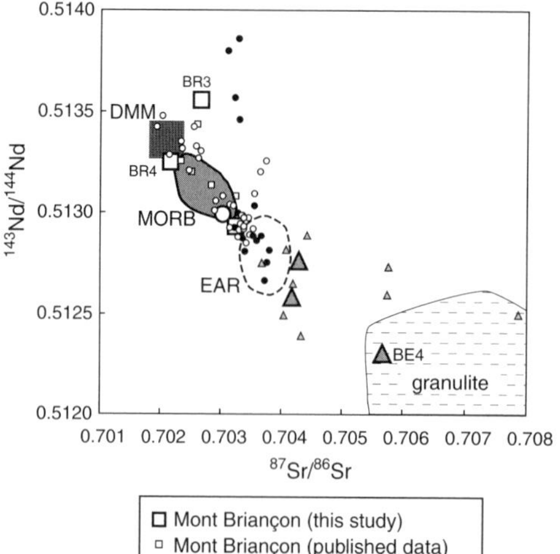

Fig. 7. $^{87}Sr/^{86}Sr$ v. $^{143}Nd/^{144}Nd$ isotopic ratios of clinopyroxene within ultramafic xenoliths from the French Massif Central. Literature data for Ray Pic (Downes & Dupuy 1987; Zangana *et al.* 1997), Mont Briançon (Downes & Dupuy 1987; Touron *et al.* 2008), other northern domain (Downes *et al.* 2003) and Puy Beaunit (Downes & Dupuy 1987; Downes *et al.* 2003) xenoliths plotted with small symbols for comparison. Fields of Pacific MORB (Hofmann 1997), EAR (European asthenospheric reservoir: Granet *et al.* 1995) and granulite of Massif Central (Downes 1984; Downes *et al.* 1990) are also shown.

pattern, contains the most enriched isotopic signature among the southern domain xenoliths investigated in the present study.

The least refractory of the investigated xenoliths from Mont Briançon is BR4, characterized by a spinel Cr-number of 9 and an olivine composition of Fo_{90}. These values approach the compositions of the most fertile mantle (spinel Cr-number 8, Fo_{88}) reported by Arai (1987), suggesting that this mantle domain was affected by a low degree of partial melting and melt extraction. BR4 also appears to be unmetasomatized ($^{87}Sr/^{86}Sr = 0.70217$; no Pb and Sr anomalies in the trace element pattern, Fig. 5a), making it an ideal sample to use in calculating a Nd model age of the depletion processes that accompanied the extraction of melts. Using values of $^{143}Nd/^{144}Nd = 0.513114$ and $^{147}Sm/^{144}Nd = 0.222$ for present-day depleted mantle (Michard *et al.* 1985), we calculated a model age (relative to depleted mantle) of 330 Ma, consistent with the partial melting age (313–377 Ma) estimated based on Hf model ages from the northern domain xenoliths by Wittig *et al.* (2006). This result indicates that xenoliths from both the northern and southern domains were affected by partial melting processes during the Variscan tectono-magmatic cycle.

Northern domain (Puy Beaunit)

The cpx of peridotites and the pyroxenite with or without phlogopite from Puy Beaunit has a relatively enriched isotopic signature compared with those of other xenoliths from the northern and southern domains (Fig. 7). The $^{87}Sr/^{86}Sr$ value obtained for cpx from sample BE4 (0.70568) is higher than that for cpx from a phlogopite-bearing xenolith from Puy Beaunit (0.70433, Bt1: Downes & Dupuy 1987). The $^{143}Nd/^{144}Nd$ ratio of cpx from the phlogopite-bearing BE4 lherzolite (0.51231) is similar to that of cpx from the phlogopite-bearing sample Bt1 (0.51239) studied by Downes & Dupuy (1987). This Nd isotopic ratio of BE4 cpx (0.51231) is lower than values published previously for cpx from phlogopite-free Puy Beaunit xenoliths ($^{143}Nd/^{144}Nd = 0.51249$–0.51289: Downes & Dupuy 1987; Downes *et al.* 2003).

Discussion

Partial melting and melt extraction resulting in chemical variations of minerals

The obtained variations in the major element chemistry, trace element patterns and Sr–Nd isotopic compositions of cpx from the investigated mantle xenoliths are indicative of partial melting and melt extraction, as well as metasomatism. The Mg- and Cr-number of spinels increases with increasing depletion, whereas the Al_2O_3 and HREE content of cpx decreases with increasing depletion (e.g. Dick & Bullen 1984; Hellebrand *et al.* 2001). The negative correlation between Al_2O_3 concentration and Mg-number for cpx, and the positive correlation between Mg- and Cr-number for spinels, within peridotite xenoliths from the French Massif Central (Fig. 3) suggest that the xenoliths are residues from various degrees of partial melting, as previously suggested by Downes *et al.* (2003).

Hellebrand *et al.* (2001) reported that HREE concentrations from cpx of abyssal peridotites show a strong correlation with the Cr-number of coexisting spinel, and proposed a factor that represents the degree of partial melting (F) based on HREE concentrations of cpx and an experimentally determined cpx–melt distribution coefficient for individual trace elements based on a fractional melting model. The authors obtained the equation $F = 10 \times \ln[\text{molar Cr}/(\text{Cr} + \text{Al}) \text{ in spinel}] + 24$

by a logarithmic fit based on a plot of F v. Cr-number of spinel. This equation is applicable to samples with Cr-numbers ranging from 10 to 60 (Hellebrand *et al.* 2001). We calculated F using the above equation, yielding values of <10%, <2% and 11–16% for peridotite xenoliths from Mont Briançon, Ray Pic and Puy Beaunit, respectively. We also applied this equation to previously published data (Werling & Altherr 1997; Zangana *et al.* 1997), obtaining F values of <17% for the northern domain and <18% for the southern domain. These results indicate that xenoliths from the French Massif Central represent residual mantle from which varying amounts of melt have been extracted. The range of F values obtained for the northern domain overlaps with that obtained for the southern domain; however, the F value calculated based on the average spinel Cr-number of xenoliths from the southern domain (*c.* 8%) is lower than that for xenoliths from the northern domain (*c.* 14%). We also calculated the Y and Yb abundances of residual peridotitic cpx in the spinel stability field using the fractional melting model of Johnson *et al.* (1990), and plotted the results, together with measured data, in Figure 8. This modelling requires less than 5% and 10% partial melting for the southern and northern domains, respectively. These F values are lower than those calculated based on the method of Cr-number in spinels. Perinelli *et al.* (2008) observed similar difference of F values between HREE contents of cpx and spinel methods in their study of Hyblean peridotite xenoliths and explained this difference in terms of slightly modified HREE contents of cpx by metasomatic process.

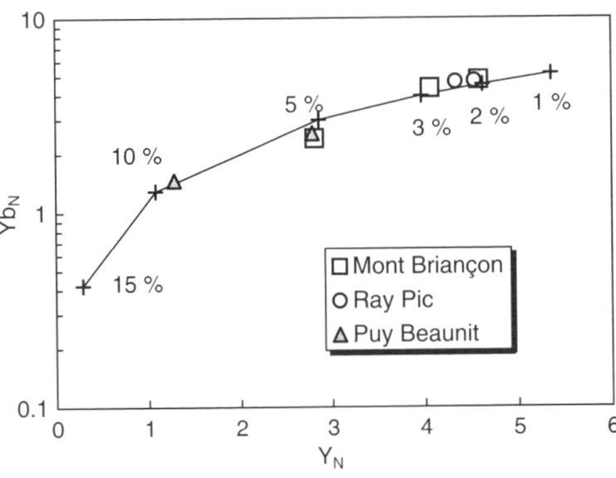

Fig. 8. Y_N v. Yb_N for clinopyroxene in the French Massif Central xenoliths. The subscript N indicates that elements are normalized to the primitive mantle values of McDonough & Sun (1995). The model melting curve of residual clinopyroxene after fractional melting in the spinel stability field presented by Johnson *et al.* (1990) is also shown.

The above results are consistent with F values reported previously for xenoliths from the southern (<8%) and northern (<25%) domains based on the concentrations of Y and Yb within cpx, using the melting model of Norman (1998; see Downes *et al.* 2003) and based on the Lu–Hf systematics of cpx (Wittig *et al.* 2006). However, the Sr–Nd isotopic ratios, as well as the LILE and LREE variations, observed in cpx from most of the Massif Central xenoliths cannot be explained solely in terms of partial melting and melt extraction, and time-integration thereafter; subsequent enrichment processes are required, as indicated in the previous studies (e.g. Downes & Dupuy 1987; Downes *et al.* 2003).

Multiple metasomatic components

Previous petrographical and geochemical studies of ultramafic xenoliths from the Massif Central have proposed several metasomatic processes (e.g. Mercier & Nicolas 1975; Downes & Dupuy 1987; Zangana *et al.* 1997; Xu *et al.* 1998; Féménias *et al.* 2004). As mentioned above, the xenoliths display at least two distinct types of LILE and LREE enrichment, as indicated by the trace element patterns (Fig. 6). We calculated the equilibrium melts with cpx based on analyses of LILE- and LREE-enriched samples (BR1, BE1, BE4 and BE11: see Fig. 9), with the aim of evaluating the nature of the metasomatic component using the partition coefficients between cpx and melt (Chalot-Prat & Boullier 1997). The melt calculated based on the cpx of BR1 has a trace element pattern similar to that of the primitive mafic magmas from the Massif Central (Fig. 9a) (Wilson & Patterson 2001). Sr–Nd isotopic compositions of BR1 cpx plot near the European asthenospheric reservoir (EAR) field (Fig. 7). The EAR was proposed as a common sub-lithospheric end member based on Pb–Nd isotopic data from primitive mafic volcanic rocks of the western and central European volcanic province (Granet *et al.* 1995). This end member is considered to represent the isotopic composition of the mantle plume source beneath the European province (Granet *et al.* 1995). Our calculation results and observations indicate that the metasomatic component of the southern domain is possibly an asthenospheric melt from the EAR.

The trace element patterns of modelled melts based on the cpx of Puy Beaunit xenoliths are generally consistent with those of glasses produced by carbonatite- and K-silicate-metasomatism (Fig. 9b) (Coltorti *et al.* 2000). This result is consistent with previous observations based on compositions of major elements of glass within Puy Beaunit xenoliths (see the earlier section on 'Major element compositions of minerals'). Such a

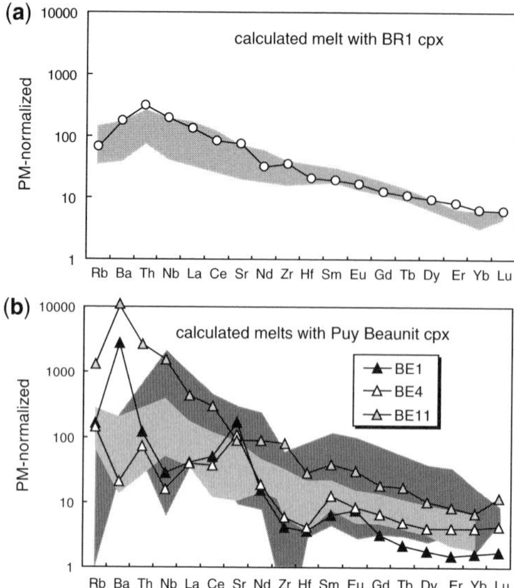

Fig. 9. Calculated trace element compositions of melts in equilibrium with clinopyroxenes in (**a**) the Mont Briançon BR1 xenolith and (**b**) the Puy Beaunit xenoliths from the French Massif Central. The compositions are calculated using the partition coefficient of Charot-Prat & Boullier (1997). The values are normalized to primitive mantle values from McDonough & Sun (1995). The middle grey field for primitive mafic magma of the Massif Central (Wilson & Patterson 2001) in (a), and the dark and light grey fields for glasses related to carbonatite- and K-silicate metasomatism (Coltorti *et al.* 2000) in (b), are shown for comparison.

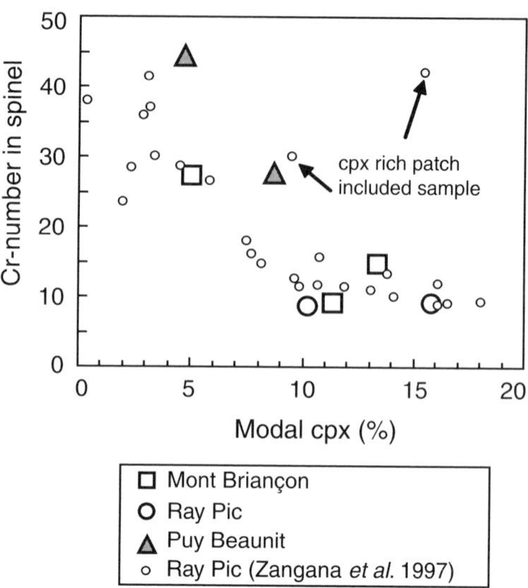

Fig. 10. Modal content of clinopyroxene v. Cr-number in spinel of the French Massif Central xenoliths compared with previously published data for Ray Pic xenoliths (Zangana *et al.* 1997).

silicate-rich carbonatite metasomatism has been proposed for peridotite xenoliths from oceanic islands by Schiano *et al.* (1994) and Neumann *et al.* (2002).

The simple partial melting and melt extraction generally result in a negative correlation between the Cr-number in spinel and modal % of cpx (Fig. 10); the Puy Beaunit xenoliths deviate from this trend, indicating an increase in the modal composition of cpx at a constant Cr-number. Similar increases in the modal composition of cpx have been reported for Ray Pic xenoliths containing cpx-rich patches (Zangana *et al.* 1997). An influx of carbonatite melt into peridotites and subsequent decarbonation reaction acts to consume opx and produce cpx and CO_2-rich fluids (e.g. Yaxley *et al.* 1991). These features of the trace element and modal composition data are not linked to the occurrence of phlogopite and glass but they do indicate, in combination with isotopic characteristics, that the Puy Beaunit xenoliths experienced modal metasomatism related to an isotopically-enriched silicate-rich carbonatite melt. Although the LILE and LREE enrichments, and HFSE depletion, of cpx of the Puy Beaunit xenoliths might suggest metasomatism by a fractionated fluid–melt from a host basanite and kimberlite (Moine *et al.* 2001; Grégoire *et al.* 2003), the relatively high Sr and low Nd isotopic compositions of the cpx suggest that the metasomatic agents are not EAR-derived magmas.

Phlogopite from Puy Beaunit xenoliths has relatively low values of $TiO_2/(TiO_2 + K_2O + Al_2O_3)$ (<0.12: Wilson & Downes 1991). This low-Ti character relative to phlogopite from other continental mantle xenoliths (>0.2: Arai & Ishimaru 2008) suggests the involvement of subduction-related aqueous fluids (e.g. Downes *et al.* 2004; Arai & Ishimaru 2008). The infiltration of fluid-rich silicate-rich melts, formed via a reaction between peridotite and basalt or carbonatite melts, has previously been proposed for explaining the formation of glass-bearing amphibole and/or phlogopite veins (e.g. Wulff-Pedersen *et al.* 1999; Moine *et al.* 2001). These earlier studies noted pronounced Ti fractionation during interaction based on the observed occurrence of high-Ti/low-Si hydrous minerals in thick veins and low-Ti/high-Si hydrous minerals in thin veins. Such a fractionation process could generate the low-Ti phlogopites found in the Puy Beaunit xenoliths. In the case of amphibole, we identified a metasomatic agent based on Nb concentration rather than Ti, because variations in Nb concentration arising from percolating melt are less pronounced than variations in Ti (Coltorti *et al.* 2007*a*, *b*). The Nb content of phlogopite in sample BE4 is 8 ppm (Table 4), clearly lower than that of veined phlogopite of intraplate xenoliths (15–180 ppm: Ionov *et al.* 1997; Coltorti *et al.* 2007*a*) and within the range of phlogopite generated by subduction-related mantle metasomatism; for example, the Finero peridotite complex and peridotite xenoliths from Papua New Guinea (0.43–55 ppm: Zanetti *et al.* 1999; Grégoire *et al.* 2001; Coltorti *et al.* 2007*a*). These findings suggest that phlogopite within sample BE4 is likely to have formed as the result of metasomatism by a slab-derived agent. The positive Sr and Pb anomalies observed in cpx from most Puy Beaunit xenoliths can be explained by the influence of a metasomatic agent derived from a subducting slab, as slab-derived fluids are enriched in Sr and Pb relative to REEs (e.g. Kogiso *et al.* 1997; Shibata & Nakamura 1997). Walter *et al.* (2008) indicated that trace-element-enriched carbonatite melt may form by a small degree of partial melting of subducted carbonated oceanic crust. Likewise mantle metasomatism via carbonated silicate melt derived from a subducting slab has also been proposed for xenoliths from Patagonia (Laurora *et al.* 2001) and the Finero complex, Italy (Morishita *et al.* 2003). The trace element and isotopic signature of Puy Beaunit xenoliths may indicate modal metasomatism by a silicate-rich carbonatite melt derived from a subducted slab, with phlogopite produced by a fluid- and silicate-rich melt separated from the silicate-rich carbonatite melt. Previous studies have also inferred that the French Massif Central was situated within the mantle wedge in a Variscan subduction zone (Shaw *et al.* 1993; Pin & Paquette 1997) and that the mantle metasomatism was caused by a hydrous component enriched in LILEs and LREEs (northern French Massif Central: Féménias *et al.* 2003, 2004), and by the partial melting of phosphatic sediments enriched in CaO, LILEs and LREEs (southern French Massif Central: Rosenbaum *et al.* 1997).

Mechanism of glass formation

Glass associated with phlogopite is found in xenoliths (including BE4) from Puy Beaunit (Wilson & Downes 1991). In previous studies, the origin of silicate glasses in mantle xenoliths has largely been discussed using mineral compositions and petrography (e.g. Frey & Green 1974; Garcia & Presti 1987; Zinngrebe & Foley 1995). Coltorti *et al.* (2000) reported the following two origins for such glasses: (1) reactions between primary minerals and an infiltrating agent in an open system; and (2) partial melting of peridotites with or without C–H–O fluids in a closed system (Draper & Green 1997). In the present study the BE4 glass is not linked to host lava. In addition, the Sr–Nd isotopic compositions of cpx of BE4 ($^{87}Sr/^{86}Sr = 0.7057$, $^{143}Nd/^{144}Nd = 0.51231$) have more enriched signatures (i.e. higher $^{87}Sr/^{86}Sr$ and lower $^{143}Nd/^{144}Nd$ ratios) than those of Massif Central magmas ($^{87}Sr/^{86}Sr = 0.7031–0.7056$, $^{143}Nd/^{144}Nd = 0.51299–0.51267$: Chauvel & Jahn 1984; Downes 1984); therefore, infiltration of the host magma played no role in the formation of the glass-bearing vein. Alternative metasomatic agents include silica-undersaturated melt and silica-oversaturated melt–fluid (Ishimaru & Arai 2005); however, percolation of a silica-undersaturated melt is unlikely to have produced the vein because the glass-bearing vein from sample BE4 is devoid of secondary cpx, which is usually precipitated during the interaction between the silica-undersaturated melt and primary opx (e.g. Yaxley *et al.* 1991). Infiltration of silica-saturated melt–fluid is also unlikely because the interaction between such a melt and peridotites results in the precipitation of opx or feldspar (e.g. Kepezhinskas *et al.* 1995; Arai & Kida 2000), neither of which is observed in the glass-bearing vein of sample BE4. The rounded shape of phlogopite, in combination with the mass-balance equation presented earlier (see the section 'Major element compositions of minerals'), suggests decompression melting of the phlogopite during eruption. Wilson & Downes (1991) suggested a mechanism of incongruent melting when explaining the formation of glasses associated with amphibole and/or phlogopite in xenoliths from the Massif Central. They also suggested that incongruent melting was possibly a result of extension-related decompression. We therefore conclude that formation of the phlogopite vein predated formation of the glass. Yaxley *et al.* (1997) reported that glass composition is related to the nature of the associated hydrous silicate phases. The extremely low Na concentration measured for the BE4 glass suggests the absence of amphibole at the time of glass formation.

Depth of origin of the investigated mantle xenoliths

Previous studies have examined the relationships between equilibration temperature, deformation textures and geochemistry of ultramafic xenoliths from a single volcanic vent at Ray Pic (Zangana *et al.* 1997) and Borée (Xu *et al.* 1998) in the French Massif Central. Zangana *et al.* (1997) reported that Ray Pic xenoliths show variations in texture from protogranular to porphyroclastic and equigranular, similar to the trends observed for other xenoliths from the Massif Central; in contrast, the Borée xenoliths show a predominant poikiloblastic texture (Xu *et al.* 1998). Zangana *et al.* (1997)

suggested that protogranular or protogranular–porphyroclastic (undeformed) xenoliths record relatively high equilibration temperatures (>900 °C), and that cpx within such xenoliths is enriched in LREEs and Sr–Nd isotopic compositions; in contrast, porphyroclastic and equigranular (deformed) xenoliths record lower equilibration temperatures (<900 °C) and have depleted LREE and isotopic signatures (Fig. 11). In contrast to the findings of Zangana *et al.* (1997), our samples show no clear relationships between texture, temperature and isotopic composition.

The estimated range of depths recorded by the Mont Briançon, Ray Pic and Puy Beaunit xenoliths is 21.2–39.8 km. In these areas, the thickness of the lithosphere is 50–70 km, as estimated by seismic tomography (Sobolev *et al.* 1996); therefore, all of the xenoliths are derived from lithospheric mantle. The present-day depth of the Moho discontinuity beneath the Massif Central is 28–29 km, as determined by seismological methods (Zeyen *et al.* 1997); accordingly, we suggest that xenoliths from the southern domain (Mont Briançon and Ray Pic) originated from the uppermost mantle, whereas the Puy Beaunit xenoliths originated from shallower depths.

A previous seismic-refraction study revealed that there is as much as 4 km of local crustal thinning beneath the western half of the Limagne graben (Zeyen *et al.* 1997). Puy Beaunit is located close to this site of crustal thinning, suggesting that the xenoliths may have been derived from a region characterized by a relatively shallow Moho. Thus, we consider that the Puy Beaunit xenoliths originated from the vicinity of the Moho discontinuity, in a zone characterized by extreme and localized crustal thinning.

Based on geochemical and textural characteristics, previous studies also concluded that Puy Beaunit xenoliths originated from near the Moho discontinuity (e.g. Mercier & Nicolas 1975; Downes & Dupuy 1987; Féménias *et al.* 2003). The depth estimates obtained in this study, based on the residual pressure of CO_2 fluid inclusions, strongly support these previous conclusions. The finding of a relationship between depth of origin and Sr–Nd isotopic composition for the investigated xenoliths (Fig. 12) suggests that, compared with deeper levels, the shallow mantle is isotopically enriched in silicate-rich carbonatite melt (Fig. 7), depleted in mineral major elements, and characterized by enriched LILE and LREE patterns

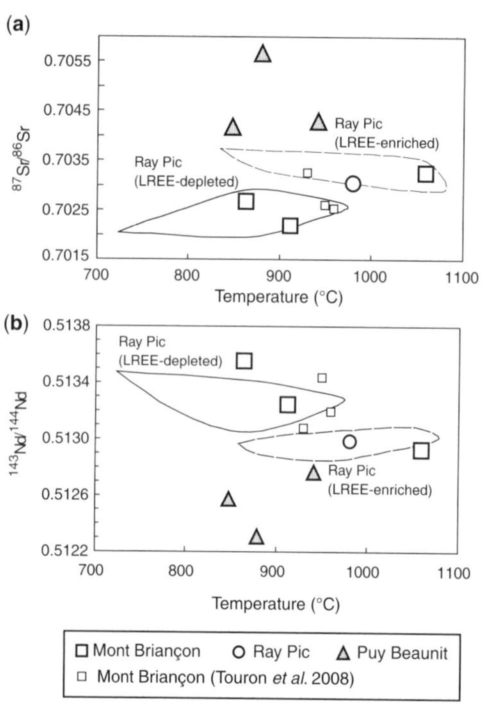

Fig. 11. Equilibrated temperature v. (**a**) $^{87}Sr/^{86}Sr$ and (**b**) $^{143}Nd/^{144}Nd$ of the French Massif Central xenoliths. Published data from Mont Briançon are reported with small symbols (Touron *et al.* 2008). Ray Pic xenoliths (Zangana *et al.* 1997) are also shown (as compositional field) for comparison.

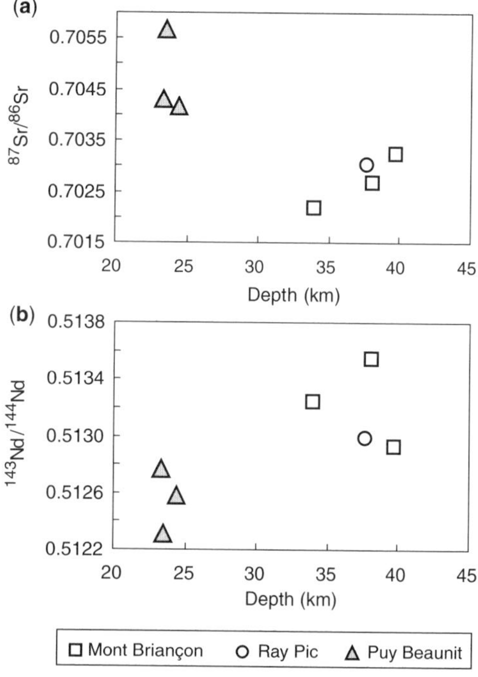

Fig. 12. Depth v. (**a**) $^{87}Sr/^{86}Sr$ and (**b**) $^{143}Nd/^{144}Nd$ of the French Massif Central xenoliths.

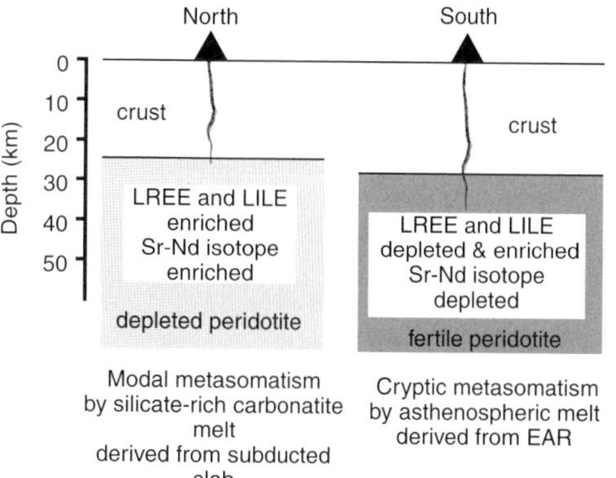

Fig. 13. Sketch showing the geochemical characteristics for the northern and southern domains beneath the French Massif Central.

(Fig. 13). The decreasing contribution of the silicate-rich carbonatite melt to greater mantle depths was influential in producing the observed chemical trends beneath the Massif Central.

Conclusions

The application of a Raman spectroscopic geobarometer and the two-pyroxene thermometer yield *P–T* conditions of 860–1060 °C at 0.92–1.10 GPa (33.3–39.8 km depth) for ultramafic xenoliths from Mont Briançon, 930–980 °C at 0.89–1.04 GPa (32.1–37.6 km depth) for those from Ray Pic and 840–940 °C at 0.59–0.71 GPa (21.2–25.6 km depth) for those from Puy Beaunit. The estimated geothermal gradient beneath the French Massif Central is steeper than that predicted by the conventional conductive model, thereby providing support for the occurrence of lithospheric thinning (Werling & Altherr 1997).

The values of Cr-number in spinel, and the abundance of Y and Yb in cpx, from xenoliths from the French Massif Central indicate various degrees of partial melting. The mineral major element chemistry, and trace element and Sr–Nd isotopic compositions of cpx, within the BR4 xenolith from Mont Briançon (the least-depleted/unmetasomatized lherzolite) yield a depleted mantle (DM) Nd model age of 330 Ma. This finding is consistent with the partial melting age based on Hf model ages obtained by Wittig *et al.* (2006), indicating that most mantle domains of the Massif Central were affected by tectono-magmatic processes during the Variscan orogenic cycle.

Trace element compositions of melt in equilibrium with cpx and Sr–Nd isotopic compositions of cpx, in combination with the modal compositional data, suggest the influence of two distinct metasomatic agents and styles: (1) cryptic metasomatism with a component related to the European asthenospheric reservoir (Granet *et al.* 1995); and (2) modal metasomatism involving a silicate-rich carbonatite melt related to the subducted slab. Overall, the relatively shallow mantle beneath the northern domain is depleted in major elements, and enriched in Sr–Nd isotopic compositions, LILEs and LREEs. These features reflect the metasomatism by a silicate-rich carbonatite melt. Relatively deep mantle beneath the southern domain consists of fertile peridotites and is isotopically depleted, reflecting metasomatism mainly by asthenospheric melts.

We thank M. Hamada for his assistance in electron microprobe analysis, H. Ishibashi, K. Matsukage, K. Nishimura and K. Niida for their useful advice, H. Zimmermann and K. Koga for their help with sampling, M. Python and G. Chazot for their valuable helps in getting references, and K. Takemura for his encouragement. We benefited from careful and constructive reviews by G. Bianchini and H. Downes, and from comments by M. Coltorti and M. Grégoire. We thank H. Downes for the thesis data by C. De Vries in the University of Utrecht and S. Arai for the presentation file. We also thank M. Coltorti and M. Grégoire for their editorial efforts. This work was supported by Kakenhi and the Institution Collaboration program of the Institute for Geothermal Sciences, a scholarship at LMV at the Université Blaise Pascal and SAKURA program to T. Kawamoto, and a JSPS fellowship to J. Yamamoto. The ICP-MS was purchased with special financial support through the President of Kyoto University.

References

Arai, S. 1987. An estimation of the least depleted spinel peridotite on the basis of olivine-spinel mantle array. *Neues Jahrbuch für Mineralogie*, **1987**, 347–354.

Arai, S. & Kida, M. 2000. Origin of fine grained peridotite xenoliths from Iraya volcano of Batan Island, Philippines: deserpentinization or metasomatism at the wedge mantle beneath an incipient arc? *Island Arc*, **9**, 458–471.

Arai, S. & Ishimaru, S. 2008. Insights into petrological characteristics of the lithosphere of mantle wedge beneath arcs through peridotite xenoliths: a Review. *Journal of Petrology*, **49**, 665–695.

Babuska, V., Plomerova, J., Vecsey, L., Granet, M. & Achauer, U. 2002. Seismic anisotropy of the French Massif Central and predisposition of Cenozoic rifting and volcanism by Variscan suture hidden in the mantle lithosphere. *Tectonics*, **21**, 11-1–11-20.

Bertrand, C. & Mercier, J. C. 1985. The mutual solubility of coexisting ortho- and clinopyroxene: toward an absolute geothermometer for the natural system? *Earth and Planetary Science Letters*, **76**, 109–122.

Brey, G. P. & Köhler, T. 1990. Geothermobarometry in four-phase lherzolites II. Thermobarometers, and practical assessment of existing thermobarometers. *Journal of Petrology*, **31**, 1353–1378.

Chalot-Prat, F. & Boullier, A.-M. 1997. Metasomatism in the subcontinental mantle beneath the Eastern Carpathians (Romania): new evidence from trace element geochemistry. *Contributions to Mineralogy and Petrology*, **129**, 284–307.

Chauvel, C. & Jahn, B.-M. 1984. Nd–Sr isotope and REE geochemistry of alkali basalts from the Massif Central, France. *Geochimica et Cosmochimica Acta*, **48**, 93–110.

Choi, S. H., Kwon, S.-T., Mukasa, S. B. & Sagon, H. 2005. Sr–Nd–Pb isotope and trace element systematics of mantle xenoliths from Late Cenozoic alkaline lavas, South Korea. *Chemical Geology*, **221**, 40–64.

Coisy, P. & Nicolas, A. 1978. Regional structure and geodynamics of the upper mantle beneath the Massif Central. *Nature*, **274**, 429–432.

Coltorti, M., Arai, S., Bonadiman, C., Faccini, B. & Ishimaru, S. 2007*a*. Nature of metasomatizing agents in suprasubduction and intraplate settings as deduced by glass and amphibole geochemistry. *Geochimica et Cosmochimica Acta*, **71**, A184.

Coltorti, M., Beccaluva, L., Bonadiman, C., Salvini, L. & Siena, F. 2000. Glasses in mantle xenoliths as geochemical indicators of metasomatic agents. *Earth and Planetary Science Letters*, **183**, 303–320.

Coltorti, M., Bonadiman, C., Faccini, B., Grégoire, M., O'Reilly, S. Y. & Powell, W. 2007*b*. Amphiboles from suprasubduction and intraplate lithospheric mantle. *Lithos*, **99**, 68–84.

De Vries, C. 2007. *A laser ablation ICP-MS study on 25 trace elements in clinopyroxenes of spinel peridotite xenoliths: a synthesis of volcanic provinces in Europe*. MSc thesis, University of Utrecht.

Delpech, G., Grégoire, M., O'Reilly, S. Y., Cottin, J. Y., Moine, B., Michon, G. & Giret, A. 2004. Feldspar from carbonate-rich silicate metasomatism in the shallow oceanic mantle under Kerguelen Islands (South Indian Ocean). *Lithos*, **75**, 209–237.

Dick, H. J. B. & Bullen, T. 1984. Chromian spinel as a petrogenetic indicator in abyssal and alpine-type peridotites and spatially associated lavas. *Contributions to Mineralogy and Petrology*, **86**, 54–76.

Downes, H. 1984. Sr and Nd isotope geochemistry of coexisting alkaline magma series, Cantal, Massif Central, France. *Earth and Planetary Science Letters*, **59**, 321–334.

Downes, H. & Dupuy, C. 1987. Textural, isotopic and REE variations in spinel peridotite xenoliths, Massif Central, France. *Earth and Planetary Science Letters*, **82**, 121–135.

Downes, H., Dupuy, C. & Leyreloup, A. F. 1990. Crustal evolution of the Hercynian belt of Western Europe: Evidence from lower-crustal granulitic xenoliths (French Massif Central). *Chemical Geology*, **83**, 209–231.

Downes, H., Macdonal, R. *et al.* 2004. Ultramafic xenoliths from the Bearpaw Mountains, Montana, USA: Evidence for multiple metasomatic events in the lithospheric mantle beneath the Wyoming craton. *Journal of Petrology*, **45**, 1631–1662.

Downes, H., Reichow, M. K., Mason, P. R. D., Beard, A. D. & Thirlwall, M. F. 2003. Mantle domains in the lithosphere beneath the French Massif Central: trace element and isotopic evidence from mantle clinopyroxenes Mantle domains in the lithosphere beneath the French Massif Central: trace element and isotopic evidence from mantle clinopyroxenes. *Chemical Geology*, **200**, 71–87.

Draper, D. S. & Green, T. H. 1997. P–T phase relations of silicic, alkaline, aluminous mantle-xenolith glasses under anhydrous and C–O–H fluid-saturated conditions. *Journal of Petrology*, **38**, 1187–1224.

Féménias, O., Coussaert, N., Berger, J., Mercier, J.-C. C. & Demaiffe, D. 2004. Metasomatism and melting history of a Variscan lithospheric mantle domain: evidence from the Puy Beaunit xenoliths (French Massif Central). *Contributions to Mineralogy and Petrology*, **148**, 13–28.

Féménias, O., Coussaert, N., Bingen, B., Whitehouse, M., Mercier, J.-C. C. & Demaiffe, D. 2003. A Permian underplating event in late- to post-orogenic tectonic setting. Evidence from the mafic–ultramafic layered xenoliths from Beaunit (French Massif Central). *Chemical Geology*, **199**, 293–315.

Frey, F. A. & Green, D. H. 1974. The mineralogy, geochemistry and origin of lherzolite inclusions in Victorian basanites. *Geochimica et Cosmochimica Acta*, **38**, 1023–1059.

Garcia, M. O. & Presti, A. A. 1987. Glass in garnet pyroxenite xenoliths from Kauia island, Hawaii: product of infiltration of host nephelinites. *Geology*, **15**, 904–906.

Gasparik, T. 1987. Orthopyroxene thermobarometry in simple and complex system. *Contributions to Mineralogy and Petrology*, **96**, 357–370.

Granet, M., Wilson, M. & Achauer, U. 1995. Imaging a mantle plume beneath the French Massif Central. *Earth and Planetary Science Letters*, **136**, 281–296.

Green, T. H. 1994. Experimental studies of trace-element partitioning applicable to igneous petrogenesis – Sedona 16 years later. *Chemical Geology*, **117**, 1–36.

Grégoire, M., Bell, D. R. & Le Roex, A. P. 2003. Garnet lherzolites from the Kaapvaal Craton (South Africa): Trace element evidence for a metasomatic history. *Journal of Petrology*, **44**, 629–657.

Grégoire, M., McInnes, B. I. A. & O'Reilly, S. Y. 2001. Hydrous metasomatism of oceanic sub-arc mantle, Lihir, Papua New Guinea Part 2. Trace element characteristics of Slab-derived fluids. *Lithos*, **59**, 91–108.

Hellebrand, E., Snow, J. E., Dick, H. J. B. & Hofmann, A. W. 2001. Coupled major and trace elements as indications of the extent of melting in mid-ocean-ridge peridotites. *Nature*, **410**, 677–681.

Hofmann, A. 1997. Mantle geochemistry: the message from oceanic volcanism. *Nature*, **385**, 219–229.

Ionov, D. A., Griffin, W. L. & O'Reilly, S. Y. 1997. Volatile-bearing minerals and lithophile trace elements in the upper mantle. *Chemical Geology*, **141**, 153–184.

Ishimaru, S. & Arai, S. 2005. Silicic glasses rapped in peridotite xenoliths: an insight into melting and metasomatism processes in mantle peridotite. *Japanese Magazine of Mineralogical and Petrological Sciences*, **34**, 205–215.

Johnson, K. T. M., Dick, H. J. B. & Shimizu, N. 1990. Melting in the oceanic upper mantle: an ion microprobe study of diopsides in abyssal peridotites. *Journal of Geophysical Research*, **95**, 2661–2678.

KEPEZHINSKAS, P. K., DEFANT, M. J. & DRUMMOND, M. S. 1995. Na Metasomatism in the Island-Arc mantle by slab melt-peridotite interaction: evidence from mantle xenoliths in the north Kamchatka arc. *Journal of Petrology*, **36**, 1505–1527.

KOGISO, T., TATSUMI, Y. & NAKANO, S. 1997. Trace element transport during dehydration processes in the subducted oceanic crust: 1. Experiments and implications for the ocean island basalt. *Earth and Planetary Science Letters*, **148**, 198–205.

KÖHLER, T. P. & BREY, G. P. 1990. Calcium exchange between olivine and clinopyroxene calibrated as a geothermobarometer for natural peridotites from 2 to 60 kb with applications. *Geochimica et Cosmochimica Acta*, **54**, 2375–2388.

LAURORA, A., MAZZUCCHELLI, M., RIVALENTI, G., VANNUCCI, R., ZANETTI, A., BARBIERI, M. A. & CINGOLANI, C. A. 2001. Metasomatism and melting in carbonated peridotite xenoliths from the mantle wedge: the Gobernador Gregores Case (southern Patagonia). *Journal of Petrology*, **42**, 69–87.

LEE, D.-C., HALLIDAY, A. N., DAVIES, G. R., ESSENE, E. J., FITTON, G. & TEMDJIM, R. 1996. Melt enrichment of shallow depleted mantle: a detailed petrological, trace element and isotopic study of mantle-derived xenoliths and megacrysts from the Cameroon Line. *Journal of Petrology*, **37**, 415–441.

LENOIR, X., GARRIDO, C. J., BODINIER, J.-L. & DAUTRIA, J.-M. 2000. Contrasting lithospheric mantle domains beneath the Massif Central (France) revealed by geochemistry of peridotite xenoliths. *Earth and Planetary Science Letters*, **181**, 359–375.

MATTE, P. 1986. Tectonics and plate tectonics model for the Variscan belt of Europe. *Tectonophysics*, **126,** 329–374.

MCDONOUGH, W. F. & SUN, S. S. 1995. The composition of the Earth. *Chemical Geology*, **120**, 223–253.

MERCIER, J.-C. & NICOLAS, A. 1975. Textures and fabrics of upper-mantle peridotites as illustrated by xenoliths from basalts. *Journal of Petrology*, **16**, 454–487.

MICHARD, A., GURRIET, P., SOUDANT, M. & ALBAREDE, F. 1985. Nd isotopes in French Phanerozoic shales: external v. internal aspects of crustal evolution. *Geochimica et Cosmochimica Acta*, **49**, 601–610.

MICHON, L. & MERLE, O. 2001. The evolution of the Massif Central rift: spatio-temporal, distribution of the volcanism. *Bulletin de la Société Géologique de France*, **172**, 201–211.

MOINE, B. N., GRÉGOIRE, M., O'REILLY, S. Y., SHEPPARD, S. M. F. & COTTIN, J. Y. 2001. High field strength element fractionation in the upper mantle: evidence from amphibole-rich composite mantle xenoliths from the Kerguelen Islands (Indian Ocean). *Journal of Petrology*, **42**, 2145–2167.

MORISHITA, T., ARAI, S. & TAMURA, A. 2003. Petrology of an apatite-rich layer in the Finero phlogopite–peridotite, Italian Western Alps; implications for evolution of a metasomatising agent. *Lithos*, **69**, 37–49.

NAKAJIMA, K. & ARIMA, M. 1998. Melting experiments on hydrous low-K tholeiite: Implications for the genesis of tonalitic crust in the Izu–Bonin–Mariana arc. *The Island Arc*, **7**, 359–373.

NEUMANN, E. R., WULFF-PEDERSEN, E., PEARSON, N. J. & SPENCER, E. A. 2002. Mantle xenoliths from Tenerife (Canary Islands): evidence for reactions betweeen mantle peridotites and silicic carbonatite melts inducing Ca metasomatism. *Journal of Petrology*, **43**, 825–857.

NORMAN, M. D. 1998. Melting and metasomatism in the continental lithosphere: laser ablation ICPMS analysis of minerals in spinel lherzolites from eastern Australia. *Contributions to Mineralogy and Petrology*, **130**, 240–255.

O'REILLY, S. & GRIFFIN, W. L. 1985. A xenolith-derived geotherm for southeastern Australia and its geophysical implications. *Tectonophysics*, **111**, 41–63.

O'REILLY, S. Y., CHEN, D., GRIFFIN, W. L. & RYAN, C. G. 1997. Minor elements in olivine from spinel lherzolite xenoliths: implications for thermobarometry. *Mineralogical Magazine*, **61**, 257–269.

PERINELLI, C., SAPIENZA, G. T., ARMIENTI, P. & MORTEN, L. 2008. Metasomatism of the upper mantle beneath the Hyblean Plateau (Sicily): evidence from pyroxenes and glass in peridotite xenoliths. *In*: COLTORTI, M. & GRÉGOIRE, M. (eds) *Metasomatism in Oceanic and Continental Lithospheric Mantle*. Geological Society, London, Special Publications, **293**, 197–221.

PIN, C. & PAQUETTE, J.-L. 1997. A mantle-derived bimodal suite in the Hercynian Belt: Nd isotope and trace element evidence for a subduction-related rift origin of the Late Devonian Brévenne metavolcanics, Massif Central (France). *Contributions to Mineralogy and Petrology*, **129**, 222–238.

POLLACK, H. H. & & CHAMPMAN, D. S. 1977. On the regional variation of heat flow, geotherms and lithospheric mantle. *Tectonophysics*, **38**, 279–296.

ROEDDER, E. 1983. Geobarometry of ultramafic xenoliths from Loihi seamount, Hawaii, on the basis of CO_2 inclusions in olivine. *Earth and Planetary Science Letters*, **66**, 369–519.

ROEDDER, E. 1984. *Fluid Inclusions. Reviews in Mineralogy*, **12**. Mineralogical Society of America.

ROSENBAUM, J. M., WILSON, M. & CONDLIFFE, E. 1997. Partial melts of subducted phosphatic sediments in the mantle. *Geology*, **25**, 77–80.

SANTOSH, M. & OMORI, S. 2008. CO_2 flushing: A plated tectonic perspective. *Gondowana Research*, **13**, 86–102.

SAPIENZA, G., HILTON, D. R. & SCRIBANOA, V. 2005. Helium isotopes in peridotite mineral phases from Hyblean Plateau xenoliths (south-eastern Sicily, Italy). *Chemical Geology*, **219**, 115–129.

SCHIANO, P., CLOCCHIATTI, R., SHIMIZU, N., WEIS, D. & MATTIELLI, N. 1994. Cogenetic silica-rich and carbonate-rich melts trapped in mantle minerals in Kerguelen ultramafic xenoliths: implications for metasomatism in the oceanic upper mantle. *Earth and Planetary Science Letters*, **123**, 167–178.

SHAW, A., DOWNES, H. & THIRLWALL, M. F. 1993. The quartz-diorites of Limousin: Elemental and isotopic evidence for Devono-Carboniferous subduction in the Hercynian belt of the French Massif Central. *Chemical Geology*, **107**, 1–18.

SHIBATA, T. & NAKAMURA, E. 1997. Across-arc variations of isotope and trace element compositions from Quaternary basaltic volcanic rocks in northeastern

Japan: implications for interaction between subducted oceanic slab and mantle wedge. *Journal of Geophysical Research*, **102**, 8051–8064.

SHIBATA, T. & YOSHIKAWA, M. 2004. Precise isotopic determination of trace amounts of Nd in magnesium-rich samples. *Journal of Mass Spectrometry*, **52**, 317–324.

SOBOLEV, S. V., ZEYEN, H., STOLL, G., WERLING, F., ALTHERR, R. & FUCHS, K. 1996. Upper mantle temperatures from teleseismic tomography of French Massif Central including effects of compositions, mineral reaction, anharmonicity, anelasticity and partial melt. *Earth and Planetary Science Letters*, **139**, 147–163.

TOURON, S., RENAC, C., O'REILLY, S. Y., COTTIN, J.-Y. & GRIFFIN, W. L. 2008. Characterization of the metasomatic agent in mantle xenoliths from Deves, Massif Central (France) using couples *in situ* trace-element and O, Sr and Nd isotopic compositions. *In*: COLTORTI, M. & GRÉGOIRE, M. (eds) *Metasomatism in Oceanic and Continental Lithospheric Mantle*. Geological Society, London, Special Publications, **293**, 177–196.

VARELA, M. E., CLOCCHIATTI, R., KURAT, G. & SCHIANO, P. 1999. Silicic glasses in hydrous and anhydrous mantle xenoliths from Western Victoria, Australia: at least two different sources. *Chemical Geology*, **153**, 151–169.

WALTER, M. J., BULANOVA, G. P. *ET AL*. 2008. Primary carbonatite melt from deeply subducted oceanic crust. *Nature*, **454**, 622–626.

WELLS, P. R. 1977. Pyroxene thermometry in simple and complex systems. *Contributions to Mineralogy and Petrology*, **62**, 129–139.

WERLING, F. & ALTHERR, R. 1997. Thermal evolution of the lithosphere beneath the French Massif Central as deduced from geothermobarometry on mantle xenoliths. *Tectonophysics*, **275**, 119–141.

WILSON, M. & DOWNES, H. 1991. Tertiary–Quaternary extension-related alkaline magmatism in Western and Central Europe. *Journal of Petrology*, **32**, 811–849.

WILSON, M. & PATTERSON, R. 2001. Intraplate magmatism related to short-wavelength convective instabilities in the upper mantle: evidence from the Tertiary–Quaternary volcanic province of western and central Europe. Geological Society of America, Special Paper, **352**, 37–58.

WITTIG, N., BAKER, J. A. & DOWNES, H. 2006. Dating the mantle roots of young continental crust. *Geology*, **34**, 237–240.

WOOD, B. J. & BANNO, S. 1973. Garnet–orthopyroxene and orthopyroxene–clinopyroxene relationships in simple and complex system. *Contributions to Mineralogy and Petrology*, **42**, 109–124.

WULFF-PEDERSEN, E., NEUMANN, E. R., VANNUCCI, R., BOTTAZZI, P. & OTTOLINI, L. 1999. Silicic melts produced by reaction between peridotite and infiltrating basaltic melts: ion probe data on glasses and minerals in veined xenoliths from La Palma, Canary Islands. *Contributions to Mineralogy and Petrology*, **137**, 59–82.

XU, X., O'REILLY, S. Y., ZHOU, X. & GRIFFIN, W. L. 1996. A xenolith-derived geotherm and the crust-mantle boundary at Qilin, southeastern China. *Lithos*, **38**, 41–62.

XU, Y.-G., MENZIES, M., BODINIER, J.-L., BEDINI, R., VROON, P. & MERCIER, J.-C. 1998. Melt percolation and reaction atop a plume: evidence from the poikiloblastic peridotite xenoliths from Boree (Massif Central, France). *Contributions to Mineralogy and Petrology*, **132**, 65–84.

YAMAMOTO, J. & KAGI, H. 2006. Extended micro-Raman densimeter for CO_2 applicable to mantle-originated fluid inclusions. *Chemistry Letters*, **35**, 610–611.

YAMAMOTO, J. & KAGI, H. 2008. Application of micro-Raman densimeter for CO_2 fluid inclusions: a probe for elastic strengths of mantle minerals. *European Journal of Mineralogy*, **20**, 529–535.

YAMAMOTO, J., ANDO, J., KAGI, H., INOUE, T., YAMADA, A., YAMAZAKI, D. & IRIFUNE, T. 2008. *In situ* strength measurements on natural upper-mantle minerals. *Physics and Chemistry of Minerals*, **35**, 249–257.

YAMAMOTO, J., KAGI, H., KANEOKA, I., LAI, Y., PRIKHOD'KO, V. S. & ARAI, S. 2002. Fossil pressures of fluid inclusions in mantle xenoliths exhibiting rheology of mantle minerals: implications for the geobarometry of mantle minerals using micro-Raman spectroscopy. *Earth and Planetary Science Letters*, **198**, 511–519.

YAMAMOTO, J., KAGI, H., KAWAKAMI, Y., HIRANO, N. & NAKAMURA, M. 2007. Paleo-Moho depth determined from the pressure of CO_2 fluid inclusions: Raman spectroscopic barometry of mantle- and crust-derived rocks. *Earth and Planetary Science Letters*, **253**, 369–377.

YAXLEY, G. M., CRAWFORD, A. J. & GREEN, D. H. 1991. Evidence for carbonatite metasomatism in spinel peridotite xenoliths from western Victoria, Australia. *Earth and Planetary Science Letters*, **107**, 305–317.

YAXLEY, G. M., KAMENETSKY, V., GREEN, D. H. & FALLOON, T. J. 1997. Glasses in mantle xenoliths from western Victoria, Australia, and their relevance to mantle processes. *Earth and Planetary Science Letters*, **148**, 433–446.

YOKOYAMA, T., MAKISHIMA, A. & NAKAMURA, E. 1999. Evaluation of the coprecipitation of incompatible trace elements with fluoride during silicate rock dissolution by acid digestion. *Chemical Geology*, **157**, 175–187.

YOSHIKAWA, M. & NAKAMURA, E. 1993. Precise isotope determination of trace amounts of Sr in magnesium-rich samples. *Journal of Mineralogy, Petrology and Economic Geology*, **88**, 548–561.

YOSHIKAWA, M. & NAKAMURA, E. 2000. Geochemical evolution of the Horoman peridotite complex: Implications for melt extraction, metasomatism and compositional layering in the mantle. *Journal of Geophysical Research*, **105**, 2879–2901.

ZANETTI, A., MAZZUCCHELLI, M., RIVALENTI, G. & VANNUCCI, R. 1999. The Finero phlogopite-peridotite massif: an example of subduction-related metasomatism. *Contributions to Mineralogy and Petrology*, **134**, 107–122.

ZANGANA, N. A., DOWNES, H., THIRLWALL, M. F. & HEGNER, E. 1997. Relationship between deformation, equilibration temperatures, REE and radiogenic isotopes in mantle xenoliths (Ray Pic, Massif Central, France): an example of plume–lithosphere interaction? *Contributions to Mineralogy and Petrology*, **127**, 187–203.

ZEYEN, H., NOVAK, O., LANDES, M., PRODEHL, C., DRIAD, L. & HIRN, A. 1997. Refraction- seismic investigations of the northern Massif Central (France). *Tectonophysics*, **275**, 99–117.

ZINNGREBE, E. & FOLEY, S. F. 1995. Metasomatism in mantle xenoliths from Gees, West Eifel, Germany: evidence for the genesis of calc-alkaline glasses and metasomatic Ca-enrichment. *Contributions to Mineralogy and Petrology*, **122**, 79–96.

Cryptic metasomatism in clino- and orthopyroxene in the upper mantle beneath the Pannonian region

G. DOBOSI[1]*, G. A. JENNER[2], A. EMBEY-ISZTIN[3] & H. DOWNES[4]

[1]*Institute for Geochemical Research, Hungarian Academy of Sciences, H-1112 Budapest, Budaörsi út 45, Hungary*

[2]*Department of Earth Sciences, Memorial University of Newfoundland, St John's, Newfoundland, Canada A1B 3X5*

[3]*Department of Mineralogy and Petrology, Hungarian Natural History Museum, H-1083 Budapest, Ludovika tér 2-6, Hungary*

[4]*School of Earth Sciences, Birkbeck College, University of London, Malet Street, London WC1E 7HX, UK*

**Corresponding author (e-mail: dobosi@geochem.hu)*

Abstract: Clino- and orthopyroxenes in anhydrous spinel peridotite xenoliths from Pliocene alkali basalts of the western Pannonian Basin have been analysed for trace elements by laser ablation inductively coupled plasma mass spectrometry (LA-ICP-MS). Clinopyroxenes show highly variable mantle normalized REE (rare earth elements) patterns but basically can be classified into three major groups: LREE-depleted, LREE-enriched and U-shaped patterns. As the REE patterns of clinopyroxenes usually reflect the REE patterns of the host peridotite, the three major REE patterns define three geochemically different groups of xenoliths. LREE-depleted xenoliths generally have undeformed protogranular textures, while the more deformed xenoliths with porphyroclastic and equigranular textures have LREE-enriched trace element patterns. The U-shaped pattern is very distinctive and is generally associated with poikilitic textures. The HREE content of the clinopyroxenes suggest that most of the xenoliths experienced less than15% partial melting, with the lowest degree occurring in the LREE-depleted xenoliths, and the highest degree in LREE-enriched xenoliths. Cryptic metasomatism frequently accompanies deformation. Metasomatic enrichment of incompatible trace elements can be observed not only in clinopyroxenes but also in coexisting orthopyroxenes. The metasomatic agents were probably alkaline mafic melts of asthenospheric origin and some may relate to upper Cretaceous alkali lamprophyre magmatism. Geochemical signatures of subduction-related melts or fluids have not been found in the anhydrous LREE-enriched xenoliths, although poikilitic xenoliths with U-shaped normalized REE patterns may indicate the influence of subduction-related melts.

The subcontinental lithospheric mantle is a rigid, non-convective layer of the upper mantle situated between the Mohorovicic discontinuity and the convective asthenospheric mantle. Different processes such as removal of melt by partial melting, addition of melt or interaction with fluids derived from subduction or asthenospheric upwelling would change its compositional characteristics and these chemical heterogeneities may be preserved for billions of years. Thus, over a long period of time the subcontinental lithospheric mantle may become chemically distinct from the rest of the Earth's mantle, especially in its trace element abundances and isotopic composition. Knowledge of the processes that changed the composition of the subcontinental lithospheric mantle is essential in understanding the geological (tectonic, igneous) evolution of a region, as well as of the long-term evolution of the mantle as a whole. Xenoliths from the mantle entrained by alkali basalts are representative samples of the lithospheric mantle. The trace element and isotope signatures of mantle xenoliths can be used to decipher the long-term chemical evolution of the subcontinental lithospheric mantle. Although the bulk trace element content of the xenoliths provides valuable information about the mantle processes, this record can be disturbed by different processes such as host magma infiltration or surficial alteration that may strongly affect the abundances of the most incompatible trace elements. *In situ* analysis of the minerals in peridotite xenoliths can help to distinguish these signatures, and provides trace element data reflecting the depletion and long-term metasomatic processes.

From: COLTORTI, M., DOWNES, H., GRÉGOIRE, M. & O'REILLY, S. Y. (eds) *Petrological Evolution of the European Lithospheric Mantle*. Geological Society, London, Special Publications, **337**, 177–194.
DOI: 10.1144/SP337.9 0305-8719/10/$15.00

This paper presents new laser ablation inductively coupled plasma mass spectrometry (LA-ICP-MS) trace element data from pyroxenes (both ortho- and clinopyroxenes) of a representative set of mantle xenoliths from the western part of the Pannonian Basin (eastern central Europe), discusses the role of melting and metasomatism in the mantle beneath the Pannonian area, and attempts to define the composition and the origin of the metasomatic agents. The representative samples have been selected from a large (>800 samples) collection of peridotite xenoliths that has been investigated petrographically during the previous decades. The present work concentrates on the anhydrous spinel peridotite xenoliths in order to study the cryptic metasomatic processes, excluding the amphibole and glass-bearing samples.

Regional distribution, geological background

The Pannonian Basin is a Tertiary extensional back-arc basin surrounded by the Eastern Alps and the Carpathian fold belt (Fig. 1). The basin shows a number of features that are characteristic of rift zones, such as high heat flow (70–110 mW m^{-2}), thin lithosphere (probably no more than 60–70 km thick), thin crust (25–30 km) and recent alkali basaltic volcanism (see Embey-Isztin *et al.* 2001 and references therein).

The formation of the Pannonian Basin is the result of two lithospheric extensional phases (Huismans *et al.* 2001). The first phase was a passive extension, which took place during the Early–Middle Miocene (18–14 Ma). The driving force of this phase was the continuous roll-back and progressive steepening of the subducted slab along the outer eastern Carpathians (Csontos *et al.* 1992; Horváth 1993; Huismans *et al.* 2001). The subduction of the European plate beneath the Carpathian area was accompanied by extensive calc-alkaline volcanism along the Carpathian Arc (Szabó *et al.* 1992; Harangi *et al.* 2007). The phase of the passive lithospheric extension and thinning was followed by a second extensional event in the Late Miocene (12–11 Ma) that affected the central part of the basin. This second extensional phase was accompanied by active asthenospheric upwelling (Huismans *et al.* 2001) and subsequent sporadic alkali basaltic volcanism (Embey-Isztin *et al.* 1993).

Several of the alkaline basalt occurrences contain ultramafic and mafic xenoliths, and discrete megacrysts of upper-mantle origin (Embey-Isztin 1978; Kurat *et al.* 1980; Embey-Isztin *et al.* 1989; Downes *et al.* 1992; Szabó *et al.* 1995; Dobosi *et al.* 1999). Large and fresh peridotite xenoliths are particularly abundant in the western part of the Pannonian Basin (Fig. 1). This study concentrates on the well-known xenolith localities of western Hungary (Szentbékkálla, Bondoróhegy and Szigliget in the Balaton Highland, and Gérce in the Lesser Hungarian Plain) and eastern Austria (Kapfenstein and Fehring in the Graz Basin). Most of the equilibrium temperature values of the studied xenoliths range between 900 and 1150 °C (Embey-Isztin *et al.* 2001), indicating that the xenolith suite represents at least a 10 km-thick vertical section of the mantle beneath the Pannonian area, assuming a uniform geothermal gradient beneath the various localities.

Petrography and textures

Peridotite xenoliths from the above-mentioned localities show a restricted mineralogical variation. Most are four-phase spinel lherzolites (diopside >5%), indicating that this is the main lithology which forms the shallow lithospheric mantle beneath the area. A smaller number of harzburgite xenoliths (diopside <5%) was also found. No trace of garnet or primary plagioclase could be detected and this strongly constrains the depth interval from which the rock fragments originate. Amphibole, the only hydrous phase, is rare and its modal abundance seldom exceeds 5% in the xenoliths. Amphibole-bearing xenoliths are relatively common in Szigliget, where 7–8% of the xenoliths may contain disseminated amphibole (Embey-Isztin *et al.* 1989, 2001), and in Kapfenstein localities (Kurat *et al.* 1980; Vaselli *et al.* 1996; Coltorti *et al.* 2007).

Texture types of peridotite xenoliths and their relative abundances at different volcanic vents have been the subject of detailed studies (Embey-Isztin 1984; Kurat *et al.* 1991). The textures show wide variations and represent the whole deformation cycle established on textural grounds by Mercier & Nicolas (1975). *Protogranular* xenoliths are coarse grained, exhibiting curvilinear grain boundaries with no or slight deformation features (Fig. 2a). They represent probably the most primitive state of the upper mantle. With increasing deformation the protogranular texture transforms into a *porphyroclastic* texture that appears to show a bimodal grain-size distribution. Large, apparently relict olivine and orthopyroxene crystals are surrounded by newly formed small grains of these same minerals, as well as those of clinopyroxene and spinel (Fig. 2b). The process is caused by dynamic recrystallization, as well as subgrain rotation, and results in a reduction in the average grain size. *Equigranular* xenoliths mark the more advanced stage of deformation. Except for some

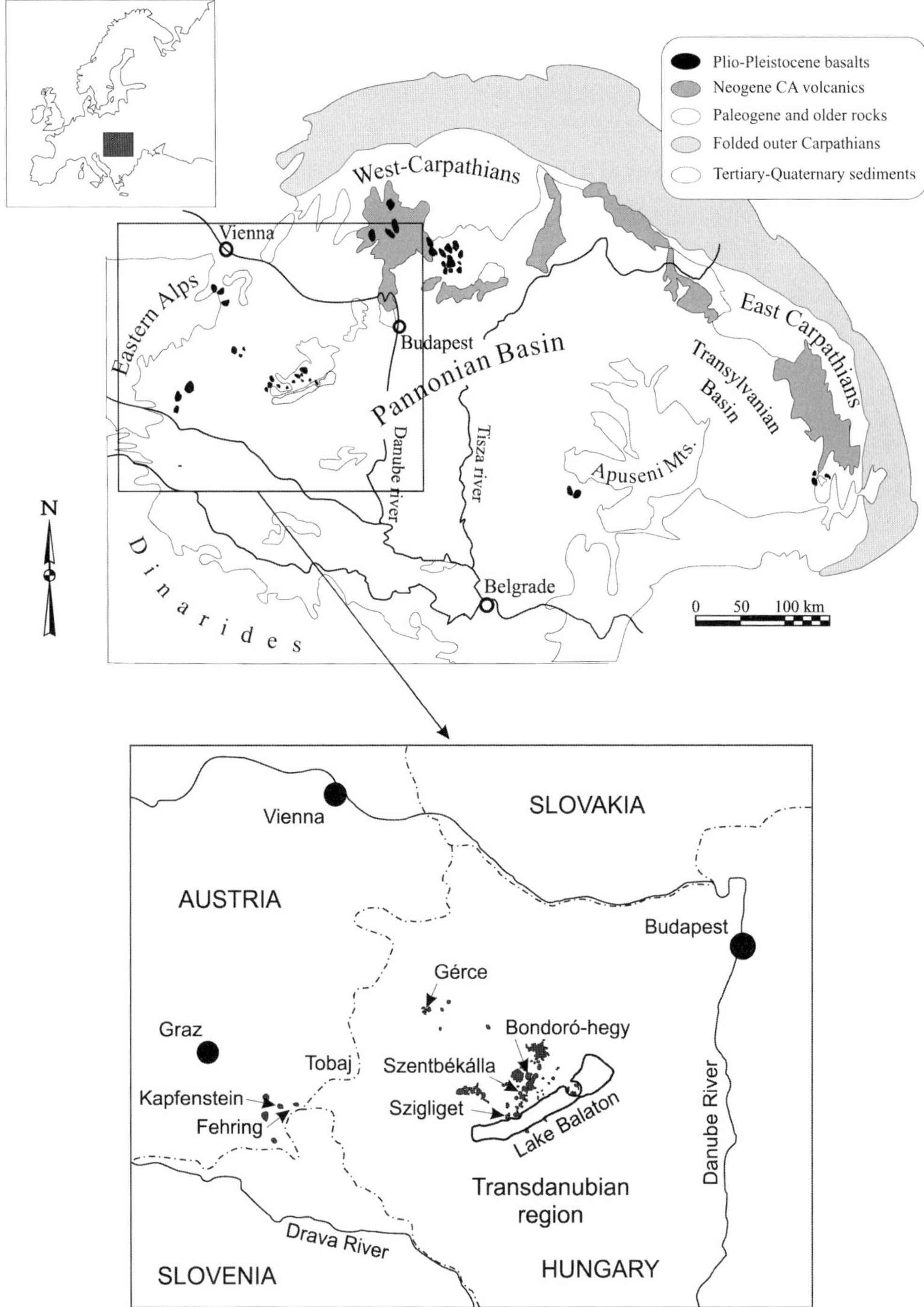

Fig. 1. Geological sketch map of the Carpathian–Pannonian region showing the alkali basalt and xenolith localities in the western Pannonian Basin.

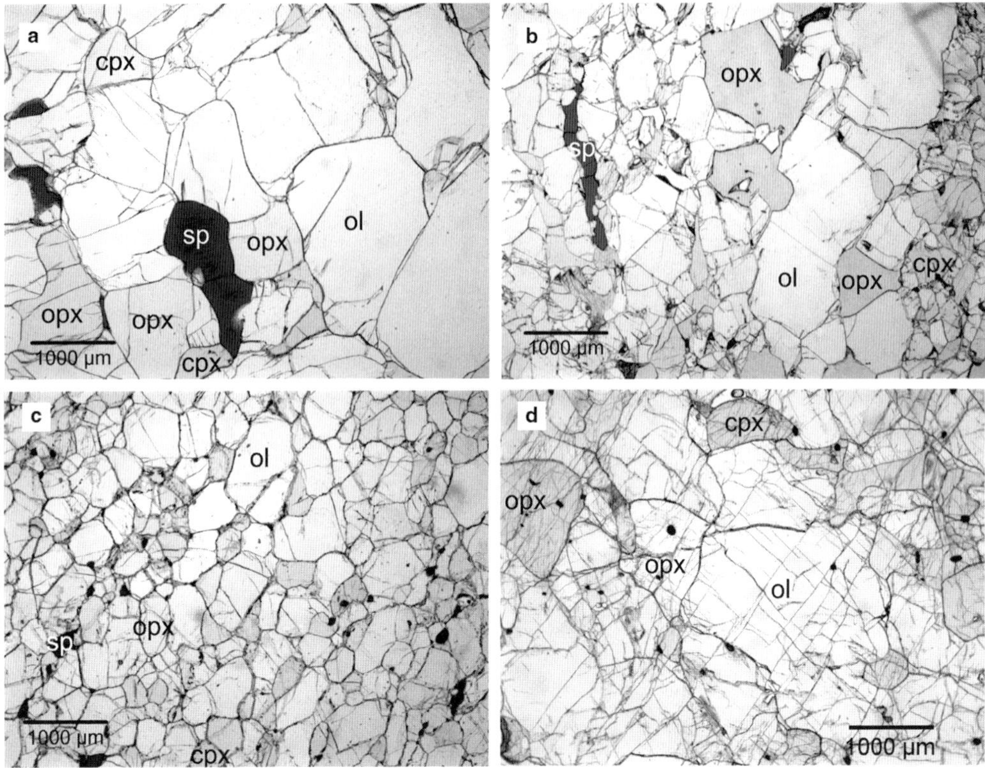

Fig. 2. Photomicrographs of the different texture types: (**a**) protogranular, (**b**) porphyroclastic, (**c**) equigranular and (**d**) poikilitic.

rare relict grains, all crystals are small and equant with typically straight grain boundaries frequently forming 120° triple junctions (Fig. 2c) pointing to a static recrystallization process. Most volcanic vents have yielded both undeformed (protogranular) and deformed (equigranular and/or porphyroclastic) peridotites; however, the proportion of these types varies between the various localities. The volcanic vent of Kapfenstein (Graz Basin) is an exception, where only untectonized protogranular xenoliths have been found. In addition to the three main textural types that constitute a series of increasing deformation from protogranular to porphyroclastic to equigranular, a fourth textural type is also common in this region. This is called poikilitic (Embey-Isztin 1984; Embey-Isztin *et al.* 1989) by virtue of having small rounded, sometimes subidiomorphic, spinel grains poikilitically enclosed in large olivine, orthopyroxene and clinopyroxene crystals (Fig. 2d). It seems that these unusual xenoliths also constitute a deformational series from a coarse-grained, undeformed to a fine-grained recrystallized variety (Embey-Isztin *et al.* 2001).

Analytical methods

Microprobe analyses of all phases in the xenoliths were carried out in three different places: at Birkbeck College in London and at the Institute for Geochemical Research in Budapest, in both places using a JEOL Superprobe 733 electron microprobe equipped with an Oxford Instruments INCA energy dispersive system), and at the University of Heidelberg, using a Cameca SX 50 equipped with five wavelength-dispersive spectrometers. The results obtained in Heidelberg were published in Embey-Isztin *et al.* (2001), but the analyses carried out in London and in Budapest are unpublished.

About 30 trace elements were determined in the ortho- and clinopyroxenes by LA-ICP-MS at the Department of Earth Sciences, Memorial University of Newfoundland in St John's. The laser ablation system used in this study is described in detail in Jackson *et al.* (1992), Jenner *et al.* (1994) and Horn *et al.* (1997). The laser beam of a Q-switched Nd:YAG laser operating 266 nm in the ultraviolet region was focused onto the sample surface through the optics of a petrographical microscope. The pulse energy of the laser beam was 0.4–0.6 mJ and the diameters of the ablation pits were about 40–60 μm. The ablated material from the sample cell was carried by an Ar gas into the plasma torch of a Fisons VG PQ2+ ICP-MS instrument. All measurements were carried out using 'time-resolved analysis' data acquisition software operating in fast, peak-jumping mode; this allowed us to monitor the progress of the laser ablation, and detect inclusions and heterogeneities. Spiked silicate glass NIST 612 was used for calibration and BCR-2 glass was used as the secondary standard. Calcium was used as an internal standard to correct the ablation yield differences between the individual analyses. Data reduction was made using the LAMTRACE© spreadsheet software written in-house by S. Jackson. In most cases the analyses were made in 100 μm-thick petrographical sections, although in some cases hand-picked clino- and orthopyroxene grains (mounted in epoxy and polished) were analysed. All spot analyses were made on clean, inclusion- and crack-free areas of the crystals, and three–four grains of each mineral were analysed (altogether between three and 10 spots). Trace element data used in this paper have not been published previously.

Mineral chemistry

Major elements

In the investigated xenoliths all minerals are homogeneous with respect to their major element content. Rare exceptions can be found in some xenoliths with porphyroclastic textures (mainly from Gérce), where the larger, relict clino- and orthopyroxene crystals may show significant zoning, especially in their Al content (Embey-Isztin *et al.* 2001). Olivine grains show rather uniform major element compositions throughout the whole xenolith suite with Mg-numbers (Fo content) between 0.88 and 0.92. In contrast, spinel compositions are highly variable and partly texture-dependent. As these two phases do not play an important role in the incompatible trace element budget of the xenoliths, they will not be regarded in the following sections.

Orthopyroxene has a similar Mg-number to the coexisting olivine, with all values lying between 0.89 and 0.92. The Al_2O_3 content varies between 1.56 and 5.65 wt%, and it is distinctly lower in orthopyroxenes of poikilitic xenoliths (1.56–4.25 wt%, average 2.93%) compared to peridotites with protogranular–porphyroclastic–equigranular textures (main series) (3.05–5.65 wt%, average 4.67%). Clinopyroxenes are chromian diopsides and they show the highest compositional variation among the silicates of the peridotites. The Mg-number varies between 0.88 and 0.94. Both Al_2O_3 and Cr_2O_3 contents are highly variable (1.20–7.70 and 0.18–2.08 wt%, respectively), and poikilitic xenoliths characteristically have a low to very low content of Al and Cr in their clinopyroxenes. In the zoned crystals Al is decreasing from core to rim in both pyroxenes, while Ca shows a smooth pattern (Embey-Isztin *et al.* 2001).

Trace elements

Trace elements show a homogeneous distribution in clino- and orthopyroxenes, no intra- and intergrain variations could be observed within the same sample. However, the incompatible trace element content of both pyroxenes displays a considerable range in variation throughout the whole xenolith suite (even within the same localities), often covering one or two orders of magnitude. Clinopyroxene has a higher content of incompatible trace elements than orthopyroxene, and in anhydrous spinel peridotites this mineral is the major host for rare earth elements Y and Sr. The primitive mantle (PM) normalized rare earth element (REE) plots of clinopyroxenes show highly variable patterns, but basically they can be classified into three major groups: LREE-depleted, LREE-enriched and U-shaped patterns. This variation is less obvious in the coexisting orthopyroxenes owing to their steep REE patterns, but the relative changes in their REE concentrations are very similar to those of the clinopyroxenes (see later). As the REE patterns of the clinopyroxenes are usually very similar to the REE patterns of the host spinel peridotite, the above-mentioned three major REE patterns define three geochemically different groups of xenoliths.

LREE-depleted xenoliths. Representative compositions of clino- and orthopyroxene from an LREE-depleted xenolith can be seen in Table 1. In clinopyroxenes the medium and heavy REEs (from Sm to Lu) show a smooth pattern in the PM normalized plot (Fig. 3a), and a slight decrease in the abundances from Nd towards La is evident. A weak negative anomaly for Ti and Zr, and a depletion in Nb and Ta, are present in all clinopyroxenes of this group (Fig. 3b). In most samples U and Th have similar low abundances as Nb and Ta, which is expected from their incompatibility, but a few LREE-depleted clinopyroxenes display an enrichment in U and Th (Fig. 3b). The PM normalized REE patterns of the coexisting orthopyroxenes show a continuous decrease from Lu to La defined by incompatibility (Fig. 2a) with strong positive anomalies at Ti, Zr, Hf, Nb and Ta (Fig. 3b), similar to those seen in mantle orthopyroxene by Rampone *et al.* (1991). Although the absolute abundances are different, almost all

Table 1. *Composition of the calculated most primitive clino- and orthopyroxene and the representative compositions of the pyroxenes from the three major types of xenoliths*

	Most primitive pyroxene		LREE-depleted (Szt 1100)		LREE-enriched (Szt 1074)		U-shaped (Szg 1087)	
	cpx	opx	cpx	opx	cpx	opx	cpx	opx
SiO_2	51.6	53.9	51.8	54.6	52.6	55.8	52.5	55.5
TiO_2	0.54	0.14	0.60	0.17	0.29	0.08	0.06	0.03
Al_2O_3	6.97	4.88	7.43	5.46	4.04	3.52	3.35	3.24
Cr_2O_3	0.90	0.44	0.82	0.48	0.55	0.35	0.75	0.54
FeO	3.02	6.16	3.23	6.23	2.72	6.41	2.80	6.08
MnO	0.08	0.14	0.09	0.13	0.08	0.15	0.08	0.14
MgO	15.6	33.0	15.7	32.2	16.5	33.5	17.1	33.5
CaO	19.8	0.80	18.4	1.07	22.3	0.63	22.2	0.80
Na_2O	1.53	0.10	1.73	0.17	0.65	0.03	0.45	0.03
Total	100.04	99.56	99.80	100.51	99.73	100.47	99.29	99.86
U	0.008		0.007		0.42		0.008	
Th	0.016		0.011		1.90		0.026	
Nb	0.19	0.045	0.25	0.068	0.075	0.005	0.12	0.048
Ta	0.015		0.013		0.008		0.012	
La	0.88	0.003	1.05	0.005	10.3	0.011	0.83	0.004
Ce	3.52	0.015	4.75	0.036	19.8	0.039	2.05	0.010
Sr	64.5	0.20	79.7	0.50	76.7	0.10	37.9	0.12
Pr	0.69	0.004	0.88	0.009	1.92	0.007	0.28	0.002
Nd	3.97	0.037	4.93	0.053	6.34	0.032	1.09	0.012
Zr	26.7	1.69	27.4	2.82	22.3	1.45	3.17	0.42
Hf	0.94	0.060	1.02	0.11	0.86	0.063	0.10	0.015
Sm	1.64	0.020	1.88	0.040	1.43	0.020	0.23	0.006
Eu	0.71	0.010	0.80	0.018	0.50	0.007	0.085	0.003
Ti	0.59	0.18	0.62	0.19	0.33	0.10	0.076	0.032
Gd	2.29	0.047	2.40	0.085	1.65	0.027	0.34	0.013
Tb	0.46	0.013	0.47	0.023	0.29	0.008	0.076	0.003
Dy	3.13	0.13	3.06	0.20	1.86	0.077	0.66	0.039
Ho	0.69	0.038	0.65	0.055	0.40	0.028	0.16	0.015
Y	17.0	1.07	15.5	1.44	10.1	0.72	4.25	0.42
Er	1.96	0.16	1.88	0.21	1.17	0.13	0.57	0.078
Tm	0.29	0.031	0.29	0.039	0.17	0.019	0.082	0.015
Yb	1.79	0.27	1.87	0.32	1.08	0.20	0.60	0.14
Lu	0.26	0.050	0.25	0.055	0.16	0.039	0.097	0.030

patterns show a more-or-less parallel arrangement in the case of both clino- and orthopyroxenes. Most of the LREE-depleted xenoliths have protogranular textures, but more deformed types such as porphyroclastic and equigranular can also be found among them.

LREE-enriched xenoliths. These xenoliths exhibit a wide range of REE patterns in their clinopyroxenes with $(Ce/Yb)_N$ often in excess of 1 (up to 13). Most of these patterns can be characterized with a more-or-less continuously increasing normalized REE content from the HREE to the LREE, although the slope of the increase may change within the pattern. More complex patterns such as convex-upwards or convex-downwards patterns with abrupt slope changes at Sm or Nd are also common (Fig. 3c). The normalized enrichment of U and Th is similar to that of the LREE, but the abundances of HFS elements (Nb, Ta, Zr, Hf, Ti) and Sr are lower than that of the neighbouring REE with similar incompatibility (Fig. 3d). The normalized incompatible trace element patterns of the coexisting orthopyroxenes are similar to that of the orthopyroxenes from LREE-depleted xenoliths. They have decreasing REE abundances from the HREE to LREEs, with positive anomalies in the HFSE and negative anomalies of Sr. Generally, the recrystallized xenoliths with equigranular textures have LREE-enriched trace element patterns.

U-shaped pattern. This third type of REE pattern is very distinctive and is generally associated with the special group of xenoliths with a poikilitic texture. The incompatible trace element content of clinopyroxene is lower than that of the LREE-depleted and

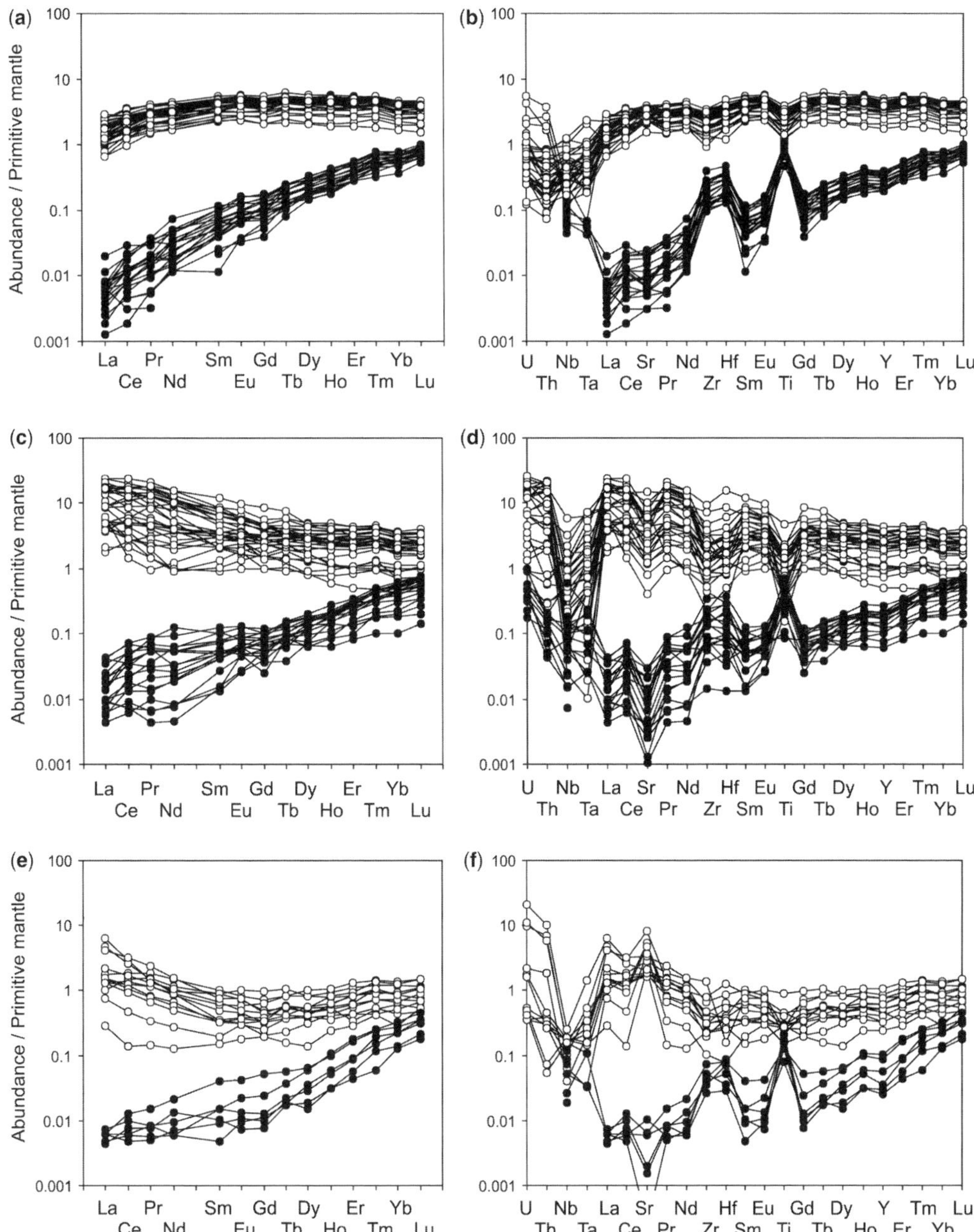

Fig. 3. Primitive mantle normalized REE and incompatible trace element patterns of the clino- and orthopyroxenes from the LREE-depleted xenoliths (**a** & **b**), LREE-enriched xenoliths (**c** & **d**) and from the xenoliths with U-shaped REE patterns (**e** & **f**). The primitive mantle values are from Sun & McDonough (1989). Clinopyroxenes are open symbols and orthopyroxenes are solid symbols.

LREE-enriched xenoliths (a representative analysis of clino- and orthopyroxene is given in Table 1). The PM normalized REE pattern of the clinopyroxene shows a depletion in MREE relative to the HREE Yb_N *c.* 0.3–1.4 × primitive mantle) and LREE (Ce_N *c.* 0.5–3 × primitive mantle) abundances, thus showing a U-shaped pattern (Fig. 3e). The most incompatible Th and U follow the slight enrichment trend defined by the LREE, having somewhat higher normalized abundances than La or Ce. As for the HFS elements (Fig. 3f), Nb and Ta show negative anomalies relative to LREE and U–Th, but no anomaly can be observed at Zr, Hf and Ti. The most conspicuous feature of the normalized trace element pattern is the high positive anomaly of Sr relative to the REEs of similar incompatibility (Fig. 3f). This pattern with the strong positive Sr anomaly strongly resembles the trace element patterns found in clinopyroxenes of poikilitic xenoliths from Beaunit, Massif Central, France (e.g. Downes & Dupuy 1987; Dobosi unpublished). A positive Sr anomaly has also been reported in both clinopyroxene and amphibole from the xenoliths of Kapfenstein (Coltorti *et al.* 2007). The PM normalized trace element patterns of orthopyroxenes (Fig. 3e, f) show a depletion from the HREE to the LREE, with positive anomalies at HFS elements and no or a negative anomaly at Sr.

The relative abundances of the different REE patterns can be visualized better if each REE pattern is represented by a single data point. REE data for mantle clinopyroxenes can be plotted by representing the LREE–MREE slope by the La/Nd ratio and the MREE–HREE slope by the Sm/Yb ratio (Rivalenti *et al.* 1996; Downes 2001). Figure 4 shows such a plot of results from the Pannonian lithospheric mantle and a comparison with the European mantle. The LREE-depleted xenoliths plot in a restricted area just below the centre of the diagram, indicating only a slight depletion in LREE abundances. Most of the LREE-enriched xenoliths plot in the upper-right quadrant, representing the LREE- and MREE-enriched xenoliths, while most of the U-shaped xenoliths plot in the upper-left quadrant, which is the field of the LREE-enriched and MREE-depleted patterns. However, the Pannonian lithospheric mantle shows a much more restricted compositional variation than the other regions (Massif Central or Eifel) of Europe. The strong LREE-depletion, as well as the strong enrichment of LREE relative to MREE (upper-left quadrant), is missing from the Pannonian xenoliths (Fig. 4).

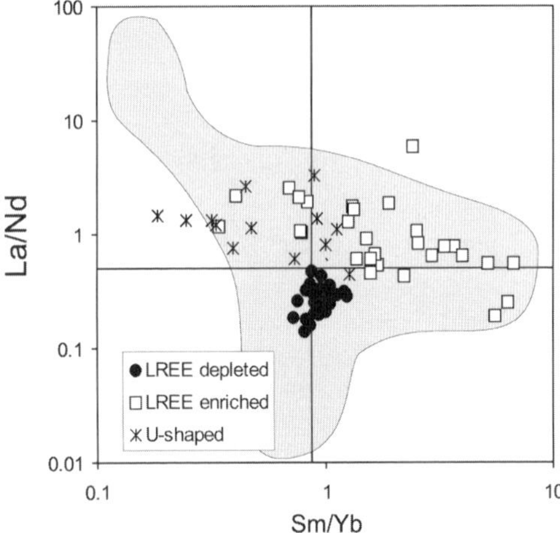

Fig. 4. La/Nd v. Sm/Yb diagram for clinopyroxenes from peridotite xenoliths of the western part of the Pannonian Basin, with the comparative field of the clinopyroxenes from the xenoliths from the European subcontinental lithospheric mantle (Downes 2001). The lines represent the ratios found in primitive mantle, according to Sun & McDonough (1989).

Discussion

Most studies of mantle spinel peridotites assume that the samples which suffered least metasomatism and also the least partial melting are those that show the least LREE depletion (Downes 2001), thus they should be the closest to the primitive mantle. Despite the fact that this view has been challenged recently by Le Roux *et al.* (2007), we consider that this approach is appropriate for the Pannonian Basin mantle xenoliths.

For the investigation of different depletion and enrichment processes in the mantle it would be useful to determine the most primitive clino- and orthopyroxene composition of the lithospheric mantle beneath the Pannonian region. These compositions can be used as starting compositions for calculations or as normalizing values to visualize enrichment and depletion in clino- and orthopyroxenes. The most primitive clinopyroxene composition has been calculated as the average of those of the LREE-depleted clinopyroxenes that have the highest HREE content (probably lowest degree of melt loss) and show little or no sign of highly incompatible trace elements, for example U- and Th- enrichment (probably not metasomatized). The composition of the most primitive orthopyroxene is the average of the orthopyroxenes coexisting with the above-mentioned clinopyroxenes. The estimated most primitive clino- and orthopyroxene trace element patterns, together with the individual samples that were averaged, can be seen in Figure 5a and b, and the calculated compositions are given in Table 1. The composition of the most primitive clinopyroxene is close to the Primitive Mantle Clinopyroxene Estimate (PMCE) given by Rivalenti *et al.* (1996), apart from the LREE abundances that are slightly depleted in our estimated clinopyroxene composition relative to the PMCE (Fig. 5a, b). The LREE element depletion may have two reasons: it may be indicative of small-scale partial melting (see later) that extracted a fraction of the most incompatible trace elements from the xenolith (and therefore from the clinopyroxene); or that

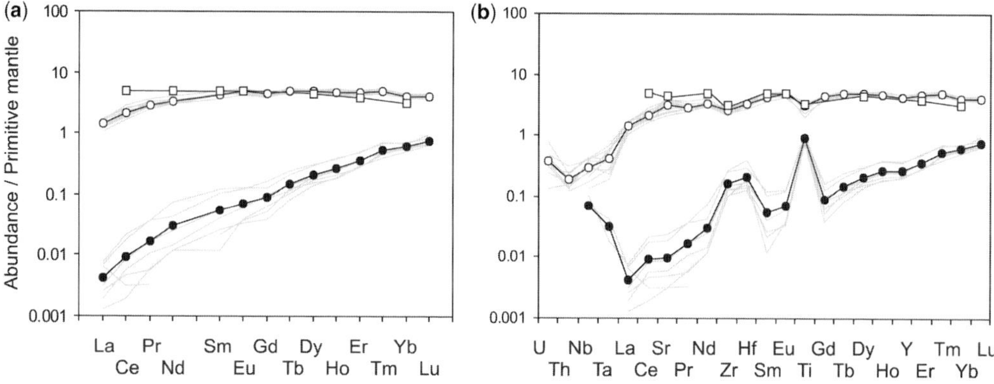

Fig. 5. Estimation of the most primitive pyroxene composition: (**a**) REE and (**b**) other incompatible elements. Open circles and filled circles connected with solid line, primitive mantle normalized composition of the most primitive clinopyroxene and orthopyroxene, respectively; grey lines, the averaged clino- and orthopyroxene compositions; open squares, PMCE given by Rivalenti *et al.* (1996). The primitive mantle values are from Sun & McDonough (1989).

the most incompatible LREE preferentially occur in fluid and melt inclusions or along grain boundaries and not in the clinopyroxene lattice.

Some representative clino- and orthopyroxene abundance patterns from the LREE-enriched xenoliths normalized to the most primitive clino- and orthopyroxene composition can be seen in Figure 6. The plots demonstrate that the normalized trace element patterns of orthopyroxenes and coexisting clinopyroxenes are very similar to each other, indicating that the depletion and enrichment processes left very similar trace element signatures on both minerals. This also means that orthopyroxene can be as sensitive an indicator of these depletion and enrichment processes in upper-mantle xenoliths as clinopyroxene. The same is true for the xenoliths, with clinopyroxenes having U-shaped REE patterns; the normalized orthopyroxene patterns are very similar to the U-shaped patterns of the coexisting clinopyroxenes (Fig. 6).

Partial melting

Extraction of basaltic magma from mantle peridotite by partial melting strongly affects the major and trace element composition of the refractory peridotites. The positive correlation between Al_2O_3 and Na_2O, and their negative correlation with MgO content observable in the pyroxenes of peridotite xenoliths (Fig. 7a, b), are the result of the removal of basalt magma from an undepleted mantle because Na and Al, as 'basaltic elements', preferably enter the melt phase during partial melting. The abundance of the incompatible trace elements in pyroxenes strongly decreases during this process, as can be seen in the Al_2O_3 v. Y plot for both clino- and orthopyroxene (Fig. 7c, d). The process of partial melting can be modelled using the incompatible trace element variations. Although there are several calculation methods for modelling clinopyroxene partial melting (e.g. Norman 1998), in this paper the fractional melting equations of Johnson *et al.* (1990) will be used. The calculated most primitive clinopyroxene composition (Table 1) is used as the starting composition for clinopyroxene, otherwise all other parameters (modal compositions, partition coefficients) used in this calculation are the same as in Johnson *et al.* (1990). The results of these calculations can be seen in Figure 8.

Figure 8a shows the variation of Er and Yb in the analysed clinopyroxene (both are less incompatible HREE) with the calculated partial melting trend. As fractional melting model was used in the calculations, the partial melting estimates indicate a minimum degree of partial melting; using equilibrium melting equations, a higher degree of melting is necessary to reach the same compositions. Most of the xenoliths, including all LREE-depleted and most of the LREE-enriched xenoliths, experienced less than 15% partial melting relative to the primitive mantle clinopyroxene. Xenoliths with LREE-depleted clinopyroxenes have experienced the lowest degree of partial melting; they are the most fertile samples as expressed in the high HREE and Al_2O_3 content of their pyroxenes. A few clinopyroxenes with LREE-enriched patterns show evidence of a higher degree of melting of up to 25%. For example, harzburgite xenolith Bo 1022 had the highest degree of partial melting from this group, which corresponds to its extremely low clinopyroxene content (1.3%: see Embey-Isztin

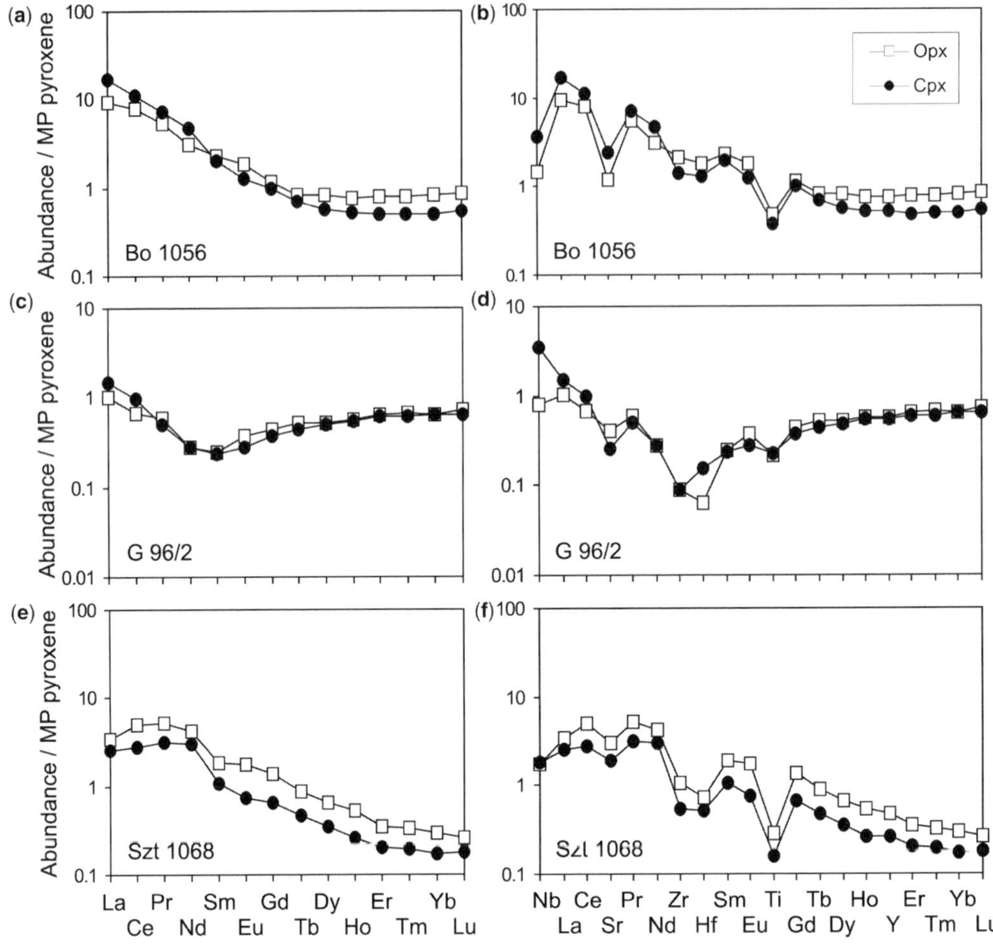

Fig. 6. Some representative REE and incompatible trace element plots of pyroxenes from the LREE-enriched xenoliths normalized to the most primitive (MP) clino- and orthopyroxene compositions (see the text for details). Note the very similar patterns in both cpx and opx.

et al. 1989). However, although all xenoliths with U-shaped REE patterns plot in this high degree partial melting field (20–25%), their modal compositions do not show such depletion in clinopyroxene. For example, the poikilitic xenolith Szg 1087 has a clinopyroxene content of 14% (Embey-Isztin *et al.* 1989). Thus, the low Al_2O_3 and HREE content of the pyroxenes in these xenoliths is not necessarily the result of partial melting but may indicate the role of other processes, such as modal metasomatism, or the cumulate origin of these xenoliths (Embey-Isztin *et al.* 1989). However, the origin of this special xenolith group is undergoing further investigation (Embey-Isztin *et al.* in prep.).

The variation of Ce (a highly incompatible REE) as a function of Yb is plotted in Figure 8b, together with the calculated partial melting curve. Owing to its high incompatibility, the concentration of Ce should decrease very quickly during the course of partial melting, as is shown from the steep slope of the calculated melting trend. It is evident from the figure that only some of the LREE-depleted samples plot on the calculated curve, suggesting that partial melting was the major factor that controlled their trace element contents. The abundance of Ce in most of the clinopyroxenes is much higher than that expected from partial melting calculations. The enrichment of LREE and other highly incompatible elements in the clinopyroxenes clearly indicates the role of other processes, such as metasomatic enrichment that overprints the effect of partial melting. Slightly elevated Ce content, as

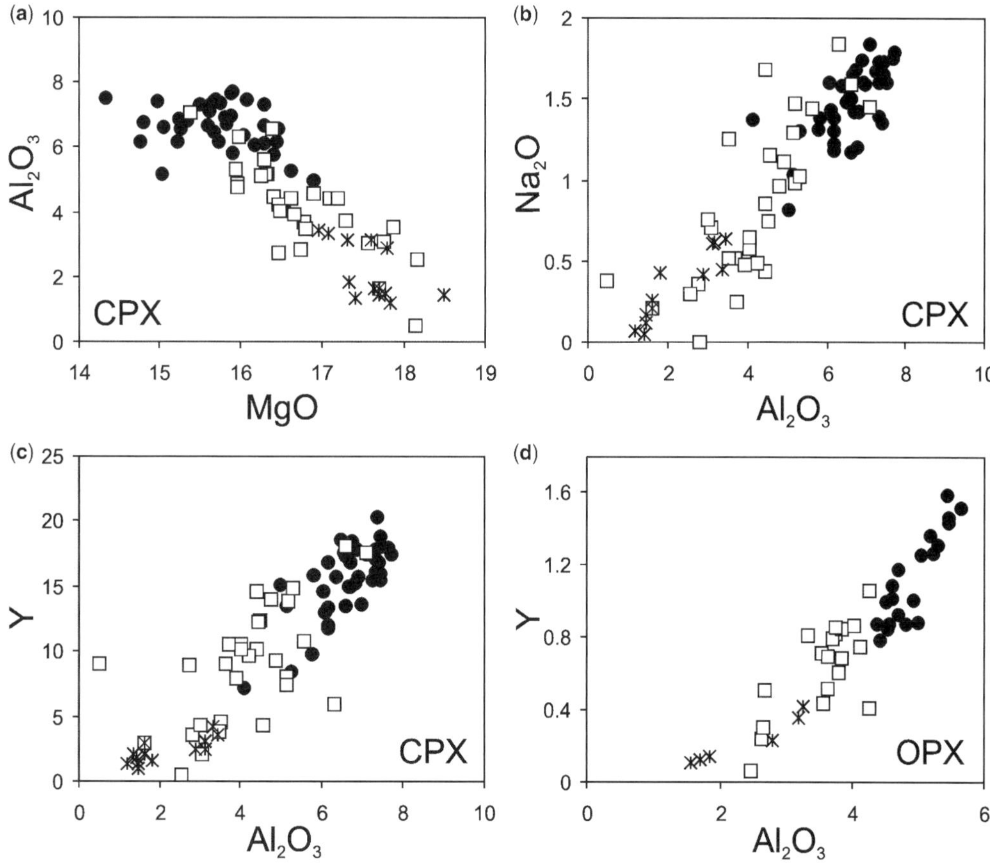

Fig. 7. Plots of (**a**) Al_2O_3 v. MgO and (**b**) Al_2O_3 v. Na_2O in clinopyroxenes, and (**c**) and (**d**) Al_2O_3 v. Y in clino- and orthopyroxenes. Symbols as in Figure 4.

well as enriched Th and U content, in clinopyroxenes from some LREE-depleted xenoliths indicate the effect of metasomatism in these relatively pristine samples.

Metasomatism

Most of the LREE-enriched xenoliths do not show evidence of a later growth of a LREE-rich accessory phase such as amphibole or apatite. These xenoliths have, therefore, been referred to as being 'cryptically' metasomatized in the sense of Dawson (1984). Metasomatic enrichment of the incompatible elements is caused by the infiltration of various incompatible trace-element-rich fluids or melts into the subcontinental lithospheric mantle. The melts and fluids would leave their geochemical signature in the mantle peridotites, thus the trace element content of the mantle xenoliths and their minerals (especially pyroxenes) may provide a key to their identification. Generally, the following melts or fluids are considered as possible agents of metasomatic enrichment: (1) alkali mafic or ultramafic silicate melts; (2) highly mobile carbonatite magmas; (3) subduction-related silicate melts such as adakites; or (4) hydrous subduction-related fluids. The possible role of subduction-related fluids is important in the Pannonian region because of the widespread subduction-related calc-alkaline volcanism during Eocene and Miocene times (see Szabó *et al.* 1992 and Harangi *et al.* 2007 for reviews).

Unfortunately, the geochemical signature left by the different fluids or melts in mantle peridotites is not unequivocal. Equilibrium between solid peridotite and the original migrating melt (metasomatizing agent) is rarely reached, especially under low melt/rock ratios, and the observed incompatible trace element patterns in the peridotites and their minerals are just snapshots of the different stages of solid matrix and percolating melt interaction. The processes during the interaction of mantle peridotite

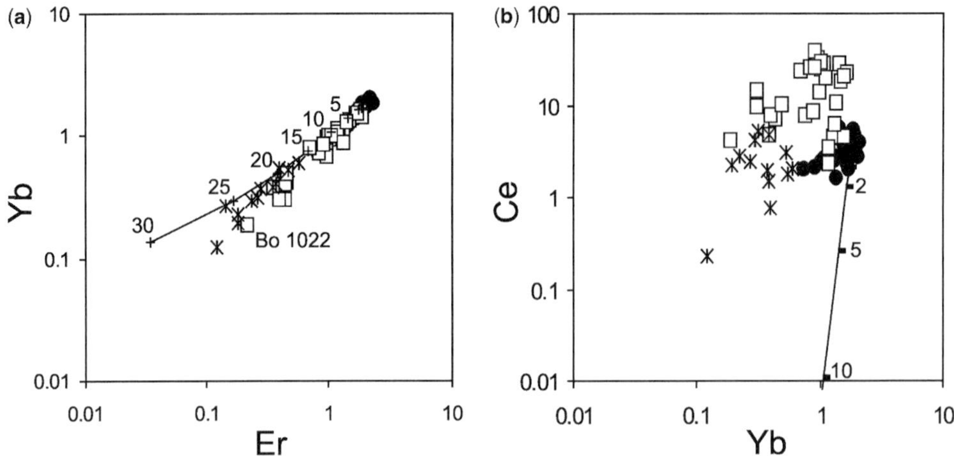

Fig. 8. Results of the partial melting calculation using (**a**) Yb and Er and (**b**) Yb and Ce abundances in clinopyroxenes. For the calculations the fractional melting formula published in Johnson *et al.* (1990) was used. The numbers indicate the degree of partial melting. Symbols as in Figure 4.

and percolating melt show a similarity to those operating in chromatographic columns (Navon & Stolper 1987; Bodinier *et al.* 1990; Vasseur *et al.* 1991). According to these models, the metasomatizing melt entering the mantle peridotite 'column matrix' is initially out of equilibrium, but as it migrates upwards it will interact with the matrix and tend to reach equilibrium with it through trace element exchange. During this solid–melt interaction each trace element behaves differently and as a consequence fractionation will occur. The chromatographic front of the more compatible elements moves more slowly than that of incompatible ones, and as a result in this transient stage of evolution the abundance ratios of elements with different compatibilities in the melt will vary considerably in space and time. If the melt is introduced continuously for a long enough time complete equilibrium can be achieved between the incoming melt and the matrix, and the minerals in mantle peridotite will have trace element compositions in equilibrium with the metasomatizing agent. However, this fully equilibrated stage is rarely achieved (in the full length of the hypothetical chromatographic column) and the various trace element patterns observed in peridotites and their minerals represent transient stages of the chromatographic fractionation. This process may explain the wide variety of LREE-enriched trace element patterns observed in the mantle xenoliths from the Pannonian region: they represent 'frozen-in stages' of the transient trace element evolution during the interaction of LREE-enriched melts and LREE-depleted solid peridotite. As can be seen in Figure 9, clinopyroxenes in xenoliths from the same locality that show similar temperatures and textures vary considerably in their LREE patterns. Moreover, the original peridotites might have varied in their degree of depletion, their mineralogical compositions and grain sizes, and all of these factors can influence their trace element evolution (see Bodinier *et al.* 1990 for detailed parameterization).

In spite of the complexity, for example, matrix inhomogeneity and the transient stages of the trace element exchange processes, an attempt at

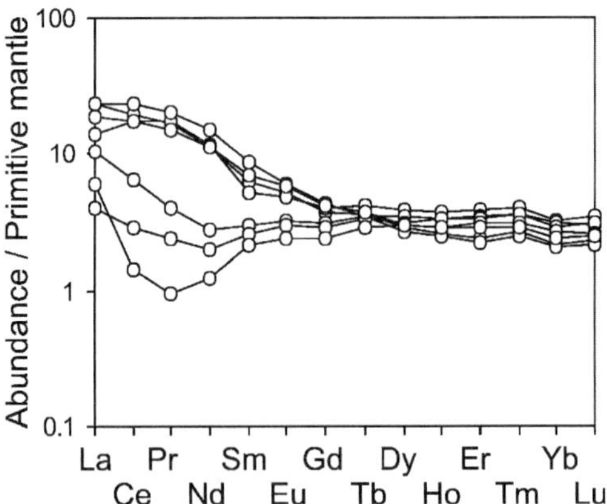

Fig. 9. REE patterns of clinopyroxenes from some LREE-enriched xenoliths showing different stages of chromatographic metasomatism – these xenoliths come from the same vent (Bo, Bondoróhegy), have the same equigranular textures and have similar equilibrium temperatures, and so they come from approximately the same mantle region. The primitive mantle values are from Sun & McDonough (1989).

estimating the metasomatizing melt composition can be made. The conversion of the clinopyroxene compositions into the hypothetical coexisting melt composition using the partition coefficients of Foley *et al.* (1996) will give a series of strongly LREE-enriched patterns (Fig. 10). These calculated abundances are not necessarily representative of the whole series of evolving melt compositions during peridotite matrix–melt interaction because of random sampling and the fact that the mantle is more complicated than a simple chromatographic column. However, one might expect that the series of calculated melt compositions converges to the original incoming melt composition. As the chromatographic fronts of the HREE move at lower velocities than those of the LREE, the latter will more readily affected, hence more depleted in the calculated melt compositions relative to the incoming melt. Thus, the calculated LREE-enrichment can be taken as a minimum concentration of the metasomatizing melt, the real concentration should be equal to or higher than that. Probably, the calculated melts in equilibrium with LREE-enriched clinopyroxenes with 'humped' concave-downwards REE patterns (e.g. Fig. 9) represent the closest approach to the metasomatizing melt composition. Calculated hypothetical melt compositions (Fig. 10) show highly fractionated LREE/HREE ratios, suggesting that the metasomatizing melt should have had a highly fractionated REE pattern with possibly more than 400 times the PM abundances of LREE.

The calculated high LREE content of the metasomatizing melts points towards the possible role of the alkali mafic magmas. Alkali basalt volcanoes (mainly Pliocene in age, 6–2 Ma) are widespread in the whole region (see Embey-Isztin *et al.* 1993; Embey-Isztin & Dobosi 1995), indicating that alkali basalt melts passed through the subcontinental lithosphere for a long (a few million years) period of time. Some of these basalts carried the xenoliths to the surface. The asthenospheric source of these basalts is indicated by their trace element and radiogenic isotope composition (Embey-Isztin *et al.* 1993, 2001; Embey-Isztin & Dobosi 1995). However, the LREE-enrichment observed in these

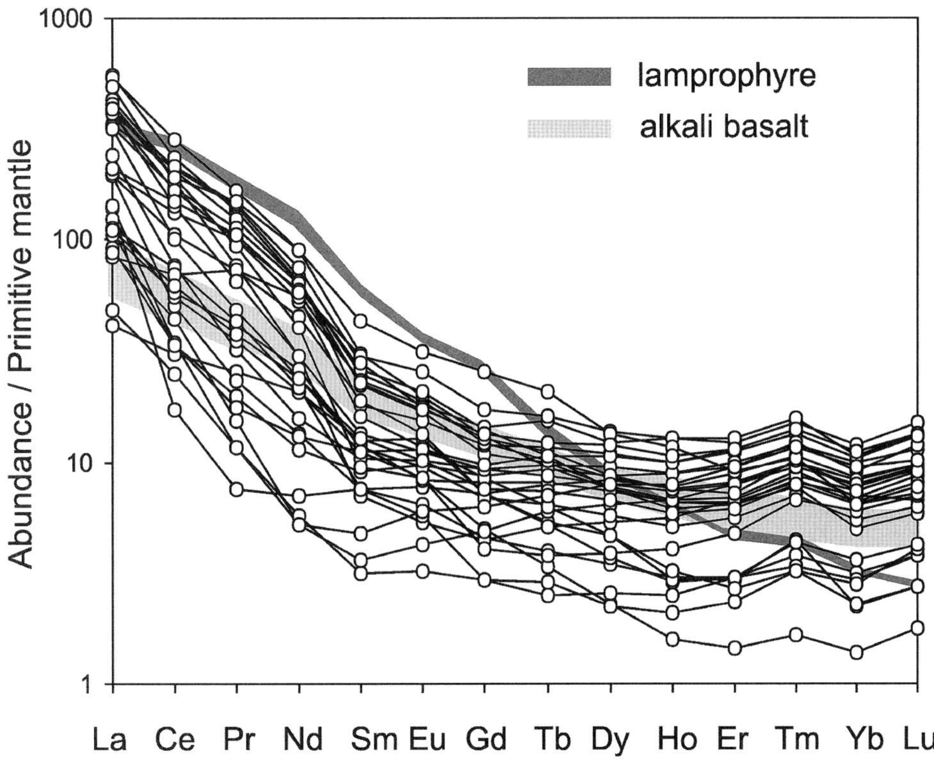

Fig. 10. Primitive mantle normalized REE patterns of hypothetical melts calculated from LREE-enriched clinopyroxene compositions together with the patterns of alkali basalts (after Embey-Isztin *et al.* 1993) and alkali lamprophyres (Dobosi unpublished) from the Pannonian Transdanubian region. The primitive mantle values are from Sun & McDonough (1989).

basalts is below the required level of enrichment to explain some of the trace element contents of clinopyroxenes from the LREE-enriched xenoliths (Fig. 10). This can have two possible explanations: (1) the metasomatizing melt was more enriched in LREE (and probably other highly incompatible elements) than the above-mentioned Pliocene alkali basalts; or (2) an 'over-enrichment' of LREE occurred during the melt–rock interaction. Such over-enrichment can be the result of 'percolative fractional crystallization', in which the mass of the melt is decreasing owing to crystallization and the residual melt becomes progressively more enriched in incompatible elements such as LREE. A model for this process was presented by Bedini *et al.* (1997).

In the case of the first explanation (i.e. the metasomatizing melts were more enriched in LREE than the Pliocene alkali basalts) some evidence for the presence of such highly LREE-enriched melts in the region should be presented. Representatives of such melts can be the Upper Cretaceous alkali lamprophyres and associated silicocarbonatites that are known as single dykes or dyke swarms in the Transdanubian Region about 50–100 km east of the xenolith localities at Lake Balaton (Horváth *et al.* 1983; Horváth & Ódor 1984; Dobosi & Horváth 1988; Szabó *et al.* 1993; Azbej *et al.* 2006). Surficial outcrops are rare; most of their occurrences are known from drilling. Recently (Szabó & Dégi pers. comm.) evidence has been found for the occurrence of alkali lamprophyre at Szigliget (see Fig. 1). The presence of these rocks suggests the passage of asthenospheric alkali lamprophyre melts through the lithospheric mantle in the Pannonian Basin. The trace element composition of these alkali lamprophyres (Figs 10 & 11) clearly indicate that they have LREE-enrichment high enough to explain some of the observed metasomatic REE patterns. From the calculated REE patterns and LREE-enrichment of the metasomatizing melts it cannot be distinguished whether alkali lamprophyres or alkali basaltic melts were the metasomatizing agents. The negative anomalies of Sr observed in some clino- and orthopyroxenes (Fig. 3d), and the negative anomalies of Zr and Hf observed in some clinopyroxenes (and bulk xenoliths), point towards the possible roles of alkali lamprophyres. However, besides alkali lamprophyres, the role of alkali basaltic melts in metasomatism cannot be excluded. The presence of alkali basalt-related pyroxenite, amphibole pyroxenite or amphibole veins in the spinel peridotite xenoliths (Embey-Isztin 1976; Embey-Isztin *et al.* 1990;

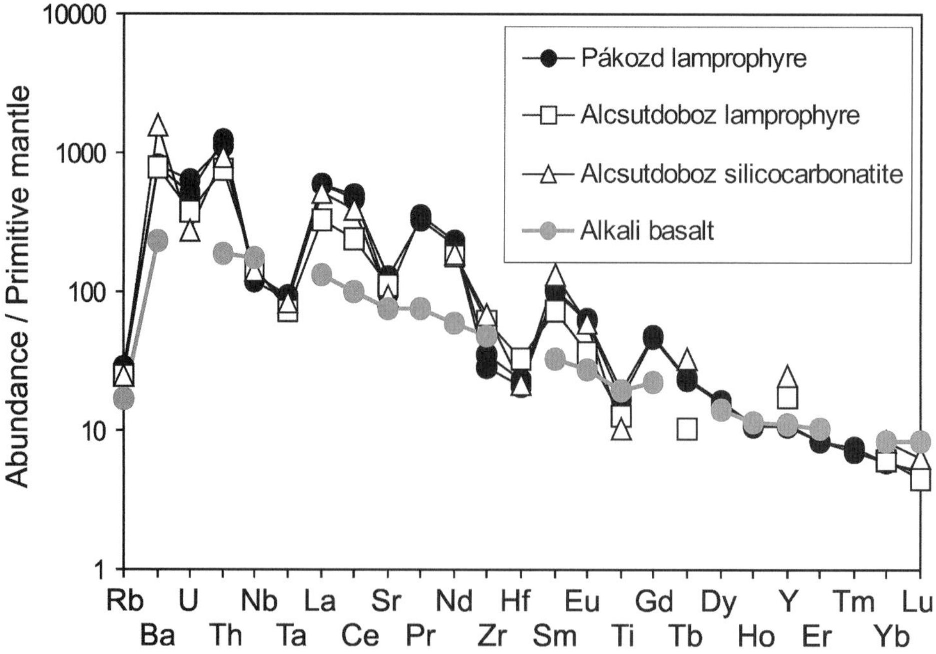

Fig. 11. Primitive mantle normalized trace element patterns for alkali basalts and some alkali lamprophyres from the Pannonian region (Pákozd surficial outcrop, Dobosi unpublished; and Alcsutdoboz drilling, Szabó *et al.* 1993). The primitive mantle values are from Sun & McDonough (1989).

Dobosi *et al.* 2003) may give evidence for the interaction between the alkali basaltic melts and the peridotitic mantle.

Unequivocal evidence for the role of subduction-related fluids and melts has not been found in the LREE-enriched xenolith group. Silicate melts deriving from the subducting slab (such as adakites) do not have high enough LREE-enrichment (Stern & Kilian 1996) to produce the observed LREE-enrichment in the clinopyroxenes. Such slab-derived melts have positive anomalies of Sr, Zr and Hf, but these characteristics have not been observed in the clinopyroxenes of the LREE-enriched group. High Nb content found in some LREE-enriched clinopyroxene (Fig. 3d) also cannot be explained by slab-derived melt metasomatism. These suggest that the xenoliths with LREE-enriched clinopyroxenes have probably been metasomatized by within-plate asthenosphere-derived mafic alkaline melts.

Recently, however, slab-melt related (proto-adakite) metasomatism was found in some xenoliths from Kapfenstein (Coltorti *et al.* 2007). Xenoliths with U-shaped REE patterns have some characteristics, such as a positive Sr anomaly in both clino- and orthopyroxenes (Fig. 3f), that may point towards the role of subduction-derived melts in their formation. However, the origin of this geochemically and texturally special group of xenoliths is beyond the scope of this paper and will be the subject of the forthcoming paper by Embey-Isztin *et al.* (in prep).

Textures and regional distribution of the xenoliths

Several studies (e.g. Downes & Dupuy 1987; Downes 1990; Downes *et al.* 1992; Zangana *et al.* 1997) suggested that there is a relationship between trace element patterns and the textural features of the xenoliths. In general, in the Pannonian Basin, the clinopyroxenes of protogranular and some of the porphyroclastic xenoliths have LREE-depleted patterns, the porphyroclastic and recrystallized equigranular xenoliths contain pyroxenes with LREE-enriched normalized patterns, while the U-shaped normalized REE pattern is restricted to the special textural group of xenoliths that have poikilitic texture.

In the three localities from the central part of the Pannonian Basin (Bondoróhegy, Szentbékálla and Szigliget, see Fig. 1) the protogranular xenoliths have the highest equilibrium temperatures of all xenoliths studied from the area (Embey-Isztin *et al.* 2001). The calculated temperatures using the method of Brey & Köhler (1990) are between 1122 and 1175 °C, which suggest that these xenoliths equilibrated close to the asthenosphere–lithosphere boundary. Clinopyroxenes from these xenoliths show the most primitive LREE-depleted patterns with the lowest inferred degree of partial melting.

In contrast, the recrystallized equigranular xenoliths have lower equilibrium temperatures, varying between 920 and 1060 °C, so they represent a shallower mantle level beneath the central part of the Pannonian area. The fact that the same volcanic vents yield low-temperature deformed and high-temperature undeformed xenoliths suggests an origin in a shear zone probably due to the uplift of hot mantle material into a cooler uppermost mantle (Embey-Isztin *et al.* 2001). The presence of rare texturally composite xenoliths containing undeformed and deformed portions indicates the presence of shear zones (Downes *et al.* 1992). The textural deformation may be owing to the rise of a mantle diapir beneath the central part of the basin and a more intense stretching of the lithosphere. Enrichment and metasomatism by asthenospheric melts could preferentially occur at shear zones (Downes 1990; Downes *et al.* 1992), which explains the enrichment of LREE in the deformed and recrystallized xenoliths. However, at the locality of Szigliget, depleted equigranular samples have also been found that indicate that deformation was not necessarily accompanied by enrichment.

The xenoliths in eastern Austria (marginal to the Pannonian Basin) have untectonized protogranular textures. Their equilibrium temperatures are lower than that of the protogranular xenoliths in the central part of the basin, that is, 988–1050 °C (Embey-Isztin *et al.* 2001) or <1100 °C (Coltorti *et al.* 2007), which may indicate a slightly cooler geotherm or higher level of xenolith extraction. Clinopyroxenes from these xenoliths have primitive LREE-depleted patterns and may show only incipient metasomatism expressed by their slightly elevated LREE, and elevated U and Th content. The inhomogeneous distribution of these elements in clinopyroxenes of a few samples indicates that metasomatic enrichment took place relatively shortly before volcanic eruption (Dobosi *et al.* 1999).

Concluding remarks

About half of the investigated Pannonian Basin mantle peridotite xenoliths show evidence for cryptic metasomatism, and this metasomatism frequently but not necessarily accompanies deformation. The calculation of the hypothetical most primitive clino- and orthopyroxene composition allowed us to visualize the effect of metasomatism on the seldom-studied orthopyroxene, which shows basically the same trends as clinopyroxenes. The agents of metasomatism were probably alkaline–highly alkaline mafic melts of asthenospheric origin, and some of these melts are probably related to the upper Cretaceous alkali lamprophyre magmatism. If this is the case, the western Pannonian lithospheric mantle might have experienced repeated intraplate alkaline metasomatism from Cretaceous to Quaternary times. Geochemical signatures of subduction-related silicic melts or hydrous fluids could not be detected in the cryptically metasomatized xenoliths, which suggests that there could not have been long-term interaction between subduction-related fluids or melts and the peridotites. However, this study does not include the amphibole-bearing xenoliths that may carry subduction-related geochemical signatures, as in the case of the hydrous xenoliths from Kapfenstein (Coltorti *et al.* 2007). The influence of subduction-related melts could be taken into account in the case of the poikilitic xenoliths that show U-shaped normalized REE patterns, sometimes with a positive anomaly of Sr.

The OTKA grants T35031 and T37382 supporting this research are gratefully acknowledged by G. Dobosi and A. Embey-Isztin, respectively. Support for G. Dobosi during his visit to Newfoundland was provided by NATO. Reviews by J.-L. Bodinier and E. R. Neumann contributed to the clarification of the manuscript. Helpful comments were received from the editors S. Y. O'Reilly and M. Coltorti.

References

AZBEJ, T., SZABÓ, CS., BODNAR, R. J. & DOBOSI, G. 2006. Genesis of carbonate aggregates in lamprophyres from the Northeastern Transdanubian Central Range; Hungary: magmatic or hydrothermal origin? *Mineralogy and Petrology*, **88**, 429–497.

BODINIER, J. L., VASSEUR, G., VERNIERES, J., DUPUY, C. & FABRIES, J. 1990. Mechanisms of mantle metasomatism: geochemical evidence from the Lherz orogenic peridotite. *Journal of Petrology*, **31**, 597–628.

BEDINI, R. M., BODINIER, J. L., DAUTRIA, J. M. & MORTEN, L. 1997. Evolution of LILE-enriched small melt fractions in the lithospheric mantle: a case study from the East African Rift. *Earth and Planetary Science Letters*, **153**, 67–83.

BREY, G. P. & KÖHLER, T. P. 1990. Geothermobarometry in four phase lherzolites II. New thermometers, and practical assessment of existing thermometers. *Journal of Petrology*, **31**, 1353–1378.

COLTORTI, M., BONADIMAN, C., FACCINI, B., NTAFLOS, T. & SIENA, F. 2007. Slab melt and intraplate metasomatism in Kapfenstein mantle xenoliths (Styrian Basin, Austria). *Lithos*, **94**, 66–89.

CSONTOS, L., NAGYMAROSY, A., HORVÁTH, F. & KOVÁCS, M. 1992. Tertiary evolution of the Intra-Carpathian area: a model. *Tectonophysics*, **208**, 221–241.

DAWSON, J. B. 1984. Kimberlites – II. The mantle and crust/mantle relationships. *In*: KORNPROBST, J. (ed.) *Developments in Petrology Series*. Elsevier, Amsterdam, 290–294.

DOBOSI, G. & HORVÁTH, I. 1988. High-and low-pressure cognate clinopyroxenes from the alkali lamprophyres of the Velence and Buda Mts., Hungary. *Neues Jahrbuch für Mineralogie Abhandlungen*, **158**, 241–256.

DOBOSI, G., DOWNES, H., EMBEY-ISZTIN, A. & JENNER, G. A. 2003. Origin of megacrysts and pyroxenite xenoliths from the Pliocene alkali basalts of the Pannonian Basin (Hungary). *Neues Jahrbuch für Mineralogie Abhandlungen*, **178**, 217–237.

DOBOSI, G., KURAT, G., JENNER, G. A. & BRANDSTÄTTER, F. 1999. Cryptic metasomatism in the upper mantle beneath Southeastern Austria: a Laser Ablation Microprobe-ICP-MS study. *Mineralogy and Petrology*, **67**, 143–161.

DOWNES, H. 1990. Shear zones in the upper mantle – Relation between geochemical enrichment and deformation in mantle peridotites. *Geology*, **18**, 374–377.

DOWNES, H. 2001. Formation and modification of the shallow sub-continental lithospheric mantle: a review of geochemical evidence from ultramafic xenolith suites and tectonically emplaced ultramafic massifs of western and central Europe. *Journal of Petrology*, **42**, 233–250.

DOWNES, H. & DUPUY, C. 1987. Textural, isotopic and REE variations in spinel peridotite xenoliths, Massif Central, France. *Earth and Planetary Science Letters*, **82**, 121–135.

DOWNES, H., EMBEY-ISZTIN, A. & THIRLWALL, M. F. 1992. Petrology and geochemistry of spinel peridotite xenoliths from the western Pannonian Basin (Hungary): evidence for an association between enrichment and texture in the upper mantle. *Contributions to Mineralogy and Petrology*, **109**, 340–354.

EMBEY-ISZTIN, A. 1976. Amphibolite/lherzolite composite xenolith from Szigliget, north of the Lake Balaton, Hungary. *Earth and Planetary Science Letters*, **31**, 297–304.

EMBEY-ISZTIN, A. 1978. On the petrology of spinel lherzolite nodules in basaltic rocks from Hungary and Auvergne, France. *Annales Hitorico-Naturales Musei Nationalis Hungarici*, **70**, 27–44.

EMBEY-ISZTIN, A. 1984. Texture types and their relative frequencies in ultramafic and mafic xenoliths from Hungarian alkali basaltic rocks. *Annales Hitorico-Naturales Musei Nationalis Hungarici*, **76**, 27–42.

EMBEY-ISZTIN, A. & DOBOSI, G. 1995. Mantle source characteristics for Miocene–Pleistocene alkali basalts, Carpathian-Pannonian Region: a review of trace elements and isotopic composition. *Acta Vulcanologica*, **7**, 155–166.

EMBEY-ISZTIN, A., DOBOSI, G., ALTHERR, R. & MEYER, H.-P. 2001. Thermal evolution of the lithosphere beneath the western Pannonian Basin: evidence from deep-seated xenoliths. *Tectonophysics*, **331**, 285–306.

EMBEY-ISZTIN, A., DOWNES, H., JAMES, D. E., UPTON, B. G. J., DOBOSI, G., SCHARBERT, H. G. & INGRAM,

G. A. 1993. The petrogenesis of Pliocene alkaline volcanic rocks from the Pannonian Basin, Eastern Central Europe. *Journal of Petrology*, **34**, 317–343.

EMBEY-ISZTIN, A., SCHARBERT, H. G., DIETRICH, H. & POULTIDIS, H. 1989. Petrology and geochemistry of peridotite xenoliths in alkali basalts from the Transdanubian volcanic region western Hungary. *Journal of Petrology*, **30**, 79–105.

EMBEY-ISZTIN, A., SCHARBERT, H. G., DIETRICH, H. & POULTIDIS, H. 1990. Mafic granulite and clinopyroxenite xenoliths from the Transdanubian Volcanic Region (Hungary): implication for the deep structure of the Pannonian Basin. *Mineralogical Magazine*, **54**, 463–483.

FOLEY, S. F., JACKSON, S. E., FRYER, B. J., GREENOUGH, J. D. & JENNER, G. A. 1996. Trace element partition coefficients for clinopyroxene and phlogopite in an alkaline lamprophyre from Newfoundland by LAM-ICP-MS. *Geochimica et Cosmochimica Acta*, **60**, 629–638.

HARANGI, SZ., DOWNES, H., THIRLWALL, M. & GMÉLING, K. 2007. Geochemistry, petrogenesis and geodynamic relationships of Miocene calc-alkaline volcanic rocks in the western Carpathian arc, eastern central Europe. *Journal of Petrology*, **48**, 2261–2287.

HORN, I., HINTON, R. W., JACKSON, S. E. & LONGERICH, H. P. 1997. Ultra-trace element analysis of NIST SRM 616 and 614 using Laser Ablation Microprobe-Inductively Coupled Plasma-Mass Spectrometry (LAM-ICP-MS): a comparison with Secondary Ion Mass Spectrometry (SIMS). *Geostandards Newsletter*, **21**, 191–203.

HORVÁTH, F. 1993. Towards a mechanical model for the formation of the Pannonian Basin. *Tectonophysics*, **226**, 333–357.

HORVÁTH, I. & ÓDOR, L. 1984. Alkaline ultrabasic rocks and associated silicocarbonatites in the NE part of the Transdanubian Mts. (Hungary). *Mineralia Slovaca*, **16**, 115–119.

HORVÁTH, I., DARIDA-TICHY, M. & ÓDOR, L. 1983. Magnesite bearing dolomitic carbonatite dike rocks from the Velence Mountains (in Hungarian). *Annual Report of the Hungarian Geological Institution*, **1981**, 41–44.

HUISMANS, R. S., PODLADCHIKOV, Y. Y. & CLOETINGH, S. 2001. Dynamic modeling of the transition from passive to active rifting, application to the Pannonian Basin. *Tectonics*, **20**, 1021–1039.

JACKSON, S. E., LONGERICH, H. P., DUNNING, G. R. & FRYER, B. J. 1992. The application of laser ablation microprobe-inductively coupled plasma-mass spectrometry (LAM-ICP-MS) to *in situ* trace element determinations in minerals. *Canadian Mineralogist*, **30**, 1049–1064.

JENNER, G. A., FOLEY, S. E., JACKSON, S. E., GREEN, T. H., FRYER, B. J. & LONGERICH, H. P. 1994. Determination of partition coefficients for trace elements in high pressure–temperature experimental run products by laser ablation microprobe-inductively coupled plasma-mass spectrometry (LAM-ICP-MS). *Geochimica et Cosmochimica Acta*, **58**, 5099–5103.

JOHNSON, K. T. M., DICK, H. J. B. & SHIMIZU, N. 1990. Melting in the oceanic upper mantle: an ion microprobe study of diopsides in abyssal peridotites. *Journal of Geophysical Research*, **95**, 2661–2678.

KURAT, G., EMBEY-ISZTIN, A., KRACHER, A. & SCHARBERT, H. 1991. The upper mantle beneath Kapfenstein and the Transdanubian volcanic region, E Austria, W Hungary: a comparison. *Mineralogy and Petrology*, **44**, 21–38.

KURAT, G., PALME, H., BADDENHAUSEN, H., HOFMEISTER, H., PALME, CH. & WANKE, H. 1980. Geochemistry of ultramafic xenoliths from Kapfenstein, Austria: evidence for a variety of upper mantle processes. *Geochimica et Cosmochimica Acta*, **44**, 45–60.

LE ROUX, V., BODINIER, J.-L., TOMMASI, A., ALARD, O., DAUTRIA, J.-M., VAUCHEZ, A. & RICHES, A. J. V. 2007. The Lherz spinel lherzolite: refertilized rather than pristine mantle. *Earth and Planetary Science Letters*, **259**, 599–612.

MERCIER, J.-C. & NICOLAS, A. 1975. Textures and fabrics of upper mantle peridotites as illustrated by xenoliths from basalts. *Journal of Petrology*, **16**, 454–487.

NAVON, O. & STOLPER, E. 1987. Geochemical consequences of melt percolation: the upper mantle as a chromatographic column. *Journal of Geology*, **95**, 285–307.

NORMAN, M. D. 1998. Melting and metasomatism in the continental lithosphere: laser ablation ICPMS analysis of minerals in spinel lherzolites from eastern Australia. *Contributions to Mineralogy and Petrology*, **130**, 240–255.

RAMPONE, E., BOTTAZZI, P. & OTTOLINI, L. 1991. Complementary Ti and Zr anomalies in orthopyroxene and clinopyroxene from mantle peridotites. *Nature*, **354**, 518–520.

RIVALENTI, G., VANNUCCI, R. *ET AL.* 1996. Peridotite clinopyroxene chemistry reflects mantle processes rather than continental versus oceanic settings. *Earth and Planetary Science Letters*, **139**, 423–437.

STERN, C. R. & KILLIAN, R. 1996. Role of the subducted slab, mantle wedge and continental crust in the generation of adakites from the Andean Austral Volcanic Zone. *Contributions to Mineralogy and Petrology*, **123**, 263–281.

SUN, S. S. & MCDONOUGH, W. F. 1989. Chemical and isotopic systematics of oceanic basalts: implications for mantle composition and processes. *In*: SAUNDERS, A. D. & NORRY, M. J. (eds) *Magmatism in the Ocean Basins*. Geological Society, London, Special Publications, **42**, 313–345.

SZABÓ, CS., HARANGI, SZ. & CSONTOS, L. 1992. Review of Neogene and Quaternary volcanism of the Carpathian–Pannonian region. *Tectonophysics*, **208**, 243–256.

SZABÓ, CS., KUBOVICS, I. & MOLNÁR, ZS. 1993. Alkaline lamprophyre and related dike rocks in NE Transdanubia, Hungary: the Alcsutdoboz-2 (AD-2) borehole. *Mineralogy and Petrology*, **47**, 127–148.

SZABÓ, CS., VASELLI, O., VANNUCCI, R., BOTTAZZI, P., OTTOLINI, L., CORADOSSI, N. & KUBOVICS, I. 1995. Ultramafic xenoliths from the Little Hungarian Plain (Western Hungary): a petrologic and geochemical study. *Acta Vulcanologica*, **7**, 249–263.

VASELLI, O., DOWNES, H., THIRLWALL, M., VANNUCCI, R. & CORADOSSI, N. 1996. Spinel peridotite xenoliths from Kapfenstein (Graz Basin, Eastern Austria): a geochemical and petrological study. *Mineralogy and Petrology*, **57**, 23–50.

VASSEUR, G., VERNIERES, J. & BODINIER, J. L. 1991. Modelling of trace element transfer between mantle melt and heterogranular peridotite matrix. *In*: MENZIES, M. A., DUPUY, C. & NICOLAS, A. (eds) *Orogenic Lherzolites and Mantle Processes. Journal of Petrology*, Special Volume, 41–54.

ZANGANA, N. A., DOWNES, H., THIRLWALL, M. F. & HEGNER, E. 1997. Relationship between deformation, equilibrium temperatures, REE and radiogenic isotopes in mantle xenoliths (Ray Pic, Massif central, France): an example of plume–lithosphere interaction? *Contributions to Mineralogy and Petrology*, **127**, 187–203.

Quantitative characterization of textures in mantle spinel peridotite xenoliths

FELICITY A. TABOR[†], BRYAN E. TABOR & HILARY DOWNES*

School of Earth Sciences, Birkbeck, University of London, Malet Street, London WC1E 7HX, UK

**Corresponding author (e-mail: h.downes@ucl.ac.uk)*

[†]Deceased 15 December 2005

Abstract: A method for quantitative characterization of grain size in thin sections has been established for mantle spinel peridotite xenoliths, using optical scanning of large areas of thin sections, skeletonization of grain-section outlines and computerized measurement of individual grain-section areas. Measurements range from 218 for the coarsest example to more than 3000 in the finest grained. Variability of the samples has been examined in relation to size and number of grain-section areas measured by using multiple and orthogonal sections from several xenoliths. The results show a linear relationship of arithmetic mean against additive standard deviation, including data from coarse-grained protogranular, through porphyroclastic to the finer-grained equigranular examples. This suggests that peridotite textures form a continuous series rather than discrete groups, as suggested by qualitative (subjective) assessment. The observed distributions of grain-section areas have been explored in relation to their description and possible mechanistic origin. By direct measurement and comparison of cumulative number and area distribution curves, we show that qualitatively assessed 'typical grain sizes' are influenced by a small number of larger grain sections. Although the arithmetic mean and standard deviation provide a convenient method for comparison, in practice grain-section area distributions show marked positive skewness more consistent with log-normal or power-law functions. Linear log-probability curves also support the existence of a continuous series of peridotite textures, suggesting that the shallow lithospheric mantle has been subject to processes of comminution and/or grain growth dependent on the Law of Proportionate Effect.

Supplementary material: Details of resolution and boundary recognition can be found at http://www.geolsoc.org.uk/SUP18398.

Peridotite xenoliths and ultramafic massifs are fragments of the Earth's lithospheric upper mantle (Harte 1983; Menzies 1983). On exposed surfaces and in thin section, mantle peridotites show a range of grain size and shape that are commonly described in the literature as 'texture' (e.g. Boullier & Nicolas 1975; Mercier & Nicolas 1975; Ave Lallement *et al.* 1980). Visual assessments are, of necessity, two-dimensional; qualitative descriptions have been applied to these two-dimensional images, and several authors have proposed systems with increasingly complex semantics (e.g. Harte 1975; Mercier & Nicolas 1975). These descriptions, however, cannot avoid the subjective nature of the qualitative criteria that are employed. The fact that there is little agreement between workers with the assignment of qualitative textural descriptions prompted us to investigate whether such assignments could be sustained by quantitative measurement of the distribution of 'grain size' in thin sections. Our investigation is mainly based on unmetasomatized spinel peridotite xenoliths in Cenozoic alkali basalts from the French Massif Central but also covers xenoliths from Germany and the USA, together with a single sample from Fontet Rouge, an ultramafic massif in the Pyrenees. The samples have largely been described in previous studies (Witt 1985; Zangana *et al.* 1997; Downes 2001; Downes *et al.* 2003). They are generally unfoliated and consist of 45–95% olivine, 10–30% orthopyroxene, 0–18% clinopyroxene and up to 3% spinel (Downes 1997), without any phlogopite or amphibole. Our method measured the size of 200–3000 individual grain-section areas of silicate minerals in thin sections. We show that mantle peridotite textures can be quantitatively assessed using a whole-slide scanning and digitization method, and form a continuum rather than being separate entities. Therefore, we cannot justify separation into the perceived categorizations of mantle texture. Although we appreciate that there may be a need for categorization (e.g. 'coarse-, medium-, fine-grained'), it is possible that the mantle textures discussed in this paper all have the same ultimate origin.

From: COLTORTI, M., DOWNES, H., GRÉGOIRE, M. & O'REILLY, S. Y. (eds) *Petrological Evolution of the European Lithospheric Mantle*. Geological Society, London, Special Publications, **337**, 195–211.
DOI: 10.1144/SP337.10 0305-8719/10/$15.00

Current descriptive textural classifications of mantle xenoliths

It has long been recognized (e.g. Collee 1963) that spinel peridotite xenoliths show wide textural variations reflecting a history of deformation, recrystallization and grain growth. However, textural nomenclature has varied between different authors. In general, textural classifications are made with little specific regard to component mineral species. Mercier & Nicolas (1975), from studies of spinel peridotite xenoliths from Western Europe and Hawaii, defined three main textural groups.

- *Protogranular* (PG) – they regarded this as the 'oldest texture', which reflected a near absence of deformation and was characterized by 'grain sizes of olivine and enstatite of *c.* 4 mm, with diopside and spinel having grain sizes of *c.* 1 mm'.
- *Porphyroclastic* (PPC) – this texture was considered to be formed by the plastic flow of protogranular material and showed large strained grains (porphyroclasts) completely surrounded by smaller unstrained grains (neoblasts).
- *Equigranular* (EG) – this texture was divided into two subgroups (tabular and mosaic) and both subtextures were fine grained, with a 'typical grain size of 0.7 mm'.

Harte (1977) reviewed and redefined textural terminology, extending the work to include garnet peridotites. He proposed the introduction of a range of eight 'General Textural Terms' (Porphyroclast, Granoblastic, Mosaic, Laminated, Fluidal, Disrupted, Equant and Tabular). These, together with subdivisions, gave rise to descriptions such as 'fluidal-mosaic-porphyroclastites', whilst 'protogranular' became 'coarse', with 'grain size' in excess of 2 mm. At the same time Nielson-Pike & Schwarzman (1977) classified textures in spinel peridotite xenoliths as igneous (igneous or pyrometamorphic), metamorphic (porphyroclastic, cataclastic, foliated and equigranular mosaic) or allotriomorphic for 'textures that are of unclear origin'. Coisy & Nicolas (1978) used the Mercier & Nicolas (1975) classification with two modifications from Harte (1977). All these descriptions, however, depend on qualitative assessment of textural features. Where 'grain sizes' were quoted there was often no explanation of what was being measured (longest dimension, equivalent diameter or some other linear feature?) or the basis for the assignment as 'typical' or 'the majority', etc. In general, there is little regard to mineral species, as olivine and orthopyroxene account for more than 80% of the modal proportions.

Numerous authors have added modifications and caveats to the basic classifications proposed by Mercier & Nicolas (1975). Lenoir *et al.* (2000) suggested that the term 'protogranular' should be restricted to samples that show specific clusters of spinel and pyroxene, and reclassified other protogranular xenoliths as 'coarse granular'. Zangana (1995) added two transitional groups ('protogranular–porphyroclastic' and 'porphyroclastic–equigranular'). The general tendency to add and redefine terminology reinforces the view of Swan & Sandilands (1995) that qualitative criteria and assessments are inevitably subjective, so that the definition should be stated in terms of numbers. Perhaps the greatest danger of qualitative descriptions is their vulnerability to interpretation by 'authoritative assertion' (Vernon 1996, 2004).

Quantification of textures in rocks

Texture in holocrystalline rocks depends on three major factors: (1) grain size and size distribution; (2) grain shape and shape variation; and (3) contact relationships of the grains. The challenge is to find physically measurable parameters that describe aspects of the 'texture' and to make sufficient measurements to have statistical significance. We have chosen grain size and size distributions (i.e. factor 1) for quantification, which have been the subject of crystal size distribution (CSD) studies (e.g. Marsh 1998; Higgins 2000, 2006).

Thin sections provide readily available two-dimensional objects for measurement. Although intuitively these can be expected to reflect the three-dimensional crystalline structure of the rock, the relationship is not simple. Distributions of grain-section areas in thin sections are a combination of the size distribution of the minerals being sectioned and a distribution produced by the random sectioning plane (Higgins 1994, 2006).

Irani & Callis (1963) stated that the 'size of a particle is the representative dimension that best describes the degree of comminution of the particle', but practical considerations influence the choice of dimension measured. If all the particles in a rock were perfect spheres, there would be no problem of the definition of size (Humphries 1969). A single dimension (e.g. diameter) would then unequivocally specify both size and shape. However, most geological materials do not approach such an ideal state. Optical methods for the determination of grain size are usually some variant of the linear intercept method (Ranalli 1984), originating in metallographic studies (Smith & Guttman 1953), which can be used to calculate the diameter of a sphere of equal volume (Etheridge & Wilkie 1979). However, Etheridge & Wilkie (1981) drew attention to the underlying shape assumptions in their calculations.

When early grain-size measurement studies were being established, measurement of individual grain-section areas in a thin section would have been prohibitively laborious. Perring *et al.* (2004) noted that 'the measurements ... are laborious to collect and typically fewer than 20 samples are analysed'. Now with readily available digital image capture and computerized pixel counting this is no longer the case. We have, therefore, developed a relatively inexpensive method of textural quantification based on digital images of whole thin sections in a photographic slide scanner, and applied the method to quantify textural variations in mantle-derived spinel peridotite xenoliths.

Methodology

To measure features of a polycrystalline rock section, the first stage is to recognize the grain boundaries and reduce the image to these basic outlines ('skeletonization'). Both intraphase and interphase boundaries need to be detected, but cleavage lines or cracks must be avoided. Most authors favour skeletonization, either by hand or by using a digitizing tablet.

Measurements can be made by hand directly on enlarged photomicrograph prints or by digital image capture with image-processing software (Perring *et al.* 2004), although automation of boundary recognition is still a major obstacle.

Petrological microscopes have a restricted field of view so that only a limited portion of the thin section can be observed at a time. Moreover, the smaller the area that can be viewed, in comparison with the grain size, the greater the proportion of grain sections that will be artificially terminated by the field of view and should be discounted in image analysis. Addressing these problems, Tarquini & Armienti (2003) put forward a method based on a film scanner with image analysis software and a thin-section holder that could be fitted with polarizing filters. Independently (Tabor 2004*a*, 2005), we devised a method using digital images of whole or selected areas of thin sections obtained using a Nikon COOLSCAN IV ED scanner fitted with a 'Slide Mount Adapter MA-20', and a Nikon Medical Slide Holder FH-G1. Pairs of mutually orthogonal Polaroid filters housed on either side of the thin section were used in different orientations so that a series of images could be obtained showing the minerals with different birefringence colours. An electron microscope specimen slot support grid was fixed to the slide (Tabor 2004*b*) to provide a 2 mm calibration bar that remained as part of the scanned image throughout subsequent processing.

The series of coloured XP (crossed polars) and PPL (plane polarized light) images were printed as A4 sheets and placed on a back-lit illuminated panel, and the boundaries traced by hand onto overlying tracing paper. An intact grain-section outline was located around the image to provide a limiting border without artificial truncations. A flat bed scanner was then used to make a black and white digital image of the grain-section outlines, limiting border and calibration bar on the tracing paper (Fig. 1). This digitized image was then fed into the Carnoy Program (Schols & Smets 2001). By counting pixels and relating these to physical dimensions by reference to the calibration bar, this program provides a measure of the area of each grain section. It assigns a number to each grain-section area, calculates the minimum and maximum grain-section areas, the mean grain-section area and standard deviation, and provides a list of the basic data for each grain section.

Magnification, resolution and mineral identity boundary recognition. We have investigated several problems relating to the recognition of grain-section boundaries, together with issues concerning magnification and resolution. Magnification must be sufficient to reveal smaller particles whilst still giving a practically achievable skeleton with sufficient grain-section areas to provide a basis for statistical assessment. In part, this will depend on the coarseness/fineness of the texture in relationship to both the area and thickness of the thin section. With increasing coarseness, to obtain comparable statistics, larger areas must be measured, which in practice means making measurements on more than one thin section. Regarding the lower limit of fineness, increased magnification does not necessarily give improved resolution and boundary discrimination. Factors militating against this are:

- physical effects from alteration at boundaries blurring the outlines;
- instrumental limitations, which give a practical lower limit to the size of grain section that can be resolved, which we have calculated for our instrumentation is 0.00035 mm^2;
- size of particle compared to thickness of thin section.

Thin sections used in this work were all of standard 30 μm thickness. Light passing through the thin section is influenced by the intersection of the particle with both surfaces of the thin section. The importance of this was recognized by Holmes (1927), Chayes (1956), Royet (1991) and Russ & Dehoff (2000), particularly in respect to 'the problem of over projection'. Etheridge & Wilkie (1981) also suggested that, in thin sections, effects of specimen thickness contribute to uncertainty of boundary recognition. An estimate of the possible influence on grain-boundary recognition can be

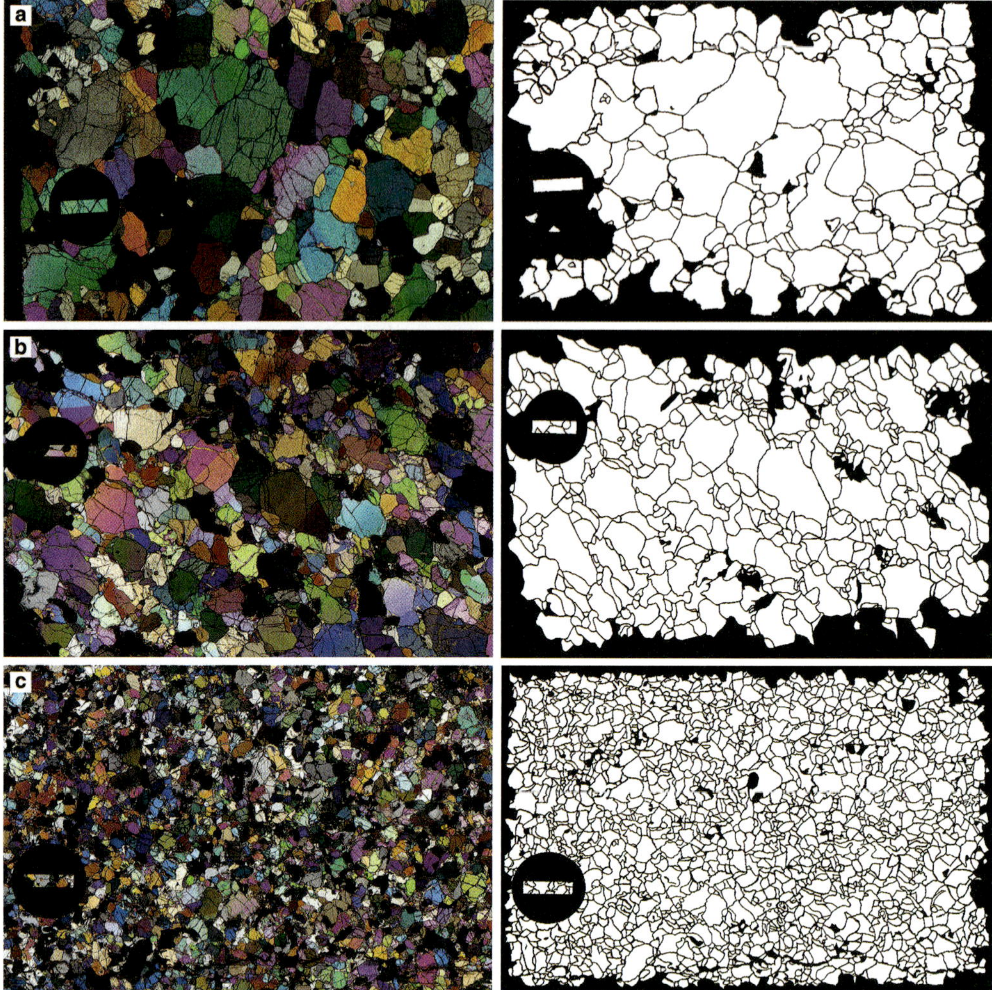

Fig. 1. Examples of scanned (XPL) thin sections, skeletonized grain-section area outlines, continuous grain-section outline borders and images of SEM (scanning electron microscopy) support slot grid providing a 2 mm calibration bar. Values from the subsequent analyses for these samples that had been designated by the original authors as follows.

Designation	No. of measurements	Minimum (mm^2)	Mean (mm^2)	Maximum (mm^2)	Standard deviation
(**a**) Protogranular	335	0.0023	0.5189	11.7526	1.1371
(**b**) Porphyroclastic	588	0.0035	0.2591	5.8676	0.5318
(**c**) Equigranular	2078	0.0015	0.0701	0.9174	0.0925

obtained from an extension of Chayes' geometry of the sectioning of a spherical particle. A general formulation of this can be shown to apply to all relative particle/slide positions and provides a basis for evaluation (see supplementary material).

Although these calculations can only approximate to the observed irregular shapes of the grain-section areas in a thin section, they support the view that there is a lower limit to measurement of individual grain-section areas, and that increasing magnification does not give greater resolution and discrimination beyond this limit. In practice, this effect only became apparent in our work with small particles at magnification of about $\times 30$ linear, when their boundaries tend to become more diffuse compared with those around larger grain sections and show coloured fringes with crossed polars in some orientations.

A spherical particle of radius 0.03 mm (comparable with thin-section thickness) can have a projected surface area, in the plane of the thin section, of 0.00283 mm^2, whilst a spherical particle of radius 0.015 mm (i.e. can just be accommodated within the thickness of the thin section) can have a projected surface area of 0.00071 mm^2. Both of these are larger than the limit imposed by instrumental resolution and should, therefore, be readily detectable.

The practical significance of all these potential effects was assessed by examining two thin sections, one from TAB23, a 'medium-grained' xenolith, and the other from RP91-7, a very 'fine-grained' example. Both thin sections, by visual subjective assessment, gave the impression of textural homogeneity. A number of scans were made from these thin sections, with increasing enlargement, imaging smaller abutting areas (Table 1). With each magnification increment there is some relative increase in the number of grain-section areas detected and measured. The results show that magnifications of more than ×12 but less than ×30 are sufficient to give adequate estimates of the grain-section size parameters without the problem of diffuse boundary recognition.

Statistical significance. It is important to establish the extent of measurement required to provide a reliable description of a particular thin section. At relatively low magnifications (*c.* ×20), the field of view in a petrological microscope is approximately 4 mm in diameter, that is, about 1.5% of the area of a typical thin section. Thus, it contains only a limited number of grain sections, particularly taking account of the proportion of grain sections that are artificially truncated by the limit of the field of view and which must, therefore, be discounted for quantification. With the coarsest 'protogranular' peridotites, there may not be an untruncated section in the field of view and even with the finest ('equigranular') there are seldom more than two dozen. A great advantage of whole-slide scanning is that it facilitates evaluation of all the grain-section areas in a thin section in a single process, rather than laboriously repeating measurements on restricted areas.

With direct measurements of grain-section areas, once the skeleton is drawn the computer program can only make the same measurement for each grain section, no matter how many times it evaluates

Table 1. *Scanning magnification effects and practical lower cut-off size limits*

Thin section proportion scanned	Magnification		Grain-section areas measured	Mean grain section (mm^2)	Standard deviation
	Linear	Areal			
TAB23					
c. 1 ≈ 600 mm^2	×8	×64	1294	0.294	0.618
c. $\frac{1}{2}$ ≈ 300 mm^2	×12	×144	660	0.325	0.668
c. $\frac{1}{2}$ ≈ 300 mm^2	×12	×144	664	0.261	0.634
Arithmetic combination			1324	0.303	0.651
Distributions combined			1324	0.303	0.662
RP91-7					
c. $\frac{1}{2}$ ≈ 250 mm^2	×15	×225	2078	0.075	0.086
c. $\frac{1}{4}$ ≈ 125 mm^2	×20	×400	1192	0.056	0.085
c. $\frac{1}{4}$ ≈ 125 mm^2	×20	×400	1314	0.052	0.081
Arithmetic combination			2506	0.054	0.083
c. $\frac{1}{8}$ ≈ 65 mm^2	×29	×841	737	0.067	0.082
c. $\frac{1}{8}$ ≈ 65 mm^2	×29	×841	783	0.038	0.065
c. $\frac{1}{8}$ ≈ 65 mm^2	×29	×841	517	0.066	0.096
c. $\frac{1}{8}$ ≈ 65 mm^2	×29	×841	545	0.065	0.088
Arithmetic combination			2582	0.059	0.083

With the medium-grained xenolith (TAB 23), the first magnification increment (×8 → ×12) gave similar increases in number of grain-section areas recognized in both areas, having small but different effects on mean grain-section area size and standard deviation for both scans, probably reflecting some degree of textural inhomogeneity even at this scale. With the very fine-grained xenolith (RP91-7), the magnification increments (×15 → ×20 → ×29) produced a more marked effect in respect to number measured, particularly at the first step. At both steps the increased numbers suggest a more significant degree of detectable textural inhomogeneity at this level. This is particularly apparent at the second step where there is a greater range of values for both the mean grain-section areas and the standard deviation, which nevertheless combine to give values close to those obtained at ×20 magnification. Furthermore, it is at the higher magnification (×29) that uncertain diffuse and fringed boundaries begin to obtrude.

the image. Clearly, two people, or one person at different times, drawing the skeleton from an image might produce slightly different interpretations of the grain boundaries. However, experiments with observers who had agreed the identification rules showed little evidence for disagreement.

If, having made an assessment on an area of a thin section, another area (e.g. the other half of the thin section) gives virtually the same result, then either or both results are characteristic of the two areas examined. However, we have found that in practice two areas of the same slide seldom give precisely the same values and concluded that in many cases there was a degree of inhomogeneity within the thin section. Therefore, it is only logical to measure the entire population of grain-section areas in a thin section as a true representation and this becomes practical with whole-slide scanning. One coincidental implication of this decision is that, if measurements are made on approximately similar areas, the number of grain-sections counted increases rapidly as their mean size gets smaller. Consequently, any parameter related to grain-section area (e.g. standard deviation) will show an apparent correlation with the number of grain-section areas analysed, solely as a result of the measurement procedure. This apparent correlation was confirmed to be spurious by experiments with subdivisions of thin sections from two particularly homogeneous xenoliths. The overall skeleton of RP91-7, a very fine-grained peridotite (*c.* 2700 grain-section areas), was first divided into approximately equal halves along a continuous grain-border outline and then each of these halves was similarly divided to give four approximately equal quarters of the original skeleton. The original whole skeleton and its six subdivisions were measured separately, at the same magnification, giving results for mean grain-section area size and standard deviation that are virtually identical (Table 2). For a coarser-grained sample (RP87-7), from which two very similar thin sections had been prepared and each scanned in two halves, the combination of the grain-section areas measurements gave results for the mean and standard deviation that are

Table 2. *Statistical parameter independence of the total number of grain-section areas measured on relatively homogeneous thin sections*

(1) A fine-grained peridotite xenolith RP91-7. Subdivision of the whole thin-section skeleton and subsequent measurement of smaller areas all scanned at the same magnification

Skeleton	Number of grain-section areas	Mean grain-section area (mm^2)	Standard deviation
Whole thin section	2683	0.0563	0.0743
Two approximately equal halves	1358	0.0538	0.0725
	1334	0.0574	0.0789
Four approximately equal quarters	654	0.0543	0.0758
	679	0.0561	0.0731
	681	0.0533	0.0752
	626	0.0558	0.0737

(2) Two thin sections, A and B, from coarser grained peridotite xenolith RP87-7. Combinations and analyses of the grain-section area measurements from the two halves of each of the thin sections, scanned at the same magnification

Skeleton	Number of grain-section areas	Mean grain-section area (mm^2)	Standard deviation
Half thin-section A	383	0.465	1.035
	371	0.454	1.059
Two halves of A combined	754	0.460	1.046
Half thin-section B	347	0.464	1.004
	377	0.397	0.818
Two halves of B combined	751	0.430	0.915
Combination of all four scans	1505	0.445	0.983

For texturally homogeneous thin sections the descriptive statistics (mean and standard deviation) are not related to the number of grain-section areas measured.

independent of the total number of grain-section areas counted (Table 2).

We also assessed the importance of border-edge effects. Drawing a rectangular frame around the area to be examined includes artificially truncated grain-section areas compared with a delineating border that follows continuous whole grain-section outlines. The influence on the statistics depends on the relative proportions and sizes of the grain-section areas being measured. The finest grained ('equigranular') samples show very little discernable influence of edge effects but they are apparent in the coarser ('protogranular') samples, where only 300 grain-section areas were measured (Table 3). This supports our intuitive assessment that at least 250–300 grain-section areas need to be measured to provide a statistically significant description of a sample.

Results

Mercier & Nicolas (1975) suggested that the textures in spinel peridotites could be divided into three main groups, protogranular, porphyroclastic and equigranular, based on grain size and shape, but the definitions of these groups were not quantified. In an early phase of our study (Tabor 2004*a*), some 30 sections or parts of sections were scanned and analysed (Table 3). All of these samples were

Table 3. *Mean grain-section areas and standard deviation for spinel peridotite mantle xenoliths from the French Massif Central (FMC) used in the initial study (Fig. 2a)*

Sample	Slide and border (see footnote)	Designated texture	Mean grain section (mm^2)	Standard deviation	Number of grain-section areas
RP91-7	1a	EG	0.06	0.08	2645
RP91-7	1b	EG	0.07	0.093	2078
RP91-7	2a	EG	0.06	0.088	2793
RP91-7	2b	EG	0.071	0.094	2097
RP91-19	1	EG	0.063	0.102	1515
RP91-20	1	EG	0.057	0.081	1985
RP91-15	1	EG	0.08	0.152	1389
RP91-14	1	EG/PPC	0.049	0.082	2521
RP91-8	1	EG/PPC	0.12	0.265	935
RP91-11	1	EG/PPC	0.065	0.164	1788
RP83-70	1	PPC	0.315	0.872	295
RP83-71	2a	PPC	0.202	0.395	853
RP83-71	2b	PPC	0.18	0.377	866
RP91-10	1a	PPC/PG	0.305	0.832	490
RP91-10	1b	PPC/PG	0.259	0.532	588
RP87-5	1	PG	0.166	0.355	553
RP91-1	1	PG	0.226	0.529	462
RP87-7R	2a	PG	0.465	1.035	383
RP87-7R	2b	PG	0.454	1.059	371
RP87-7P	1a	PG	0.519	1.137	335
RP87-7P	2a	PG	0.464	1.004	374
RP87-7P	2b	PG	0.397	0.818	377
RP87-7P	3a	PG	0.476	1.054	300
Mb4	1	PG	0.245	0.754	564
Mb-1	2a	PG	0.546	1.326	328
Mb-1	2b	PG	0.383	0.789	428
Mb-9	1	PG	0.392	0.947	571
Mb-57	1	PG	0.415	0.972	507
CH-11	1	PG	0.655	1.68	218
CH-11	2	PG	0.577	1.471	247

Abbreviations: EG, equigranular; PPC, porphyroclastic; PG, protogranular.
R and P are two separate thin sections from the one xenolith. a and b are two parts (approximate halves) of the same thin section.
The second column indicates the methodology employed with respect to the outline border.

1. Natural outline of intact grain-section areas within the area scanned. No truncated grain-section areas in the area being measured.
2. Forced border representing the scanned area but which truncates some grain-section areas in the area being measured.
3. Forced border like 2 but with truncated grain-section areas 'blacked out'. Should approximate to 1 but might be easier to achieve in practice.

from the French Massif Central, where Mercier & Nicolas (1975) set up their original textural classification. Many were from the Ray Pic locality (Zangana 1995; Zangana *et al.* 1997) and cover the common textural range. Other samples, described by Downes *et al.* (2003), were chosen specifically to include very coarse samples described as 'protogranular' by Lenoir *et al.* (2000). The samples were divided into textural types according to the classification assigned by the previous authors, using the criteria of Mercier & Nicolas (1975). These textures are illustrated in the thin sections in Figure 1, together with their skeletonized images.

The number of grain-section areas measured depended on the grain size of the sample, and varied from nearly 3000 for very fine-grained samples to more than 200 for very coarse-grained ones (Table 3). The mean areas occupied by individual grain sections range from 0.06 to 0.66 mm^2. The standard deviations likewise vary from 0.08 for the fine-grained samples to in excess of 1 for the coarsest. Xenoliths from Ray Pic cover almost the entire size range of grain-section areas, whereas the coarsest texture is seen in sample CH-11 as described by Lenoir *et al.* (2000).

Reproducibility of the technique is shown by the results for four separate analyses of fine-grained sample RP91-7 (Table 3), showing variation in mean grain-section area from 0.06 to 0.07 mm^2. Similarly, for a coarse-grained xenolith (RP87-7), on which six separate analyses were performed on two thin sections, the range of variation in mean grain-section area is from 0.40 to 0.52 mm^2. The larger variation may be due to genuine differences in grain size in different areas of this sample.

When mean grain-section area sizes are plotted against the standard deviation of these grain-section area sizes, a straight-line relationship is found (Fig. 2a). Peridotites previously designated as protogranular (PG) have mean grain-section areas greater than approximately 0.2 mm^2 and standard deviations of more than 0.4, whereas peridotites described as equigranular (EG) display much smaller grain-section areas (<0.1 mm^2) and smaller standard deviations (<0.2). There is strong overlap between the results for samples previously identified as porphyroclastic (PPC) and those considered to be protogranular (Fig. 2a). Xenoliths previously designated as 'equigranular' all plot in a small region of Figure 2a, so this textural type can be easily and consistently identified using our digitization method. They consistently have grain-section areas of less than 0.1 mm^2 and concomitantly small standard deviations of grain-section area. However, they are completely overlapped by some samples that had been designated as 'porphyroclastic–equigranular' by Zangana (1995). A subgroup of coarse-grained samples previously recognized as 'protogranular' have grain-section areas of more than 0.35 mm^2, which are clearly different from the equigranular group. However, difficulties arise with those protogranular (PG) samples that overlap the porphyroclastic (PPC) and transitional PPC–PG groups, in the grain-section area size range of 0.2–0.4 mm^2. We therefore propose that a minimum mean grain-section area of 0.35 mm^2 be used to delimit the protogranular texture, and a maximum mean grain-section area of 0.1 mm^2 be used to delimit the equigranular texture. In this classification, porphyroclastic textures would have mean grain-section areas of between 0.1 and 0.35 mm^2.

We have extended the study by examination of 30 thin sections of unfoliated, unmetasomatized spinel peridotite xenoliths from additional locations including the Massif Central and the Herault regions (France), the Eifel region (Germany), Cima (California, USA) and Kilbourne Hole (New Mexico, USA), and a single sample from an ultramafic massif from the Pyrenees (Table 4 and Fig. 2b). The results for these samples are entirely consistent with the previously established linear trend, suggesting that the linear trend is a general relationship and that peridotite textures form an overlapping continuum rather than separate groups.

Xenolith sample variability

Whilst whole-slide scanning provides an effective assessment of the population characteristics of grain-section areas in a thin section, it is not clear how characteristic the results from a single thin section are in relation to the whole xenolith from which it was cut. To explore this aspect, orthogonal sections were cut from three xenoliths (MR1-a,b,c; MR2-a,b,c; and KH3-a,b,c) and quantified (Table 4 & Fig. 3a). Four equal areas, which together accounted for the whole slide, from a thin section that showed significant grain-section size banding (DW83-20-a,b,c,d) were also quantified (Table 4 & Fig. 3a). Three successive sections cut from xenolith TE189 were also quantified (Figs 2a & 3a).

Whilst the orthogonal sections from sample MR1 cluster together about the established trend line, those from MR2 spread along the line to an extent that, taken individually, they might be given different textural designations. Values for the three orthogonal sections of xenolith KH3 also spread along the trend line (Fig. 3a). Even with the successive sections from xenolith TE189, there are minor differences. Whilst the results for three different parts of the single DW83-20 slide fall very closely together, a much coarser band across the centre of the thin section gave a very different result. If this banding continued into the xenolith

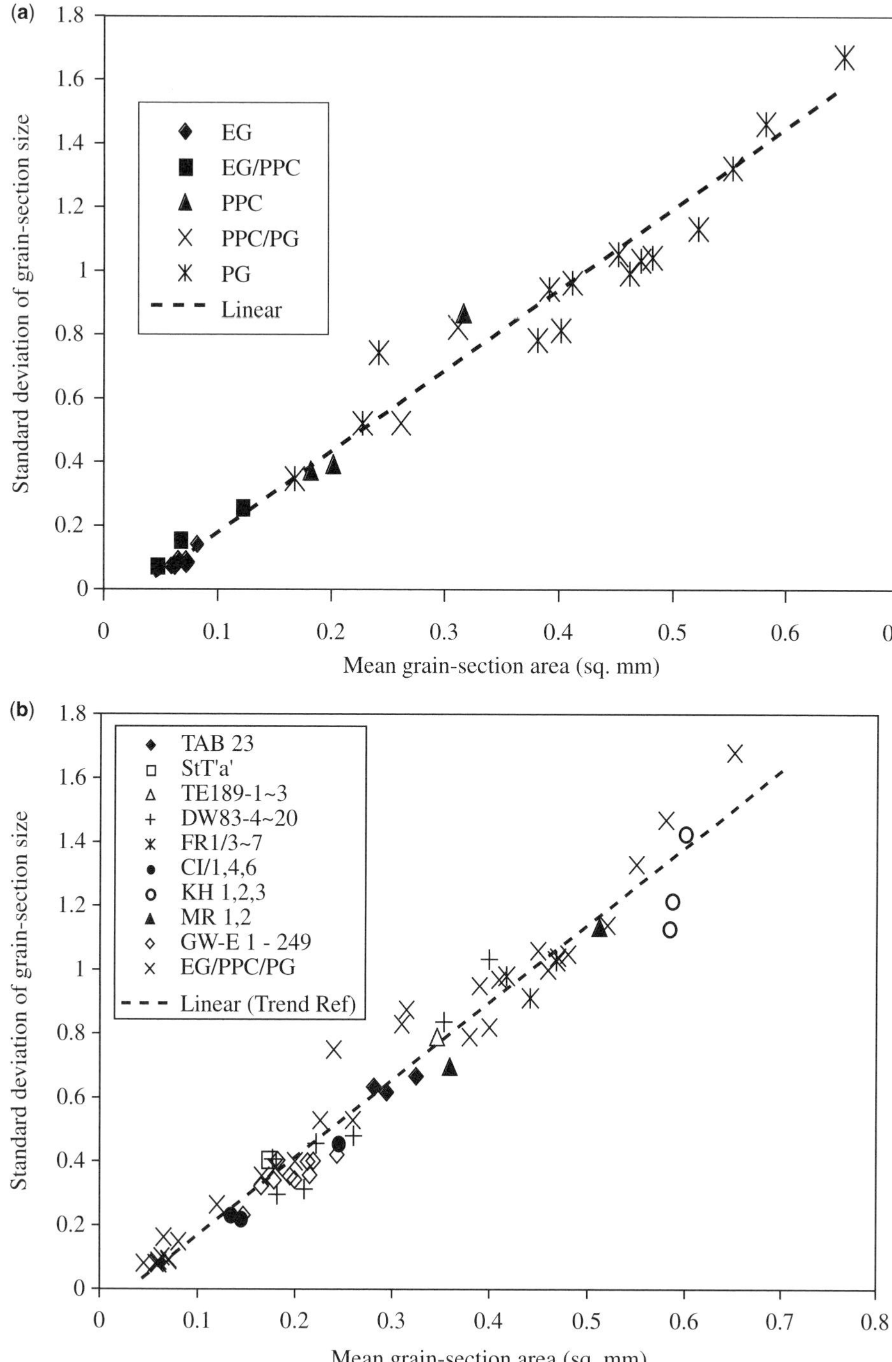

Fig. 2. Standard deviation v. mean grain-section area size for mantle peridotites. (**a**) From the French Massif Central, used in the initial study (data from Table 3). (**b**) For all samples, including those from the initial study (EG/PPC/PG), together with samples from other geographical locations and one sample from an ultramafic massif (data from Table 4).

Table 4. *Mean grain-section areas and standard deviation for mantle spinel peridotites used for the geographical extension of the study (Fig. 2b)*

Sample	Origin	Mean grain-section (mm^2)	Standard deviation	Number of grain-section areas
TAB 23 a	French	0.294	0.618	1294
TAB 23 b	Massif	0.325	0.668	667
TAB 23 c	Central	0.281	0.634	664
MR 1-1,2,3	Boree	0.512	1.134	2232
MR 2-1,2,3	Region	0.360	0.698	3148
TE189 = Bo83-73		0.347	0.791	773
St T 1 St. Thibery	Herault	0.173	0.406	1578
FR 1/3	Pyrenees	0.468	1.029	949
FR 1/5	Fontet	0.442	0.912	930
FR 1/7	Rouge	0.418	0.981	887
DW 83-4	Germany	0.177	0.407	1942
DW 83-5		0.353	0.837	906
DW 83-6		0.222	0.456	1236
DW 83-12		0.400	1.034	684
DW 83-13		0.182	0.296	1590
DW 83-15		0.209	0.314	1250
DW 83-20		0.261	0.480	433
GW-E 1Dreis		0.243	0.421	819
GW-E 6Dreis		0.147	0.232	2811
GW-E 8Dreis		0.165	0.323	2429
GW-E 10Dreis		0.182	0.404	2102
GW-E 15Dreis		0.200	0.343	1454
GW-E 16Dreis		0.213	0.400	1902
GW-E 18Dreis		0.215	0.358	1917
GW-E 171		0.195	0.354	1676
GW-E 195		0.179	0.341	2351
GW-E 249		0.219	0.402	2092
CI/1	SW USA	0.133	0.234	1623
CI/4	Cima	0.245	0.457	928
CI/6		0.144	0.222	1797
KH 1-MD	Kilbourne	0.586	1.217	642
KH 2-MD	Hole	0.584	1.131	688
KH 3-1,2,3		0.600	1.429	1750

TAB 23 a,b,c are measurements from a single thin section scanned whole and in two approximately equal halves. FR 1/3, FR 1/5, FR 1/7 are three separate thin sections taken from a collection randomly cut from the same sample. KH 3-1,2,3, MR 1-1,2,3 and MR 2-1,2,3 are the averaged values of three orthogonally cut thin sections from xenoliths KH 3, MR 1 and MR 2. The separate orthogonal values have been plotted in Figure 3a.

and another thin section had been cut normal to the studied thin section, a very different assessment of the texture of the xenolith would have been made. However, the calculated parameters for each individual area still fit on the line established for the plot of standard deviation v. mean grain-section area (Figs 2 & 3a). Xenoliths are often too small to make multiple and orthogonal thin sections. Nevertheless, interpretations based on single thin sections may not represent the entire sample.

Qualitative visual size assessment

Authors commonly make estimates of 'typical' grain size to accompany their qualitative descriptions of rock textures, although what constitutes 'typical' is seldom clearly defined. For example, Mercier & Nicolas (1975) gave linear measurements, 'typically coarse (4 mm)', '8 × 2 mm' and 'rather fine grained (0.7 mm)' in reference to features of their textural groups. Significantly these dimensions are generally greater than the mean grain-section areas determined in our quantitative study. Mercier & Nicolas (1975) showed photomicrographs (crossed polars) and derived mineral identity outlines (skeletons) that we have digitally scanned and analysed, on the basis of the scale bar dimensions given in each figure. Using this methodology, four of the five skeletons drawn by Mercier & Nicholas (1975)

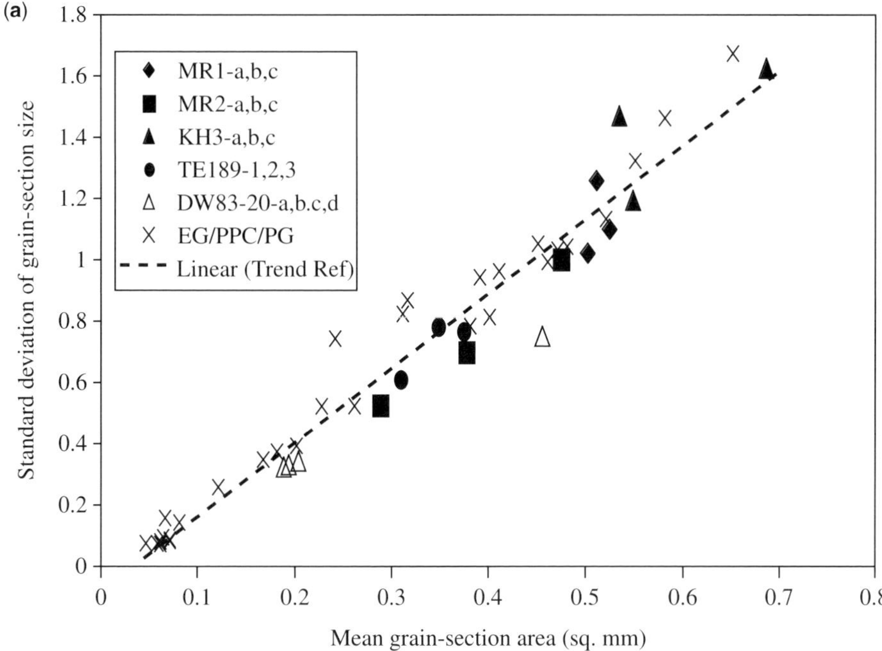

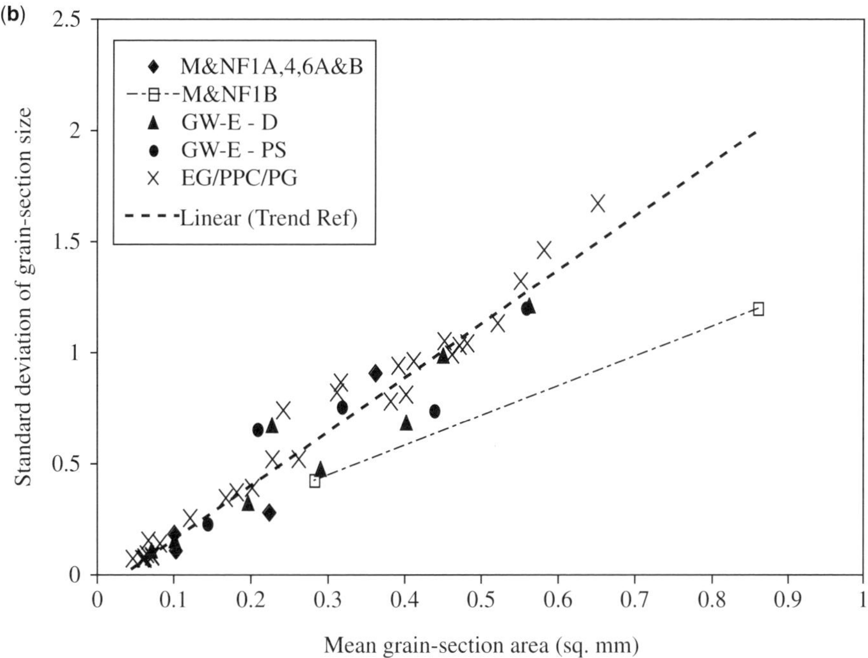

Fig. 3. Standard deviation v. mean grain-section area size for mantle peridotites. (**a**) For samples from which orthogonal (MR1, MR2, KH3) and successive (TE189) thin sections had been prepared, together with the results for a clearly banded thin section (DW83-20) (data from Table 4). (**b**) Mineral identity outlines (skeletons) prepared by other (earlier) authors. Results from the initial study and the linear trend line (Fig. 2a) are included for reference. Two points for M & N F1B (Mercier & Nicolas 1975) represent computations at two different scale sizes (see section on 'Qualitative visual assessment' for details).

plot consistently with our observed linear relationship (Fig. 3b) but one sample (shown in their fig. 1b) falls well away from the general trend line. This skeleton, although nominally representing a similar area to that shown in their figure 1a, contains only about half the number of grain sections. However, if its scale bar is taken to represent 1 mm rather than the 2 mm shown, then the resultant mean and standard deviation from our digital image analysis are consistent with all our other observations (Fig. 3b).

Similarly, Witt (1985) provided eight mineral outline skeletons from spinel peridotite xenoliths from Germany, prepared by hand tracing around the individual grain-section areas on a projected image of the thin sections. Digital scanning and computer analysis of these illustrations (GW-E – D) gave values for the mean grain-section area and standard deviation consistent with our observed linear relationship (Fig. 3b). Five of the original thin sections were scanned, skeletonized and analysed. These gave substantially similar results, again consistent with the linear relationship (Fig. 3b, GW-E – PS). These results show that comparable skeletons prepared independently by other authors give results consistent with our observed linear relationship between grain-section area size and standard deviation, suggesting that this linear relationship is generally applicable to spinel peridotite xenoliths worldwide.

Discussion

Cumulative distribution plots

The discrepancy between qualitative visual size assessments and actual grain-section area measurements (digitization) is explained by a comparison of cumulative number and area distributions (Fig. 4a), in which 95% of the number of grain-section areas only account for 50% of the total thin-section area. The remaining 5% of the grain-section areas occupy the rest of the scanned area and so attract disproportionate visual attention. This is a well-known effect shown by Randolph & Larson (1971).

The cumulative distribution curves (Fig. 4a) also show that the smallest grain-section areas measured (0.0015 mm^2) are significantly greater than the practical lower resolution limit of our equipment (0.00035 mm^2). Furthermore, the smallest grain-section areas amount to only about 1% of the number of grain-section areas measured and represent no more than 0.007% of the total area scanned. The shape of these cumulative distribution curves, plotted on a logarithmic size axis, reflects the highly skewed size distributions that are commonly observed in geological materials (Size 1987) as a result of natural processes.

The nature of grain-section area size distributions in thin sections

The term 'statistics' covers both 'descriptive' and 'inductive' applications. Descriptive statistics usually relate only to the calculation or presentation of figures to summarize or characterize a set of data. No assumptions are made or implied (Gibbons 1985). The calculation of mean, median, variance, range, etc., does not depend on any distribution within the data. If such data constitute a random sample from a certain population, the sample represents that population in miniature (i.e. a proxy) (Herdan 1953) and descriptive statistical parameters provide information regarding the description of the population. Even if the distribution of the data is not 'normal', the description and characterization are still valid (Borradaile 2003). In contrast, inductive statistics or statistical inferences (evaluations and tests of hypotheses, etc.) cannot be made without some information regarding the probability function of the variable relating to the sample description used in the inference procedure (Gibbons 1985).

In common with many geological observations (e.g. Koch & Link 1970; Borradaile 2003), the grain-section area size distributions in our study are positively skewed, with long tails towards higher values. Most of the observations are of small values, with the peak near the lower size limit of grain-section area. It is common practice to visualize such distributions by log-probability plots (Hatch & Choate 1929), and to interpret highly peaked and skewed results as the result of log-normal distributions. Log-probability plots (e.g. Fig 4b), and non-parametric Kolmogorov–Smirnov tests at the 5% significance level both suggest that the grain-section area size distributions found in the present work are not statistically inconsistent with a 'log-normal' model.

The assumption of a log-normal distribution is common in geological applications (e.g. Pettijohn 1957; Irani & Callis 1963; Aitchison & Brown 1969). However, our probability plots tend to depart from linearity at both ends of the size range (Fig. 4b). This characteristic has been observed with other geological systems (e.g. Bagnold 1954), and is often attributed to the effects of limited numbers of larger particles and truncation at the smaller end of the distributions. However, this explanation does not accord with the cumulative distribution curves observed in our study. In many industrial (e.g. Prasher 1987; Bye 1999) and geological applications (e.g. Kittleman 1964; Kotov &

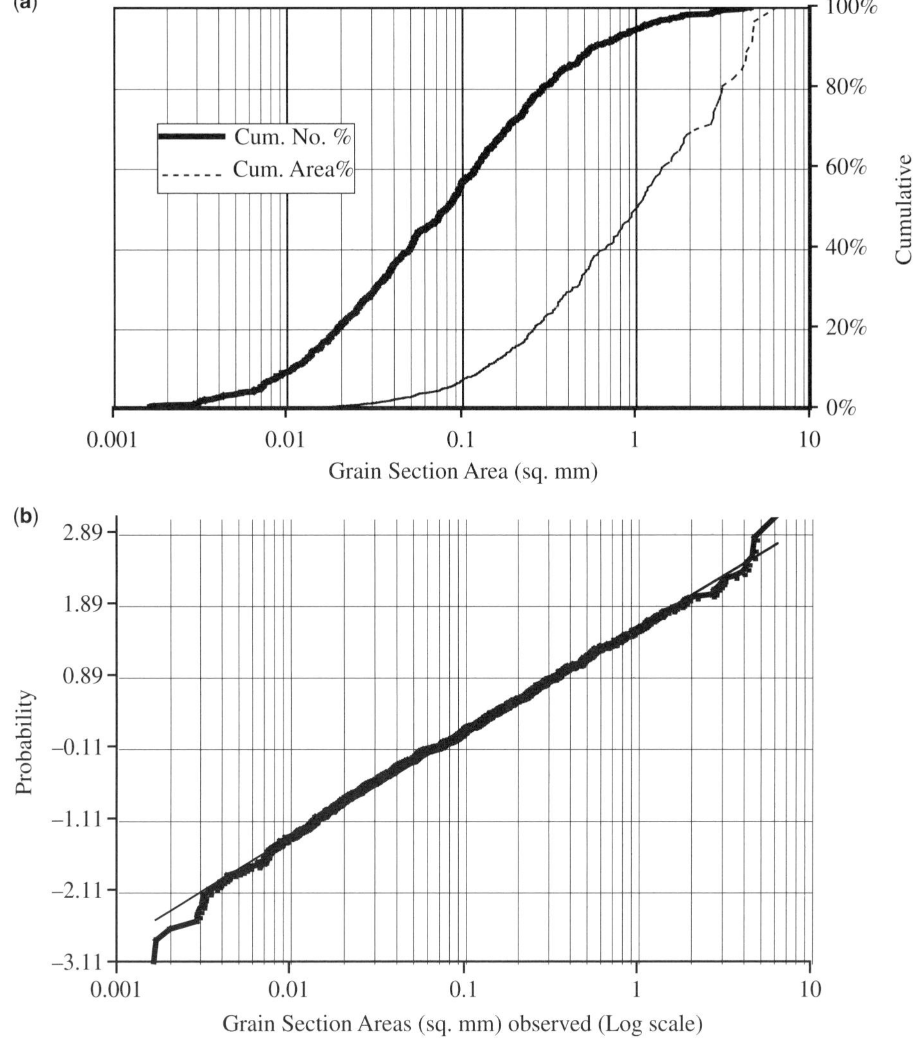

Fig. 4. Grain-section area size distributions. (**a**) Cumulative plots (number of grain-section areas and area occupied by grain-section areas). All other xenoliths examined show similar characteristics. (**b**) Log-probability plot of grain-section areas from the xenolith TAB 23, compared with the linear relationship expected for a log-normal distribution.

Berendsen 2002) better fits have been found using a 'power-law type' equation originally proposed by Rosin & Rammler (1933). A power-law relationship for olivine grain-size distributions in mantle xenoliths was also proposed by Armienti & Tarquini (2002). Like the log-normal distribution, the power-law distribution can be deduced from a model of grinding and crushing (Bennett 1936; Herdan 1953) or crystallization (Kottler 1950). A mechanism that gives rise to positively skewed frequency distributions can be based on the Law of Proportionate Effect (Herdan 1953), in which each stage of the process has a proportionate effect on the subsequent outcome (Vistelius 1960; Agterberg 1974). Many textural studies are based on nucleation and growth processes that inherently assume constant growth rates. However, Eberl *et al.* (2002) argued that observed size distributions may be better explained by size-dependent proportionate models.

This characteristic behavior can be demonstrated with a simplified model, described in Figure 5. For a body/particle subjected to fragmentation the resulting subdivision distribution will only be uniform if, at each successive stage, each component is divided into two equal parts. Unless

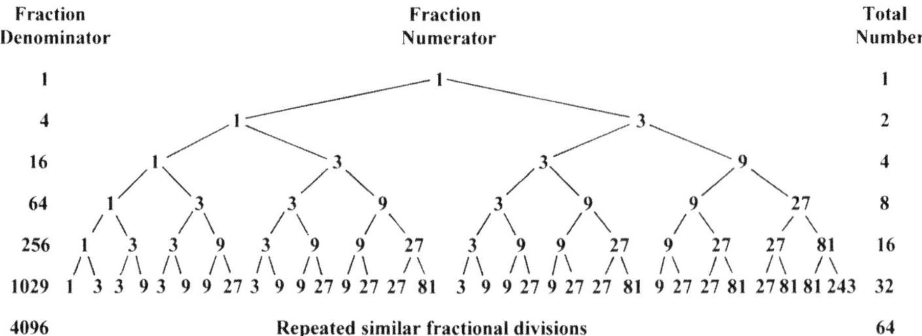

Fig. 5. Fragmentation patterns derived from the progressive $\frac{1}{4}$–$\frac{3}{4}$ division. Subsequent similar fractional divisions produce rapidly increased numbers of fragments (1024 at the 10th division) with increasingly positive skewed distributions (Fig. 6a).

there are special features of the material being fragmented, it is more likely that two unequal parts (a larger and a smaller) will occur in a random fashion. As such, this is difficult to model but the general behaviour can be seen with a simplified case in which the proportional division is the same at each stage (e.g. as in Fig. 5, $\frac{1}{4}$ and $\frac{3}{4}$).

Frequency distributions generated by this model (Fig. 6a) are increasingly positively skewed with each successive stage. Similar fragmentation patterns (e.g. $\frac{1}{3}$–$\frac{2}{3}$, $\frac{1}{5}$–$\frac{4}{5}$ and $\frac{1}{6}$–$\frac{5}{6}$) generate analogous distributions that become more rapidly skewed at each successive fragmentation stage with increasing disparity between the sizes of the fragments (i.e. $\frac{1}{3}$–$\frac{2}{3}$ shows a less skewed distribution than $\frac{1}{5}$–$\frac{4}{5}$). Plots of the standard deviation v. mean fragment size for each of these distributions are gently curved and over a limited range could approximate to a linear relationship. However, the fragmentation patterns generate some stages that give rise to values for skewness and kurtosis that are either greater than or less than those found for the samples examined in our study. Selecting only those fragmentation stages that gave skewness and kurtosis values found within the observed range for mantle peridotites gave a plot of standard deviation v. mean fragment size (Fig. 6b) consistent with the linear relationship we have discovered (e.g. Fig. 2).

From our results it might be suggested that, as a result of plate tectonics, the shallow lithospheric mantle represented by the xenoliths has been subject to tectonic and metamorphic processes of comminution and/or grain growth that are dependent on the Law of Proportionate Effect. It is beyond the scope of this paper to speculate on the nature of these geological processes, other than to say that both mechanical grinding and recrystallization can give rise to very similar mathematics (Kottler 1950).

Conclusions

This study shows that textures in mantle-derived spinel peridotite xenoliths can be quantified using manual skeletonization and digital image-analysis software. There is a linear relationship between the mean grain-section area size and standard deviation of the grain-section area size for all the samples studied. This method provides a means for consistent comparison of samples from different localities that is independent of the observer.

Protogranular and equigranular peridotites that form the two end members of textures (Mercier & Nicolas 1975) can be clearly distinguished using our method. Porphyroclastic peridotites tend to overlap protogranular ones at larger grain-section areas and equigranular ones at smaller grain-section areas, suggesting that spinel peridotite textures form a continuum rather than being discrete entities. This is further supported by the observation that with some thin sections, although different parts may show a variation in mean grain-section area and standard deviation, they still plot on the same linear relationship.

With markedly skewed and peaked distributions of the type observed in our study, it is difficult to assign distribution functions with any certainty. It is probably better to use non-parametric methods to examine and compare measured grain-section areas, and to plot them on the graph of the linear mean grain-section area v. standard deviation of mean grain-section area (Fig. 2). Since all the samples examined have the marked skewed peak and long right-hand (larger size) tail in common, it is reasonable to suppose that this is characteristic of the population from which they were drawn. It is, therefore, possible that the apparent variations in texture might be no more than the statistical fluctuation to be expected when drawing small

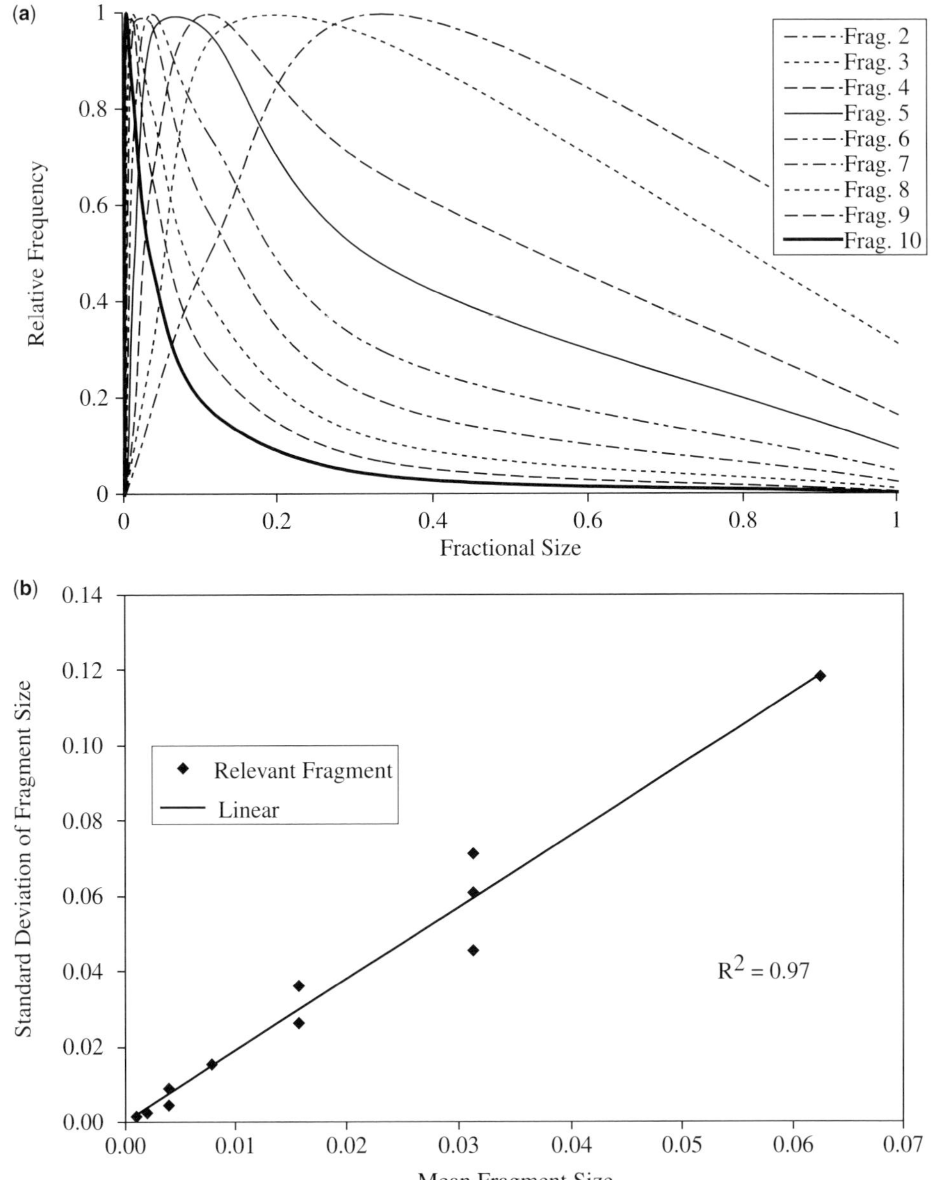

Fig. 6. 'Proportionate Effect' simulation. (**a**) Computed model relative frequency distributions based on the simplified $\frac{1}{4}$–$\frac{3}{4}$ fragmentation pattern (Fig. 5). Frag., successive stages of fragmentation. (**b**) Standard deviation v. mean fragment size plot for relevant distribution fragment stages that have skewness and kurtosis parameters within the range found experimentally.

(random) samples from the very much larger population representing the lithospheric upper mantle. In these circumstances it is probably even more important to avoid terminology that might be thought to imply separate/discrete states and formation mechanisms.

We thank G. Witt-Eickschen, P. Kempton and E. Rall for providing samples from Germany and the USA. We also gratefully acknowledge the skill of A. Beard in preparing thin sections of friable xenoliths and of S. Hirons for help with preparing the figures. The authors wish to acknowledge that this paper was significantly improved by suggestions from M. Holness, P. Armienti, M. Wilson, M. Higgins and M. Mazzucchelli.

References

AGTERBERG, F. P. 1974. *Geomathematics*. Elsevier, New York, 206–207.

AITCHISON, J. & BROWN, J. A. C. 1969. *The Lognormal Distribution*. Cambridge University Press, Cambridge.

ARMIENTI, P. & TARQUINI, S. 2002. Power law olivine crystal size distributions in lithospheric mantle xenoliths. *Lithos*, **65**, 273–285.

AVE LALLEMANT, H. G., MERCIER, J.-C. C., CARTER, N. L. & ROSS, J. V. 1980. Rheology of the upper mantle: inferences from peridotite xenoliths. *Tectonophysics*, **70**, 85–113.

BAGNOLD, R. A. 1954. *The Physics of Blown Sand and Desert Dunes*. Methuen, London.

BENNETT, J. G. 1936. Broken coal. *Journal of the Institute of Fuel*, **10**, 22–39.

BINDEMAN, I. N. 2005. Fragmentation phenomena in populations of magmatic crystals. *American Mineralogist*, **90**, 1801–1815.

BORRADAILE, G. J. 2003. *Statistics of Earth Science Data*. Springer, Berlin.

BOULLIER, A. M. & NICOLAS, A. 1975. Classification of textures and fabrics of peridotite xenoliths from South African kimberlites. *In*: AHRENS, L. H., DAWSON, J. B., DUNCAN, A. R. & ERLANK, A. J. (eds) *Physics and Chemistry of the Earth*, Volume 9. Pergamon, London, 97–105.

BYE, G. C. 1999. *Portland Cement*, 2nd edn. Thomas Telford, London, 76–78.

COISY, P. & NICOLAS, A. 1978. Regional structure and geodynamics of the upper mantle beneath the Massif Central. *Nature*, **274**, 429–432.

CHAYES, F. 1956. *Petrographic Modal Analysis*. Wiley, Chichester.

COLLEE, A. L. G. 1963. A fabric study of lherzolites with special reference to ultrabasic nodular inclusions in the lavas of Auvergne (France). *Leidse Geologische Mededelingen*, **28**, 1–102.

DOWNES, H. 1997. Shallow continental lithospheric mantle heterogeneity – petrological constraints. *In*: FUCHS, K. (ed.) *Upper Mantle Heterogeneities from Active and Passive Seismology*. Kluwer, Dordrecht, 295–308.

DOWNES, H. 2001. Formation and modification of the shallow sub-continental lithospheric mantle: a review of geochemical evidence from ultramafic xenolith suites and tectonically emplaced ultramafic massifs of Western and Central Europe. *Journal of Petrology*, **42**(1), 233–250.

DOWNES, H., REICHOW, M. K., MASON, P. R. D., BEARD, A. D. & THIRLWALL, M. F. 2003. Mantle domains in the lithosphere beneath the French Massif Central: trace element and isotopic evidence from mantle clinopyroxenes. *Chemical Geology*, **200**, 71–87.

EBERL, D. D., KILE, D. E. & DRITS, V. A. 2002. On geological interpretations of crystal size distributions: Constant vs. proportionate growth. *American Mineralogist*, **87**, 1235–1241.

ETHERIDGE, M. A. & WILKIE, J. C. 1979. The geometry and microstructure of a range of QP-mylonite zones. US Geological Survey, Open File Report, **79-1239**, 448–504.

ETHERIDGE, M. A. & WILKIE, J. C. 1981. An assessment of dynamically recrystallized grain size as a palaeopiezometer in quartz-bearing mylonite zones. *Tectonophysics*, **78**, 475–508.

GIBBONS, J. D. 1985. *Nonparametric Statistical Inference*, 2nd edn. Marcel Dekker, New York.

HARTE, B. 1977. Rock nomenclature with particular relation to deformation and recrystallisation textures in olivine-bearing xenoliths. *Journal of Geology*, **85**, 279–288.

HARTE, B. 1983. Mantle peridotites and processes – the kimberlite sample. *In*: HAWKESWORTH, C. J. & NORRY, M. J. (eds) *Continental Basalts and Mantle Xenoliths*. Shiva, Nantwich, 46–91.

HATCH, T. & CHOATE, S. P. 1929. Statistical description of the size properties of non-uniform particulate substances. *Journal of the Franklin Institute*, **207**, 369–386.

HERDAN, G. 1953. *Small Particle Statistics*. Elsevier, Amsterdam.

HIGGINS, M. D. 1994. Determination of crystal morphology and size from bulk measurements on thin sections: numerical modeling. *American Mineralogist*, **79**, 113–119.

HIGGINS, M. D. 2000. Measurement of crystal size distributions. *American Mineralogist*, **85**, 1105–1116.

HIGGINS, M. D. 2006. *Quantitative Textural Measurements in Igneous and Metamorphic Petrology*. Cambridge University Press, Cambridge.

HOLMES, A. 1927. *Petrographic Methods and Calculations*. Thomas Murby and Co., London.

HUMPHRIES, D. W. 1969. Mensuration methods in optical microscopy. *In*: BARER, R. & COSSLETT, V. E. (eds) *Advances in Optical and Electron Microscopy*, Volume 3. Academic Press, London, 33–99.

IRANI, R. R. & CALLIS, C. F. 1963. *Particle Size: Measurement, Interpretation and Application*. Wiley, New York.

KITTLEMAN, L. R. 1964. Application of Rosin's distribution in size–frequency analysis of clastic rocks. *Journal of Sedimentary Petrology*, **34**(3), 483–502.

KOCH, G. S. & LINK, R. F. 1970. *Statistical Analysis of Geological Data*, (Dover Edition 1980.) Dover Publications, New York.

KOTTLER, F. 1950. The distribution of particle size. *Journal of the Franklin Institute*, **250**, 339–356.

KOTOV, S. & BERENDSEN, P. 2002. Statistical characteristics of xenoliths in the Antioch kimberlite pipe, Marshall county, northeastern Kansas. *Natural Resources Research*, **11**, 289–297.

LENOIR, X., GARRIDO, C. J., BODINIER, J.-L. & DAUTRIA, J.-M. 2000. Contrasting lithospheric mantle domains beneath the Massif Central (France) revealed by geochemistry of peridotite xenoliths. *Earth and Planetary Science Letters*, **181**, 359–375.

MARSH, B. D. 1998. On the interpretation of crystal size distributions in magmatic systems. *Journal of Petrology*, **39**(4), 553–599.

MENZIES, M. 1983. Mantle ultramafic xenoliths in alkaline magmas: Evidence for mantle heterogeneity modified by magmatic activity. *In*: HAWKESWORTH, C. J. &

NORRY, M. J. (eds) *Continental Basalts and Mantle Xenoliths*. Shiva, Nantwich, 92–110.

MERCIER, J.-C. C. & NICOLAS, A. 1975. Textures and fabrics of upper-mantle peridotites as illustrated by xenoliths from basalts. *Journal of Petrology*, **16**(2), 454–487.

NIELSON-PIKE, J. E. & SCHWARZMAN, E. C. 1977. Classification of textures in ultramafic xenoliths. *Journal of Geology*, **85**, 49–61.

PERRING, C. S., BARNES, S. J., VERRALL, M. & HILL, R. E. T. 2004. Using automated digital image analysis to provide quantitative petrographic data on olivine–phyric basalts. *Computers & Geosciences*, **30**, 183–195.

PETTIJOHN, F. P. 1957. *Sedimentary Rocks*, 2nd edn. Harper & Row, New York.

PRASHER, C. L. 1987. *Crushing and Grinding Process Handbook*. Wiley, Chichester, 87–88.

RANALLI, G. 1984. Grain size distribution and flow stress in tectonites. *Journal of Structural Geology*, **6**(4), 443–447.

RANDOLPH, A. D. & LARSON, M. A. 1971. *Theory of Particulate Processes*, 2nd edn. Academic Press, New York, 6–7.

ROSIN, P. & RAMMLER, E. 1933. Laws governing the fineness of powdered coal. *Journal of the Institute of Fuel*, **7**, 29–36.

ROYET, J.-P. 1991. Stereology: a method for analyzing images. *Progress in Neurobiology*, **37**, 433–474.

RUSS, J. C. & DEHOFF, R. T. 2000. *Practical Stereology*. 2nd edition. Kluwer Academic/Plenum Publishers, New York.

SCHOLS, P. & SMETS, E. 2001. Carnoy; analysis software for LM, SEM and TEM images. Leuven; distributed by the authors. Web address: http://www.carnoy.org.

SIZE, W. B. 1987. Use of representative samples and sampling plans in describing geologic variability and trends. *In*: SIZE, W. B. (ed.) *Use and Abuse of Statistical Methods in the Earth Sciences*. Oxford University Press, Oxford.

SMITH, C. S. & GUTTMAN, L. 1953. Measurement of internal boundaries in three-dimensional structures by random sectioning. *Transactions of the American Institute of Mineral & Metals Engineers*, **197**, 81–87 and 1561.

SWAN, A. R. H. & SANDILANDS, M. 1995. *Introduction to Geological Data Analysis*. Blackwell Science, Oxford.

TABOR, F. A. 2004*a*. *Mantle Xenoliths. A Quantitative Characterisation of Textures*. MRes thesis, Birkbeck College, University of London.

TABOR, B. E. 2004*b*. If you can't count it, it doesn't count: Quantitative Microscopy. *Quekett Journal of Microscopy*, **39**, 715–722.

TABOR, F. A. 2005. Spinel lherzolite xenoliths, quantitative characterization of textures (Abstract). *Ofioliti*, **30**, 221.

TARQUINI, S. & ARMIENTI, P. 2003. Quick determination of crystal size distributions of rocks by means of a color scanner. *Image Analysis Stereology*, **22**, 27–34.

VERNON, R. H. 1996. Observation versus argument by authority – the origin of enclaves in granites. *Journal of Geoscience Education*, **44**, 57–64.

VERNON, R. H. 2004. *A Practical Guide to Rock Microstructure. 1.10 Importance of Evidence*. Cambridge University Press, Cambridge, 8–9.

VISTELIUS, A. B. 1960. The skew frequency distributions and the fundamental law of the geochemical processes. *Journal of Geology*, **68**(1), 1–22.

WITT, G. 1985. *Die junge Evolution des Erdmantels unter der rheinischen Vulkanprovinz*. PhD thesis, Universitat zu Koln.

ZANGANA, N. A. H. 1995. *Geochemical Variations in Mantle Xenoliths from Ray Pic, Massif Central, France*. PhD thesis, University of London.

ZANGANA, N. A., DOWNES, H., THIRLWALL, M. F. & HEGNER, E. 1997. Relationship between deformation, equilibration temperatures, REE and radiogenic isotopes in mantle xenoliths (Ray Pic, Massif Central, France): an example of plume–lithosphere interaction? *Contributions to Mineralogy and Petrology*, **127**, 187–203.

Mafic alkaline metasomatism in the lithosphere underneath East Serbia: evidence from the study of xenoliths and the host alkali basalts

V. CVETKOVIĆ[1]*, H. DOWNES[2], V. HÖCK[3], D. PRELEVIĆ[4] & M. LAZAROV[5]

[1]*Faculty of Mining and Geology, University of Belgrade, Đušina 7, 11000 Belgrade, Serbia*

[2]*School of Earth Sciences, Birkbeck University of London, Malet Street, London WC1E 7HX, UK*

[3]*Fachbereich für Geographie, Geologie und Mineralogie, Fakultät für Naturwissenschaften, Universität Salzburg, Hellbrunnerstraße 34/III, A-5020 Salzburg, Austria*

[4]*Johannes Gutenberg-Universität, Institut für Geowissenschaften FB 22 Mineralogie, Becherweg 21, 55099 Mainz, Germany*

[5]*Institut für Mineralogie, Johann Wolfgang Goethe-Universität, Seckenberganlage 28, 60054 Frankfurt am Main, Germany*

**Corresponding author (e-mail: cvladica@rgf.bg.ac.rs)*

Abstract: Effects of mafic alkaline metasomatism have been investigated by a combined study of the East Serbian mantle xenoliths and their host alkaline rocks. Fertile xenoliths and tiny mineral assemblages found in depleted xenoliths have been investigated. Fertile lithologies are represented by clinopyroxene (cpx)-rich lherzolite and spinel (sp)-rich olivine websterite containing Ti–Al-rich Cr-augite, Fe-rich olivine, Fe–Al-rich orthopyroxene and Al-rich spinel. Depleted xenoliths, which are the predominant lithology in the suite of East Serbian xenoliths, are harzburgite, cpx-poor lherzolite and rare Mg-rich dunite. They contain small-scale assemblages occurring as pocket-like, symplectitic or irregular, deformation-assisted accumulations of metasomatic phases, generally composed of Ti–Al- and incompatible element-rich Cr-diopside, Cr–Fe–Ti-rich spinel, altered glass, olivine, apatite, ilmenite, carbonate, feldspar, and a high-TiO_2 (*c.* 11 wt%) phlogopite. The fertile xenoliths are too rich in Al, Ca and Fe to simply represent undepleted mantle. By contrast, their composition can be reproduced by the addition of 5–20 wt% of a basanitic melt to refractory mantle. However, textural relationships found in tiny mineral assemblages inside depleted xenoliths imply the following reaction: opx + sp1 (primary mantle Cr-spinel) ± phlogopite + Si-poor alkaline melt = Ti–Al-cpx + sp2 (metasomatic Ti-rich spinel) ± ol ± other minor phases. Inversion modelling, performed on the least contaminated and most isotopically uniform host basanites ($^{87}Sr/^{86}Sr$ = *c.* 0.7031; $^{143}Nd/^{144}Nd$ = *c.* 0.5129), implies a source that was enriched in highly and moderately incompatible elements (*c.* 35–40× chondrite for U–Th–Nb–Ta, 2× chondrite for heavy rare earth elements (HREE), made up of clinopyroxene, carbonate (*c.* 5%), and traces of ilmenite (*c.* 1%) and apatite (*c.* 0.05%). A schematic model involves: first, percolation of CO_2- and H_2O-rich fluids and precipitation of metasomatic hydrous minerals; and, second, the subsequent breakdown of these hydrous minerals due to the further uplift of hot asthenospheric mantle. This model links intraplate alkaline magmatism to lithospheric mantle sources enriched by sublithospheric melts at some time in the past.

Mantle metasomatism is generally believed to be responsible for small-scale heterogeneities within the lithosphere (Dawson 1984; Hart 1984; Harte 1987; Menzies *et al.* 1987). Evidence from the study of mantle xenoliths entrained in alkali basalts proved to be particularly important for understanding the origin of such heterogeneities. Previous studies have shown that the major metasomatic agents are: (i) carbonatitic melts (Baker *et al.* 1998; Yaxley *et al.* 1998; Gorring & Kay 2000; Wang & Gasparik 2001); (ii) silicate melts (Menzies *et al.* 1987; Kepezhinskas *et al.* 1995, 1996; Vannucci *et al.* 1998; Zangana *et al.* 1999; Grégoire *et al.* 2000; Schiano *et al.* 2002); and (iii) fluids (O'Reilly & Griffin 1988; Baker *et al.* 1998; Gorring & Kay 2000; Larsen *et al.* 2003) – the latter two having a wide range of compositions (Coltorti *et al.* 2004). All of these agents produce different metasomatic styles recognized by changes in primary mantle mineralogy and specific enrichments in trace element composition.

These metasomatic agents dissolve primary orthopyroxene (±spinel) and precipitate clinopyroxene and spinel, along with other minor phases such as apatite, ilmenite or carbonate. The metasomatic phases are sometimes found filling veinlets

From: Coltorti, M., Downes, H., Grégoire, M. & O'Reilly, S. Y. (eds) *Petrological Evolution of the European Lithospheric Mantle*. Geological Society, London, Special Publications, **337**, 213–239.
DOI: 10.1144/SP337.11 0305-8719/10/$15.00

and patchy accumulations within xenoliths. In addition, the same xenolith suites commonly contain granular clinopyroxene-rich lherzolite or wehrlite xenoliths, interpreted as products of similar melt–rock interactions (e.g. Carpenter *et al.* 2002). It is widely accepted that mafic alkaline metasomatism is produced by infiltration of small amounts of alkaline melts genetically related to the host alkaline magmas (e.g. Dawson 2002).

A detailed understanding of the effects of mafic alkaline metasomatism is usually difficult to achieve because of the following problems: (i) primary metasomatic products and textural relationships may often be destroyed or/and superimposed by those resulting from interaction with the host magma during magma ascent or immediately before eruption (e.g. Klügel 1998); and (ii) compositional changes caused by carbonatitic melts/fluids usually occur roughly concomitantly with silicate metasomatism, producing an overlap in geochemical effects (Neumann *et al.* 2002; Rivalenti *et al.* 2004; Bonadiman *et al.* 2008). In this context, the information provided by a careful study of textural relationships and mineral compositional variations within the presumed metasomatic associations is essential. Recognition of late-stage associations that texturally and compositionally appear to be unrelated to the host magma is very important for proving that these metasomatic associations genuinely originate in the mantle. In addition, significant information about alkali silicate metasomatism can be potentially provided from the study of the host rocks. It is generally accepted that alkaline rocks that host mantle xenoliths, although having a strong asthenospheric signature, usually require a metasomatic component believed to reside in the lithosphere (e.g. Wilson & Downes 2006 and references therein). Because these rocks are fairly homogeneous isotopically, it is believed that this metasomatism occurred shortly before the eruption and xenolith capture. It is, therefore, logical to suppose that these lithospheric domains, which are believed to have played a role in the petrogenesis of the host alkaline rocks, might be compositionally (and texturally?) similar to metasomatic associations found in xenoliths.

In this paper we revisit the problem of mafic alkaline metasomatic effects in mantle xenoliths from East Serbia. The xenoliths have been studied for mineral major [electron probe microanalysis (EPMA)] and trace element [laser ablation inductively coupled mass spectrometry (LA-ICP-MS)] compositions, and the new data are discussed together with some previously published results (Cvetković *et al.* 2004*a*). We present new textural and mineral chemistry evidence that provides better insight into these metasomatic events. In addition, we report a new set of whole rock ICP-MS trace element data on host basanites and perform quantitative inversion modelling in order to infer the mineralogy of their sources. These results were used to compare the inferred mantle source with the metasomatic associations observed in the studied xenoliths. Consequently, we offer a model in which mafic alkaline metasomatic associations, similar to those observed in the studied xenoliths, play an important role in the petrogenesis of the host rocks, putting forward the idea that metasomatism may be considered a precursor of the magmatism.

Characteristics of the local upper mantle and volcanism

The studied mantle xenoliths are found in Palaeogene basanites of East Serbia (SE Europe). The characteristics of the main xenolith lithologies representing the East Serbian lithospheric mantle (ESLM) are reported by Cvetković *et al.* (2004*a*) and summarized in Table 1. The predominant lithology is composed of very depleted harzburgite and clinopyroxene-poor lherzolites, as well as rare Mg-rich dunites. Cvetković *et al.* (2007*a*) discussed the problem of the very high degree of depletion, and suggested that the mantle segment underneath East Serbia is similar to sub-arc oceanic mantle and may represent a slice of Tethyan oceanic lithosphere accreted during convergence. A subordinate lithology of spinel-poor, orthopyroxene-rich olivine websterite xenoliths is interpreted to have originated by crystallization of high-Mg and silica-saturated magmas (Cvetković *et al.* 2007*b*). The existence of such magmas is independent confirmation of the existence of ultra-depleted mantle underneath the region. Various subordinate xenolith lithologies containing clinopyroxene-rich and altered glass-bearing pocket assemblages were attributed to the effects of mafic alkaline metasomatism. Cvetković *et al.* (2004*b*) described some textural relationships of these presumed metasomatic assemblages and reported trace element patterns of clinopyroxene.

The East Serbian Palaeogene alkaline rocks occur as relicts of small eroded monogenetic volcanoes along the western margin of the Carpatho-Balkanides (Jovanović *et al.* 2001; Cvetković *et al.* 2004*a*). This alkaline magmatism originated within an orogenic setting as small pulses of magma between 60 and 40 Ma ago (Jovanović *et al.* 2001; Cvetković *et al.* 2004*a*). Ten localities of these rocks are known and at two of them, Sokobanja and Striževac, mantle xenoliths are present (Cvetković *et al.* 2004*b*). The rocks are olivine- and clinopyroxene-phyric, and range in composition from basanites and olivine tephrites to tephriphonolites. Their isotope characteristics and trace element patterns are generally similar to alkaline rocks of the

Table 1. *Main characteristics of xenoliths from East Serbian Palaeogene basanites*

Xenolith type, equilibration temperature	Abundance, size, texture	Olivine	Orthopyroxene	Clinopyroxene	Spinel	Other minerals (in pockets and veins)
Mg-dunite/harzburgite/lherzolite, 950–1100 °C	Most abundant, *c.* 90% of the suite; few mm to 10 cm, protogranular, undeformed or slightly deformed	*c.* 75 vol.% $Fo_{89.6-92}$	*c.* 20 vol.% Mg# = 90.5–92.5; Al_2O_3 = 1–2 wt%	Mostly <5 vol.% Mg# = 91–93 Al_2O_3 = 1.4–3.4 wt%; Cr_2O_3 = 0.5–2 wt%	*c.* 1 vol.% Cr# = 0.5–0.7	Clinopyroxene, spinel, ±apatite, ± ilmenite, ±carbonate, ±glass (see text for detailed description)
Sp-poor, opx-rich olivine websterite*, 800–1200 °C	Rare; 1–4 cm, protogranular to cumulitic, undeformed	<20 vol.% Fo_{85-88}	*c.* 70–80 vol.% Mg# = 86–87; Al_2O_3 = 1.5–2.5 wt%	*c.* 5–10 vol.% Mg# = 85–92 TiO_2 = 0.2–0.7 wt%; Al_2O_3 = <1–4.4 wt%; Cr_2O_3 >1 wt%	≪1 vol.% TiO_2 = 0.3–11.5 wt%; Cr# = 0.8–0.9	Similar secondary minerals as in the first group, occurring in the interstices
Sp-rich olivine, websterite*, 1050–1150 °C	Rare; 1–2 cm, protogranular, undeformed	<20 vol.% Fo_{86-88}	*c.* 70–75 vol.% Mg# *c.* 89; Al_2O_3 = 3–6 wt%	*c.* 10–20 vol.% Mg# = 87–88 TiO_2 < 0.5 wt%; Al_2O_3 *c.* 5 wt%; Cr_2O_3 > 1.5 wt%	>2 vol.% TiO_2 >1 wt%; Cr# = 0.3–0.35	
Cpx-rich lherzolite, 1000–1200 °C	Rare; 1–3 cm, protogranular, undeformed	*c.* 60 vol.% Fo_{86-88}	*c.* 10–20 vol.% Mg# = 87–88; Al_2O_3 > 4 wt%	*c.* 20 vol.% Mg# = 85–92 TiO_2 = ≥1 wt%; Al_2O_3 = 5–7 wt%; Cr_2O_3 ≥ 1 wt%	*c.* 2–3 vol.% TiO_2 *c.* 0.5 wt%; Cr# = 0.15–0.2	
Cpx (±olivine) megacrysts*	Abundant, from <0.5 cm to >5 cm	Fo_{86}		Mg# = 85–86 TiO_2 = *c.* 1 wt%; Al_2O_3 = 6–10 wt%; Cr_2O_3 *c.* 0.5 wt%		
Fe-rich dunite*	Rare, 1–3 cm	>95 vol.% Fo_{85-86}		<5 vol.%; Composition similar to clinopyroxene megacrysts		

*Not addressed in this study (see Cvetković *et al.* 2004*a*, 2007*b*).

Central European Cenozoic volcanism (e.g. Wedepohl & Baumann 1999), which are usually interpreted as derived from partial melting of mixed asthenospheric and lithospheric sources (Wilson & Downes 1991). However, mantle xenolith-bearing basanites from Sokobanja are characterized by uniformly low initial strontium isotope ratios ($^{87}Sr/^{86}Sr = 0.7029-0.7035$: Cvetković *et al.* 2004*b*) and can be potentially considered as derived from nearly pure asthenospheric melts. This assumption will be addressed in part where the mantle source characteristics of these lavas are discussed (see below).

Analytical methods

Major element microprobe data of minerals from these metasomatic associations are given in Table 2. The data were obtained using a wavelength-dispersive (WDS) electron microprobe at the University of Frankfurt. The analytical conditions are summarized by Cvetković *et al.* (2007*b*). Additional mineral chemical data used here can be found in Cvetković *et al.* (2004*b*, 2007*a*, *b*).

Trace element concentrations in minerals were determined at the University of Frankfurt using an *in situ* LA-ICP-MS technique using a New Wave Research LUV213™ ultraviolet Nd_Yag laser coupled with a Finnigan MAT ELEMENT2™. The pulse frequency of 10 Hz, pulse energy in the range of 03–1.2 mJ, and spot size between 30 and 150 μm were used. The USGS BIR 1-G glass (Eggins *et al.* 1997) and NIST612 glass (Pearce *et al.* 1997) were used as external standards. Silica was used as an internal standard. Trace element concentrations in olivines from a symplectite-bearing Mg-rich dunite xenolith (SB-M-3) and from a Fe-rich dunite xenolith (SB-M-1), as well as in olivines from refractory peridotite xenoliths (an average analysis from SB-M-4 and SB-M-5), are reported in Table 3. Only three clinopyroxene crystals from metasomatic pockets were successfully ablated and their analyses are also given in Table 3, along with compositions of discrete clinopyroxene and clinopyroxene in a lenticular symplectite from the Mg-rich dunite xenolith.

Trace element concentrations in whole-rock samples of host Sokobanja basanites (Table 4), which were used for geochemical modelling, were obtained using an ICP-MS technique in the ACME Analytical Laboratories Ltd (Vancouver, BC, Canada).

Petrography and major element mineral chemistry

Here we present characteristics of the mantle lithologies that we believe are related to enrichments by mafic silicate alkaline melts and describe newly observed textural relationships. For more general information on the xenolith petrography the reader is referred to Cvetković *et al.* (2004*b*). These supposed metasomatic lithologies are recognized as fertile xenoliths and as tiny mineral + glass assemblages situated within depleted xenoliths. Clinopyroxene (±olivine) megacrysts and Fe-rich dunite xenoliths are also present but they are not addressed here, except for comparison. Cvetković *et al.* (2004*b*) demonstrated that these represent deep-seated magmatic cumulates.

The fertile xenoliths are represented by undeformed and protogranular clinopyroxene-rich lherzolite and spinel-rich olivine websterite xenoliths. The lherzolites have the highest modal content of clinopyroxene (*c.* 20 vol%) in the ESLM xenolith suite. The clinopyroxenes show a spongy texture (Fig. 1a), suggesting incipient melting. Similar textures are found in many xenolith suites and are especially frequent in wehrlite xenoliths (e.g. Yaxley *et al.* 1991; Xu *et al.* 1996; Weichert *et al.* 1997; Bonadiman *et al.* 2001; Carpenter *et al.* 2002). The olivine websterite xenoliths are predominantly composed of tabular and undeformed orthopyroxene, smaller and less abundant olivine, and interstitial spongy greenish clinopyroxene. Spinel is particularly abundant (>2 vol%) and always appears as coarse-grained accumulations associated with clinopyroxene.

The clinopyroxene-rich lherzolite and spinel-rich websterite xenoliths are very similar in terms of mineral chemistry (Cvetković *et al.* 2004*b*). They both contain high-Ti–Al-rich Cr augite (*c.* 1 wt% TiO_2, 5–7 wt% Al_2O_3, 1–1.5 wt% Cr_2O_3), Fe-rich olivine (Mg# <88), Fe- and Al-rich orthopyroxene (Mg# *c.* 88, Al_2O_3 up to 6 wt%) and Al-rich spinel (Cr# 0.14–0.4).

In depleted peridotite xenoliths, patchy and pocket-like mineral accumulations are also very common, while they are typically lacking in fertile lithologies. They are composed of euhedral/subhedral clinopyroxene, anhedral spinel and altered glass, but substantial amounts of olivine, apatite, ilmenite, carbonate and feldspar are often present. A few high-TiO_2 (*c.* 11 wt%) phlogopite crystals were found in one harzburgite xenolith (Cvetković *et al.* 2004*b*). They are represented by very small flakes with diffuse boundaries to the adjacent clinopyroxene, carbonate and altered glass (Fig. 1b). Textural evidence of orthopyroxene dissolution and formation of clinopyroxene selvages (Fig. 1c), precipitation of new spinel (Fig. 1d, e), and spinel dissolution and overgrowths is also present (Fig. 1e, analyses 12–14 in Table 2). In some pockets clinopyroxene occurs as fairly euhedral phenocryst-like crystals associated with tiny spinel and completely enclosed by carbonate

Table 2. *(**a**) Selected microprobe data of clinopyroxene occurring in various metasomatic associations in East Serbian xenoliths. (**b**) Selected microprobe data of spinel occurring in various metasomatic associations in East Serbian xenoliths. (**c**) Selected microprobe data of carbonate and apatite occurring in patchy metasomatic assemblages within East Serbian xenoliths. (**d**) Selected microprobe data of feldspar, glass and ilmenite occurring in patchy metasomatic assemblages within East Serbian xenoliths*

(a) **Clinopyroxene**

	Patchy metasomatic assemblages within depleted xenoliths									Clinopyroxene-spinel-olivine symplectites				
	SB-M-5					STZ-20-4			X-11/2	SB-M-3				
	1	2	3	4	5	6	7	8	9	10	11	12	13	14
SiO_2	49.72	49.30	50.55	47.44	48.37	47.20	48.29	47.51	48.82	52.19	51.82	50.18	50.84	50.76
TiO_2	0.40	0.28	0.85	1.33	0.28	0.44	0.35	0.42	1.48	0.25	0.29	0.81	0.61	0.74
Al_2O_3	6.61	7.12	5.20	7.83	7.42	8.66	8.08	8.57	6.49	5.32	5.80	5.75	6.18	6.28
FeO	2.98	2.95	3.05	3.52	3.02	2.74	2.57	2.56	2.92	3.08	3.06	3.24	3.14	2.48
MnO	0.07	0.03	0.08	0.09	0.08	0.06	0.03	0.07	0.05	0.11	0.12	0.06	0.09	0.05
MgO	15.46	15.12	15.77	13.78	15.06	13.78	14.25	13.80	14.84	16.43	16.23	15.09	15.77	15.79
CaO	20.60	21.07	21.12	20.02	21.14	21.14	21.31	21.01	22.47	20.20	20.00	23.32	21.12	21.81
Na_2O	0.73	0.82	0.87	0.98	0.94	1.00	0.90	0.92	0.57	1.10	1.11	0.45	0.89	0.70
Cr_2O_3	2.41	2.60	1.60	4.10	2.78	4.14	3.59	4.36	1.62	0.89	1.05	1.03	1.18	1.03
Total	99.04	99.39	99.19	99.17	99.15	99.21	99.46	99.27	99.28	99.65	99.54	99.97	99.87	99.71
Mg#	0.90	0.90	0.90	0.87	0.90	0.90	0.91	0.91	0.90	0.90	0.90	0.89	0.90	0.92

(b) **Spinel**

	Patchy metasomatic assemblages within depleted xenoliths							Clinopyroxene-spinel-olivine symplectites				Spinel-rich olivine websterite		
	X-6	X-11/2			K-19			SB-M-3				SB-3		
	1	2	3	4	5	6	7	8	9	10	11	12	13	14
SiO_2	0.40	0.72	0.12	0.57	0.38	0.34	0.25	0.08	0.09	0.45	0.11	0.10	0.18	0.04
TiO_2	1.78	2.59	0.49	2.83	0.10	0.29	0.22	0.13	0.27	0.20	0.28	1.32	2.11	8.02
Al_2O_3	26.10	19.07	37.10	30.87	25.36	36.52	37.60	49.29	50.82	51.33	43.59	30.69	24.61	16.70
FeO^t	18.47	23.48	18.38	22.22	15.23	11.94	11.61	12.28	13.07	13.41	13.53	27.81	36.67	49.52
MnO	0.22	0.25	0.23	0.24	n.a.	n.a.	n.a.	0.12	0.17	0.16	0.13	0.39	0.52	0.60
MgO	14.19	12.47	16.21	13.40	14.82	17.53	17.84	20.73	19.74	20.32	19.44	12.47	9.59	7.23
Cr_2O_3	36.91	41.19	26.85	29.89	44.44	33.84	30.67	17.52	15.90	13.98	23.17	26.43	24.85	15.33
Total	98.09	99.78	99.39	100.02	100.33	100.46	98.18	100.14	100.06	99.84	100.25	99.21	98.53	97.44
Mg#	0.64	0.58	0.72	0.58	0.67	0.74	0.77	0.84	0.80	0.83	0.81	0.59	0.49	0.35
Cr#	0.49	0.59	0.33	0.39	0.54	0.38	0.35	0.19	0.17	0.15	0.26	0.37	0.40	0.38

(Continued)

Table 2. *Continued*

(c) Carbonate and apatite

	Carbonate								Apatite				
	STZ-20-1				SB-M-5		K-19	X-20/10	STZ-20-1				
	1	2	3	4	5	6	7	8	1	2	3	4	5
SiO_2	bdl	bdl	bdl	bdl	bdl	bdl	bdl	bdl	0.45	0.93	1.70	0.44	0.92
Al_2O_3	bdl	bdl	bdl	bdl	bdl	bdl	bdl	bdl	bdl	bdl	bdl	bdl	bdl
FeO	0.11	0.07	0.14	0.51	0.12	0.08	0.31	0.11	0.31	0.28	0.58	0.24	0.17
MgO	1.42	1.20	1.32	13.14	2.34	1.42	1.38	2.08	0.19	0.31	0.97	0.20	0.23
CaO	57.10	58.30	57.85	44.72	56.73	57.65	55.59	55.26	54.38	53.67	53.10	54.96	54.93
P_2O_5	bdl	bdl	bdl	bdl	bdl	bdl	bdl	bdl	39.55	40.60	35.64	40.31	40.00
Total	58.71	59.83	59.48	58.49	59.27	59.26	57.63	57.90	94.95	95.92	92.40	96.19	96.44

(d) Feldspar, glass and ilmenite

	Feldspars				Glass					Ilmenite			
	K-19	X-11/2	X-20/10	X-20/11	K-19	X-11/2		X-20/5		STZ-20-1			
	1	2	3	4	5	6	7	8	9	1	2	3	4
SiO_2	56.98	65.69	65.14	53.80	61.57	52.82	61.61	59.24	53.27	0.06	0.05	0.07	0.44
TiO_2	0.08	0.10	0.65	0.10	0.25	1.38	2.28	0.24	0.10	55.16	55.45	53.24	50.68
Al_2O_3	27.07	20.38	19.80	28.45	16.77	18.32	21.45	24.83	28.15	0.04	0.05	0.06	0.16
FeO	0.37	0.20	0.15	0.41	0.38	1.58	0.22	0.31	0.41	35.67	34.90	37.37	38.12
MnO	0.02	0.00	0.02	0.03	0.03	0.08	0.11	0.03	0.02	0.66	0.64	0.90	0.90
MgO	0.06	0.02	0.01	0.09	3.36	11.31	0.00	5.96	10.63	8.36	8.71	7.67	6.09
CaO	8.67	0.58	0.67	10.79	4.74	0.65	0.93	5.96	10.63	0.38	0.96	0.04	0.10
Na_2O	5.72	5.62	5.65	4.72	4.47	1.64	2.73	6.30	3.95	bdl	bdl	bdl	bdl
K_2O	0.39	6.67	7.59	0.30	4.23	6.16	8.48	0.63	0.26	bdl	bdl	bdl	bdl
Cr_2O_3	bdl	bdl	bdl	bdl	bdl	bdl	bdl	bdl	bdl	0.07	0.13	0.82	0.45
NiO	bdl	bdl	bdl	bdl	bdl	bdl	bdl	bdl	bdl	0.14	0.16	0.07	0.09
Total	99.48	99.49	99.71	98.69	96.33	93.94	97.81	97.62	96.91	100.57	101.09	100.28	97.26

Abbreviations: bdl, below detection limit; n.a., not available.

(Fig. 1f). Euhedral terminations of crystal faces (Fig. 1b, f) suggest that clinopyroxene had crystallized in an open space.

Following the classification of Morimoto *et al.* (1988), clinopyroxene in all these metasomatic assemblages is Cr-diopside (Table 2a, analyses 1–9). Cr-spinels show variable composition but have generally higher Cr# and Fe and Ti content (Table 2b, 1–7) in comparison to discrete primary spinel grains. Carbonate is Mg-bearing calcite with 1–3 wt% MgO, and apatite is characterized by low totals (<96.5 wt%). Ilmenite has high MgO content (*c.* 8 wt%). Feldspars vary from Ca-rich plagioclase to Na-rich K-feldspar (Table 2d), and they are compositionally similar to feldspars found in other xenoliths from both oceanic and continental mantle (e.g. Delpech *et al.* 2004; Bonadiman *et al.* 2005).

Apart from such accumulations, dominated by freely crystallized euhedral clinopyroxene crystals, there is evidence of metasomatic reactions that were probably assisted by deformation. These assemblages were observed only in rare, slightly deformed xenoliths. Secondary minerals occupy the space between partially disrupted and rearranged spinel grains and surrounding silicates. Figure 2a shows a sheared zone in a harzburgite xenolith containing spinel aggregates composed of scattered irregular grains, which sometimes display a jigsaw-puzzle texture (Fig. 2b, e). Backscattered electron (BSE) images show that the spinel has spongy rims that suggest melting and resorption (Fig. 2c, d). An assemblage of clinopyroxene and olivine neoblasts, orthopyroxene relicts, and glass altered into a non-stoichiometric mixture of phyllosilicates (Table 2d, analysis 5) is found adjacent to the spinel rims (Fig. 2d). Similar spinel decomposition is found, for instance, in metasomatized Yitong xenoliths (Xu *et al.* 1996, fig. 3a, p. 409; Bonadiman *et al.* 2001, fig. 2a, b). Clinopyroxene in these associations is Al–Ti-rich and cannot be distinguished from that in the above-described patchy pockets. The fact that similar mineral phases were found in the above described interstitial patches and in deformed xenoliths is a very important observation implying that infiltration of a melt of similar composition, and not the host–melt–xenolith interaction, is responsible for the origin of both secondary assemblages.

A particular textural type that belongs to the metasomatic assemblages (see discussion below) is seen in clinopyroxene–spinel–olivine symplectites. These were found only in two Mg-rich dunite xenoliths and occur in two textural forms. Lenticular or nest-like symplectites are around 1–2 mm in length and approximately 1 mm wide and, according to the size and shape, they resemble the above-mentioned metasomatic patches (Fig. 3a). By contrast to the mentioned patches, the symplectites lack glass and feldspar, and consist of fine-grained intergrowths of clinopyroxene, olivine and spinel in modal proportions of roughly 3:2:1. In a two-dimensional profile (Fig. 3a) clinopyroxene forms the basis of the symplectite aggregate, while olivine and spinel display a variety of shapes: rod-like, cylindrical, semi-circular, vermicular, etc. The other textural forms are intergrowths that do not show the well-confined outer forms seen in the lenticular symplectites. They rather appear as aggregates of tiny vermicular spinels in association with coarser olivine and clinopyroxene (Fig. 3b). Olivine from symplectites and intergrowths has generally the same composition (*c.* Fo_{90}) as adjacent olivine crystals unrelated to symplectitic forms. However, they are slightly less magnesian in comparison to olivine in other depleted xenoliths. Spinel in the symplectite is slightly richer in FeO, TiO_2 and Al_2O_3 in comparison to discrete spinel grains from the same xenolith, and differs from secondary spinels from pocket accumulations in having lower Cr# values (Table 2b, analyses 8–11). Clinopyroxene from symplectite shows high Al_2O_3 (4.5–7 wt%), Na_2O (0.7–1.1 wt%), along with irregular but high TiO_2 (0.2–1.1 wt%) and Cr_2O_3 (0.9–3.6 wt%) content (Table 2a, analyses 10–14), and is generally similar in composition to clinopyroxene from metasomatic pockets (see later).

Minor and trace element mineral chemistry

Olivine from a Fe-rich dunite (Table 3a, sample SB-M-1) is richer in Li (>2 ppm), Ca (*c.* 1500 ppm), Sc (>5.5 ppm), Ti (*c.* 50 ppm), Mn (*c.* 1500 ppm) and Cr (*c.* 300 ppm), and poorer in Ni (*c.* 2600 ppm), than olivine from the symplectite-bearing Mg-rich dunite (SB-M-3) (Li < 1.5 ppm, Ca *c.* 1000 ppm, Sc > 5.5 ppm, Ti < 10 ppm, Mn *c.* 1200 ppm, Cr *c.* 150 ppm, Ni *c.* 3500 ppm). Olivines from harzburgite xenoliths (SB-M-4 and SB-M-5) have the lowest content of all measurable elements (e.g. Li < 2 ppm, Ca *c.* 750 ppm, Sc < 4.5 ppm, Ti < 5 ppm, Mn *c.* 1000 ppm and Cr *c.* 70 ppm) coupled with the highest Ni content (>3700 ppm). Given that most elements behave incompatibly in olivine in terrestrial olivine–melt systems (e.g. Karner *et al.* 2003), it can be concluded that the olivine from the symplectite-bearing dunite is less enriched than olivine from the Fe-rich dunite but comparatively more enriched than olivine from depleted xenoliths.

Trace element concentrations and multi-element patterns of clinopyroxenes are shown in Table 3b and Figure 4, respectively. Clinopyroxene from pocket accumulations displays the highest contents

Table 3. *(**a**) LA-ICP-MS trace element analyses of olivines from Fe-rich dunite (SB-M-1), clinopyroxene–spinel–olivine symplectite-bearing Mg-rich dunite (SB-M-3) and an average analysis of olivine from depleted harzburgite xenoliths (HZ: SB-M-4 and SB-M-5); the forsterite contents are given in brackets. (**b**) LA-ICP-MS trace element analyses of clinopyroxenes; data for clinopyroxene from spinel-poor olivine websterite xenoliths are from Cvetković* et al. *(2007*b*) and are referring to clinopyroxene produced by orthopyroxene replacement (Cpx-1) and euhedral metasomatic clinopyroxene (Cpx-2)*

(a) Olivines

	SB-M-1 (*c.* Fo_{88})					SB-M-3 (*c.* Fo_{91})					HZ (*c.* $Fo_{>91}$)
ppm	Ol-1	Ol-2	Ol-3	Ol-4	Ol-5	Ol-1	Ol-2	Ol-3	Ol-4	Ol-5	Avr Ol
Li	2.01	2.04	2.16	2.19	2.11	1.07	1.22	1.22	1.20	1.24	1.78
Ca	1592	1583	1593	1591	1599	1075	998	1014	969	1005	771
Sc	5.89	5.78	5.89	5.79	5.95	5.6	5.4	5.5	5.6	6.1	3.96
Ti	48.6	48.7	52.5	49.7	49.9	7	6.07	5.7	8.57	11.7	4.26
Cr	296	295	288	294	289	167	164	155	155	158	73
Mn	1458	1497	1504	1532	1511	1164	1211	1237	1249	1251	1038
Ni	2703	2683	2642	2662	2589	3556	3506	3466	3452	3425	3713

(b) Clinopyroxenes

	Discrete clinopyroxene				Cpx from sympl.		Cpx from pockets			Cpx from sp-poor olivine websterites	
	SB-M-3						SB-M-4	SB-M-5			
ppm	Cpx-1	Cpx-2	Cpx-3	Cpx-4	Cpx-6	Cpx-7	Cpx-4	Cpx-3	Cpx-5	Type1	Type 2
Li	37.1	33.8	21.6	38.6	11.8	15.5	39.7	12.0	28.7	11.4	5.9
B	n.a.	n.a.	n.a.	n.a.	13.1	10.9	5.1	n.a.	4.1	3.9	4.0
Sc	86	96	97	96	84	98	94	81	105	93	94
V	n.a.	n.a.	n.a.	n.a.	682	370	281	n.a.	225	283	387
Cr	4979	5119	4628	4899	145 779	24 617	n.a.	4292	3480	n.a.	9368
Mn	648	666	695	727	1413	1221	1964	1186	1337	1528	939
Co	n.a.	n.a.	n.a.	n.a.	139	87	75	n.a.	35	26	40
Ni	517	479	446	525	2342	2168	1770	1084	1613	580	822

Cu	n.a.	n.a.	n.a.	n.a.	16.9	58.5	206	n.a.	206	35	11
Zn	n.a.	n.a.	n.a.	n.a.	310	32	103	n.a.	115	26	29
Rb	0.08	bdl	0.06	0.11	0.64	1.07	1.22	4.04	2.92	11.00	3.90
Sr	258	193	136	218	400	197	208	382	212	103	98
Y	7.3	6.7	7.0	6.6	8.8	8.7	10.7	30.0	25.1	11.0	9.4
Zr	10.51	9.21	8.09	9.51	17.25	12.47	39.18	226.00	79.00	28.00	6.78
Nb	0.05	bdl	0.06	0.17	1.16	1.37	7.93	60.60	27.40	8.85	1.14
Cs	n.a.	n.a.	n.a.	n.a.	0.04	0.06	0.14	n.a.	0.16	0.07	0.05
Ba	0.77	0.11	0.07	1.75	2.84	1.96	11.04	341.00	79.00	40.70	19.50
La	1.49	1.00	0.63	1.17	1.63	1.43	8.95	36.70	25.10	6.35	1.24
Ce	2.57	1.80	1.33	2.18	3.99	4.47	20.80	81.60	35.80	15.30	3.40
Pr	0.37	0.27	0.21	0.27	0.47	0.58	2.82	10.09	4.84	1.87	0.47
Nd	1.29	1.28	1.01	1.12	2.01	2.30	12.8	26.0	14.55	7.97	1.79
Sm	0.57	0.56	0.44	0.39	0.71	0.48	3.04	8.45	4.02	1.67	0.94
Eu	0.17	0.14	0.11	0.19	0.32	0.34	0.94	2.29	1.56	0.59	0.26
Gd	0.64	0.71	0.70	0.69	0.64	1.10	2.52	6.37	2.91	1.86	1.02
Tb	0.19	0.14	0.14	0.16	0.22	0.24	0.33	1.04	0.48	0.27	0.19
Dy	1.24	1.11	0.95	1.03	1.30	1.63	2.19	5.67	4.40	1.97	1.65
Ho	0.32	0.28	0.33	0.23	0.35	0.35	0.48	1.12	0.64	0.43	0.35
Er	0.88	0.80	0.68	0.81	0.88	1.07	1.26	2.86	1.67	1.02	0.95
Tm	0.14	0.13	0.09	0.13	0.17	0.18	0.18	0.41	0.36	0.15	0.20
Yb	0.73	0.99	0.88	0.88	1.21	1.23	0.91	2.46	1.26	1.10	1.39
Lu	0.16	0.18	0.21	0.15	0.17	0.18	0.13	0.45	0.35	0.20	0.23
Hf	0.45	0.36	0.30	0.26	0.56	0.51	0.94	4.10	1.44	0.85	0.20
Ta	bdl	bdl	bdl	bdl	0.10	0.08	0.61	3.35	1.39	0.43	0.20
Pb	0.30	0.22	0.27	0.47	2.45	3.80	2.97	2.28	3.67	0.57	bdl
Th	0.09	0.08	0.08	0.12	0.21	0.16	1.04	10.15	4.47	0.68	0.96
U	bdl	bdl	bdl	0.01	0.11	0.08	0.83	0.49	0.74	0.14	0.15

Abbreviations: bdl, below detection limit; n.a., not available.

Table 4. *Major and trace element analyses of the Sokobanja mafic alkaline rocks used for modelling*

	X-2 1	X-12 2	SB-61 9	SB-62 10	SB-63 11	SB-64 12	SB-65 13	SB-66 14	SB-67 15
SiO_2	42.37	42.49	42.49	42.7	42.57	42.24	43.04	41.87	42.7
TiO_2	1.78	1.88	1.83	1.82	1.84	1.87	1.72	1.68	1.66
Al_2O_3	13.01	13.72	13.74	13.51	13.50	12.99	13.60	12.73	12.71
Fe_2O_3	10.08	10.46	10.11	10.25	10.18	10.12	10.00	10.08	10.16
MnO	0.16	0.17	0.17	0.17	0.17	0.18	0.16	0.16	0.17
MgO	12.55	10.99	11.40	11.14	12.03	11.83	11.11	13.22	13.45
CaO	11.28	11.39	11.47	11.43	10.98	11.42	11.74	11.12	10.77
Na_2O	3.20	3.63	3.74	3.78	3.49	3.58	3.70	2.98	3.08
K_2O	0.64	0.81	0.73	0.76	0.71	0.81	0.77	0.90	0.74
P_2O_5	0.76	0.83	0.79	0.81	0.79	0.71	0.76	0.74	0.76
Mg#	0.71	0.68	0.69	0.68	0.70	0.70	0.69	0.72	0.72
Cr	567.9	417.4	458.4	458.4	478.9	521.5	492.6	622.6	608.9
Ni	303.1	218.5	213.7	197.7	230.2	241.2	218.2	295.1	293.9
Sc	23	22	20	20	20	22	21	20	20
Ba	710.8	745.5	761.0	774.0	805.9	837.0	826.8	839.6	790.9
Co	46.5	43.8	45.2	44.0	48.5	49.7	47.5	49.1	49.6
Cs	1.1	1.4	1.4	1.3	1.1	1.1	1.1	0.7	1.1
Ga	14.8	16.7	16.5	17.5	16.8	16.1	17.0	16.8	16.0
Hf	3.8	4.4	4.4	4.3	4.6	4.2	4.3	3.9	4.1
Nb	73.2	81.3	74.8	78.5	75.1	69.0	75.9	67.2	68.5
Rb	7.5	10.4	7.1	7.2	6.5	7.3	8.4	21.4	11.7
Sr	889	949	972	973	957	899	971	929	892
Ta	4.0	4.5	4.7	4.9	4.7	4.3	4.5	4.4	4.1
Th	9.9	9.3	11.2	10.5	9.9	8.6	9.6	9.3	8.9
U	2.2	2.4	2.5	2.4	2.6	2.2	2.4	2.3	2.5
V	196	206	197	206	206	203	212	194	188
W	1.1	1.2	1.7	2.3	1.4	0.9	2.6	0.8	1.0
Zr	170	189	178	182	177	166	176	161	159
Y	27.2	28.4	27.6	27.9	27.1	24.3	28.0	25.1	24.7
La	56.2	59.8	62.4	62.8	63.2	54.0	62.0	55.0	57.3
Ce	101.8	111.0	109.7	108.8	111.0	93.0	108.6	96.8	99.9
Pr	10.51	11.09	11.89	12.17	12.06	10.9	11.94	10.73	11.0
Nd	43.0	46.7	44.9	44.1	43.9	37.4	44.0	39.5	38.9
Sm	7.9	8.5	8.1	8.1	8.0	7.1	7.8	7.1	7.2
Eu	2.31	2.45	2.48	2.41	2.48	2.18	2.51	2.23	2.26
Gd	5.98	6.43	6.47	6.46	6.40	5.91	6.72	6.03	6.03
Tb	0.87	0.98	1.00	0.98	1.05	0.93	0.95	0.88	0.93
Dy	4.32	5.11	5.20	4.92	5.07	4.46	5.21	4.69	4.87
Ho	0.87	0.96	0.97	0.94	0.95	0.89	0.96	0.88	0.88
Er	2.28	2.51	2.46	2.40	2.60	2.03	2.54	2.28	2.37
Tm	0.36	0.39	0.35	0.36	0.37	0.30	0.33	0.32	0.32
Yb	2.11	2.33	2.04	2.16	2.21	1.96	2.23	2.02	1.90
Lu	0.31	0.32	0.34	0.33	0.34	0.30	0.36	0.30	0.30
Mo	2.9	3.4	2.8	2.9	2.7	2.6	2.3	1.7	1.7
Cu	52.5	47.8	47.6	45.0	46.7	54.0	55.3	51.1	51.9
Pb	3.7	3.5	3.4	2.9	3.2	3.4	3.0	3.5	3.5
Zn	61	64	59	59	59	63	52	62	61

of all trace elements and has the most LREE-enriched pattern ($[La/Yb]_N$ *c.* 10). On the primitive mantle normalized diagram it shows a steady decrease in concentration from the most incompatible towards the less incompatible elements. The exceptions are Rb and Ba, which have normalized values an order of magnitude lower than adjacent Th. Clinopyroxene from clinopyroxene–spinel–olivine symplectites displays a flat REE pattern ($[La/Yb]_N$ *c.* 1) with a slight increase in concentration from Gd to Lu. Mantle-normalized values of the highly incompatible elements are low, with the exception of positive spikes at U, Pb and Sr. Discrete interstitial clinopyroxene from the same dunite and the symplectitite clinopyroxene have roughly similar patterns. However, the former has

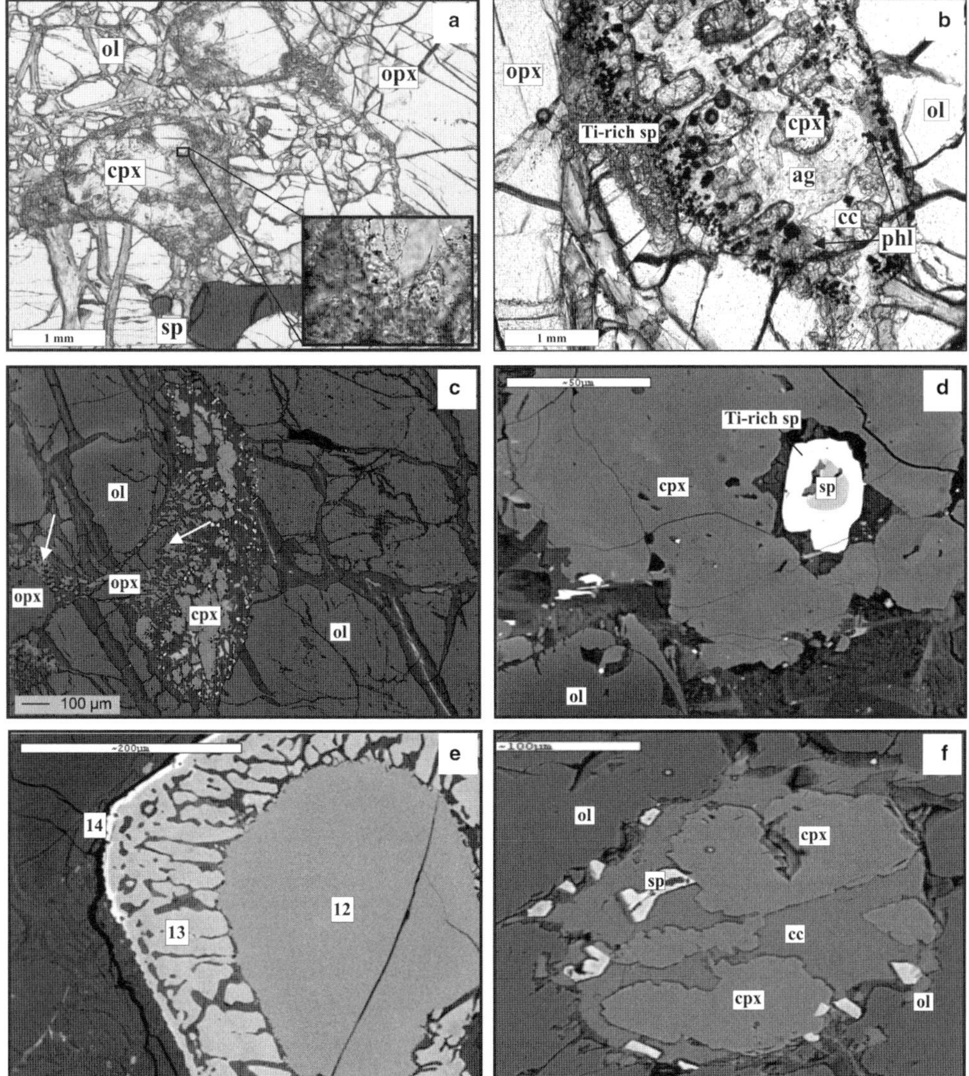

Fig. 1. (**a**) Plane polarized (PPL) photomicrograph of a clinopyroxene-rich lherzolite; two clinopyroxene crystals with spongy zones are situated within the central part of the image; note the different appearance of replacement selvage around the orthopyroxene (upper right); the inset shows a spongy area in clinopyroxene. (**b**) PPL photomicrograph of a pocket containing euhedral clinopyroxene, Ti-rich spinel, altered glas, carbonate and two tiny phlogopite relicts; the phlogopite flakes display diffuse boundaries towards adjacent phases. (**c**) Back-scattered electron (BSE) image of a metasomatic pocket in a harzburgite xenolith; the arrows show replacement selvages around the orthopyroxene. (**d**) BSE image displaying a spinel included in metasomatic clinopyroxene; the spinel shows Ti-rich overgrowth. (**e**) BSE image of a coarse-grained spinel with a reaction pattern of increasing Fe–Ti and slightly increasing Cr content (analyses 1–5 are shown in Table 2). (**f**) BSE image of a rounded pocket with euhedral clinopyroxene and Al-rich spinel enclosed by Mg-calcite. Abbreviations: sp, spinel; ol, olivine; opx, orthopyroxene; cpx, clinopyroxene; cc, carbonate; phl, phlogopite; ag, altered glass.

slightly lower concentrations of LREE, and lacks positive anomalies at U and Pb. For comparison, trace element patterns of discrete clinopyroxenes from harzburgite xenoliths are also shown. They exhibit moderately LREE-enriched ($[La/Lu]_N$ = 3.5–10) patterns with low concentrations of HREE (e.g. Yb = 1.4–2.2× chondrite) and a pronounced peak at Sr.

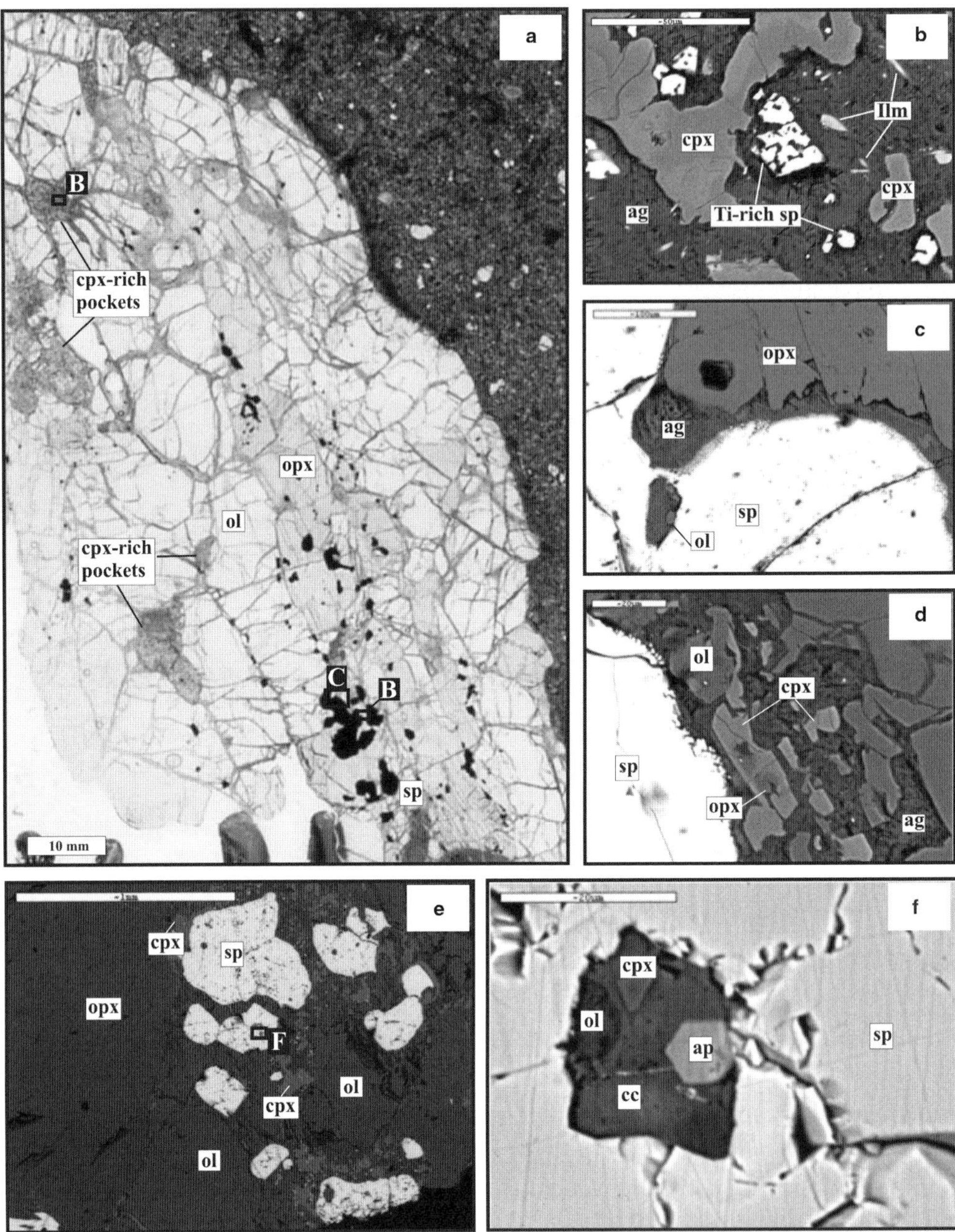

Fig. 2. (**a**) PPL scanned thin-section image of a slightly sheared harzburgite xenolith; note that BSE images shown in (b), (c) and (d) represent small areas in this xenolith. (**b**) BSE image of a pocket containing clinopyroxene-, Ti-rich spinel-, altered glas and tiny ilmenite crystals. (**c**), (**d**) BSE images showing spongy rims of a partially disrupted spinel crystal as well as clinopyroxene and olivine neoblasts, orthopyroxene relicts and altered glass (?) within the reaction assemblage. (**e**) BSE image displaying another spinel aggregate with a jigsaw-fit puzzle texture. (**f**) BSE image of a detail from (e) showing a pocket inside a spinel crystal; the pocket contains a perfectly idiomorphic apatite crystal as well as subhedral tiny grains of clinoypyroxene and altered olivine, all enclosed in a carbonate. Abbreviations: ap, apatite; ilm, ilmenite; others as in Figure 1.

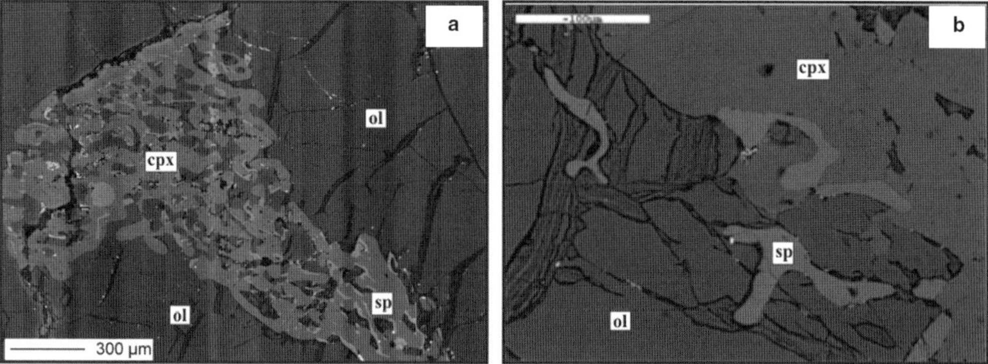

Fig. 3. (**a**) BSE image showing a lenticular clinopyroxene–olivine–spinel symplectite found in a Mg-rich dunite xenolith. (**b**) BSE image of isolated clinopyroxene–spinel and olivine–spinel intergrowths without having any form of concentrated symplectite aggregates. Abbreviations as in Figure 1.

Petrochemistry of the host volcanics

In the modelling (see below) only the Sokobanja mantle xenolith-bearing rocks were used. The rock samples of other localities where xenoliths were found (Striževac) are more heterogeneous, some having a K_2O content of up to 1.66 wt% and these were omitted. The chemical analyses of Sokobanja mantle xenolith-bearing mafic alkaline rocks are given in Table 4. All samples are olivine (Fo_{76-86}) $\pm$clinopyroxene (Mg# 75–85) phyric with a groundmass composed of clinopyroxene (Mg# >70), nepheline, plagioclase (An *c.* 60) and Ti-bearing magnetite (TiO_2 *c.* 15 wt%). They have high Mg# (*c.* 70) and high concentrations of incompatible elements (Ni mostly >200 ppm and Cr > 450 ppm). The whole-rock MgO v. Fe_2O_3 diagram with the equilibrium line of primitive mantle melts is shown in Figure 5. Although some samples show effects of small olivine accumulation (<6–7 vol.%), this has only negligible effects on trace element concentrations used in the modelling ($\ll$1% relative). The samples have low Sr (*c.* 0.7031) and high Nd (*c.* 0.5129) initial isotope ratios (Jovanović *et al.* 2001) indicating that they represent fairly uncontaminated magmas. Therefore, no correction for low-pressure processes was necessary.

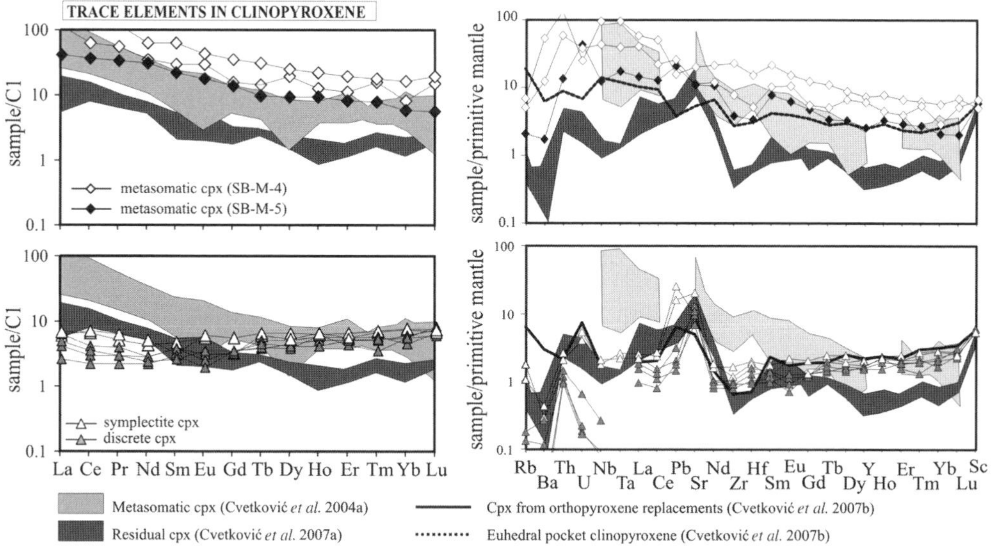

Fig. 4. Trace element patterns of clinopyroxene from metasomatic and residual xenolith assemblages. Chondrite and primitive mantle normalization after McDonough & Sun (1995).

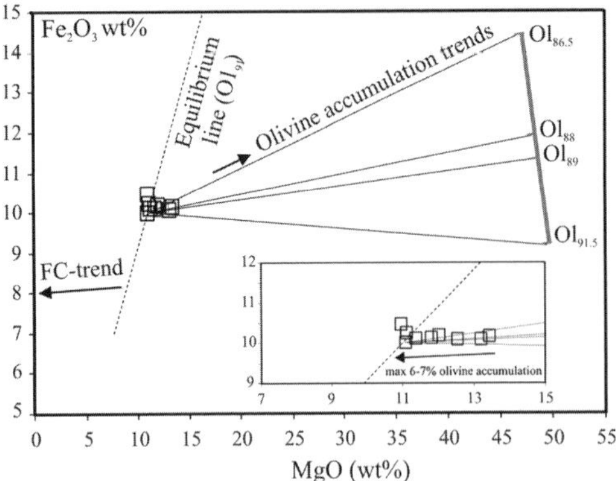

Fig. 5. MgO v. Fe_2O_3 diagram for whole-rock samples of mafic alkaline rocks from Sokobanja showing the effects of the accumulation of olivine in the studied samples.

Discussion

Origin of fertile xenoliths

Clinopyroxene-rich lherzolite and olivine websterite xenoliths show similar textures to depleted xenoliths but their modal composition and mineral chemistry are much too fertile to represent fragments of residual, very fertile mantle. Their clinopyroxenes generally exhibit trace element patterns similar to those of the clinopyroxene megacrysts (Cvetković *et al.* 2004*b*), suggesting that they are also related to alkaline mafic magmas. However, it is unlikely that these pyroxene-rich lithologies formed by direct magmatic crystallization, as suggested above for the megacrysts, because they contain orthopyroxene – a phase that is not on the liquidus in Si-undersaturated magmas (Ghiorso & Carmichael 1987). The possibility that they represent mixtures of residual mantle peridotites and crystal cumulates from mafic alkaline magmas is also unlikely. Their orthopyroxene is too rich in Al and Ca to be accounted for by the simple addition of Fe-rich olivine and clinopyroxene to a harzburgite. By contrast, the composition of the fertile xenoliths can be explained by melt–rock interactions. We recalculated bulk composition of the xenoliths using their modal analyses and chemistry of present minerals. The residual mantle can be represented by the average composition of harzburgite and lherzolite of this xenolith suite while the average major element analysis of the host basanites can be used as a proxy for the composition of presumed mafic alkaline metasomatic melts. Accordingly, the calculated bulk Al_2O_3, CaO and TiO_2 content of the fertile xenoliths can be obtained by the addition of 5–20 wt% of a basanitic melt to a refractory mantle. It is noteworthy that the addition of pure cumulitic material (80% clinopyroxene and 20% olivine) cannot produce a decrease in bulk CaO/Al_2O_3 ratio of the residual mantle lithology.

Although the approach of recalculating bulk compositions on the basis of modal analysis has serious limitations, the above-presented calculations at least indicate that the fertile xenolith lithologies could have originated by refertilization of residual mantle by relatively Fe-rich alkaline basaltic melts. Similar compositional and textural changes in mantle peridotites are generally believed to be confined to the wall rocks along magmatic conduits, that is, in the area of high melt/wall rock ratio (Kempton 1987). Xu *et al.* (1996) reported a linear change in composition from lherzolite to orthopyroxene-bearing wehrlite and finally to orthopyroxene-free wehrlite, and interpret this as resulting from advancing melt–peridotite reactions. This reaction is usually referred to as carbonatite metasomatism, which stabilizes clinopyroxene with respect to orthopyroxene (e.g. Coltorti *et al.* 1999 among many others). Such a scenario proposes an infiltration of melts and their subsequent equilibration with the surrounding peridotite. Moreover, it also suggests that, along with precipitation of new Ti–Al-clinopyroxene, the already existing mantle minerals underwent detectable changes in composition; for example, a decrease in Mg# in olivine and orthopyroxene, and an increase in Al_2O_3 and CaO in orthopyroxene and Al_2O_3 for spinel. Such additions of clinopyroxene and equilibration of primary mantle phases require time and provide evidence that the reactions occurred *in situ* in the mantle rather than due to direct contact with the host magma. The occurrence of spongy clinopyroxene (see Fig. 1a) also argues against the second possibility. These spongy textures are developed on the presumed metasomatic clinopyroxene. Consequently, if the spongy textures formed by a later thermal event, possibly during entrapment of the xenoliths, previous additions of clinopyroxene must have had occurred earlier.

Origin of glassy patches, deformation-assisted and symplectitic assemblages

Many authors (e.g. Yaxley *et al.* 1998; Coltorti *et al.* 1999 among others) argued that the glassy pockets found in mantle xenoliths originated *in situ* in the lithosphere. Following their arguments it can be emphasized that the secondary assemblages found in ESLM xenoliths are different from the phase associations found in host-related veins because: (i) the latter often cut the whole xenolith; (ii) they contain Fe-rich and Cr-poor pinkish clinopyroxene, pure Ti-oxides and nepheline; and (iii) they usually produce Fe-rich rims on surrounding olivines. By contrast, the *in situ* metasomatic aggregates

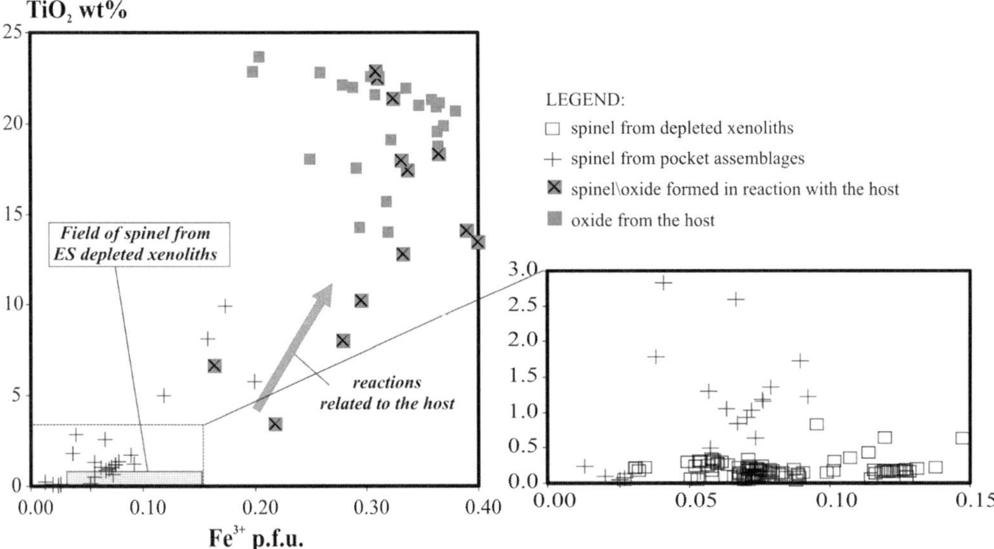

Fig. 6. Fe^{3+} [p.f.u. (per formula unit)] v. TiO_2 (wt%) for spinel and Fe–Ti-oxide from various mineral associations found in East Serbian xenoliths; field of spinel from ES-depleted xenoliths after Cvetković *et al.* (2004*a*).

contain greenish Cr-rich clinopyroxene and usually do not produce any effect on surrounding olivine. Figure 6 shows TiO_2 wt% v. Fe^{3+} variations for spinel and Fe–Ti-oxides in various associations of ESLM xenoliths. Ferric iron was calculated by accepting that spinel represents solid solution between $R^{2+}R^{3+}O_4$ and $R^{2+}R^{4+}O_4$. Most spinels in metasomatic pockets display an increase in TiO_2 that is not accompanied by a substantial increase in Fe^{3+} concentrations, as seen for spinels that reacted with the host. Only a few pocket spinels show higher Fe^{3+} content that can be related to shallow-level interaction between the host magma and spinel, which was not recognized texturally.

In general, these metasomatic assemblages are texturally similar to xenolith-hosted secondary mineral associations usually attributed to mantle metasomatism (e.g. Ionov *et al.* 1994, 1999; Zinngrebe & Foley 1995; Wulff-Pedersen *et al.* 1996, 1999). It is, however, highly possible that the observed textural relationships do not represent the original textures inherited from the mantle. Some aggregates in pockets could have formed due to decompressional or/and fluid-induced melting of earlier metasomatic phases followed by rapid crystallization of the melt, as, for instance, reported by Bali *et al.* (2002, 2008) for silicate melt pockets found in mantle xenoliths from western Hungary. That is indicated by the occurrence of euhedral clinopyroxene and quench-like spinel crystals in some pockets. Decomposition of pre-existing metasomatic minerals is also suggested by the presence of phlogopite (relict?) crystals in one sample. Given that all present metasomatic minerals are anhydrous, it can be concluded that this decomposition was an open-system process in which fluids released from phlogopite could leave the system. This implies that a metasomatic phase could have formed in the lithosphere before the xenoliths were captured by basanitic magma.

Textures of frozen reactions found in slightly deformed xenoliths further suggest that at least some metasomatic reactions occurred *in situ* in the ESLM. Puzzle-like disintegrated spinels with spongy/recrystallized Ti-rich rims found along sheared zones in such xenoliths indicate that infiltration of metasomatic melt was assisted by deformation. Evidence for similar magmatic processes related to localized deformation was reported from a xenolith from the Pannonian Basin (Falus *et al.* 2004). It is highly unlikely that such textural relationships could originate after the xenolith capture. Post-entrapment deformation within xenoliths is related to thermal stress and the release of internal fluids upon decompression, and commonly gives rise exclusively to brittle cracking (Nicolas 1986; Wilshire & Kirby 1989; Klügel 1998).

Clinopyroxene–spinel–olivine symplectites are texturally similar to symplectites in the Horoman peridotite (Morishita & Arai 2003). However, the symplectites in ESLM xenoliths lack orthopyroxene and cannot be attributed to garnet breakdown reactions (Smith 1977). Zinngrebe & Foley (1995) mentioned conspicuous, millimetre-sized symplectic intergrowths of clinopyroxene and spinel surrounded by thick margins of phlogopite in Gees xenoliths, and they speculated that they may have formed after the previous amphibole. To characterize the symplectites occurring in ESLM xenoliths, we use the composition of clinopyroxene because this mineral shows relatively large compositional variations and occurs in a variety of textural assemblages. Major element compositional variations of texturally different clinopyroxenes are illustrated in Figure 7. The observed compositional variations suggest that the symplectite-related clinopyroxene is clearly compositionally different from clinopyroxene megacrysts and from those related to reactions with the host, and that it generally follows the compositional trends of clinopyroxenes believed to be melt–peridotite reaction products. In addition, the symplectitic clinopyroxene is more similar in composition to small pocket clinopyroxene than to coarser clinopyroxene from fertile lherzolite and websterite xenoliths. Both symplectite clinopyroxene and clinopyroxene from metasomatic pockets have high Mg# and very high Cr_2O_3 content. Similarities in trace element patterns of clinopyroxene from symplectites and those from overgrowths around orthopyroxene (Cvetković *et al.* 2007*b*) indicate that dissolution of orthopyroxene may have been involved in their genesis.

The above characteristics suggest that metasomatic pockets and symplectites probably originated in a similar way, involving a interaction between mantle peridotite and infiltrating melts. In contrast to the reactions that produced fertile pyroxene-rich lithologies, these metasomatic processes were probably associated with a low melt/rock ratio. During low melt/rock ratio interactions the MgO and Cr_2O_3 content in clinopyroxene is buffered by peridotite rather than by infiltrating melts. This can explain why clinopyroxenes from metasomatic pockets and symplectites have high metasomatic components (Al, Ti and Na_2O), along with very high Mg# values that are comparable to those of interstitial residual clinopyroxene.

The foregoing discussion offers textural and compositional evidence that the metasomatic pockets and symplectites found in ESLM xenoliths most probably represent products of mantle metasomatism. Textural and mineral compositional characteristics indicate the role of the following reaction: opx + Cr-rich sp ± phlogopite + Si-undersaturated alkaline melt = Ti–Al-cpx + Ti-rich spinel ± olivine ± other minor phases. The presence of tiny phlogopite crystals in metasomatic pockets is taken as an indication of the possible role of pre-existing hydrous phases and their subsequent breakdown. However, even if the present phlogopite crystal are relicts, this reaction should be taken as an oversimplification because it cannot account for all of the observed textural assemblages. There are open questions as regards the following. (i) Had

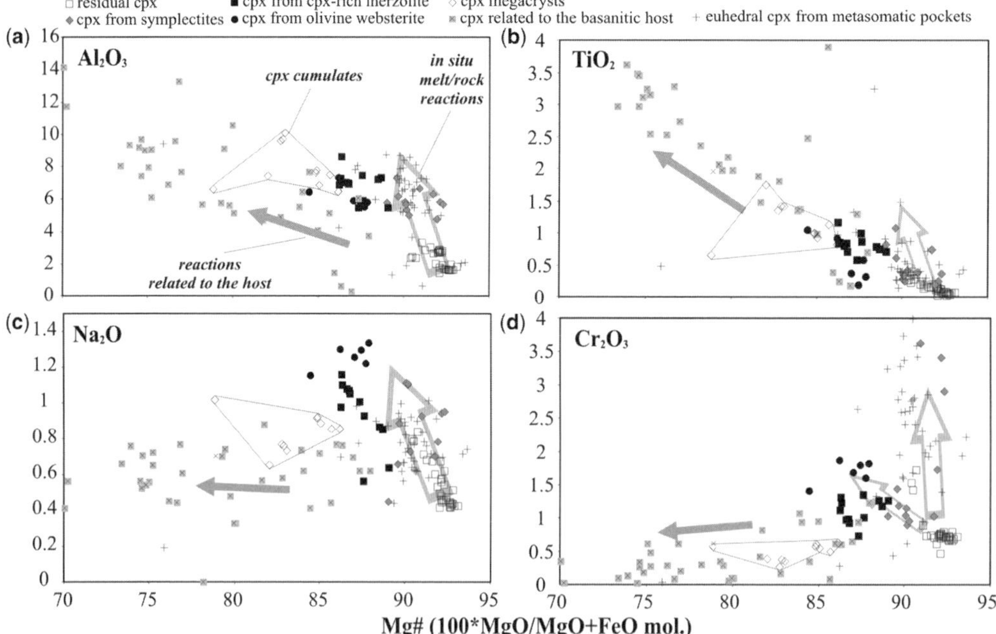

Fig. 7. Mg# (100 × MgO/MgO + FeO) v. (**a**) Al_2O_3, (**b**) TiO_2, (**c**) Na_2O and (**d**) Cr_2O_3 for clinopyroxene from various associations found in East Serbian xenoliths.

metasomatic phases other than phlogopite also been present and then entirely decomposed? (ii) Was phlogopite (±other phases) related to a compositionally different metasomatic event? (iii) When (and why?) exactly did the breakdown reactions occur? Very high TiO_2 content found in phlogopite crystals in this xenolith suite lend support to the conclusion that the mica was related to compositionally similar metasomatic agents. Similar high-Ti hydrous minerals are reported from different xenolith suites and are often interpreted as products of alkaline metasomatism (e.g. Stern *et al.* 1999; Kogarko *et al.* 2007; Wang *et al.* 2008 among others). However, to constrain the exact time and cause of the breakdown reactions is very difficult, if not impossible. Generally, it could have occurred within the mantle due to thermal (± new infiltration) events subsequent to the main enrichment phase or it might have been related to post-entrainment decompression melting due to the fast uplift of host lavas. Consequently, a proposed scenario derived from this section is that the ESLM has undergone addition of a silica-undersaturated alkaline melt rich in Fe–Ti–Al, CO_2 and H_2O. Similar metasomatic assemblages have been known from many xenolith suites and peridotite massifs, and was also interpreted in terms of carbonate-rich alkaline silicate metasomatism (e.g. Harte 1987; Wilshire 1987; Canil & Scarfe 1989; Xu *et al.* 1996; McGuire & Mukasa 1997; Witt Eickschen & Kramm 1998; Delpech *et al.* 2004, Rivalenti *et al.* 2004, etc.).

Implications from the geochemistry of host basanites

Presented textural and mineral chemistry evidence suggests that prior to the onset of the Palaeogene mafic alkaline magmatism, the ESLM already contained metasomatized domains. If this is correct, it raises a logical question: could these metasomatized lithospheric regions have played any role in petrogenesis of the host basanites? In this section we try to demonstrate that this is a likely hypothesis, namely that the host lavas could have originated by melting of a mantle source containing metasomatic assemblages similar to those observed in xenoliths.

We have used the following twofold approach: (a) performing inversion modelling based on the whole-rock trace element data of basanites adopting the method of Cebria & Lopez-Ruiz (1996) for constraining the source composition and bulk partition coefficients of incompatible elements during the partial melting; and (b) inferring possible mineralogical characteristics of the source and comparing them to the observed mineralogy of metasomatic associations found in xenoliths. As mentioned above, for the modelling we used the Sokobanja basanites (Table 4) because their composition appears to be unaffected by fractionation and contamination processes, and because they have the most uniform isotopic composition of all the East Serbian Palaeogene alkaline rocks.

The analyses used for modelling, the general explanation of the procedure and the equations used are given in Appendix, while the step-by-step modelling procedure can be found in Cebria & Lopez-Ruiz (1996). The results of the modelling are presented in Table 5. The modelled source is enriched in highly and moderately incompatible elements (e.g. average estimates between *c.* 35–40× chondrite for U–Th–Nb–Ta and around 2× chondrite for HREE). The trace element pattern of the source is compared to the pattern of the MORB source (Salters & Stracke 2004) in Figure 8. With respect to the MORB-source, the calculated source is more than two orders of magnitude more enriched in concentrations of highly incompatible elements, and from one to two orders of magnitude more enriched in content of highly–moderate incompatible elements. The concentrations in the range Dy–Lu are either very similar or slightly lower in comparison to the MORB-source.

The results of the modelling imply that the primary magmas of Sokobanja basanites were not derived by the melting of a homogeneous asthenospheric source, provided that it is MORB-like. Even extremely small degrees of partial melting ($\ll 1\%$) are not capable of producing the observed enrichment in most incompatible elements. Partial melting of deep-mantle (plume) sources can be excluded as there is no geological evidence that mantle plume could have been related to the petrogenesis of East Serbian Palaeogene alkaline rocks (e.g. Cvetković *et al.* 2004*a*). Cvetković *et al.* (2005) performed partial melting modelling using the approach of McKenzie & O'Nions (1995). The model involving 5–10% of melting of a garnet- and amphibole-bearing REE-enriched mantle (MORB-source enriched by 8% of melts formed by extraction of 0.3% of fractional melting of the same source) could roughly reproduce the observed REE patterns, but still had difficulties in matching the observed concentrations of La and Ce. Importantly, the model did not take into consideration the Sr and HFSE content, which are generally very high in alkali basalts.

The modelling presented here has an advantage in determining not only the trace element concentrations in the source but also their bulk distribution coefficients. These can be used as additional constraints on the source mineralogy. The estimated ranges of bulk partition coefficients for the source phases (D_0) and for the phases entering the melt

Table 5. *Results of the inverse modelling of the source composition of the host basanites; the minimal (MIN), maximal (MAX) and average (AVR) estimates of the bulk partition coefficient of the phases in the source* (D_0) *and of the phases entering the melt* (P_L)*, as well as of the concentrations in the source; xCH are the enrichment factors with respect to the chondrite composition (McDonough & Sun 1995)*

	D_0^{MIN}	D_0^{MAX}	D_0^{AVR}	P_L^{MIN}	P_L^{MAX}	P_L^{AVR}	C_0^{MIN}	C_0^{MAX}	C_0^{AVR}	xCH
Th	0.0088	0.0093	0.0091	0.0711	0.0184	0.0448	1.01	1.07	1.04	35.82
U	0.0340	0.0440	0.0390	0.2730	0.0591	0.1660	0.26	0.34	0.30	40.60
Nb	0.0270	0.0310	0.0290	0.2077	0.0904	0.1490	8.26	9.48	8.87	36.95
Ta	0.0200	0.0220	0.0210	0.1229	0.0352	0.0791	0.53	0.58	0.55	40.70
La	0.0020	0.0020	0.0020	0.0035	0.0035	0.0035	6.40	6.40	6.40	27.00
Ce	0.0022	0.0022	0.0022	0.0040	0.0040	0.0040	11.08	11.08	11.08	18.08
Sr	0.0570	0.0880	0.0725	0.4156	0.0978	0.2567	112.69	173.98	143.34	19.77
Nd	0.0065	0.0067	0.0066	0.0418	0.0123	0.0271	4.43	4.57	4.50	9.85
P	0.0500	0.0770	0.0635	0.4060	0.0852	0.2456	0.08	0.13	0.11	0.00
Zr	0.0360	0.0480	0.0420	0.3021	0.0695	0.1858	19.53	26.04	22.78	5.96
Hf	0.0280	0.0340	0.0310	0.2069	0.0370	0.1220	0.51	0.62	0.56	5.45
Sm	0.0250	0.0300	0.0275	0.1953	0.0344	0.1148	0.85	1.01	0.93	6.29
Eu	0.0240	0.0280	0.0260	0.1936	0.0591	0.1263	0.26	0.30	0.28	4.96
Gd	0.0390	0.0540	0.0465	0.3207	0.0594	0.1900	0.70	0.97	0.83	4.19
Tb	0.0240	0.0270	0.0255	0.1486	0.0422	0.0954	0.11	0.13	0.12	3.38
Dy	0.0300	0.0370	0.0335	0.2408	0.0637	0.1523	0.53	0.65	0.59	2.41
Y	0.0290	0.0360	0.0325	0.2434	0.0608	0.1521	2.86	3.54	3.20	2.04
Ho	0.0420	0.0600	0.0510	0.3511	0.0730	0.2120	0.10	0.15	0.13	2.33
Er	0.0380	0.0530	0.0455	0.3214	0.0536	0.1875	0.27	0.38	0.32	2.02
Yb	0.0605	0.0605	0.0605	0.0640	0.0640	0.0640	0.34	0.34	0.34	2.08
Lu	0.0540	0.0540	0.0540	0.0600	0.0600	0.0600	0.05	0.05	0.05	1.96

(P_L) are also given in Table 5. The obtained D_0 values confirmed that La and Ce (D_0 *c.* 0.002) behaved most incompatibly during partial melting, as expected. This was inferred from covariation diagrams and was used as a pre-assumption in the modelling (see the Appendix). D_0 values for Nd (*c.* 0.007) and Th (*c.* 0.009) suggest slightly more compatible behaviour of these two elements. All other trace elements show $D_0 > 0.02$. P_2O_5 and Sr are least incompatible among the elements used in modelling, with average D_0 values of 0.063 and 0.07, respectively. Among the incompatible elements only Ti behaved more compatibly but due to very poor correlation ($r_{[La\ v\ TiO_2]}$ *c.* 0.2) it was excluded from the modelling. K, Rb and Ba were also excluded from the modelling because of their near-compatible behaviour. This characteristic is observed in most alkaline provinces and was usually attributed to the presence of amphibole and/or phlogopite in the source (e.g. Wilson & Downes 1991).

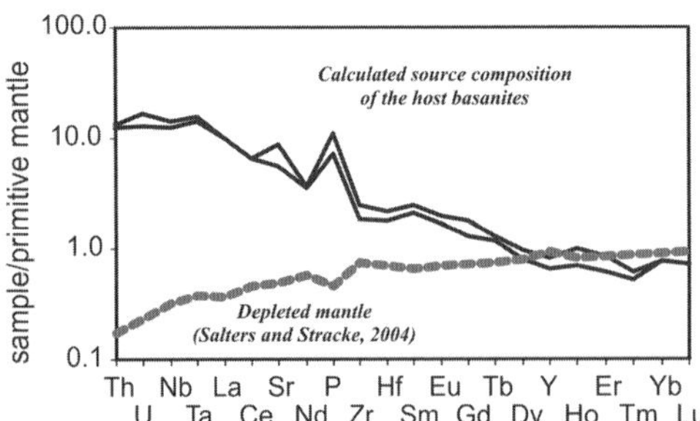

Fig. 8. Primitive mantle-normalized trace element pattern showing the source composition inferred from inverse modelling. The compositional pattern of depleted mantle is shown for comparison (Salters & Stracke 2004).

To infer possible source mineralogy of the host basanites we compared D_0 values of the basanitic mantle source obtained by inversion modelling with the D_0 values of mantle lithologies differing in their mineral composition using mineral–melt partition coefficients from the literature (see Table 6). The results are illustrated in Figure 9, and mineral–melt partition coefficients used in calculations are given in Table 5. Note that the mineralogy of unmetasomatized mantle (shaded area, shown in all diagrams) can roughly reproduce the D_0 values obtained by inversion modelling for the REE (except for Sm, to a certain extent) and for Y. However, there is strong discrepancy in D_0 values for Th, U, Nb, Ta, Sr, P, Zr and Hf, which appear to be much more compatible in the source calculated by the modelling. Consequently, if a metasomatic assemblage was present in the source of Sokobanja basanites, it must be an association that can substantially increase D_0 for the mentioned elements without affecting the bulk partition

Table 6. *Bulk mineral–melt partition coefficients used for the calculation of source mineralogy for host basanites*

	Ol	Opx	Cpx	Sp	Amph	Phl	Ilm	cc	Ap
Th	0.00005	0.0009	0.002		0.04	0.0014	0.0006		19
U	0.00005	0.0032	0.0006		0.15	0.13	0.0082		16
Nb	0.0004	0.0033	0.0052	0.02	0.196	0.088	2		
Ta	0.0004	0.0049	0.0035	0.02	0.03	0.1096	1.7		
La	0.0001	0.001	0.0123	0.01	0.17	0.0413	0.000029		2.5
Ce	0.0003	0.0039	0.01	0.001	0.26	0.0455	0.000054		11.2
Sr	0.001	0.003	0.05		0.12	0.183	0	1	1.1
Nd	0.001	0.012	0.031	0.01	0.44	0.0293	0.0005		14
P	0.01	0.03	0.03				0.05		50
Zr	0.0033	0.0167	0.016	0.03	1.2	0.2806	3.002		0.986
Hf	0.001	0.033	0.035	0.03	0.92	0.2703	1.18		0.878
Sm	0.001	0.019	0.082	0.01	0.76	0.0255	0.0006		14.6
Eu	0.0015	0.03	0.1	0.01	0.88	0.0218	0.0011		9.6
Gd	0.0015	0.016	0.3	0.01	0.86	0.0205	0.0034		15.8
Tb	0.0015	0.019	0.31	0.01	0.83	0.2	0.0067		15.4
Dy	0.0017	0.022	0.33	0.01	0.78	0.0281	0.01		3.9
Y	0.0099	0.0455	0.15		0.396	0.007	0.0045		
Ho	0.0016	0.026	0.31	0.01	0.73		0.011		13.4
Er	0.0132	0.0717	0.182	0.01	0.68	0.0303	0.0275		41.6
Yb	0.0305	0.1013	0.18	0.01	0.59	0.0484	0.17		8.1
Lu	0.043	0.127	0.18	0.01	0.51	0.0471	0.084		3.8

Explanation and data sources: $D_{mineral-melt}$ for olivine (Ol), orthopyroxene (Opx) and clinopyroxene (Cpx) from Keshav *et al.* (2005) and references therein, except for Gd, Tb, Dy and Ho (McKenzie & O'Nions 1991) and for P (Ulmer 1989); $D_{mineral-melt}$ for spinel (Sp) are after McKenzie & O'Nions (1991) except for Nb, Ta, Zr and Hf (Horn *et al.* 1994); $D_{mineral-melt}$ for amphibole (Amph) from McKenzie & O'Nions (1991) except for Th (La Tourrette *et al.* 1995), U, Zr and Hf (Villemant *et al.* 1981), Nb and Y (Chazot *et al.* 1996) and Ta (Green *et al.* 1993); $D_{mineral-melt}$ for phlogopite (Phl) from Fujimaki *et al.* (1984) except for Th and Nb (La Tourrette *et al.* 1995), U and Tb (Villemant *et al.* 1981), and for Ta, Sr and Y (Foley *et al.* 1996); $D_{mineral-melt}$ for ilmenite (Ilm) from Zack & Brumm (1998) except for P (Anderson & Greenland 1969) and for Zr and Hf (Fujimaki *et al.* 1984); The values for $D_{mineral-melt}$ for calcite (cc) are considered too low for all elements except for Sr, which is considered to be unity (e.g. Ionov & Harmer 2002); $D_{mineral-melt}$ for apatite (Ap) from Paster *et al.* (1974), except for Th, U (average values from Luhr *et al.* 1984; Mahood & Stimac 1990; Bea *et al.* 1994), Sr, Dy and Lu (Watson & Green 1981), Zr, Hf and Er (Fujimaki 1986) and for P for which $D_{mineral-melt}$ was calculated on the basis of P_2O_5 concentrations in apatite from ESLM xenoliths and in host basanites. Missing $D_{mineral-melt}$ values are considered not important for the calculations because of low element partition into given minerals.

coefficients for REEs. Amphibole-bearing metasomatized mantle (with or without phlogopite) is characterized by a very high $^{REE}D_0$, but the D_0 values for Nb, Ta, Sr and P are still too low (Fig. 9b, c). The presence of phlogopite alone is also unable to account for the high D_0 values for Nb, Ta, Sr and P (Fig. 9d). The best fit with D_0 of the calculated basanitic source was produced using anhydrous metasomatized mantle with small additions of metasomatic clinopyroxene and carbonate (*c.* 5%), and with traces of ilmenite (*c.* 1%) and apatite (*c.* 0.05%) (Fig. 9e). The pattern of D_0 values is almost completely matched along with still higher $^{U}D_0$ and $^{LREE}D_0$ ($\times 0.5 - \times 4$) values than in the mantle source calculated by the inversion technique. Note that any addition of phlogopite and, especially, amphibole will produce an increase in $^{LREE}D_0$ and thereby a larger misfit with the pattern of the modelling.

The presented model strongly depends on the mineral–melt partition used, and for some mineral–element pairs these values have a wide range or are poorly constrained, or both, and for these reasons this should be taken as an indication rather than as a straightforward conclusion. However, the model strongly argues for the presence of similar minerals in the basanitic mantle source and in metasomatic associations in the xenoliths. Hence, we may assume that the presence of carbonate and apatite in the source of host rocks is a robust hypothesis. In the absence of carbonate the calculated $^{Sr}D_0$ of 0.073 can be matched only if relatively large proportions (>15 vol%.) of amphibole are present. However, as shown above, this would make the middle rare earth elements (Sm–Dy) four–five times more compatible than calculated. If we theoretically assume much lower $^{REE}D_{Amph}$ values we would potentially solve the problem of D_0 values in the source. However, even if amphibole enters the melt totally, it cannot account for the $^{Sr}P_L > 0.2$ estimated by inversion calculations. However, because of their high K_D^{Sr}, traces of carbonate and apatite in the mantle could account for the calculated low $^{Sr}D_0$ and relatively

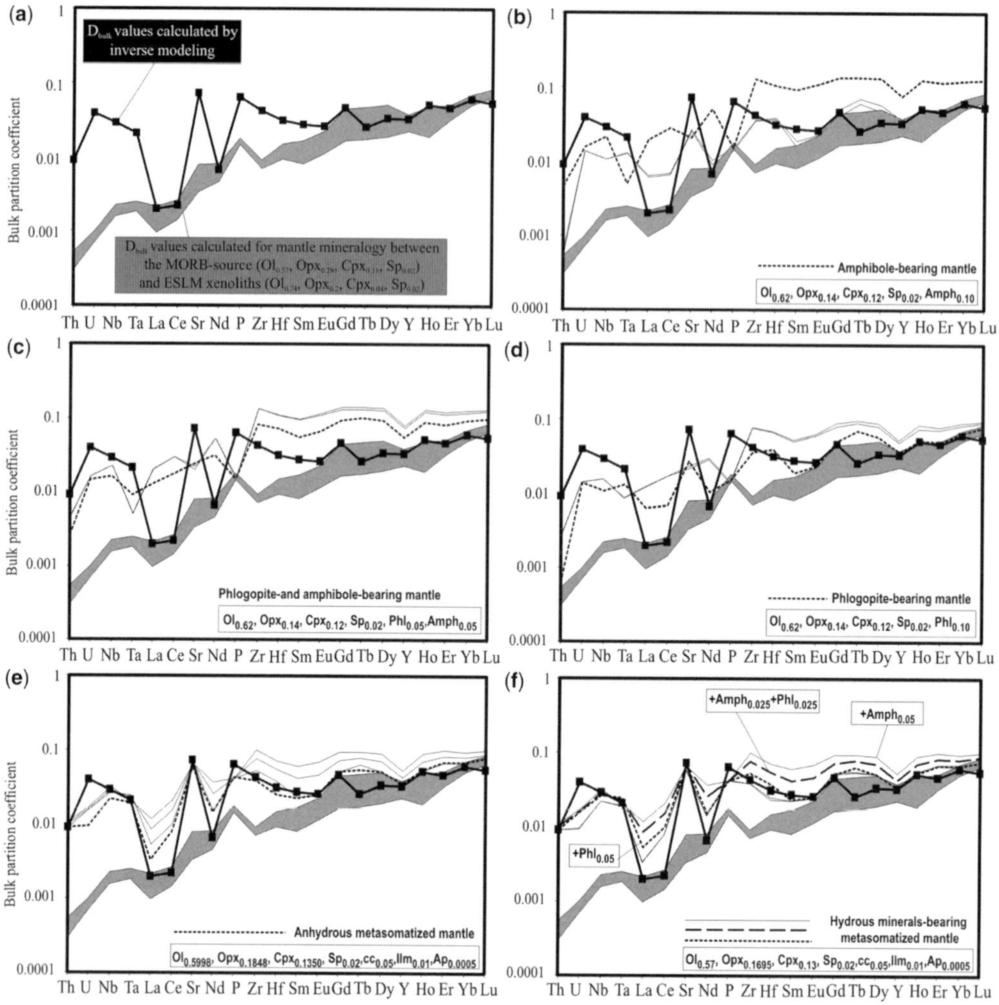

Fig. 9. A comparison between $D_{mineral-melt}$ values for the source of basanite primary melts obtained by inverse modelling (full squares) and $D_{mineral-melt}$ values for unmetasomatized mantle (shaded area, shown in all diagrams), which varies in modal composition between depleted mantle (57% olivine, 28% orthopyroxene, 13% clinopyroxene and 2% spinel: Workman & Hart 2005) and the average of ESLM depleted xenoliths (74% olivine, 20% orthopyroxene, 4% clinopyroxene and 2% spinel: Cvetković *et al.* 2004*a*), as well as with various modally metasomatized mantle (see the figure for explanation).

high ${}^{Sr}P_L$. This conclusion is in accordance with the observation that Mg-bearing calcite and apatite are frequently found in metasomatic associations found in ESLM xenoliths (also Cvetković *et al.* 2007*b*). Similar reasoning can be followed to infer the presence of Fe–Ti-oxides as repositories of Nb and Ta, whose estimated concentration in the source is more than 35× chondrite. Ilmenite is observed in many metasomatic associations in ESLM xenoliths (Cvetković *et al.* 2004*a*, 2007*b*), and this mineral has high partition coefficients for Nb and Ta (e.g. Zack & Brumm 1998). Hence, it can account for high Nb and Ta concentrations in the source, and also for their relatively high partitioning into the melt.

Asthenosphere–lithosphere interaction: the order of events

The foregoing discussion implies that there is agreement in mineral composition, at least qualitatively, between the observed secondary assemblages in the studied xenoliths and the inferred metasomatic phases in the source of host magmas. The only discrepancy is the presence is phlogopite crystals in the observed assemblages while, on the other hand, the

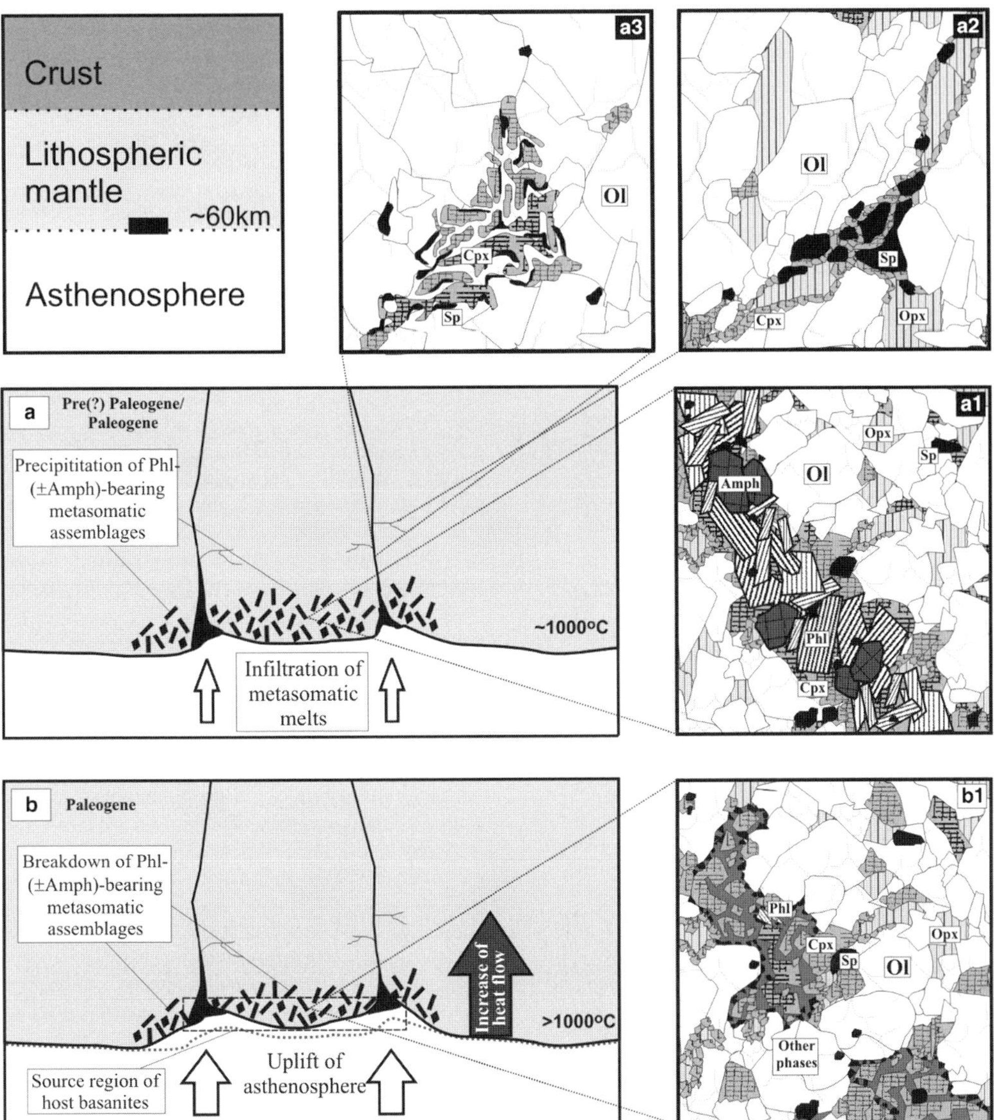

Fig. 10. A schematic model of metasomatic processes within the ESLM. The first phase (**a**) is related to the precipitation of primary phlogopite and possibly (?) amphibole, producing the various metasomatic assemblages illustrated in inset (**a**1). The percolations that were facilitated by local deformation zones gave rise to the disintegration of spinel grains (inset **a**2), whereas small portions of alkaline melts solidified as metasomatic symplectites (**a**3). This phase must have occurred before the Palaeogene, at least before East Serbian mafic alkaline magmatism occurred. The second phase (**b**) comprises a succession of events: the uplift of hot asthenospheric mantle, the increase in heat flow at the base of the lithosphere, the breakdown of previously formed phlogopite and amphibole leaving behind anhydrous metasomatic associations (inset **b**1), and the subsequent partial melting of the basanite source (hatched area). The inferences about temperature values and depth of the lithosphere–asthenosphere bottom are after Cvetković *et al.* (2007*b*).

modelling suggests an anhydrous source for the host basanites. In Figure 10 we offer a schematic two-phase model of metasomatic processes that might have occurred at the base of the ESLM. We do not suggest that the metasomatized ESLM xenoliths represent the actual source for host magmas, only that they may represent counterparts of similar metasomatic regions situated deeper in the lithosphere. The first phase (Fig. 10a) involves the percolation of CO_2- and H_2O-rich mafic alkaline melts and the precipitation of primary metasomatic assemblages possibly containing phlogopite and amphibole (Fig. 10a1). The original presence of amphibole is highly speculative because no amphibole is found in the studied xenoliths. Despite this fact amphibole is included in the model mainly because it was found in many other xenolith suites (e.g. Witt-Eickschen *et al.* 1998), and its role in the source processes of alkali basalts and basanites has already been reported by many studies (Jung & Masberg 1998; Jung & Hoernes 2000). The percolation–infiltration processes were most probably facilitated by local deformation and fault zones, as also suggested by textural assemblages with rotated, disrupted and partially resorbed spinel grains formed (Fig. 10a2). Simultaneously, small portions of the alkaline melt were trapped by wall-rock peridotite and solidified as metasomatic symplectites (Fig. 10a3). The second phase of the model is related to the uplift of hot asthenospheric mantle and this should have occurred at the very onset of the host basanitic magmatism. The asthenospheric uplift produced an increase in heat flow and caused a temperature rise at the base of the lithosphere. Decompression and high temperature had caused the breakdown of previously formed phlogopite and amphibole, leaving behind anhydrous metasomatic associations in the lithosphere bottom that are texturally and compositionally similar to the assemblages found in the ESLM xenoliths (Fig. 10b1). The lithosphere bottom, along with convective mantle underneath, could have represented the source for basanites (hatched area, Fig. 10b).

This model uses a general idea of many studies that have linked intraplate alkaline magmatism to lithospheric mantle sources enriched by melts derived from sublithospheric convecting mantle at some time in the past (Stein & Hofmann 1992; Baker *et al.* 1998; Beccaluva *et al.* 2001; Bonadiman *et al.* 2001; Coltorti *et al.* 2004; Jung *et al.* 2005). We have presented evidence of Nb–Ta enrichment in the ESLM and in agreement with Pilet *et al.* (2004), who attributed high Nb and Ta concentrations to short-term metasomatic events in the lithospheric mantle. The model also suggests that the host basanites probably originate from an anhydrous metasomatized peridotite source and that the phlogopite–amphibole signature in these rocks is due to a previous phase of dehydration that occurred prior to magmatism. Assuming an open system, it can be expected that, after dehydration, the source was depleted in H_2O, Rb and K with respect to other incompatible elements, especially HFSE. This scenario may be of significance for the petrogenesis of at least some alkaline mafic rocks. A similar idea was very recently reported and modelled by Weinstein *et al.* (2006) for the example of the Harrat As Shaam basanites in Israel.

Conclusions

Prior to the onset of the Palaeogene, the East Serbian mantle was already metasomatized. The metasomatism produced domains that are represented by fertile xenoliths and small-scale metasomatic associations found in depleted xenoliths.

The composition of the fertile xenoliths is modelled by the addition of 5–20 wt% of a basanite melt to a refractory mantle. The model suggests that this lithology could have originated by the refertilization of residual mantle regions by mafic alkaline basaltic melts. However, the small-scale metasomatic associations comprise: (i) metasomatic phases crystallized in an open space; (ii) those formed owing to simultaneous deformation events; and (iii) rare symplectitic assemblages. They all show a similar mineral composition and imply the following reaction: opx + Cr-rich sp $\pm$ phlogopite + Si-undersaturated alkaline melt = Ti–Al-cpx + Ti-rich spinel $\pm$ olivine $\pm$ other minor phases.

There is agreement in modal composition, at least qualitatively, between the observed secondary metasomatic assemblages in the xenoliths and the source of host magmas, the latter being inferred from inversion modelling. In this context, the host basanites probably originated from a peridotitic source that contained metasomatic domains similar to those observed in xenoliths. The phlogopite–amphibole signature in these rocks is probably owing to a dehydration phase caused by further asthenospheric uplift that occurred prior to alkaline magmatism.

We wish to thank Dr H. Höfer and Dr A. Gerdes for help with EPMA and LA-ICP-MS analyses. Careful and encouraging reviews by C. Bonadiman and Cs. Szabo, and the editorial handling by M. Coltorti, have been very helpful and are gratefully acknowledged. The authors also thank M. Grégoire for constructive reading of an earlier version of the manuscript. This study was supported by the Austrian Science Foundation FWF as a Lise Meitner project no. M832-N10 granted to V. Cvetković. The Serbian Ministry of Science and Technological Development project no. 146013 is acknowledged.

Appendix

Inversion modelling

This modelling is based on mathematical expressions that govern the geochemical behaviour of trace elements

Table A1. *Values for slope* (A_j) *and intercept* (B_j) *in La v. trace element and La v. La/trace element diagrams;* r *is the correlation coefficient for La v. trace element diagrams*

	A_j	B_j	r		A_j	B_j
Th	0.156	0.485	0.701	La/Th	0.0067	5.732
U	0.029	0.699	0.757	La/U	0.1224	17.548
Nb	0.958	17.047	0.729	La/Nb	0.0031	0.62
Ta	0.063	0.727	0.780	La/Ta	0.0362	11.158
Ce	1.743	1.340	0.931	La/Ce	0.00002	0.56526
Sr	8.444	437.360	0.862	La/Sr	0.0005	0.0336
Nd	0.671	2.789	0.775	La/Nd	0.001	1.3334
P_2O_5	0.008	0.329	0.738	La/P2O5	0.5582	43.612
Zr	1.996	55.287	0.720	La/Zr	0.0018	0.233
Hf	0.054	1.027	0.769	La/Hf	0.0557	10.73
Sm	0.102	1.720	0.730	La/Sm	0.0273	6.025
Eu	0.032	0.479	0.930	La/Eu	0.0861	19.896
TiO_2	0.006	1.438	0.261	La/TiO2	0.4403	7.1075
Gd	0.071	2.083	0.898	La/Gd	0.0539	6.2438
Tb	0.013	0.182	0.820	La/Tb	0.2045	50.087
Dy	0.062	1.256	0.871	La/Dy	0.0523	8.9451
Y	0.334	6.934	0.773	La/Y	0.0094	1.6603
Ho	0.010	0.335	0.872	La/Ho	0.3972	40.649
Er	0.028	0.780	0.842	La/Er	0.1376	16.355
Tm	0.005	0.064	0.596	La/Tm	0.4837	143.8
Yb	0.023	0.761	0.591	La/Yb	0.1736	17.87
Lu	0.004	0.101	0.754	La/Lu	1.0765	123.93

during non-modal partial melting processes (Shaw 1970):

$$C_L^i = \frac{C_0^i}{D_0^i + F(1 - P_L^i)} \tag{1}$$

$$C_{RS}^i = \frac{C_0^i}{(1 - F)}\left[\frac{D_0^i - P_L^i F}{D_0^i + F(1 - P_L^i)}\right] \tag{2}$$

where C_0^i, concentration of the element i in the source; C_{RS}^i, concentration of the element i in the residual solid; C_L^i, concentration of the element i in the partial melt; D_0^i, bulk distribution coefficient of the element i for the phases present in the source; P_L^i, bulk distribution coefficient of the element i for the phases entering the melt; F, amount of partial melt.

It is combined with the graphical procedure of Treuil & Joron (1975) for the determination of the relative degree of incompatibility of the trace elements using C_L^i–C_L^j and C_L^i–C_L^i/C_L^j diagrams. In the element-ratio diagram, when C_L^i is concentration of the most incompatible and C_L^j the concentration of a less incompatible element, the slope (A) and intercept (B) have direct relations to geochemical parameters of partial melting processes, which can be mathematically expressed as:

$$A^j = \frac{D_0^j - D_0^i \dfrac{1 - P_L^j}{1 - P_L^i}}{C_0^j} \tag{3}$$

$$B^j = \frac{C_0^i(1 - P_L^j)}{C_0^j(1 - P_L^i)} \tag{4}$$

In the studied suite of samples lanthanum (La) is the most incompatible element. The values of the slope, intercept and correlation coefficient for the La-element, and the slope and intercept for La–La/element diagrams are given in Table A1. The graphical expression of the relative incompatibility of the studied trace elements is given in Figure A1, a diagram displaying the ratio of intercept on La-element v. slope on La-ratio diagrams (both recalculated on maximal concentrations) in which the origin is La. Using the relationship between intercepts on La-ratio diagrams for two elements suggested by Clague & Frey (1982)

$$\frac{B^k}{B^j} = \frac{C_0^j(1 - P_L^k)}{C_0^k(1 - P_L^j)} \tag{5}$$

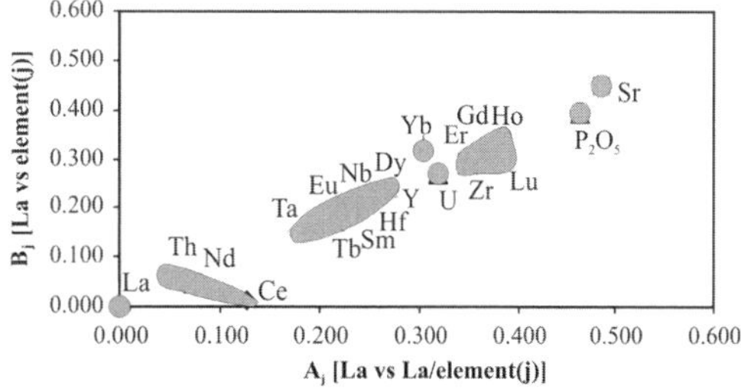

Fig. A1. Diagram of A_j v. B_j displaying relative incompatibility of the studied elements, where A_j is the slope on La v. La/element (j) diagram and B_j is the intercept on La v. element (j) diagram.

it was possible to calculate ratios of Ce/La, Ce/Nd, Ce/Th and Ce/Lu, and ranges of La, Ce, Nd and Th concentrations, namely their minimal enrichment in the source. Because for equation (5) it was necessary to constrain the P_L^j and P_L^k values, the assumptions are that: (a) P_L values for Ce, Nd and Th are negligible against 1, which is inferred above; and (b) that the source concentrations of Lu are in range of 2×–4× chondrite, as suggested by McDonough & Frey (1989). In addition, using the above-constrained minimal concentrations of Ce in the source and the least enriched rock sample with La = 54 ppm (to satisfy the condition that $D_0 \ll F$), the maximum degree of partial melting was estimated at $F > 0.1$, by the equation:

$$F = \frac{B^j C_0^j}{C_L^i}. \quad (6)$$

The calculation of C_0, D_0 and P_L values for La, Ce, Yb and Lu was performed using a best-fit solution of a system of non-linear equations combining equations (1) and (2), and assuming other constraints used in the initial approach: $F_{max} < 0.12$; $D_0^i < P_L^i$; $D_0^{La} \approx D_0^{Ce} \approx 0$; $D_0^{La} < D_0^{Ce} < D_0^{Yb} < D_0^{Lu}$. The combination of the obtained parameters was used to estimate that the maximum percentage of melting is $F = 0.117$. Finally, the C_0, D_0 and P_L for other elements were calculated using the equations:

$$1 - P_L^j = D_0^j \frac{1 - P_L^i}{\frac{A^j}{B^j} C_0^i + D_0^i} \quad (7)$$

$$\frac{C_0^j}{C_L^j} = D_0^j \left[1 + \frac{F(1 - P_L^i)}{\frac{A^j}{B^j} C_0^i + D_0^i} \right] \quad (8)$$

which include the above-estimated values for D_0^{La}, P_L^{La}, C_0^{La} and F. The assumptions that $D_0^i < P_L^i$ and that C_{RS}^i must be <0 were satisfied.

References

ANDERSON, A. T. & GREENLAND, L. P. 1969. Phosphorous fractionation diagrams as a quantitative indicator of crystallization differentiation of basaltic liquids. *Geochimica et Cosmochimica Acta*, **33**, 493–505.

BAKER, J., CHAZOT, G., MENZIES, M. & THIRLWALL, M. 1998. Metasomatism of the shallow mantle beneath Yemen by the Afar Plume; implications for mantle plumes, flood volcanism, and intraplate volcanism. *Geology*, **26**(5), 431–434.

BALI, E., SZABÓ, C., VASELLI, O. & TÖRÖK, K. 2002. Significance of silicate melt pockets in upper mantle xenoliths from the Bakony–Balaton Highland Volcanic Field, Western Hungary. *Lithos*, **61**, 79–102.

BALI, E., ZANETTI, A., SZABÓ, CS., PEATE, D. & WAIGHT, T. E. 2008. Evolution of the subcontinental lithospheric mantle beneath the Central Pannonian Basin: trace element evidence from silicate melt pockets in mantle xenoliths from the Bakony-Balaton Highland Volcanic Field (western Hungary). *Contributions to Mineralogy and Petrology*, **155**, 165–179.

BEA, F., PEREIRA, M. D. & STROH, A. 1994. Mineral/leucosome trace-element partitioning in a peraluminous migmatite (a laser ablation-ICP-MS study). *Chemical Geology*, **117**, 291–312.

BECCALUVA, L., BONADIMAN, C., COLTORTI, M., SALVINI, L. & SIENA, F. 2001. Depletion events, nature of metasomatizing agent and timing of enrichment processes in lithospheric mantle xenoliths from the Veneto Volcanic Province. *Journal of Petrology*, **42**, 173–188.

BONADIMAN, C., BECCALUVA, L., COLTRTI, M. & SIENA, F. 2005. Kimberlite-like metasomatism and 'garnet signature' in spinel-peridotite xenoliths from Sal, Cape Verde Archipelago: relics of a subcontinental mantle domain within the atlantic oceanic lithosphere? *Journal of Petrology*, **46**, 2465–2493.

BONADIMAN, C., COLTORTI, M., DUGGEN, S., PALUDETTI, L., SIENA, F., THIRLWALL, M. F. & UPTON, B. G. J. 2008. Palaeozoic subduction-related and kimberlite or carbonatite metasomatism in the Scottish lithospheric mantle. *In*: COLTORTI, M. & GRÉGOIRE, M. (eds) *Metasomatism in Oceanic and Continental Lithospheric Mantle*. Geological Society, London, Special Publications, **293**, 303–333.

BONADIMAN, C., COLTORTI, M., MILANI, L., SALVINI, L., SIENA, F. & TASSINARI, R. 2001. Metasomatism in the lithospheric mantle and its relationships to magmatism in the Veneto Volcanic Province, Italy. *Peridico di Mineralogia*, **LXX/3**, 333–357.

CANIL, D. & SCARFE, C. M. 1989. Origin of phlogopite in mantle xenoliths from Kostal Lake Wells Grey Park, British Columbia. *Journal of Petrology*, **30**, 1159–1179.

CARPENTER, R. L., EDGAR, A. D. & THIBAULT, Y. 2002. Origin of spongy textures in clinopyroxene and spinel from mantle xenoliths, Hessian Depression, Germany. *Mineralogy and Petrology*, **74**, 149–162.

CEBRIA, J. M. & LOPEZ-RUIZ, J. 1996. A refined method for trace element modelling of nonmodal batch partial melting processes; the Cenozoic continental volcanism of Calatrava, central Spain. *Geochimica et Cosmochimica Acta*, **60**(8), 1355–1366.

CHAZOT, G., MENZIES, M. A. & HARTE, B. 1996. Determination of partition coefficients between apatite, clinopyroxene, amphibole, and melt in natural spinel lherzolites from Yemen: implications for wet melting of the lithospheric mantle. *Geochimica et Cosmochimica Acta*, **60**(3), 423–437.

CLAGUE, D. A. & FREY, F. A. 1982. Petrology and trace element geochemistry of the Honolulu volcanics, Oahu: implications for the oceanic mantle below Hawaii. *Journal of Petrology*, **23**, 447–504.

COLTORTI, M., BECCALUVA, L., BONADIMAN, C., FACCINI, F., NTAFLOS, T. & SIENA, F. 2004. Amphibole genesis via metasomatic reaction with clinopyroxene in mantle xenoliths from Victoria Land, Antarctica. *Lithos*, **75**, 115–139.

COLTORTI, M., BONADIMAN, C., HINTON, R. W., SIENA, F. & UPTON, B. G. J. 1999. Carbonatite metasomatism of the oceanic upper mantle: evidence from clinopyroxenes and glasses in ultramafic xenoliths of Grande

Comore, Indian Ocean. *Journal of Petrology*, **40**, 133–165.

Cvetković, V., Downes, H., Prelević, D., Jovanović, M. & Lazarov, M. 2004*a*. Characteristics of the lithospheric mantle beneath east Serbia inferred from ultramafic xenoliths in Palaeogene basanites. *Contributions to Mineralogy and Petrology*, **148**(3), 335–357.

Cvetković, V., Downes, H., Prelević, D., Lazarov, M. & Resimić-Sarić, K. 2007*a*. Geodynamic significance of ultramafic xenoliths from Eastern Serbia: relics of sub-arc oceanic mantle? *Journal of Geodynamics*, **43**, 504–527.

Cvetković, V., Höck, V. & Prelević, D. 2005. Mantle source characteristics of the Palaeogene mafic alkaline rocks of the East Serbian Carpatho-Balkanides. *Mitteilungen der Österreichischen Mineralogischen Gesellschaft*, **151**, 36.

Cvetković, V., Lazarov, M., Downes, H. & Prelević, D. 2007*b*. Modification of the subcontinental mantle beneath East Serbia: evidence from orthopyroxene-rich xenoliths. *Lithos*, **94**, 90–110.

Cvetković, V., Prelević, D., Downes, H., Jovanović, M., Vaselli, O. & Pecskay, Z. 2004*b*. Origin and geodynamic significance of Tertiary postcollisional basaltic magmatism in Serbia (central Balkan Peninsula). *Lithos*, **73**, 161–186.

Dawson, J. B. 1984. Carbonate tuff cones in northern Tanganyika. *Geological Magazine*, **101**, 129–137.

Dawson, J. B. 2002. Metasomatism and partial melting in upper-mantle peridotite xenoliths from the Lashaine Volcano, northern Tanzania. *Journal of Petrology*, **43**, 1749–1777.

Delpech, G., Gregoire, M., O'Reilly, S. Y., Cottin, J. Y., Moine, B., MIchon, G. & Giret, A. 2004. Feldspar from carbonate-rich silicate metasomatism in the shallow oceanic mantle under Kerguelen Islands (South Indian Ocean). *Lithos*, **75**, 209–237.

Eggins, S. M., Woodhead, J. D. *et al.* 1997. A simple method for the precise determination of ≥40 trace elements in geological samples by ICP-MS using enriched isotope internal standardisation. *Chemical Geology*, **134**, 311–326.

Falus, G., Drury, M. R., Van Roermund, H. L. M. & Szabo, C. S. 2004. Magmatism-related localized deformation in the mantle: a case study. *Contributions to Mineralogy and Petrology*, **146**, 493–505.

Foley, S. F., Jackson, S. E., Fryer, B. J., Greenough, J. D. & Jenner, G. A. 1996. Trace element partition coefficients for clinopyroxene and phlogopite in an alkaline lamprophyre from Newfoundland by LAM-ICP-MS. *Geochimica et Cosmochimica Acta*, **60**(4), 629–638.

Fujimaki, H. 1986. Partition-coefficients of Hf, Zr, and REE between zircon, apatite, and liquid. *Contributions to Mineralogy and Petrology*, **94**(1), 42–45.

Fujimaki, H., Tatsumoto, M. & Aoki, K.-I. 1984. Partition coefficients of Hf, Zr, and REE between phenocrysts and groundmasses. *Journal of Geophysical Research*, **89**, 662–672.

Ghiorso, M. S. & Carmichael, I. S. E. 1987. Modeling magmatic systems: petrologic applications. *Reviews in Mineralogy*, **17**, 467–499.

Gorring, M. L. & Kay, S. M. 2000. Carbonatite metasomatized peridotite xenoliths from suthern Patagonia: implications for lithospheric processes and Neogene plateau magmatism. *Contributions to Mineralogy and Petrology*, **140**, 55–72.

Green, T. H., Adam, J. & Site, S. H. 1993. Proton microprobe determined trace element partition coefficients between pargasite, augite and silicate or carbonatitic melts. *Eos, Transactions of the American Geophysical Union*, **74**, 340.

Gregoire, M., Moine, B. N., O'Reilly, S. Y., Cottin, J. Y. & Giret, A. 2000. Trace element residence and partitioning in mantle xenoliths metasomatized by highly alkaline, silicate-and carbonate-rich melts (Kerguelen Islands, Indian Ocean). *Journal of Petrology*, **41**, 477–509.

Hart, S. R. 1984. A large scale isotope anomaly in the southern hemisphere mantle. *Nature*, **309**, 753–757.

Harte, B. 1987. Metasomatic events recorded in mantle xenoliths: an overview. *In*: Nixon, P. (ed.) *Mantle Xenoliths*. Wiley, Hoboken, NJ, 625–640.

Horn, I., Foley, S. F., Jackson, S. E. & Jenner, G. A. 1994. Experimentally determined partitioning of high field strength- and selected transition elements between spinel and basaltic melt. *Chemical Geology*, **117**, 193–218.

Ionov, D. & Harmer, R. E. 2002. Trace element distribution in calcite–dolomite carbonatites from Spitskop: inferences for differentiation of carbonatite magmas and the origin of carbonates in mantle xenoliths. *Earth and Planetary Science Letters*, **198**, 495–510.

Ionov, D. A., Grégoire, M. & Prikhod'ko, V. S. 1999. Feldspar–Ti-oxide metasomatism in off-cratonic continental and oceanic upper mantle. *Earth and Planetary Science Letters*, **165**(1), 37–44.

Ionov, D., Hofmann, A. W. & Shimizu, N. 1994. Metasomatism-induced melting in mantle xenoliths from Mongolia. *Journal of Petrology*, **35**(3), 753–785.

Jovanović, M., Downes, H., Vaselli, O., Cvetković, V., Prelević, D. & Pecskay, Z. 2001. Palaeogene mafic alkaline volcanic rocks of East Serbia. *Acta Vulcanologica*, **13**(1–2), 159–173.

Jung, S. & Hoernes, S. 2000. The major- and trace-element and isotope (Sr, Nd, O) geochemistry of Cenozoic alkaline rift-type volcanic rocks from the Rhoen area (central Germany): petrology, mantle source characteristics and implications for asthenosphere–lithosphere interactions. *Journal of Volcanology and Geothermal Research*, **99**(1–4), 27–53.

Jung, S. & Masberg, P. 1998. Major- and trace-element systematics and isotope geochemistry of Cenozoic mafic volcanic rocks from the Vogelsberg (central Germany); constraints on the origin of continental alkaline and tholeiitic basalts and their mantle sources. *Journal of Volcanology and Geothermal Research*, **86**(1–4), 151–177.

Jung, S., Pfaender, J. A., Bruegmann, G. & Stracke, A. 2005. Sources of primitive alkaline volcanic rocks from the Central European Volcanic Province (Rhoen, Germany) inferred from Hf, Os and Pb isotopes. *Contributions to Mineralogy and Petrology*, **150**, 546–559.

Kogarko, L. N., Kurat, G. & Ntaflos, T. 2007. Henrymeyerite in the metasomatized upper mantle of Eastern Antarctica. *Canadian Mineralogist*, **45**(3), 497–501.

KARNER, J. M., PAPIKE, J. J. & SHEARER, C. K. 2003. Olivine from planetary basalts; chemic signatures that indicate planetary parentage and those that record igneous setting and process. *American Mineralogist*, **88**(5–6), 806–816.

KEMPTON, P. D. 1987. Mineralogic and geochemical evidence for differing styles of metasomatism in spinel lherzolite xenoliths: enriched mantle source regions of basalts? *In*: MENZIES, M. A. & HAWKESWORTH, C. J. (eds) *Mantle Metasomatism*. Academic Press, London, 45–89.

KEPEZHINSKAS, P. K., DEFANT, M. J. & DRUMMOND, M. S. 1995. Na Metasomatism in the island-arc mantle by slab melt–peridotite interaction: evidence from mantle xenoliths in the North Kamchatka Arc. *Journal of Petrology*, **36**(6), 1505–1527.

KEPEZHINSKAS, P. K., DEFANT, M. J. & DRUMMOND, M. S. 1996. Progressive enrichment of island arc mantle by melt peridotite interaction inferred from Kamchatka xenoliths. *Geochemica et Cosmochimica Acta*, **60**, 1217–1229.

KESHAV, S., CORGNE, A., GUDFINNSSON, G. H., BIZIMIS, M., MCDONOUGH, W. F. & FEI, Y. W. 2005. Kimberlite petrogenesis; insights from clinopyroxene–melt partitioning experiments at 6 GPa in the CaO–MgO–Al_2O_3–SiO_2–CO_2 system. *Geochimica et Cosmochimica Acta*, **69**(11), 2829–2845.

KLÜGEL, A. 1998. Reactions between mantle xenoliths and host magma beneath La Palma (Canary Islands); constraints on magma ascent rates and crustal reservoirs. *Contributions to Mineralogy and Petrology*, **131**(2–3), 237–257.

LARSEN, L. M., PEDERSEN, A. K., SUNDVOLL, B. & FREI, R. 2003. Alkali picrites formed by melting of old metasomatized lithospheric mantle: Mantdlat Member, Vaigat Formation, Palaeocene of West Green-land. *Journal of Petrology*, **44**, 3–38.

LA TOURRETTE, T., HERVIG, R. L. & HOLLOWAY, J. R. 1995. Trace-element partitioning between amphibole, phlogopite, and basanite melt. *Earth and Planetary Science Letters*, **135**(1–4), 13–30.

LUHR, J. F., CARMICHAEL, I. S. E. & VAREKAMP, J. C. 1984. The 1982 eruptions of El Chichon volcano, Chiapas, Mexico: mineralogy and petrology of the anhydrite-bearing pumices. *Journal of Volcanology and Geothermal Research*, **23**, 69–108.

MAHOOD, G. A. & STIMAC, J. A. 1990. Trace-element partitioning in pantellerites and trachytes. *Geochimica et Cosmochimica Acta*, **54**, 2257–2276.

MCDONOUGH, W. H. & FREY, F. A. 1989. Rare earth elements in upper mantle rocks. *In*: LIPIN, B. R. & MCKAY, G. A. (eds) *Geochemistry and Mineralogy of Rare Earth Elements. Reviews in Mineralogy*, **21**, 99–145.

MCDONOUGH, W. F. & SUN, S.-S. 1950. The composition of the Earth. *Chemical Geology*, **120**, 223–253.

MCGUIRE, A. V. & MUKASA, S. B. 1997. Magmatic modification of the uppermost mantle beneath the Basin and Range to Colorado Plateau transition zone; evidence from xenoliths, Wikieup, Arizona. *Contributions to Mineralogy and Petrology*, **128**(1), 52–65.

MCKENZIE, D. & O'NIONS, R. K. 1991. Partial melt distributions from inversion of rare earth element concentrations. *Journal of Petrology*, **32**(5), 1021–1091.

MCKENZIE, D. & O'NIONS, R. K. 1995. The source regions of ocean island basalts. *Journal of Petrology*, **36**(1), 133–159.

MENZIES, M. A., RODGERS, N., TINDLE, A. & HAWKESWORTH, C. J. 1987. Metasomatic and enrichment processes in lithospheric peridotites, and effect of asthenosphere–lithosphere interaction. *In*: MENZIES, M. A. & HAWKESWORTH, C. J. (eds) *Mantle Metasomatism*. Academic Press, London, 313–361.

MORIMOTO, N., FABRIES, J. ET AL. 1988. Nomenclature of pyroxenes. *American Mineralogist*, **62**, 53–62.

MORISHITA, T. & ARAI, S. 2003. Evolution of spynel pyroxene symplectite in spinel-lherzolites from the Horoman Complex, Japan. *Contributions to Mineralogy and Petrology*, **144**, 509–522.

NEUMANN, E.-R., WULFF-PEDERSEN, E., PEARSON, N. & SPENCER, E. A. 2002. Mantle xenoliths from Tenerife (Canary Islands); evidence for reactions between mantle peridotites and silicic carbonatite melts inducing Ca metasomatism. *Journal of Petrology*, **43**(5), 825–857.

NICOLAS, A. 1986. A melt extraction model based on structural studies in mantle peridotites. *Journal of Petrology*, **27**, 999–1022.

O'REILLY, S. Y. & GRIFFIN, W. L. 1988. Mantle metasomatism beneath western Victoria, Australia: I. Metasomatic processes in Cr-diopside lherzolites. *Geochimica et Cosmochimica Acta*, **52**, 433–447.

PASTER, T. P., SCHAUWECKER, D. S. & HASKIN, L. A. 1974. The behavior of some trace elements during solidification of the Skaergaard layered series. *Geochimica et Cosmochimica Acta*, **38**(10), 1549–1577.

PEARCE, N. J. G., PERKINS, W. T., WESTGATE, J. A., GOTRON, M. P., JACKSON, S. E., NEAL, C. R. & CHENERY, S. P. 1997. A compilation of new and published major and trace element data for NIST SRM 610 and NIST SRM 612 glass reference materials. *Geostandards Newsletter – The Journal of Geostandards and Geoanalysis*, **21**(1), 115–144.

PILET, S., HERNANDEZ, J., BUSSY, F. & SYLVESTER, P. 2004. Short-term metasomatic control of Nb/Th ratios in the mantle sources of intraplate basalts. *Geology (Boulder)*, **32**(2), 113–116.

RIVALENTI, G., ZANETTI, A., MAZZUCCHELLI, M., VANNUCCI, R. & CINGOLANI, C. A. 2004. Equivocal carbonatite markers in the mantle xenoliths of the Patagonia backarc; the Gobernador Gregores case; Santa Cruz Province, Argentina. *Contributions to Mineralogy and Petrology*, **147**(6), 647–670.

SALTERS, V. J. M. & STRACKE, A. 2004. Composition of the depleted mantle. *Geochemistry, Geophysics, Geosystems*, **5**(5), 1–27.

SCHIANO, P., CLOCCHIATTI, R., SHIMIZU, N. C., MAURY, R., JOCHUM, K. P. & HOFMANN, A. 2002. Hydrous, silica-rich melts in the sub-arc mantle and their relationship with erupted arc lavas. *Nature*, **377**, 595–600.

SHAW, D. M. 1970. Trace element fractionation during anatexis. *Geochemica Cosmochimica Acta*, **34**, 237–243.

SMITH, D. 1977. The origin and interpretation of spinel–pyroxene clusters in peridotite. *Journal of Geology*, **85**, 476–482.

STEIN, M. & HOFMANN, A. W. 1992. Fossil plume head beneath the Arabian lithosphere? *Earth and Planetary Science Letters*, **114**, 193–209.

STERN, C. R., KILIAN, R., OLKER, B., HAURI, E. H. & KURTIS KYSER, T. 1999. Evidence from mantle xenoliths for relatively thin (<100 km) continental lithosphere below the Phanerozoic crust of southernmost South America. *Lithos*, **48**(1–4), 217–235.

TREUIL, M. & JORON, J. L. 1975. Utilisation des elements hygromagmatophyles pour la simplification de la modelisation quantitative des processus magmatiques. Examples de l' Afar et de la Dorsale medioatlantique. *Societá Italiana di Mineralogia e Petrografia*, **131**, 125–174.

ULMER, P. 1989. Partitioning of high field strength elements among olivine, pyroxenes, garnet and calc alkaline picrobasalt: experimental results and an application. *International Journal of Mass Spectrometry and Ion Physics*, **89**, 42–47.

VANNUCCI, R., BOTTAZZI, P., WULFF-PEDERSEN, E. & NEUMANN, E.-R. 1998. Partitioning of REE, Y, Sr, Zr and Ti between clinopyroxene and silicate melts in the mantle under La Palma (Canary Islands); implications for the nature of the metasomatic agents. *Earth and Planetary Science Letters*, **158**(1–2), 39–51.

VILLEMANT, B., JAFFREZIC, H., JORON, J. L. & TREUIL, M. 1981. Distribution coefficients of major and trace-elements – fractional crystallization in the alkali basalt series of Chaine-Des-Puys (Massif Central, France). *Geochimica et Cosmochimica Acta*, **45**(11), 1997–2016.

WANG, W. & GASPARIK, T. 2001. Metasomatic clinopyroxenes in diamonds from Lianing province, China. *Geochimica et Cosmochimica Acta*, **65**, 611–620.

WANG, J., HATTORI, K. H., LI, J. & STERN, C. R. 2008. Oxidation state of Paleozoic subcontinental lithospheric mantle below the Pali Aike volcanic field in southernmost Patagonia. *Lithos*, **105**, 98–110.

WATSON, E. B. & GREEN, T. H. 1981. Apatite/liquid partition coefficients for the rare-earth elements and strontium. *Earth and Planetary Science Letters*, **56**, 405–421.

WEDEPOHL, K. H. & BAUMANN, A. 1999. Central European Cenozoic plume volcanism with OIB characteristics and indications of a lower mantle source. *Contributions to Mineralogy and Petrology*, **136**(3), 225–239.

WIECHERT, U., IONOV, D. A. & WEDEPOHL, K. H. 1997. Spinel peridotite xenoliths from the Atsagin-Dush Volcano, Dariganga lava plateau, Mongolia; a record of partial melting and cryptic metasomatism in the upper mantle. *Contributions to Mineralogy and Petrology*, **126**(4), 345–364.

WEINSTEIN, Y., NAVON, O., ALTHERR, R. & STEIN, M. 2006. The role of lithospheric mantle heterogeneity in the generation of Plio-Pleistocene alkali basaltic suites from NW Harrat Ash Shaam (Israel). *Journal of Petrology*, **47**(5), 1017–1050.

WILSHIRE, H. G. 1987. A model of mantle metasomatism. *In*: MORISS, E. M. & PASTERIS, J. D. (eds) *Mantle Metasomatism and Alkaline Magmatism*. Geological Society of America, Special Paper, **215**, 47–60.

WILSHIRE, H. G. & KIRBY, S. H. 1989. Dikes, joints and faults in the upper mantle. *Tectonophysics*, **161**, 23–31.

WILSON, M. & DOWNES, H. 1991. Tertiary–Quaternary intra-plate magmatism in Europe and its relationship to mantle dynamics. *Journal of Petrology*, **32**, 811–849.

WILSON, M. & DOWNES, H. 2006. Tertiary–Quaternary intra-plate magmatism in Europe and its relationship to mantle dynamics. *In*: GEE, D. & STEPHENSON, R. (eds) *European Lithosphere Dynamics*. Geological Society, London, Memoirs, **32**, 147–166.

WITT-EICKSCHEN, G. & KRAMM, U. 1998. Evidence for the multiple stage evolution of the subcontinental lithospheric mantle beneath the Eifel (Germany) from pyroxenite and composite pyroxenite/peridotite xenoliths. *Contributions to Mineralogy and Petrology*, **131**(2–3), 258–272.

WITT-EICKSCHEN, G., KAMINSKY, W., KRAMM, U. & HARTE, B. 1998. The nature of young vein metasomatism in the lithosphere of the West Eifel (Germany); geochemical and isotopic constraints from composite mantle xenoliths from the Meerfelder Maar. *Journal of Petrology*, **39**(1), 155–185.

WORKMAN, R. K. & HART, S. R. 2005. Major and trace element composition of the depleted MORB mantle (DMM). *Earth and Planetary Science Letters*, **231**, 53–72.

WULFF-PEDERSEN, E., NEUMANN, E. R. & JENSEN, B. J. 1996. The upper mantle under La Palma, Canary Islands; formation of Si–K–Na-rich melt and its importance as a metasomatic agent. *Contributions to Mineralogy and Petrology*, **125**(2–3), 113–139.

WULFF-PEDERSEN, E., NEUMANN, E.-R., VANNUCCI, R., BOTTAZZI, P. & OTTOLINI, L. 1999. Silicic melts produced by reaction between peridotite and infiltrating basaltic melts: ion probe data on glasses and minerals in veined xenoliths from La Palma, Canary Islands. *Contributions to Mineralogy and Petrology*, **137**, 59–82.

XU, Y., MERCIER, J.-C. C., MENZIES, M., ROSS, J. V., HARTE, B., LIN, C. & SHI, L. 1996. K-rich glass-bearing wehrlite xenoliths from Yitong, Northeastern China: petrological and chemical evidence for mantle metasomatism. *Contributions to Mineralogy and Petrology*, **125**, 406–420.

YAXLEY, G. M., CRAWFORD, A. J. & GREEN, D. H. 1991. Evidence for carbonatite metasomatism in spinel peridotite xenoliths from western Victoria, Australia. *Earth and Planetary Science Letters*, **107**(2), 305–317.

YAXLEY, G. M., GREEN, D. H. & KAMENETSKY, V. 1998. Carbonatite metasomatism in the southern Australian lithosphere. *Journal of Petrology*, **39**, 1917–1939.

ZACK, T. & BRUMM, R. 1998. Ilmenite/liquid partition coefficients of 26 trace elements determined through ilmenite/clinopyroxene partitioning in garnet pyroxene. *In*: GURNEY, J. J., GURNEY, J. L., PASCOE, M. D. & RICHARDSON, S. H. (eds) *7th International Kimberlite Conference*. Red Roof Design, Cape Town, 986–988.

ZANGANA, N. A., DOWNES, H., THIRWALL, M. F., MARRINER, G. F. & BEA, F. 1999. Geochemical variation on peridotite xenoliths and their constituent clinopyroxenes from Ray Pic (French Massif Central): implications for the composition of the shallow lithospheric mantle. *Chemical Geology*, **153**, 11–35.

ZINNGREBE, E. & FOLEY, S. F. 1995. Metasomatism in mantle xenoliths from Gees, West Eifel, Germany; evidence for the genesis of calc-alkaline glasses and metasomatic Ca-enrichment. *Contributions to Mineralogy and Petrology*, **122**(1–2), 79–96.

Index

Note: Page numbers in *italics* refer to Figures; page numbers in **bold** refer to Tables.